Operator Theory: Advances and Applications
Vol. 152

Editor:
I. Gohberg

Analytic Methods in the Theory of Differential and Pseudo-Differential Equations of Parabolic Type

Samuil D. Eidelman
Stepan D. Ivasyshen
Anatoly N. Kochubei

Springer Basel AG

Authors:

Samuil D. Eidelman
Anatoly N. Kochubei
Institute of Mathematics
National Academy of Sciences of Ukraine
Tereshchenkivska 3
Kiev 01601
Ukraine
e-mail: seidelman@amath.pp.kiev.ua
e-mail: kochubei@imath.kiev.ua

Stepan D. Ivasyshen
Department of Mathematical Physics
National Technical University of Ukraine
Kiev Polytechnical Institute
Prosp. Peremogy 37
Kiev 03056
Ukraine

2000 Mathematics Subject Classification 39A12, 47B39, 65M06, 65N06, 65Q05

A CIP catalogue record for this book is available from the
Library of Congress, Washington D.C., USA

Bibliographic information published by Die Deutsche Bibliothek
Die Deutsche Bibliothek lists this publication in the Deutsche Nationalbibliografie; detailed bibliographic data is available in the Internet at <http://dnb.ddb.de>.

ISBN 978-3-0348-9592-7 ISBN 978-3-0348-7844-9 (eBook)
DOI 10.1007/978-3-0348-7844-9

Originally published by Birkhäuser Verlag, Basel – Boston – Berlin in 2004
Softcover reprint of the hardcover 1st edition 2004

Printed on acid-free paper produced from chlorine-free pulp. TCF ∞
Cover design: Heinz Hiltbrunner, Basel

ISBN 978-3-0348-9592-7

9 8 7 6 5 4 3 2 1

www.birkhauser-science.com

Contents

Preface

The theory of parabolic equations, a well-developed part of the contemporary theory of partial differential equations and mathematical physics, is the subject of an immense research activity. A continuing interest in parabolic equations is caused both by the depth and complexity of mathematical problems emerging here, and by its importance in specific applied problems of natural science, technology, and economics.

This book aims at a consistent and, as far as possible, a complete exposition of analytic methods of constructing, investigating, and using fundamental solutions of the Cauchy problem for the following four classes of linear parabolic equations with coefficients depending on all variables:

$\mathbf{E}_1$: $\overrightarrow{2b}$-parabolic partial differential equations (parabolic equations of a quasi-homogeneous structure), in which every spatial variable may have its own weight with respect to the time variable.

$\mathbf{E}_2$: degenerate partial differential equations of Kolmogorov's structure, which generalize classical Kolmogorov equations of diffusion with inertia.

$\mathbf{E}_3$: pseudo-differential equations with non-smooth quasi-homogeneous symbols.

$\mathbf{E}_4$: fractional diffusion equations.

These classes of equations generalize in various directions the classical equations and systems parabolic in the Petrovsky sense, which were defined in [180] and studied in a number of monographs [83, 45, 146, 107, 76] and survey articles [102, 1, 215, 70, 46].

Note that the principal quantitative characteristics of equations from the classes $\mathbf{E}_1$, $\mathbf{E}_2$, and $\mathbf{E}_4$ are the weights of the time variable with respect to spatial variables. The weight equals $1/(2b_j)$, with respect to the j-th spatial variable, for the class $\mathbf{E}_1$, $1/(2b)$ with respect to the main spatial variables, and $1 + 1/(2b)$, $2 + 1/(2b)$, with respect to the two groups of spatial variables corresponding to the degeneration. The weight for the class $\mathbf{E}_4$ equals $\alpha/2$.

The above classes of equations are investigated by the same general scheme. As the first step, a fundamental solution of the Cauchy problem is constructed and studied. Then it is used for finding classes of existence, uniqueness, and correctness.

Fundamental solutions of the Cauchy problem are constructed through various modifications of the classical Levi method. Its essence is as follows. A fundamental solution is looked for as the sum of the main term (called a parametrix), and an integral with the parametrix as a kernel and an unknown density. The density is determined from an integral equation solved by the iteration method. Usually the parametrix is a fundamental solution of the Cauchy problem for a model equation with certain parameters. The model equation contains only the principal (in a certain sense) terms whose coefficients depend only on the time variable and parameters, or solely on parameters.

For model equations, the fundamental solutions of the Cauchy problem are constructed and investigated by the following stages:

- The Fourier transform in spatial variables is applied to a model equation. For the classes $\mathbf{E}_1$ and $\mathbf{E}_3$ this leads to ordinary differential equations, while for the class $\mathbf{E}_2$ a first order partial differential equation with respect to time and some of the spatial variables is obtained. For the class $\mathbf{E}_4$ the Fourier transform results in a fractional differential equation with respect to the time variable.
- For the equations obtained this way, properties of fundamental solutions of the Cauchy problem as functions of parameters are investigated.
- The above properties, together with appropriate theorems about the Fourier transform, are used to obtain estimates (as precise as possible) and a complete analytical description of a fundamental solution of the Cauchy problem for a model equation.

Implementation of the procedure of constructing a parametrix, and then a fundamental solution of the Cauchy problem, requires a variety of analytic tools like theorems about the Fourier transforms of analytic functions, theory of distributions, theory of hypersingular integrals, various special functions (the Mittag-Leffler functions, Fox's H-functions etc).

As a rule, differences in the ways of applying Levi's method appear in its first stage, as a parametrix is chosen in order to take into account specific properties of an equation. The subsequent stages are standard though the necessary techniques guaranteeing sufficient precision of estimates can be very complicated.

The simultaneous treatment of four substantially different generalizations of the classical Petrovsky's definition of parabolic equations makes it possible to give an interesting comparative analysis of the methods and results. In spite of the fact that the structure of each equation under consideration has its own peculiarities, and the estimates of a parametrix differ in certain details, each time transparent and sufficiently precise information is obtained regarding analytic properties of a parametrix. That results in a very precise description of properties of fundamental solutions of the Cauchy problem for equations whose coefficients depend on all independent variables.

The estimates of fundamental solutions, the existence, uniqueness, and correctness classes for equations of the types $\mathbf{E}_1$, $\mathbf{E}_2$, and $\mathbf{E}_4$ are similar – the fundamental solutions decay exponentially, as the spatial variables grow, and the existence, uniqueness, and correctness classes for the Cauchy problem consist of rapidly increasing functions. On the other hand, fundamental solutions of the

Cauchy problem for equations of the class $\mathbf{E}_3$ have only a power-like decay (and this result cannot be improved).

Parallels and differences in properties of equations of the above classes can be interpreted in physical terms. All of them provide mathematical models for various diffusion phenomena. For a diffusion described by the heat equation (and more general second order parabolic partial differential equations), the mean square displacement $\overline{\Delta x^2}$ of a diffusive particle for the time t behaves like Ct as $t \to \infty$. The anomalous diffusion, for which $\overline{\Delta x^2} \sim Ct^\alpha$, often appears in more complicated physical systems. The subdiffusion ($\alpha < 1$) is modelled by the fractal diffusion equation $\mathbf{E}_4$, while the superdiffusion ($\alpha > 1$) is described by equations of class $\mathbf{E}_3$.

Mathematically, it is perhaps the most interesting (probably, also important) to compare the results of Chapter 5 regarding fundamental solutions of the Cauchy problem for fractional diffusion equations, with corresponding results for the usual diffusion.

It appears that certain important properties of solutions, like the maximum principle, positivity, and some other properties of the fundamental solution, are similar. However the estimates of fundamental solutions, classes of existence and correctness for the Cauchy problem for fractional diffusion equations resemble, to a greater extent, the results known for equations of an arbitrary order, parabolic in the sense of Petrovsky (while, for instance, the positivity property is absent for equations of an order greater than two). The scheme of Levi's method for fractional diffusion equations is carried out as usual, but the estimates of the parametrix are qualitatively different, which necessitates a modification in the procedures for obtaining the estimates for this case.

Chapter 1

Equations. Problems. Results. Methods. Examples

This chapter is introductory. It contains definitions and examples of equations from the classes considered in this book, setting the context of the problems to be solved. We describe main results and techniques. In addition, we have included in this chapter several lemmas essential for this book.

Note that the term "equation" is used both in the case of a scalar equation and in the case of a vector equation, that is a system of equations.

1.1 Differential equations

1.1.1. Equations. Let us define the classes $\mathbf{E}_1$ and $\mathbf{E}_2$ of partial differential equations which will be considered in this book. These classes generalize the class $\mathbf{E}_0$ of equations parabolic in the Petrovsky sense. Therefore we begin with the definition of the class $\mathbf{E}_0$.

$\mathbf{E}_0$. Parabolic equations in the sense of Petrovsky

The initial general definition of parabolicity of equations (or systems of equations) was given by Petrovsky [180] who was also the first to investigate deeply such equations. This class of equations is wide enough and possesses many properties (with natural modifications) typical for the heat equation, the simplest and best studied representative of this class.

Let $n \in \mathbb{N}$ and $N \in \mathbb{N}$ be natural numbers. Below we shall consider a one-dimensional variable t and an n-dimensional variable $x := (x_1, \ldots, x_n)$; t and $x_1, \ldots, x_n$ will be interpreted as the time and spatial variables respectively.

Denote by Q a certain set of points (t, x) of the space $\mathbb{R}^{n+1}$. Let us consider the equation

$$L_N(t, x, \partial_t, \partial_x)u(t, x) = f(t, x), \quad (t, x) \in Q. \tag{1.1.1}$$

Here $L_N(t, x, \partial_t, \partial_x)$ is a linear differential expression, scalar if $N = 1$ or an $N \times N$-matrix if $N > 1$, with complex-valued coefficients depending on t and x, $(t, x) \in Q$; $f : Q \to \mathbb{C}_{N1}$ is a given function, $u : Q \to \mathbb{C}_{N1}$ is an unknown function.

Suppose first that $N = 1$. For any point $(t, x) \in Q$, the function $L_1(t, x, p, i\sigma)$, where $p \in \mathbb{C}$, $\sigma \in \mathbb{R}^n$, is a polynomial with respect to p and σ. Suppose that the degree of the polynomial $L_1(t, x, \lambda^m p, i\lambda\sigma)$ with respect to λ equals mr where m and r are natural numbers. Denote by L_1^0 the principal part of the polynomial L_1, that is the sum of all its terms satisfying the generalized homogeneity condition, so that

$$L_1^0(t, x, \lambda^m p, i\lambda\sigma) = \lambda^{mr} L_1^0(t, x, p, i\sigma), \quad \lambda \in \mathbb{R}.$$

Thus, the principal part L_1^0 is a generalized homogeneous polynomial of the degree mr.

Definition 1.1. *A differential expression L_1 is called parabolic in the sense of Petrovsky (with the weight m) at a point (t, x), if there exists such a constant $\delta > 0$ that for all $\sigma \in \mathbb{R}^n$, p-roots of the polynomial $L_1^0(t, x, p, i\sigma)$ satisfy the inequality*

$$\operatorname{Re} p(t, x, \sigma) \leq -\delta|\sigma|^m. \tag{1.1.2}$$

A differential expression L_1 is called uniformly parabolic in the sense of Petrovsky (with the weight m) on the set Q, if it is parabolic in the sense of Petrovsky at each point of the set Q, and the inequality (1.1.2) *holds with the same constant δ for every point $(t, x) \in Q$.*

Note that for a parabolic in the sense of Petrovsky differential expression L_1 the weight m is necessarily an even number (below it will be denoted $2b$), and the numbers b and r are defined in a unique way – r is the degree of the polynomial $L_1(t, x, p, i\sigma)$ in p, while $2br$ is its degree in $\sigma_1, \dots, \sigma_n$.

Let us consider the case $N > 1$.

Definition 1.2. *A matrix differential expression $L_N(t, x, \partial_t, \partial_x)$ with elements $L_{Njl}(t, x, \partial_t, \partial_x)$, $\{j, l\} \subset \{1, \dots, N\}$, is called parabolic in the sense of Petrovsky at a point (t, x) or uniformly on Q, if:*

1) *the differential expression* $\det L_N(t, x, \partial_t, \partial x)$ *is (respectively) parabolic in the sense of Definition* 1.1*;*

2) *the degree of each polynomial $L_{Njl}(t, x, \lambda^{2b} p, i\lambda\sigma)$, $\{j, l\} \subset \{1, \dots, N\}$, with respect to λ does not exceed $2br_l$ and*

$$L_{Njl}(t, x, p, i\sigma) = \delta_{jl} p^{r_l} + \tilde{L}_{Njl}(t, x, p, i\sigma),$$

where δ_{jl} is a Kronecker symbol, r_l is a certain number from $\mathbb{N}$, and $\tilde{L}_{Njl}$ is a polynomial which does not contain p^{r_l}.

Note that the degree of generalized homogeneity of the principal part L_{Njl}^0 of the polynomial L_{Njl} equals $2br_l$ and does not depend on j.

Definition 1.3. *The equation* (1.1.1) *is called parabolic in the sense of Petrovsky at a point (t, x) or uniformly on the set Q, if such is the differential expression L_N.*

Thus, an equation (1.1.1), parabolic in the sense of Petrovsky with the weight $2b$, in the scalar form, has the following structure:

$$\partial_t^{r_j} u_j(t,x) - \sum_{l=1}^{N} \sum_{\substack{2bk_0+|k|\le 2br_l \\ (k_0<r_l)}} a_{k_0 k}^{jl}(t,x)\partial_t^{k_0}\partial_x^k u_l(t,x)$$
$$= f_j(t,x), \quad (t,x)\in Q, \quad j\in\{1,\dots,N\}, \tag{1.1.3}$$

where u_j and $f_j, j \in \{1,\dots,N\}$, are respectively elements of the columns u and f from $\mathbb{C}_{N1}$, $|k| := k_1 + \cdots + k_n$ for $k := (k_1,\dots,k_n) \in \mathbb{Z}_+^n$. In particular, if $r_l = 1$, $l \in \{1,\dots,N\}$, then such an equation can be written as

$$\left(I\partial_t - \sum_{|k|\le 2b} a_k(t,x)\partial_x^k\right)u(t,x) = f(t,x), \quad (t,x)\in Q, \tag{1.1.4}$$

where I is a unit matrix of the order N, and $a_k : Q \to \mathbb{C}_{NN}$, $|k| \le 2b$.

We see that the definition of parabolicity in the sense of Petrovsky assumes a certain structure of an equation, separation of its principal part satisfying an algebraic condition. All the spatial variables $x_1,\dots,x_n$ have the same strength, in the sense that the differentiation in each of them has the same weight $1/(2b)$ with respect to the differentiation in t.

If $N = b = 1$, then the equation (1.1.1) is a scalar second order equation

$$\left(\partial_t - \sum_{j,l=1}^{n} a_{jl}(t,x)\partial_{x_j}\partial_{x_l} - \sum_{j=1}^{n} a_j(t,x)\partial_{x_j} - a(t,x)\right)u(t,x)$$
$$= f(t,x), \quad (t,x)\in Q. \tag{1.1.5}$$

Its uniform parabolicity in the sense of Petrovsky on Q means the existence of such a constant $\delta > 0$ that for any $(t,x) \in Q$ and $\sigma \in \mathbb{R}^n$,

$$\operatorname{Re}\sum_{j,l=1}^{n} a_{jl}(t,x)\sigma_j\sigma_l \ge \delta|\sigma|^2.$$

If the coefficients of the equation (1.1.5) are real, this condition means that the quadratic form

$$\sum_{j,l=1}^{n} a_{jl}(t,x)\sigma_j\sigma_l, \quad \sigma\in\mathbb{R}^n \tag{1.1.6}$$

is positive definite, uniformly with respect to $(t,x) \in Q$.

Let us give the simplest important examples of equations which are uniformly parabolic in the sense of Petrovsky on $\mathbb{R}^{n+1}$. In these examples $a > 0, \alpha > 0, \beta \in \mathbb{R}, b \in \mathbb{N}$, and $\Delta \equiv \sum_{j=1}^{n} \partial_{x_j}^2$.

Example 1.1. The classical heat equation

$$(\partial_t - a\Delta)u = f.$$

Example 1.2. Higher order equations with constant coefficients

$$(\partial_t + (-1)^b a\Delta^b)u = f, \quad (\partial_t + \alpha\Delta^2 + \beta\Delta)u = f.$$

Example 1.3. The polycaloric equation

$$(\partial_t - a\Delta)^b u = f,$$

the simplest example of an equation of an arbitrary order in t, which is parabolic in the sense of Petrovsky.

Example 1.4. The bi-parabolic heat equation

$$(\alpha(\partial_t - a\Delta)^2 + \beta(\partial_t - a\Delta))u = f,$$

proposed in [85] for a more adequate mathematical description of heat and diffusion processes than the classical heat equation.

Example 1.5. The equation

$$(\partial_t - a\Delta)u - \beta\sum_{j=1}^{n}\partial_{x_j}(x_j u) = f,$$

generalizing the Kolmogorov equation

$$(\partial_t - a\partial_x^2)u - \beta\partial_x(xu) = f \tag{1.1.7}$$

for a one-dimensional Ornstein-Uhlenbeck random process [18]. Equations of this type with coefficients, which increase with the growth of spatial variables, appear, in particular, in some problems of statistical radiophysics [220].

The definition of parabolicity in the sense of Petrovsky can be generalized in various directions. In particular, there exist some definitions of parabolicity of the equation (1.1.1), in which the condition 1) of Definition 1.2 is preserved, but the degrees of generalized homogeneity of the polynomials L^0_{Njl} may depend both on l and j. The latter means that not only different unknown functions in a system of equations have different weights, as it is in the definition of parabolicity in the sense of Petrovsky, but different equations too. The widest generalization in this direction was proposed by Solonnikov [215, 146] who defined and studied a very wide class of systems of equations called parabolic in the sense of Solonnikov.

Let us consider now some generalizations of Petrovsky's parabolicity related to giving up the equal "strength" of all spatial variables and the positive definiteness of appropriate algebraic forms like the quadratic form (1.1.6) for the equation (1.1.5).

$\mathbf{E}_1$. $\overrightarrow{2b}$-parabolic equations

This class of equations (also called parabolic equations of a quasi-homogeneous structure) was defined and investigated by Eidelman [51]. Its study was continued [53] by Eidelman and Ivasyshen, and then in [108, 15, 114, 115, 178, 109] by Ivasyshen and his collaborators. The difference between this class and the class $\mathbf{E}_0$ is that the differentiations in different variables $x_j, j \in \{1, \dots, n\}$, may have different weights $1/(2b_j)$ with respect to the differentiation in t. Here $b_1, \dots, b_n$ are some numbers from $\mathbb{N}$.

Denote by $\overrightarrow{2b}$ the vector $(2b_1, \dots, 2b_n)$, by b the least common multiple of the numbers $b_1, \dots, b_n$, and by m_j the number b/b_j, $j \in \{1, \dots, n\}$.

The next definitions are similar to Definitions 1.1–1.3. We shall denote by $L_1(t,x,p,i\sigma)$ and $L_{Njl}(t,x,p,i\sigma)$ such polynomials with respect to $p \in \mathbb{C}$ and $\sigma \in \mathbb{R}^n$ that the degrees of the polynomials $L_1(t,x,\lambda^{2b}p, i(\lambda^{m_1}\sigma_1, \dots, \lambda^{m_n}\sigma_n))$ and $L_{Njl}(t,x,\lambda^{2b}p, i(\lambda^{m_1}\sigma_1, \dots, \lambda^{m_n}\sigma_n))$ with respect to λ are equal to $2br$ and does not exceed $2br_l$, respectively, where r and r_l are some numbers from $\mathbb{N}$; the principal part L_1^0 of a polynomial L_1 is defined by the equality

$$L_1^0(t,x,\lambda^{2b}p, i(\lambda^{m_1}\sigma_1, \dots, \lambda^{m_n}\sigma_n)) = \lambda^{2br} L_1^0(t,x,p,i\sigma), \quad \lambda \in \mathbb{R};$$

the polynomials L_{Njl} satisfy the equality

$$L_{Njl}(t,x,p,i\sigma) = \delta_{jl}p^{r_l} + \tilde{L}_{Njl}(t,x,p,i\sigma),$$

where $\tilde{L}_{Njl}$ is a polynomial not containing p^{r_l}.

Definition 1.4.

1) *A differential expression L_1 is called $\overrightarrow{2b}$-parabolic at a point (t,x) if p-roots of the polynomial $L_1^0(t,x,p,i\sigma)$ satisfy the inequality*

$$\mathrm{Re}\, p(t,x,\sigma) \leq -\delta \sum_{j=1}^{n} \sigma_j^{2b_j} \tag{1.1.8}$$

for any $\sigma \in \mathbb{R}^n$, with some constant $\delta > 0$ which may depend on the point (t,x).

2) *A differential expression L_1 is called uniformly $\overrightarrow{2b}$-parabolic on a set Q, if it is $\overrightarrow{2b}$-parabolic at every point of Q, and the inequality* (1.1.8) *holds for any point $(t,x) \in Q$, with the same constant δ.*

3) *A matrix differential expression L_N with elements L_{Njl}, $\{j,l\} \subset \{1,\dots,N\}$, is called $\overrightarrow{2b}$-parabolic at a point, or uniformly on Q, if the differential expression $\det L_N(t,x,\partial_t,\partial_x)$ is $\overrightarrow{2b}$-parabolic at the point, or uniformly on Q, respectively.*

4) *The equation* (1.1.1) *is called $\overrightarrow{2b}$-parabolic at a point, or uniformly on Q, if the differential expression L_N has the same property.*

For multi-indices $k := (k_1, \dots, k_n) \in \mathbb{Z}_+^n$ set $\|k\| := \sum_{j=1}^{n} m_j k_j$. Then a $\overrightarrow{2b}$-parabolic equation (1.1.1) in the general case, and in the case when $r_l = 1$, $l \in \{1, \dots, N\}$, has the following structure, similar to (1.1.3) and (1.1.4), respectively:

$$\partial_t^{r_j} u_j(t,x) - \sum_{l=1}^{N} \sum_{\substack{2bk_0 + \|k\| \le 2br_l \\ (k_0 < r_l)}} a_{k_0 k}^{jl}(t,x) \partial_t^{k_0} \partial_x^k u_l(t,x)$$
$$= f_j(t,x), \quad (t,x) \in Q, \quad j \in \{1, \dots, N\}, \tag{1.1.9}$$

and

$$\left(I\partial_t - \sum_{\|k\| \le 2b} a_k(t,x) \partial_x^k \right) u(t,x) = f(t,x), \quad (t,x) \in Q. \tag{1.1.10}$$

In particular, the uniform on Q $\overrightarrow{2b}$-parabolicity of the equation (1.1.10) means that p-roots of the equation

$$\det \left(Ip - \sum_{\|k\| = 2b} a_k(t,x)(i\sigma)^k \right) = 0$$

satisfy for all $(t,x) \in Q$ and $\sigma \in \mathbb{R}^n$ the inequality (1.1.8) with some constant $\delta > 0$. This constant will be called the constant of $\overrightarrow{2b}$-parabolicity of the equation (1.1.10).

For an illustration, here are the simplest $\overrightarrow{2b}$-parabolic equations with constant coefficients.

Example 1.6. Analogs of the simplest higher order equations, parabolic in the sense of Petrovsky, from Example 1.2:

$$\left(\partial_t + a \sum_{j=1}^{n} (-1)^{b_j} \partial_{x_j}^{2b_j} \right) u = f,$$

$$\left(\partial_t - a_1 \sum_{j=1}^{n_1} \partial_{x_1}^2 + a_2 \left(\sum_{j=1}^{n_2} \partial_{x_{n_1+j}}^2 \right)^2 - a_3 \left(\sum_{j=1}^{n_3} \partial_{x_{n_1+n_2+j}}^2 \right)^3 \right) u = f,$$

where $a > 0$, $a_j > 0$, $j \in \{1,2,3\}$, $\{n_1, n_2, n_3\} \subset \mathbb{N}$, $n_1 + n_2 + n_3 = n$, in the first equation b_j, $j \in \{1, \dots, n\}$, are arbitrary numbers from $\mathbb{N}$, in the second equation $b_j = 1$ for $j \in \{1, \dots, n_1\}$, $b_j = 2$ for $j \in \{n_1 + 1, \dots, n_1 + n_2\}$, and $b_j = 3$ as $j \in \{n_1 + n_2 + 1, \dots, n\}$.

Hypothetically, $\overrightarrow{2b}$-parabolic equations can emerge if the process described by the equation is considerably heterogeneous with respect to different spatial variables. However it looks more realistic to expect that such a heterogeneity will be described by pseudo-differential equations, not by differential ones.

Below we shall consider the classes $\mathbf{E}_{1j}$, $j \geq 1$, subclasses of the above class $\mathbf{E}_1$. The basic class is $\mathbf{E}_{11}$, the class of $\overrightarrow{2b}$-parabolic equations, of the form (1.1.10), uniform in the layer $\Pi_{[0,T]} \equiv [0,T] \times \mathbb{R}^n$, $T > 0$, with bounded Hölder continuous coefficients. Other subclasses consist of equations which are in a certain sense degenerate in comparison to $\mathbf{E}_{11}$.

In particular, $\mathbf{E}_{12}$ is the class of $\overrightarrow{2b}$-parabolic equations of the form (1.1.10), uniform in $\Pi_{[0,T]}$, with bounded coefficients from the group of principal terms and increasing coefficients, as $|x| \to \infty$, from the group of lower terms. The main notion for this class is the dissipativity of an equation. Dissipative equations, parabolic in the sense of Petrovsky, were introduced and studied by Eidelman [50, 45, 46], Eidelman and Porper [68]. Such equations generalize the equation

$$\left(\partial_t - \sum_{j=1}^{n} \partial_{x_j}^2 + (a(x))^2\right) u = f,$$

where the function $a : \mathbb{R}^n \to [1, \infty)$ tends to infinity as $|x| \to \infty$. The latter means increasing heat emission.

The notion of dissipativity for $\overrightarrow{2b}$-parabolic equations was defined by Ivasyshen and Pasichnyk [114, 115, 178], as follows.

Definition 1.5. *The equation* (1.1.10) *is called dissipative $\overrightarrow{2b}$-parabolic in $Q = \Pi_{[0,T]}$, if there exists a continuous function $D : \mathbb{R}^n \to [1, \infty)$, satisfying the following conditions:*

1) $D(x) \to \infty$ *as* $|x| \to \infty$;
2) *the functions* $b_k(t,x) := a_k(t,x)(D(x))^{\|k\|-2b}$, $(t,x) \in \Pi_{[0,T]}$, $\|k\| \leq 2b$, *are bounded;*
3) *the equation* $\left(I\partial_t - \sum_{\|k\|+k_{n+1}=2b} b_k(t,x)\partial_x^k(-i\partial_{x_{n+1}})^{k_n+1}\right) v(t,x,x_{n+1}) = 0$

with bounded coefficients and an additional spatial variable x_{n+1} is, uniformly on $\Pi_{[0,T]} \times \mathbb{R}$, $\overrightarrow{2B}$-parabolic, where $\overrightarrow{2B} := (2b_1, \ldots, 2b_n, 2b)$, that is the p-roots of the equation

$$\det\left(Ip - \sum_{\|k\|+k_{n+1}=2b} b_k(t,x)(i\sigma)^k \mu^{k_{n+1}}\right) = 0$$

satisfy the inequality

$$\operatorname{Re} p(t,x,\sigma,\mu) \leq -\delta\left(\sum_{j=1}^{n} \sigma_j^{2b_j} + \mu^{2b}\right)$$

for any $(t,x) \in \Pi_{[0,T]}$, $\sigma \in \mathbb{R}^n$, $\mu \in \mathbb{R}$ and some constant $\delta > 0$.

D is called the characteristic of dissipation for the equation (1.1.10).

Example 1.7. Let a function $a : \mathbb{R}^3 \to [1,\infty)$ be continuous and such that $a(x) \to \infty$ as $|x| \to \infty$. Then the equation

$$(\partial_t - \partial_{x_1}^2 + \partial_{x_2}^4 - \partial_{x_3}^6 + (a(x))^{12})u = f$$

is a dissipative $\overrightarrow{2b}$-parabolic equation on $[0,T] \times \mathbb{R}^3$ with the characteristic of dissipation $D = a$, where $\overrightarrow{2b} = (2,4,6)$, T is an arbitrary positive number. Here $2b = 12$.

Apart from dissipative equations, the class E_{12} contains some $\overrightarrow{2b}$-parabolic equations with increasing coefficients which can be transformed into dissipative ones.

The next class of equations we consider, is the class $\mathbf{E}_{13}$ of $\overrightarrow{2b}$-parabolic equations with certain degenerations on the initial hyperplane, of the form

$$\left(\alpha(t)I\partial_t - \beta(t) \sum_{0<\|k\|\leq 2b} a_k(t,x)\partial_x^k - a_0(t,x)\right)u(t,x) = f(t,x),$$
$$(t,x) \in \Pi_{(0,T]}, \tag{1.1.11}$$

under the following assumptions: α and β are continuous on $[0,T]$ functions, for which $\alpha(t) > 0$, $\beta(t) > 0$ for $t \in (0,T]$, $\alpha(0)\beta(0) = 0$, and β is monotonically non-decreasing; the differential expression

$$I\partial_t - \sum_{\|k\|\leq 2b} a_k(t,x)\partial_x^k$$

is uniformly $\overrightarrow{2b}$-parabolic on $\Pi_{[0,T]}$.

Equations of the form (1.1.11), parabolic in the sense of Petrovsky, were studied by Ivasyshen, Voznyak, and Medynsky [118, 111], $\overrightarrow{2b}$-parabolic ones were investigated by Berezan and Ivasyshen [15].

Example 1.8. The equation

$$\left(t^{\alpha_0}\partial_t + t^{\beta_0}\sum_{j=1}^{n}(-1)^{b_j}\partial_{x_j}^{2b_j} + a_0\right)u = f,$$

with $\alpha_0 \geq 0$, $\beta_0 \geq 0$, $\alpha_0 + \beta_0 > 0$ and $a_0 \in \mathbb{R}$, is a typical example of an equation from the class $\mathbf{E}_{13}$.

Let us introduce also the class $\mathbf{E}_{14}$, which can be seen as a synthesis of the classes $\mathbf{E}_{12}$ and $\mathbf{E}_{13}$. It consists of $\overrightarrow{2b}$-parabolic equations of the form (1.1.11), with degenerations on the initial hyperplane and possible arbitrary growth, as $|x| \to \infty$, of coefficients at lower derivatives. The investigation of such a class was initiated by Ivasyshen and Pasichnyk [114, 115, 178].

Example 1.9. The equation

$$(t^{\alpha_0}\partial_t - t^{\beta_0}(\partial^2_{x_1} - \partial^4_{x_2} + \partial^6_{x_3}) + (a(x))^{12})u = f,$$

with α_0, β_0, and a from Examples 1.7 and 1.8, belongs to the class $\mathbf{E}_{14}$.

$\mathbf{E}_2$. Degenerate Equations of the Kolmogorov Type

We shall define a class of equations constituting a natural generalization of the equation of diffusion with inertia used by Kolmogorov [135] as a mathematical model of the Brownian motion for a system with n degrees of freedom taking into account the inertia of the system. The latter is achieved by describing the state of a system by values of its n coordinates $y_1, \ldots, y_n$ and their time derivatives $\dot{y}_1, \ldots, \dot{y}_n$. The simplest Kolmogorov equation for $n = 1$ has the form

$$(\partial_t + \dot{y}_1\partial_{y_1} - \partial^2_{\dot{y}_1})u = f. \tag{1.1.12}$$

This equation is, degenerate in y_1, parabolic in the sense of Petrovsky with respect to the variables t and $\dot{y}_1$. It belongs to the class of ultra-parabolic equations. Numerous generalizations of the equation (1.1.12) were introduced and studied by many authors [227, 101, 216, 209, 142, 143, 144, 145, 64, 65, 73, 155, 207, 158, 100, 2, 159, 62, 3, 74, 161, 119, 55, 38, 61, 120, 56, 57, 36, 58, 59, 60]. The degeneration may happen not only with respect to a single group of variables; there may be two or more of them.

We shall consider mostly the case of degeneration in two groups of variables. Thus we shall assume that the n-dimensional spatial variable x consists of the n_1-dimensional variable $x_1 := (x_{11}, \ldots, x_{1n_1})$, n_2-dimensional variable $x_2 := (x_{21}, \ldots, x_{2n_2})$, and n_3-dimensional variable $x_3 := (x_{31}, \ldots, x_{3n_3})$, that is $x := (x_1, x_2, x_3)$. Here n_1, n_2, and n_3 are such numbers from $\mathbb{N}$, that $n_3 \le n_2 \le n_1$, and $n_1 + n_2 + n_3 = n$. Correspondingly, a multi-index $k \in \mathbb{Z}^n_+$ will be written as $k := (k_1, k_2, k_3)$, where $k_l := (k_{l1}, \ldots, k_{ln_l}) \in \mathbb{Z}^{n_l}_+$, $l \in \{1, 2, 3\}$.

Let us consider a scalar equation of the form

$$\left(\partial_t - \sum_{j=1}^{n_2} x_{1j}\partial_{x_{2j}} - \sum_{j=1}^{n_3} x_{2j}\partial_{x_{3j}} - A(t, x, \partial_{x_1})\right)u(t,x) = f(t,x),$$
$$(t,x) \in Q, \tag{1.1.13}$$

where $A(t, x, \partial_{x_1})$ is a differential expression in x_1 with coefficients depending on t and x. The variables t and x_1 will be called basic.

Definition 1.6.

1) *An equation* (1.1.13) *is called a degenerate equation of the Kolmogorov type, if the differential expression* $\partial_t - A(t, x, \partial_{x_1})$ *is parabolic (in a certain sense) with respect to the basic variables.*

2) *If a differential expression* $\partial_t - A(t, x, \partial_{x_1})$ *is parabolic, uniformly on* Q, *in the sense of Petrovsky with a weight* $2b$ *in the basic variables, then the equation*

(1.1.13) *is called a degenerate parabolic equation of the Kolmogorov type of the order* $2b$*; in the case* $b = 1$ *it is called an ultra-parabolic equation of the Kolmogorov type.*

3) *If an expression* $\partial_t - A(t, x, \partial_{x_1})$ *is* $\overrightarrow{2b} := (2b_1, \dots, 2b_{n_1})$*-parabolic uniformly on* Q*, then the equation* (1.1.13) *is called a degenerate equation of the Kolmogorov type with a* $\overrightarrow{2b}$*-parabolic part with respect to the basic variables.*

The class of equations defined in 1) will be denoted by $\mathbf{E}_2$, $\mathbf{E}_{21}$ and $\mathbf{E}_{22}$ will mean the classes of 2), and $\mathbf{E}_{23}$ will denote the class defined in 3).

Thus, an equation from the class $\mathbf{E}_{21}$ has the form

$$\left(\partial_t - \sum_{j=1}^{n_2} x_{1j}\partial_{x_{2j}} - \sum_{j=1}^{n_3} x_{2j}\partial_{x_{3j}} - \sum_{|k_1|\le 2b} a_{k_1}(t,x)\partial_{x_1}^{k_1}\right)u(t,x) = f(t,x),$$
$$(t,x) \in Q, \tag{1.1.14}$$

an equation from $\mathbf{E}_{22}$ can be written as

$$\left(\partial_t - \sum_{j=1}^{n_2} x_{1j}\partial_{x_{2j}} - \sum_{j=1}^{n_3} x_{2j}\partial_{x_{3j}} - \sum_{j,l=1}^{n_1} a_{jl}(t,x)\partial_{x_{1j}}\partial_{x_{1l}}\right.$$
$$\left. - \sum_{j=1}^{n_1} a_j(t,x)\partial_{x_{1j}} - a_0(t,x)\right)u(t,x) = f(t,x), \quad (t,x) \in Q; \tag{1.1.15}$$

the form of an equation from $\mathbf{E}_{23}$ is

$$\left(\partial_t - \sum_{j=1}^{n_2} x_{1j}\partial_{x_{2j}} - \sum_{j=1}^{n_3} x_{2j}\partial_{x_{3j}} - \sum_{\|k_1\|\le 2b} a_{k_1}(t,x)\partial_{x_1}^{k_1}\right)u(t,x) = f(t,x),$$
$$(t,x) \in Q. \tag{1.1.16}$$

In (1.1.14) b is a given number from $\mathbb{N}$, $|k_1| := \sum_{j=1}^{n_1} k_{1j}$; in (1.1.16) b is the least common multiple of the given numbers $b_1, \dots, b_{n_1}$ from $\mathbb{N}$, $\|k_1\| := \sum_{j=1}^{n_1} m_j k_{1j}$, where $m_j := b/b_j, j \in \{1, \dots, n_1\}$.

Appropriate parabolicity conditions hold for the equations (1.1.14)–(1.1.16), that is there exists such a constant $\delta > 0$ that for any $(t,x) \in Q$ and $\sigma_1 := (\sigma_{11}, \dots, \sigma_{1n_1}) \in \mathbb{R}^{n_1}$ the respective inequalities

$$\operatorname{Re} \sum_{|k_1|=2b} a_{k_1}(t,x)(i\sigma_1)^{k_1} \le -\delta|\sigma_1|^{2b}, \tag{1.1.17}$$

$$\operatorname{Re} \sum_{j,l=1}^{n_1} a_{jl}(t,x)\sigma_{1j}\sigma_{1l} \ge \delta|\sigma_1|^2, \tag{1.1.18}$$

and

$$\operatorname{Re} \sum_{\|k_1\|=2b} a_{k_1}(t,x)(i\sigma_1)^{k_1} \leq -\delta \sum_{j=1}^{n_1} \sigma_{1j}^{2b_j} \tag{1.1.19}$$

hold.

The equations (1.1.13)–(1.1.16) contain three groups x_1, x_2, and x_3 of spatial variables. If there are only two groups x_1 and x_2, then the second sum is absent from the corresponding equations. We shall assume that these equations are special cases of (1.1.13)–(1.1.16) corresponding formally to $n_3 = 0$.

Equations from the class $\mathbf{E}_{22}$ have been investigated in many papers, in particular, in [227, 101, 216, 209, 142, 143, 144, 145, 64, 207, 62, 74, 161, 38, 61, 36]. The class $\mathbf{E}_{21}$ was defined by Eidelman and Malitskaya [65]. Equations from this class were investigated by Eidelman, Malitskaya, Tychinskaya, Ivasyshen, Androsova, and Voznyak [65, 73, 155, 158, 159, 2, 3, 119]. The class $\mathbf{E}_{23}$ was defined by Eidelman and Ivasyshen who began its investigation in [56, 57, 58, 59, 60].

Let us give some examples of equations from the above classes. In these examples a and a_j are positive constants, b, b_j, and n_1' are natural numbers with $n_1' < n_1$.

Example 1.10. An equation from the class $\mathbf{E}_{22}$

$$\left(\partial_t - \sum_{j=1}^{n_2} x_{1j}\partial_{x_{2j}} - \sum_{j=1}^{n_3} x_{2j}\partial_{x_{3j}} - a\sum_{j=1}^{n_1} \partial^2_{x_{1j}}\right)u = f.$$

Example 1.11. An equation from the class $\mathbf{E}_{21}$

$$\left(\partial_t - \sum_{j=1}^{n_2} x_{1j}\partial_{x_{2j}} - \sum_{j=1}^{n_3} x_{2j}\partial_{x_{3j}} + (-1)^b a\left(\sum_{j=1}^{n_1} \partial^2_{x_{1j}}\right)^b\right)u = f.$$

Example 1.12. Equations from the class $\mathbf{E}_{23}$

$$\left(\partial_t - \sum_{j=1}^{n_2} x_{1j}\partial_{x_{2j}} - \sum_{j=1}^{n_3} x_{2j}\partial_{x_{3j}} - a_1\sum_{j=1}^{n_1'} \partial^2_{x_{1j}} + (-1)^b a_2\left(\sum_{j=n_1'+1}^{n_1} \partial^2_{x_{1j}}\right)^b\right)u = f,$$

$$\left(\partial_t - \sum_{j=1}^{n_2} x_{1j}\partial_{x_{2j}} - \sum_{j=1}^{n_3} x_{2j}\partial_{x_{3j}} + \sum_{j=1}^{n_1} (-1)^{b_j} a_j \partial^{2b_j}_{x_{1j}}\right)u = f.$$

1.1.2. Problems. For the above classes of equations, it is natural to consider the Cauchy problem, that is the problem of finding a solution of an equation in the layer $\Pi_{(0,T]} := (0,T] \times \mathbb{R}^n$ satisfying, in the case of the equations (1.1.3) and (1.1.9) the initial conditions

$$\partial_t^{s-1} u_j(t,x)\Big|_{t=0} = \varphi_j^s(x), \quad x \in \mathbb{R}^n,$$
$$s \in \{1,\dots,r_j\}, \quad j \in \{1,\dots,N\}; \tag{1.1.20}$$

for equations of the first order in t the initial condition has the form

$$u(t,x)\Big|_{t=0} = \varphi(x), \quad x \in \mathbb{R}^n. \tag{1.1.21}$$

We assume in (1.1.20) and (1.1.21) that $\partial_t^{s-1} u_j(t,\cdot)$ and $u(t,\cdot)$ tend, as $t \to 0$, to φ_j^s and φ respectively, in the topologies of those spaces, in which the solvability of the problems is studied. In the simplest case (1.1.20) and (1.1.21) may be understood as the limits for any fixed x, or as the uniform limits on each compact subset of $\mathbb{R}^n$.

The basic notion in the approach to the Cauchy problem used in this book is a fundamental solution (or a fundamental matrix of solutions) of the Cauchy problem. If it is constructed and thoroughly studied, then it is possible to obtain exact, sometimes the best possible, results regarding intrinsic properties of solutions, correct solvability of the Cauchy problem in various function spaces, integral representations of solutions, local solvability of the Cauchy problem for nonlinear equations, extendability of solutions onto a larger time interval etc [83, 45, 146].

The fundamental solution of the Cauchy problem (FSCP) is defined as follows.

Definition 1.7.

1) *A FSCP for an equation of the form*

$$(L_N u)(t,x) := (I\partial_t - A(t,x,\partial_x))u(t,x) = f(t,x), \quad (t,x) \in \Pi_{(0,T]},$$

is such a function

$$Z(\cdot,\cdot;\tau,\xi) : \Pi_{(\tau,T]} \to \mathbb{C}_{NN}, \tag{1.1.22}$$

depending on the parametric point $(\tau,\xi) \in \Pi_{[0,T)}$, *that the formula*

$$u(t,x) = \int_{\mathbb{R}^n} Z(t,x;\tau,\xi)\varphi(\xi)d\xi, \quad (t,x) \in \Pi_{(\tau,T]},$$

determines a solution of the homogeneous equation $L_N u = 0$ *on* $\Pi_{(\tau,T]}$, *satisfying the initial condition*

$$u|_{t=\tau} = \varphi \tag{1.1.23}$$

for any $\tau \in [0,T)$ *and an arbitrary continuous bounded function* φ.

In other words, FSCP is a function (1.1.22) *solving for any* $(\tau,\xi)\in\Pi_{[0,T)}$ *the Cauchy problem*

$$LZ=0,\quad Z|_{t=\tau}=I\delta_\xi,$$

where δ_ξ *is the Dirac delta-function concentrated in the point* ξ.

2) *For equations of arbitrary orders in* t *of the form* (1.1.3) *or* (1.1.9), *the FSCP is a function* (1.1.22) *which, for any* $(\tau,\xi)\in\Pi_{[0,T)}$, *is a solution on* $\Pi_{(\tau,T]}$ *of the Cauchy problem for the corresponding homogeneous equation with the following initial conditions:*

$$\partial_t^{s-1}Z_{jl}\Big|_{t=\tau}=\delta_{jl}\delta_{sr_j}\delta_\xi,\quad s\in\{1,\dots,r_j\},\quad \{j,l\}\subset\{1,\dots,N\},$$

where Z_{jl} *are elements of the matrix* Z.

Using only a FSCP, one can obtain an integral representation of a solution of the Cauchy problem for equations of the first order in t, and also for higher order equations, if the functions φ_j^s, $s\le r_j-1$, from the condition (1.1.20) equal zero. In the case of general initial conditions an integral representation of solutions of the Cauchy problem (1.1.9), (1.1.20) can be made using *a Green matrix of the Cauchy problem* [109]. For the scalar equation (1.1.1) ($N=1$) of an order $r>1$ in t, a Green matrix of the Cauchy problem is such a row-matrix $G:=(G_0,G_1,\dots,G_r)$ that the solution of this equation with the initial conditions

$$\partial_t^{s-1}u(t,x)|_{t=0}=\varphi_s(x),\quad x\in\mathbb{R}^n,\quad s\in\{1,\dots,r\},$$

for arbitrary smooth functions f and φ_s with compact supports has the representation

$$u(t,x)=\int_0^t d\tau\int_{\mathbb{R}^n}G_0(t,x;\tau,\xi)f(\tau,\xi)d\xi$$
$$+\sum_{s=1}^r\int_{\mathbb{R}^n}G_s(t,x;\xi)\varphi_s(\xi)d\xi,\quad (t,x)\in\Pi_{(0,T]}.$$

The FSCP Z generates generalized heat potentials – the Poisson integral

$$(P\varphi)(t,x):=\int_{\mathbb{R}^n}Z(t,x;0,\xi)\varphi(\xi)d\xi,\quad (t,x)\in\Pi_{(0,T]},\tag{1.1.24}$$

of a function φ, the Poisson integral

$$(P\varphi)(t,x):=\int_{\mathbb{R}^n}Z(t,x;0,\xi)d\varphi(\xi),\quad (t,x)\in\Pi_{(0,T]},\tag{1.1.25}$$

of a generalized measure φ, and the volume potential

$$(Vf)(t,x) := \int_0^t d\tau \int_{\mathbb{R}^n} Z(t,x;\tau,\xi)f(\tau,\xi)d\xi, \quad (t,x) \in \Pi_{(0,T]}$$

of a density f.

Studying the Cauchy problem, it is important to investigate properties of these potentials, especially to give a precise description of the ranges of the operators P and V, and, in particular, to investigate the behaviour of $P\varphi$ as $t \to 0$. These properties are necessary for proving existence, uniqueness, and correct solvability of the Cauchy problem in various function spaces, for obtaining an integral representation of solutions, for studying the behaviour, as $t \to \pm\infty$, of solutions defined in domains unbounded with respect to the time variable.

In this book we solve the following principal problems of the classes of equations defined in Section 1.1.1:

$\mathbf{P}_1$. Construction and study of FSCPs.

$\mathbf{P}_2$. Investigation of properties of generalized heat potentials.

$\mathbf{P}_3$. Finding the classes of existence, uniqueness, and correctness of the Cauchy problem.

$\mathbf{P}_4$. Finding conditions for the existence of weighted L_p-traces of solutions, defined on the open layer $\Pi_{(0,T]}$, on the initial hyperplane $\{t = 0\}$, and the reconstruction of the solutions from their traces.

1.1.3. Results. Completeness and precision of the investigation of correct solvability and properties of solutions of the Cauchy problem depends crucially on the completeness of the study of properties of the FSCP including properties of the corresponding potentials. Therefore the central problems of Chapters 2 and 3 are $\mathbf{P}_1$ and $\mathbf{P}_2$. In the course of solving them, we implement the procedures for constructing the FSCP Z and obtaining sufficiently precise estimates of Z and its derivatives. We obtain estimates for the Poisson integrals of functions and generalized measures from special weighted classes and study the behaviour of potentials for $t \to 0$. Conditions are found upon densities of the volume potentials, under which they are solutions (from appropriate spaces) of inhomogeneous equations.

Note that the most complete and precise of the above results from Chapters 2 and 3 are those for equations of the class $\mathbf{E}_{11}$ (with bounded and Hölder continuous coefficients). For that class of equations, not only precise estimates of the function Z and its derivatives, but also the exact estimates of differences in t and x of its higher derivatives are obtained.

The exact estimates of the FSCP make it possible to prove precise theorems of the correct solvability of the Cauchy problem. The construction given below is a general framework for such theorems, and also for the theorems on the representation of solutions defined on an open layer $\Pi_{(0,T]}$, in terms of their limit values on the hyperplane $\{t = 0\}$. The essence of the construction is a characterization of the evolution of a solution of the Cauchy problem in t by their belonging to a family of Banach spaces (a specific space for each value of t).

Let us describe the general assumptions on the families of the Banach spaces, and the results based on these families, and also the basic definitions and results for parabolic equations in the sense of Petrovsky.

Consider a parabolic (in a sense to be specified later) equation of the form

$$(\partial_t - A(t,x,\partial_x))u(t,x) = 0, \quad (t,x) \in \Pi_{(0,T]}, \tag{1.1.26}$$

with the condition on the lower boundary of the layer $\Pi_{(0,T]}$ (the initial condition)

$$u|_{t=0} = \varphi. \tag{1.1.27}$$

Denote by Φ such a space of initial data that for any Φ there exists a classical solution u of the Cauchy problem (1.1.26), (1.1.27). Let U be the space of all such solutions u.

In the 1950s Eidelman [49, 52] noticed for the case of parabolic equations in the sense of Petrovsky (1.1.26) that if Φ is the space of functions increasing exponentially as $|x| \to \infty$, then the appropriate space U consists of functions which increase exponentially for $|x| \to \infty$ as functions of x, for every fixed $t \in (0,T]$. The type of growth may depend on t. That means that a solution $u(t,x)$, $(t,x) \in \Pi_{(0,T]}$, as a function of x for each fixed $t \in (0,T]$, belongs to a certain space U_t. Hence the temporal evolution of solutions of the problem (1.1.26), (1.1.27) can be described by their belonging to the appropriate spaces U_t. Ivasyshen [108] has proved that the space U can be defined as the space of classical solutions u of the equation (1.1.26), possessing the following properties – for any $t \in (0,T]$ $u(t,\cdot) \in U_t$, and there exists such a constant $C > 0$ that for all $t \in (0,T]$,

$$\|u(t,\cdot)\|_{U_t} \leq C.$$

The above space U can be characterized as the range of the Poisson operator P defined by the formulas (1.1.24) and (1.1.25) on the space Φ. Now the operator $P : \Phi \to U$ is an isomorphism.

Thus, if Φ is chosen appropriately, and $U_t, t \in (0,T]$, are the corresponding spaces, as above, then the following statements are valid (see [60]).

Metatheorem A. *If $\varphi \in \Phi$, then there exists a unique solution u of the problem* (1.1.26), (1.1.27) *which belongs to the space U and has the form*

$$u = P\varphi. \tag{1.1.28}$$

The exact meaning of the initial condition (1.1.27) *depends on the choice of the space Φ and should be specified.*

Metatheorem B. *If u is a solution of the equation* (1.1.26) *from the space U, then there exists a unique element $\varphi \in \Phi$, such that the representation* (1.1.28) *is valid.*

Hence Metatheorems A and B are valid if we choose appropriate spaces Φ and find the corresponding families of spaces $U_t, t \in (0,T]$.

Wide enough families of spaces Φ and $U_t, t \in (0,T]$, are described in the monograph [45] by Eidelman for equations parabolic in the sense of Petrovsky. Metatheorem 1.1 for such equations was also validated there. Ivasyshen [108] extended the class of families essentially, and substantiated Metatheorem B. Let us give the definitions for the case of equations

$$\left(\partial_t - \sum_{|k|\le 2b} a_k(t,x)\partial_x^k\right)u(t,x) = 0, \quad (t,x) \in \Pi_{(0,T]}. \tag{1.1.29}$$

Our assumptions are as follows:

1) the equation (1.1.29) is uniformly parabolic in the sense of Petrovsky on the layer $\Pi_{[0,T]}$;
2) the coefficients $a_k, |k| \le 2b$, are bounded on $\Pi_{[0,T]}$, continuous in t, uniformly in $x \in \mathbb{R}^n$, and satisfy the Hölder conditions in $x \in \mathbb{R}^n$, uniformly with respect to t;
3) there exists a Lagrange adjoint equation to the equation (1.1.29), and its coefficients satisfy the condition 2).

It is known [45] that the conditions 1) and 2) imply the existence of a FSCP Z for the equation (1.1.29), with the estimates

$$\begin{gathered}|\partial_t^{k_0}\partial_x^k Z(t,x;\tau,\xi)| \le C(t-\tau)^{-k_0-(n+|k|)/(2b)}\exp\{-c\rho(t-\tau,x,\xi)\},\\ 0 \le \tau < t \le T, \quad \{x,\xi\} \subset \mathbb{R}^n, \quad 2bk_0 + |k| \le 2b,\end{gathered} \tag{1.1.30}$$

where $C > 0$, $c > 0$, and $\rho(t,x,\xi) := t^{1-q}|x-\xi|^q$, $q := 2b/(2b-1)$.

If the condition 3) holds too, then the FSCP is normal, and the convolution formula is valid for it.

In order to define the spaces Φ and U_t, $t \in (0,T]$, for the equation (1.1.29), we use the following arguments. If we assume that the initial function φ satisfies the inequality

$$|\varphi(x)| \le C\exp\{a|x|^q\}, \quad x \in \mathbb{R}^n, \quad a \ge 0, \tag{1.1.31}$$

then, as it has been mentioned, the solution u of the equation (1.1.29) corresponding to the function φ by (1.1.24), has the bound

$$|u(t,x)| \le C\exp\{k(t,a)|x|^q\}, \quad (t,x) \in \Pi_{(0,T]}.$$

Using (1.1.24) and the estimates (1.1.30) and (1.1.31), we see that in order to find the function k, one has to find an upper bound for the function

$$f(\xi) := -c_0 t^{1-q}|x-\xi|^q + a|\xi|^q, \quad \xi \in \mathbb{R}^n,$$

for fixed $(t,x) \in \Pi_{(0,T]}$, $a \ge 0$, and $c_0 \in (0,c)$, where c is the constant from the estimate (1.1.30). Since $f(\xi) \le -c_0 t^{1-q}||x| - |\xi||^q$, it suffices to find the maximum of the function $f_0(r) := -c_0 t^{1-q}||x| - r|^q + ar^q$, $r \geq 0$. This maximum equals $k(t,a)|x|^q$ where

$$k(t,a) := c_0 a (c_0^{2b-1} - a^{2b-1} t)^{1-q}, \quad t \in [0,T], \tag{1.1.32}$$

if $a \geq 0$ is chosen in such a way that $T < (c_0/a)^{2b-1}$. Hence we have the estimate

$$-c_0 t^{1-q}|x-\xi|^q + a|\xi|^q \leq k(t,a)|x|^q, \quad t \in (0,T], \quad \{x,\xi\} \subset \mathbb{R}^n.$$

Note that $k(0,a) = a$, and the function (1.1.32) possesses the following important semigroup property:

$$k(t-\tau, k(\tau,a)) = k(t,a), \quad 0 \leq \tau \leq t \leq T. \tag{1.1.33}$$

This equality can be verified by a direct calculation.

As the spaces Φ we shall choose the spaces Φ_p^a, $1 \leq p \leq \infty$, $a \geq 0$.

If $1 < p \leq \infty$, the space Φ_p^a consists of the Lebesgue measurable functions $\varphi : \mathbb{R}^n \to \mathbb{C}$, for which the norm

$$\|\varphi\|_p^a := \|\varphi(\cdot)\exp\{-a|\cdot|^q\}\|_{L_p(\mathbb{R}^n)}$$

is finite.

Let $\mathcal{B}_n$ be the σ-algebra of Borel subsets of $\mathbb{R}^n$. Define Φ_1^a as the space of all countably additive functions $\varphi : \mathcal{B}_n \to \mathbb{C}$ (generalized Borel measures on $\mathbb{R}^n$) satisfying the condition

$$\|\varphi\|_1^a := \int_{\mathbb{R}^n} \exp\{-a|x|^q\} d|\varphi|(x) < \infty$$

where $|\varphi|$ is the full variation of φ.

The spaces $U_t, t \in (0,T]$, corresponding to Φ_p^a, are the spaces $L_p^{k(t,a)}$ of the Lebesgue measurable functions $v : \mathbb{R}^n \to \mathbb{C}$, for which the norm

$$\|v\|_p^{k(t,a)} := \|v(\cdot)\exp\{-k(t,a)|\cdot|^q\}\|_{L_p(\mathbb{R}^n)}$$

is finite.

Let P be the Poisson operator corresponding to (1.1.29). On elements $\varphi \in \Phi_p^a$, P is defined by the formula (1.1.24) if $p > 1$, or by the formula (1.1.25) if $p = 1$.

In order to formulate the main results, we introduce also the spaces Ψ_1^a and Ψ_0^a of all the Lebesgue measurable or, respectively, continuous functions $\psi : \mathbb{R}^n \to \mathbb{C}$, having, respectively, the properties

$$\|\psi(\cdot)\exp\{k(T,a)|\cdot|^q\}\|_{L_1(\mathbb{R}^n)} < \infty \qquad \text{and} \qquad |\psi(x)|\exp\{k(T,a)|x|^q\} \xrightarrow[|x|\to\infty]{} 0.$$

For the equation (1.1.29) satisfying 1)–3), Metatheorems A and B have the following specific forms.

Theorem 1.1.

1. *Let* $\varphi \in \Phi_p^a$, $1 \le p \le \infty$. *Then the function* (1.1.28) *is the unique solution of the equation* (1.1.29) *on the layer* $\Pi_{(0,T]}$, *with the following properties: for any* $t \in (0,T]$ $u(t,\cdot) \in L_p^{k(t,a)}$; *there exists such a constant* $C > 0$ *that for an arbitrary* $t \in (0,T]$,
$$\|u(t,\cdot)\|_p^{k(t,a)} \le C\|\varphi\|_p^a;$$
if $1 < p < \infty$, *then* $\lim_{t\to 0} \|u(t,\cdot) - \varphi(\cdot)\|_p^{k(t,a)} = 0$, *while for* $p = 1$ *and* $p = \infty$ $u(t,\cdot) \to \varphi$, $t \to 0$ *weakly, that is for each* ψ *from the spaces* Ψ_0^a *and* Ψ_1^a *respectively*
$$\lim_{t\to 0} \int_{\mathbb{R}^n} \psi(x)u(t,x)dx = \begin{cases} \displaystyle\int_{\mathbb{R}^n} \psi(x)d\varphi(x), & p = 1, \\ \displaystyle\int_{\mathbb{R}^n} \psi(x)\varphi(x)dx, & p = \infty. \end{cases}$$

2. *Let* u *be a solution of the equation* (1.1.29) *satisfying the condition*
$$\|u(t,\cdot)\|_p^{k(t,a)} \le C$$
for any $t \in (0,T]$, *with some* $C > 0$ *and* $p \in [1,\infty]$. *Then there exists a unique element* $\varphi \in \Phi_p^a$, *such that the solution* u *can be represented in the form* (1.1.28).

Theorems similar to Theorem 1.1 are proved in Chapters 2 and 3 for $\overrightarrow{2b}$-parabolic equations and degenerate parabolic equations of the Kolmogorov type.

In particular, for an equation from the class $\mathbf{E}_{11}$ solutions may have different types of growth with respect to different spatial variables. Therefore in this case the growth type is a vector-function $\vec{k}(t,\vec{a}) := (k_1(t,a_1),\dots,k_n(t,a_n))$, where $\vec{a} := (a_1,\dots,a_n)$,
$$k_j(t,a_j) := c_0 a_j (c_0^{2b_j-1} - a_j^{2b_j-1}t)^{1-q_j}, \quad j \in \{1,\dots,n\}. \tag{1.1.34}$$

Here $c_0, a_1, \dots, a_n$ are given numbers, such that $c_0 \in (0,c)$, $a_j \ge 0$, $j \in \{1,\dots,n\}$, $T < \min_{j\in\{1,\dots,n\}} (c_0/a_j)^{2b_j-1}$, where c is the constant from estimates like (1.1.30) for the FSCP; $q_j := 2b_j/(2b_j-1)$, $j \in \{1,\dots,n\}$. Note that for each function (1.1.34) the identity (1.1.33) holds.

For an equation from the class $\mathbf{E}_{21}$ the growth type of solutions is different for different groups of spatial variables. The formulas like (1.1.32) and (1.1.34) are much more complicated. Here we have to consider two sets of functions determining the type of a possible growth of solutions.

The above approach makes it possible to determine precisely a possible behaviour of solutions of the Cauchy problem, as $|x| \to \infty$.

Chapter 2 contains also the Schauder type theorems about the correct solvability of the Cauchy problem. Such theorems establish a precise dependence, in terms of the Hölder spaces, of the smoothness of a solution on the smoothness of initial functions, coefficients, and right-hand sides of equations.

1.1.4. Techniques. As it was mentioned in the preface, FSCPs are constructed and studied by means of Levi's parametrix method [152].

In this method a FSCP $Z(t,x;\tau,\xi)$ is sought as a sum of the principal term $Z_0(t,x;\tau,\xi)$ called a parametrix, and an additional summand $W(t,x;\tau,\xi)$ having the form of the integral with the kernel Z_0 and some unknown density Q. The parametrix Z_0 has a desired singularity at $(t,x)=(\tau,\xi)$. Usually it is defined on the basis of the FSCP $G_0(t,x;\tau,\xi;y)$ or $G_0(t,x;\tau,\xi;\beta,y)$ for model equations containing only principal (in a certain sense) groups of terms. In their coefficients the variables x or t,x are fixed in the points y or (β,y) respectively. In order to obtain the parametrix, in the expression for G_0 one should substitute, instead of y or β,y, appropriate functions which may depend on the main variables t,x and the parametric ones τ,ξ. The unknown density Q is determined from an integral equation which has the Volterra type in the variable t and can be solved by successive approximations.

Let us explain, in a little greater detail, the scheme of the Levi method for the equations (1.1.10) with $N=1$, and (1.1.14), from the classes $\mathbf{E}_{11}$ and $\mathbf{E}_{21}$ respectively.

The above scheme consists of the following successive steps.

Step 1. *Construction and study of FSCPs for equations whose coefficients may depend only on the time variable t.*

Consider the Cauchy problem with the initial condition (1.1.23) where τ is an arbitrary fixed number from the interval $[0,T)$, for the homogeneous equations (1.1.10), with $N=1$, and (1.1.14), in which the coefficients a_k, $\|k\|\le 2b$, and a_{k_1}, $|k_1|\le 2b$, may depend only on t. Applying the Fourier transform $F_{x\to\sigma}$ in the spatial variables, we transform the problem into the Cauchy problem with the initial condition

$$v|_{t=\tau}=\psi:=F_{x\to\sigma}[\varphi], \tag{1.1.35}$$

for the ordinary differential equation

$$\left(\frac{d}{dt}-\sum_{\|k\|\le 2b}a_k(t)(i\sigma)^k\right)v=0, \tag{1.1.36}$$

and the first order partial differential equation

$$\left(\partial_t+\sum_{j=1}^{n_2}\sigma_{2j}\partial_{\sigma_{1j}}+\sum_{j=1}^{n_3}\sigma_{3j}\partial_{\sigma_{2j}}-\sum_{|k_1|\le 2b}a_{k_1}(t)(i\sigma_1)^{k_1}\right)v=0, \tag{1.1.37}$$

respectively.

Solving the problems (1.1.35), (1.1.36) and (1.1.35), (1.1.37) (the latter is solved by the method of characteristics), and then applying the inverse Fourier transform $F^{-1}_{\sigma\to x}$, we get the following formula for the FSCP G corresponding to the equations (1.1.10) and (1.1.14) for the cases under consideration:

$$G(t,x;\tau,\xi) := (t-\tau)^{-M} F^{-1}_{\sigma\to y}[\tilde{V}(t,\tau,\sigma)](t,\tau,y)|_{y=y(t-\tau,x,\xi)},$$
$$0\le\tau<t\le T,\quad \{x,\xi\}\subset\mathbb{R}^n, \tag{1.1.38}$$

where in the case of the equation (1.1.10) $M := \sum_{j=1}^{n}(1/(2b_j))$,

$$\tilde{V}(t,\tau,\sigma) := \exp\left\{\sum_{\|k\|\le 2b} i^{|k|}(t-\tau)^{1-\|k\|/(2b)}\int_0^1 a_k(\tau+(t-\tau)\beta)d\beta\sigma^k\right\},$$
$$y(t,x,\xi) := (t^{-1/(2b_1)}(x_1-\xi_1),\dots,t^{-1/(2b_n)}(x_n-\xi_n)),$$

whereas for the equation (1.1.14) $M := \sum_{j=1}^{3}(j-1+1/(2b))n_j$,

$$\tilde{V}(t,\tau,\sigma) := \exp\left\{\sum_{|k_1|\le 2b} i^{|k_1|}(t-\tau)^{1-|k_1|/(2b)}\int_0^1 a_{k_1}(\tau+(t-\tau)\beta)\right.$$
$$\left.\times\,(\sigma_1'+\beta\sigma_2'+\frac{\beta^2}{2}\sigma_3)^{k_1'}(\sigma_1''+\beta\sigma_2'')^{k_1''}(\sigma_1''')^{k_1'''}\,d\beta\right\},$$

$$y(t,x,\xi) := (t^{-1/(2b)}(x_1-\xi_1), t^{-1-1/(2b)}(x_2+t\hat{x}_1-\xi_2),$$
$$t^{-2-1/(2b)}(x_3+tx_2'+\frac{t^2}{2}x_1'-\xi_3)),$$

$$x_1' := (x_{11},\dots,x_{1n_3}),\qquad x_1'' := (x_{1(n_3+1)},\dots,x_{1n_2}),$$
$$x_1''' := (x_{1(n_2+1)},\dots,x_{1n_1}),\qquad \hat{x}_1 := (x_{11},\dots,x_{1n_2}).$$

Next we study properties of the function $\tilde{V}$ and use a theorem from [86] on the Fourier transform of entire functions of a finite order (see also Lemma 1.1). As a result we obtain a complete analytical description of the function (1.1.38) and all of its derivatives, and their precise estimates.

Step 2. *Investigation of properties of FSCPs for equations with coefficients depending on the parameters.*

We consider equations of the same form as in Step 1, but with coefficients depending on the variable t and the parameter $y\in\mathbb{R}^n$, or solely on the parameters

$(\beta, y) \in \Pi_{[0,T]}$. As before, we construct the FSCPs $G(\cdot,\cdot;\cdot,\cdot;y)$ and $G(\cdot,\cdot;\cdot,\cdot;\beta,y)$, and study their properties as functions of y or β, y (their differentiability, estimates of derivatives and their increments in y or β, y etc).

Note that for the Levi method itself it is sufficient to consider in Steps 1 and 2 the equations containing only the principal terms. For the classes $\mathbf{E}_{11}$ and $\mathbf{E}_{21}$ such equations with parameters have the form

$$\left(\partial_t - \sum_{\|k\|=2b} a_k(t,y)\partial_x^k\right)u = 0 \tag{1.1.39}$$

and

$$\left(\partial_t - \sum_{j=1}^{n_2} x_{1j}\partial_{x_{2j}} - \sum_{j=1}^{n_3} x_{2j}\partial_{x_{3j}} - \sum_{|k_1|=2b} a_{k_1}(t,y)\partial_{x_1}^{k_1}\right)u = 0, \tag{1.1.40}$$

or the equation

$$\left(\partial_t - \sum_{\|k\|=2b} a_k(\beta,y)\partial_x^k\right)u = 0 \tag{1.1.41}$$

and

$$\left(\partial_t - \sum_{j=1}^{n_2} x_{1j}\partial_{x_{2j}} - \sum_{j=1}^{n_3} x_{2j}\partial_{x_{3j}} - \sum_{|k_1|=2b} a_{k_1}(\beta,y)\partial_{x_1}^{k_1}\right)u = 0, \tag{1.1.42}$$

respectively.

Step 3. *Determination of a parametrix and investigation of its properties.*

Let $G_0(\cdot,\cdot;\cdot,\cdot;y)$ be a FSCP for the equation (1.1.39) or (1.1.40). We choose the function

$$Z_0(t,x;\tau,\xi) := G_0(t,x;\tau,\xi;\xi), \quad 0 \le \tau < t \le T, \quad \{x,\xi\} \subset \mathbb{R}^n \tag{1.1.43}$$

to be the parametrix for the equation (1.1.10). The parametrix for the equation (1.1.14) can be chosen also in the form (1.1.43), but this may cause complications in the analysis of the iteration procedure of the Levi method, as well as inaccurate estimates for the FSCP and its derivatives.

It appears that the choice of a parametrix can happen to be better (under an additional condition upon coefficients of the equation) if we start from the FSCP $G(\cdot,\cdot;\cdot,\cdot;\beta,y)$ for the equation (1.1.42). In this case the parametrix is defined by the equality

$$Z_1(t,x;\tau,\xi) := G_0(t,x;\tau,\xi;\tau,\xi), \quad 0 \le \tau < t \le T, \quad \{x,\xi\} \subset \mathbb{R}^n. \tag{1.1.44}$$

After the parametrix is defined, its properties are studied. Most of them follow from properties of the FSCP for equations with parameters found in Step 2. In addition we have to study only properties of the integrals with the kernels (1.1.43) and (1.1.44).

Step 4. *Construction and investigation of FSCPs for equations with coefficients depending on all the variables.*

By the Levi method, FSCPs for the equations (1.1.10) and (1.1.14) are sought respectively in the form

$$Z(t,x;\tau,\xi) = Z_0(t,x;\tau,\xi) + \int_\tau^t d\lambda \int_{\mathbb{R}^n} Z_0(t,x;\lambda,y)Q(\lambda,y;\tau,\xi)dy,$$
$$0 \le \tau < t \le T, \quad \{x,\xi\} \subset \mathbb{R}^n, \tag{1.1.45}$$

and

$$Z(t,x;\tau,\xi) = Z_1(t,x;\tau,\xi) + \int_\tau^t d\lambda \int_{\mathbb{R}^n} Z_1(t,x;\lambda,y)Q(\lambda,y;\tau,\xi)dy,$$
$$0 \le \tau < t \le T, \quad \{x,\xi\} \subset \mathbb{R}^n, \tag{1.1.46}$$

where Z_0 and Z_1 are the parametrices, Q is an unknown function.

Using the definition of the FSCP and properties of integrals like the ones from (1.1.45) and (1.1.46) studied in the previous step, we obtain an integral equation of the Volterra type for the function Q:

$$Q(t,x;\tau,\xi) = K(t,x;\tau,\xi) + \int_\tau^t d\lambda \int_{\mathbb{R}^n} K(t,x;\lambda,y)Q(\lambda,y;\tau,\xi)dy,$$
$$0 \le \tau < t \le T, \quad \{x,\xi\} \subset \mathbb{R}^n. \tag{1.1.47}$$

This equation has a quasi-regular kernel and can be solved by iterations. Investigating properties of the solution of (1.1.47) and using properties of the volume potentials from (1.1.45) and (1.1.46), we prove the existence of the FSCP and its bounds similar to those from Step 1.

If coefficients of the above equations are smooth enough, then another choice of a parametrix is possible. For example, we can choose as a parametrix for the equation (1.1.10) any of the functions $G_0(t,x;\tau,\xi;x)$, $G_0(t,x;\tau,\xi;\tau,\xi)$, or $G_0(t,x;\tau,\xi;t,x)$, $0 \le \tau < t \le T$, $\{x,\xi\} \subset \mathbb{R}^n$, where the function $G_0(\cdot,\cdot;\cdot,\cdot;y)$ is the same as in Step 3, while $G_0(\cdot,\cdot;\cdot,\cdot;\beta,y)$ is the FSCP for the equation (1.1.41).

The application of a FSCP to the study of the Cauchy problem and the investigation of properties of solutions is based on traditional techniques requiring scrupulous estimates and analysis of properties of various potentials in special function spaces which reflect a complicated structure of equations. It is the exactness of estimates that make it possible not only to find precise classes of correctness of the Cauchy problem, but also to prove the inverse theorems mentioned in Section 1.1.3.

In the above description we dealt with equations from the classes $\mathbf{E}_{11}$ and $\mathbf{E}_{21}$ with bounded coefficients. If the coefficients are allowed to grow with a growth of spatial variables, then the above methods should be modified. In particular, for equations from the class $\mathbf{E}_{12}$ the following approach is used. First we consider the class of dissipative equations. For them, depending on properties of the characteristic of dissipation, there are two different possibilities for choosing a parametrix in the Levi method. Note that in this case the parametrix depends on all the coefficients, not just the principal ones, as above. Next we find a class of equations with increasing coefficients which can be reduced to dissipative equations by an appropriate change of the unknown function. This leads to the proof of existence of a FSCP Z, estimates for Z and its derivatives, theorems of solvability and, in some cases, well-posedness of the Cauchy problem.

The study of equations from the classes $\mathbf{E}_{13}$ and $\mathbf{E}_{14}$, which degenerate on the initial hyperplane, also requires an appropriate modification of the above method. For the equation (1.1.11) with bounded coefficients the parametrix is defined by the formula (1.1.43), in which $G_0(\cdot,\cdot;\cdot,\cdot;y)$ is the FSCP for the equation

$$\left(\alpha(t)I\partial_t - \beta(t)\sum_{\|k\|=2b} a_k(t,y)\partial_x^k - a_0(t,y)\right)u = 0. \tag{1.1.48}$$

1.2 Pseudo-differential equations

1.2.1. Equations. We shall consider two classes of equations – those with a quasi-homogeneous non-smooth pseudo-differential operator with respect to spatial variables, and those with a fractional derivative with respect to the time variable.

$\mathbf{E}_3$. Pseudo-differential equations with non-smooth quasi-homogeneous symbols

We consider equations of the form

$$\partial_t u(t,x) + (Au)(t,x) + \sum_{k=1}^{m}(A_k u)(t,x) = f(t,x), \quad (t,x) \in \Pi_{[0,T]}, \tag{1.2.1}$$

where A and $A_1,\dots,A_m$ are pseudo-differential operators (ΨDOs) with symbols $a(t,x,\sigma)$ and $a_1(t,x,\sigma),\dots,a_m(t,x,\sigma)$ homogeneous in the "dual" variable σ. The principal symbol $a(t,x,\sigma)$ is assumed to be homogeneous of an order $\gamma \geq 1$ and uniformly elliptic, that is

$$\operatorname{Re} a(t,x,\sigma) \geq a_0 > 0, \quad (t,x) \in \Pi_{[0,T]}, \ |\sigma| = 1,$$

the symbols $a_k(t,x,\sigma)$ have orders of homogeneity γ_k, $0 < \gamma_k < \gamma$; f is a bounded continuous function.

The investigation of the equation (1.2.1) was initiated by Eidelman and Drin' [32, 31, 33]. A thorough theory was developed by Kochubei [128].

The existence of a point $\sigma = 0$, where the symbols are not smooth, prevents us from using the standard ΨDO calculus in studying the equation (1.2.1). Moreover, the usual formula

$$(Au)(t,x) = (2\pi)^{-n} \int\limits_{\mathbb{R}^n} e^{i(x,\sigma)} a(t,x,\sigma)\widehat{u}(t,\sigma)\, d\sigma,$$

$$\widehat{u}(t,\sigma) = \int\limits_{\mathbb{R}^n} e^{-i(y,\sigma)} u(t,y)\, dy, \tag{1.2.2}$$

determines the ΨDO only on smooth rapidly decreasing functions. Since (just as for parabolic differential equations) it is natural to consider solutions of (1.2.1) which are only bounded, the operator defined initially by (1.2.2) should be extended to wider classes of functions. Such extension is given by representing a ΨDO with a homogeneous symbol as a hyper-singular integral operator.

Suppose that $f(x)$ and $\Omega(x,\theta)$ are bounded continuous complex-valued functions on $\mathbb{R}^n$ and $\mathbb{R}^n \times S^{n-1}$ respectively. An expression

$$\left(\mathbf{D}^\alpha_\Omega f\right)(x) = \frac{1}{d_{n,l}(\alpha)} \int\limits_{\mathbb{R}^n} \Omega\left(x, \frac{h}{|h|}\right) \frac{\left(\mathbf{\Delta}^l_h f\right)(x)}{|h|^{n+\alpha}}\, dh, \quad x \in \mathbb{R}^n, \tag{1.2.3}$$

where $\alpha > 0$,

$$\left(\mathbf{\Delta}^l_h f\right)(x) = \sum_{k=0}^{l} (-1)^k C^k_l f(x - kh),$$

l is a positive integer, and $d_{n,l}(\alpha)$ is a normalization constant, is called a *hyper-singular integral* (HSI) of order α with characteristic Ω (note that Ω will often depend also on the time parameter t). Along with the HSI (1.2.3) we shall use the *restricted HSI* $\mathbf{D}^\alpha_{\Omega,\varepsilon} f$ obtained by replacing the domain of integration with $\{h \in \mathbb{R}^n : |h| > \varepsilon\}$, $\varepsilon > 0$. The theory of HSIs was developed by Samko [200, 202].

The HSI (1.2.3) is absolutely convergent if, for example, α is not an integer, $n > \alpha$, and f has bounded derivatives up to the order $[\alpha]+1$ ($[\alpha]$ is the integer part of α). Often the HSI makes sense in more general situations, if it is understood as conditionally convergent, that is

$$\left(\mathbf{D}^\alpha_\Omega f\right)(x) = \lim_{\varepsilon \to 0} \left(\mathbf{D}^\alpha_{\Omega,\varepsilon} f\right)(x).$$

If $f \in \mathcal{S}(\mathbb{R}^n)$, then the HSI-operator $f \mapsto \mathbf{D}^\alpha_\Omega f$ can be represented as a ΨDO with the symbol

$$\widetilde{\Omega}(x,\sigma) = \frac{1}{d_{n,l}(\alpha)} \int\limits_{\mathbb{R}^n} \frac{\left(1 - e^{-i(\sigma,h)}\right)^l}{|h|^{n+\alpha}} \Omega\left(x, \frac{h}{|h|}\right) dh.$$

Conversely, under some natural assumptions a ΨDO (1.2.2) can be represented as a HSI-operator (1.2.3), with the one-to-one correspondence between symbols and characteristics.

Therefore in the equation (1.2.1) we shall understand the operators A, A_1, $\dots$, A_m as conditionally convergent HSI-operators.

Example 1.13. A ΨDO (1.2.2) with the symbol $|\sigma|^\gamma$, $\gamma > 0$ (independent of t and x) can be represented as a HSI-operator with the characteristic $\Omega(x,\theta) \equiv 1$. This operator coincides on $\mathcal{S}(\mathbb{R}^n)$ with the fractional power $(-\Delta)^{\gamma/2}$ of the Laplacian. The corresponding equation (1.2.1) (with $a(t,x,\sigma) = |\sigma|^\gamma$, $1 < \gamma < 2$, $a_k \equiv 0$) has been used in physics for modeling anomalous superdiffusion in single molecule spectroscopy, the diffusion on a polymer chain in chemical space etc. (the "Lévy flights" model; see the survey [170] and references there). These applications are connected with the probabilistic interpretation of this equation (with $\gamma < 2$) and more general ones, which will be discussed below.

E_4. Fractional diffusion equations

Equations of this class contain fractional derivatives with respect to the time variable, and elliptic operators in the spatial variables:

$$\left(\mathbb{D}_t^{(\alpha)} u\right)(t,x) = Bu(t,x), \quad (t,x) \in \Pi_{[0,T]}, \tag{1.2.4}$$

where

$$\left(\mathbb{D}_t^{(\alpha)} u\right)(t,x) = \frac{1}{\Gamma(1-\alpha)}\left[\partial_t \int_0^t (t-\tau)^{-\alpha} u(\tau,x)\,d\tau - t^{-\alpha} u(0,x)\right], \quad 0 < \alpha < 1, \tag{1.2.5}$$

is a regularized fractional derivative,

$$B = \sum_{i,j=1}^n a_{ij}(t,x)\partial_{x_i}\partial_{x_j} + \sum_{j=1}^n b_j(t,x)\partial_{x_j} + c(t,x)$$

is an elliptic second order differential operator with bounded continuous real-valued coefficients.

Such equations were introduced and studied by Wyss [231], Schneider and Wyss [206] (for $B = \Delta$), Kochubei [129, 130], Eidelman and Kochubei [63] (for the general case). A strong motivation for investigating such equations comes from physics – fractional diffusion equations describe anomalous (sub-)diffusion on fractals (physical objects of a fractional dimension, like some amorphous semiconductors or strongly porous materials; see [4, 170, 169, 92] and references therein). By this time, the mathematical theory of fractional diffusion equations has made only its first steps.

1.2.2. Problems. Just as for the parabolic equations, we shall consider the Cauchy problem (1.1.20) for the equations (1.2.1) and (1.2.4). Note that for the fractional diffusion equation (1.2.4) it is the setting of the Cauchy problem (1.1.20) (which has a direct physical meaning of recovering the evolution from initial data) that determines the necessity of using the regularized fractional derivative (1.2.5). If, for example, one considers the Riemann-Liouville fractional derivative defined similarly, but without subtracting $t^{-\alpha}u(0,x)$, then the appropriate initial data will be the limit value ($t \to 0$) of the fractional integral of a solution of the order $1-\alpha$, not the limit value of the solution itself (see [202, 140]).

A FSCP is defined for both our classes of ΨDO's in a way similar to that of the differential equations (Definition 1.7 with $N = 1$). The basic problems are the same problems P_1–P_3 and P_5, as above. In addition, an important problem is to prove properties of the FSCP needed for the probabilistic interpretation of the equation (1.2.1) with $\gamma \leq 2$, and the equation (1.2.4).

Note that the ΨDO with the symbol $|\sigma|^{\gamma}$, $\gamma < 2$ (see Example 1.13), is the generator of the symmetric stable process, the most important representative of the class of Lévy processes (stochastic processes with independent increments) interesting both for various applications (from physics to finance), and as an object of the theory of stochastic processes.

More generally, if the Cauchy problem for the equation (1.2.1) is well-posed, in particular, if there exists a FSCP, and it is non-negative, then the FSCP is the transition density of a Markov process whose path properties can be deduced from appropriate estimates of the FSCP. In contrast to the case of the Brownian motion or processes corresponding to parabolic second order differential equations with sufficiently "regular" coefficients, the processes with pseudo-differential generators are discontinuous – their paths may have jumps.

There is a considerable literature employing either stochastic differential equations or the theory of semigroups of operators to construct discontinuous Markov processes from integro-differential and, in particular, pseudo-differential generators; see, for example, [104, 212, 219, 151, 139, 122, 123], and this list is very far from being complete. Both these approaches do not give information about the analytic properties of the transition probabilities of a Markov process; such information can be obtained by studying the FSCP. On the other hand, the methods of the theory of parabolic pseudo-differential equations have been used successfully for general Lévy-type generators [137].

The fractional diffusion equation (1.2.4) also admits a probabilistic interpretation [204, 205, 138, 232]. This time the resulting process is not a Markov one; it can be obtained from a Markov process by a random time change.

1.2.3. Results. In Chapter 4 we construct and investigate a FSCP for the equation (1.2.1). For equations of an order ≤ 2 we find sufficient conditions for the non-negativity of the FSCP, a crucial property for the probabilistic applications. Within this class of equations, a uniqueness theorem for solutions of the Cauchy problem, tending to zero at infinity, is obtained. We also study conditions, under which a solution stabilizes as $t \to \infty$.

For the equation (1.2.4) considered in Chapter 5, we also give a construction and analysis of properties of a FSCP. In this case, the uniqueness theorem is proved for the class of bounded solutions of the Cauchy problem. For the equation with one spatial variable a precise uniqueness class of solutions with exponential growth is also found.

1.2.4. Techniques. Just as for parabolic differential equations, the basic approach is the Levi method. For pseudo-differential equations (1.2.1), the most difficult task is to obtain precise estimates of the parametrix. That is done with the use of the theory of spherical harmonics and rather delicate techniques of analytic continuation of homogeneous distributions. The machinery of HSIs [200, 202] is also essential.

In contrast to all types of classical parabolic equations, the FSCP for the fractional diffusion equation (1.2.4) (even for $B = \Delta$) has, if $n > 1$, a singularity not only at $t = 0$, but also at the diagonal $x = \xi$ with respect to the spatial variables. This causes serious complications for the implementation of the Levi method. The estimates of the iterated kernels require a series of additional regularization procedures absent in the classical case. Moreover, while the classical heat kernel has a simple expression in elementary functions, in the fractional case we have to deal with Fox's H-functions and their asymptotics, which makes even simple, in principle, transformations and estimates difficult technical tasks. Note also that, in order to represent the solution of the Cauchy problem for the equation (1.2.4) by the usual formula

$$u(t,x) = \int_{\mathbb{R}^n} Z(t,x;0,\xi)u_0(\xi)\,d\xi + \int_0^t d\lambda \int_{\mathbb{R}^n} Y(t,x;\lambda,y)f(\lambda,y)\,dy$$

where $u_0(x) = u(0,x)$, we have to carry out the Levi method twice – first for the FSCP Z, and then separately for the function Y.

1.3 Main lemmas

1.3.1. On the Fourier transform of some entire functions. Let $s := (s_1, \dots, s_n) \in \mathbb{C}^n$, $s := \sigma + i\gamma$, $\{\sigma := (\sigma_1, \dots, \sigma_n), \gamma := (\gamma_1, \dots, \gamma_n)\} \subset \mathbb{R}^n$. We shall consider the functions $f(s), s \in \mathbb{C}^n$, with values in $\mathbb{C}_{NN}$, $N \geq 1$ ($\mathbb{C}_{11} := \mathbb{C}$), assuming them to be entire, with finite orders of growth on $\mathbb{C}^n$ and decrease on $\mathbb{R}^n$. Such functions belong to the space $L_1(\mathbb{R}^n)$ and admit the classical direct and inverse Fourier transforms $F_{\sigma\to x}$ and $F^{-1}_{\sigma\to x}$ defined as

$$F_{\sigma\to x}[f](x) := \int_{\mathbb{R}^n} \exp\{-i(x,\sigma)\}f(\sigma)d\sigma, \quad x \in \mathbb{R}^n, \tag{1.3.1}$$

and

$$F^{-1}_{\sigma\to x}[f](x) := (2\pi)^{-n} \int_{\mathbb{R}^n} \exp\{i(x,\sigma)\}f(\sigma)d\sigma, \quad x \in \mathbb{R}^n, \tag{1.3.2}$$

where $(x,\sigma) := \sum_{j=1}^n x_j\sigma_j$.

Lemma 1.1. *Let $f : \mathbb{C}^n \to \mathbb{C}_{NN}$ be an entire function satisfying the inequality*

$$|f(s)| \le C_0 \exp\left\{ -\sum_{j=1}^{n} a_j|\sigma_j|^{p_j} + \sum_{j=1}^{n} b_j|\gamma_j|^{q_j} \right\}, \quad s \in \mathbb{C}^n, \tag{1.3.3}$$

with the constants $C_0 > 0$, $a_j > 0$, $b_j > 0$, $p_j > 1$, and $q_j \ge p_j$, $j \in \{1, \dots, n\}$. Then its inverse Fourier transform $g(x) := F^{-1}_{\sigma \to x}[f](x)$, $x \in \mathbb{R}^n$, can be extended onto $\mathbb{C}^n$ to an entire function $g(z), z := (z_1, \dots, z_n) \in \mathbb{C}^n$, $z := x + iy, \{x := (x_1, \dots, x_n), y := (y_1, \dots, y_n)\} \subset \mathbb{R}^n$, with the estimate

$$|g(z)| \le C \exp\left\{ -\sum_{j=1}^{n} \alpha_j|x_j|^{q_j'} + \sum_{j=1}^{n} \beta_j|y_j|^{p_j'} \right\}, \quad z \in \mathbb{C}^n, \tag{1.3.4}$$

where $C > 0, \alpha_j > 0, \beta_j > 0$, and the numbers p_j' and q_j' are determined from the equalities

$$\frac{1}{p_j} + \frac{1}{p_j'} = 1, \quad \frac{1}{q_j} + \frac{1}{q_j'} = 1.$$

Proof. It follows from the estimate (1.3.3) that the function $g(x), x \in \mathbb{R}^n$, can be extended onto $\mathbb{C}^n$ by the formula (1.3.2) with x replaced by z. Using the analyticity of the function f, the estimate (1.3.3), and using the Cauchy theorem we obtain that

$$g(z) = (2\pi)^{-n} \int_{\mathbb{R}^n} \exp\{i(x + iy, \sigma + i\gamma)\} f(\sigma + i\gamma) d\sigma,$$

$$z := x + iy \in \mathbb{C}^n,$$

for an arbitrary fixed point $\gamma \in \mathbb{R}^n$. Together with (1.3.3), this implies the inequality

$$|g(z)| \le (2\pi)^{-n} C_0 \exp\left\{ -(x, \gamma) + \sum_{j=1}^{n} b_j|\gamma_j|^{q_j} \right\}$$
$$\times \prod_{j=1}^{n} \int_{\mathbb{R}} \exp\left\{ -a_j|\sigma_j|^{p_j} + |y_j\sigma_j| \right\} d\sigma_j. \tag{1.3.5}$$

Choose $\gamma \in \mathbb{R}^n$ in such a way that $\gamma_j = \delta_j (\operatorname{sign} x_j)^{q_j'} x_j^{q_j'-1}, \delta_j > 0, j \in \{1, \dots, n\}$. Using the equality $(q_j' - 1)q_j = q_j'$ we find that

$$-(x, \gamma) + \sum_{j=1}^{n} b_j|\gamma_j|^{q_j} = -\sum_{j=1}^{n} \alpha_j|x_j|^{q_j'}, \tag{1.3.6}$$

where $\alpha_j := \delta_j - b_j\delta_j^{q_j} > 0$, if δ_j is small enough, since $q_j > 1$.

Next we use the Young inequality

$$|y_j\sigma_j| \le \frac{\eta_j^{p_j}|\sigma_j|^{p_j}}{p_j} + \frac{|y_j|^{p_j'}}{\eta_j^{p_j'}p_j'},$$

where $\eta_j := ((1-\varepsilon)a_j p_j)^{1/p_j}, \varepsilon \in (0,1)$. Then

$$\int\limits_{\mathbb{R}} \exp\{-a_j|\sigma_j|^{p_j} + |y_j\sigma_j|\}\, d\sigma_j$$
$$\le (\varepsilon a_j)^{-1/p_j} \int\limits_{\mathbb{R}} \exp\{-|\sigma_j|^{p_j}\}\, d\sigma_j \exp\left\{\beta_j|y_j|^{p_j'}\right\}, \tag{1.3.7}$$

where $\beta_j := \dfrac{1}{p_j'}((1-\varepsilon)a_j p_j)^{1-p_j'}$.

The inequalities (1.3.5)–(1.3.7) imply the estimate (1.3.4), in which

$$C := (2\pi)^{-n} C_0 \prod_{j=1}^{n} (\varepsilon a_j)^{-1/p_j} \int\limits_{\mathbb{R}} \exp\{-|\sigma_j|^{p_j}\} d\sigma_j. \qquad \square$$

1.3.2. The reference functions and their properties. Consider the functions

$$\rho_{j\beta}(t,r) := (t^{-\beta-j}r)^{1/(1-\beta)}, \quad t>0, \quad r\ge 0, \quad j\in\{0,1,2\}, \tag{1.3.8}$$

where $\beta \in (0,1)$. These functions will appear in estimates of FSCPs.

The investigation of properties of the function (1.3.8) will be preceded by the following lemma about some elementary inequalities.

Lemma 1.2. *Let $n \in \mathbb{N}$, $p \in [1,\infty)$ and $\lambda \in (-1,1)$. For any $\{a,b,c\} \subset \mathbb{R}^n$,*

$$2^{1-p}|a|^p - |b|^p \le |a+b|^p \le 2^{p-1}\Big(|a|^p + |b|^p\Big), \tag{1.3.9}$$

$$4^{1-p}|a|^p - 2^{1-p}|b|^p - |c|^p \le |a+b+c|^p \le 4^{p-1}\Big(|a|^p + |b|^p + |c|^p\Big), \tag{1.3.10}$$

$$|a+\lambda b|^p + |b|^p \ge \delta\Big(|a|^p + |b|^p\Big), \qquad ((1.3.11))$$

where $\delta := \min\Big(1 - |\lambda|^p, 2^{1-p}\Big)$.

Proof. The inequalities (1.3.9) are obvious for $p=1$. If $p>1$, the function $y = x^p$, $x \ge 0$, is convex downwards on $[0,\infty)$. Therefore

$$2^{-p}\Big(|a| + |b|\Big)^p \le 2^{-1}\Big(|a|^p + |b|^p\Big).$$

Hence

$$|a+b|^p \le \Big(|a|+|b|\Big)^p \le 2^{p-1}\Big(|a|^p+|b|^p\Big).$$

We have proved the second inequality from (1.3.9). It implies all other inequalities (1.3.9)–(1.3.11).

Indeed, we have

$$\begin{aligned}
|a+b+c|^p &\le 2^{p-1}\Big(|a+b|^p+|c|^p\Big)\\
&\le 2^{p-1}\Big(2^{p-1}(|a|^p+|b|^p)+|c|^p\Big)\\
&\le 4^{p-1}\Big(|a|^p+|b|^p+|c|^p\Big),\\
|a|^p = |a+b-b|^p &\le 2^{p-1}\Big(|a+b|^p+|b^p|\Big),\\
|a+b|^p = |a+b+c-c|^p &\le 2^{p-1}\Big(|a+b+c|^p+|c|^p\Big),\\
|a|^p = |a+\lambda b-\lambda b|^p &\le 2^{p-1}\Big(|a+\lambda b|^p+|\lambda b|^p\Big),
\end{aligned}$$

and we obtain the second inequality from (1.3.10), the first inequality from (1.3.9), and also the inequalities

$$|a+b+c|^p \ge 2^{1-p}|a+b|^p-|c|^p \ge 2^{1-p}\Big(2^{1-p}|a|^p-|b|^p\Big)-|c|^p,$$

$$|a+\lambda b|^p+|b|^p \ge 2^{1-p}|a|^p-|\lambda b|^p+|b|^p = 2^{1-p}|a|^p+\Big(1-|\lambda|^p\Big)|b|^p.$$

The latter inequalities imply the first inequality from (1.3.10) and the inequality (1.3.11). □

Lemma 1.3. *Let* $n\in\mathbb{N}$, $\beta\in(0,1)$, *and*

$$\begin{aligned}
\rho_\beta^{(0)}(t,x;\tau,\xi) &:= \rho_{0\beta}(t-\tau,|x-\xi|),\\
\rho_\beta^{(1)}(t,x,y;\tau,\xi,\eta) &:= \rho_\beta^{(0)}(t,x;\tau,\xi)+\rho_{1\beta}(t-\tau,|y-\eta+(t-\tau)x|),\\
\rho_\beta^{(2)}(t,x,y,z;\tau,\xi,\eta,\zeta) &:= \rho_\beta^{(1)}(t,x,y;\tau,\xi,\eta)\\
&\quad+\rho_{2\beta}(t-\tau,|z-\zeta+(t-\tau)y+2^{-1}(t-\tau)^2x|),\\
\{t,\tau\}\subset\mathbb{R},\quad \tau<t,&\quad \{x,y,z,\xi,\eta,\zeta\}\subset\mathbb{R}^n.
\end{aligned}\tag{1.3.12}$$

For any

$$\{t,\lambda,\tau\}\subset\mathbb{R},\quad \tau<\lambda<t,\quad \text{and}\quad \{x,x',y,y',z,z',\xi,\eta,\zeta\}\subset\mathbb{R}^n$$

the inequalities

$$\begin{aligned} A_0 &:= \rho_\beta^{(0)}(t,x;\lambda,x') + \rho_\beta^{(0)}(\lambda,x';\tau,\xi) \\ &\geq \rho_\beta^{(0)}(t,x;\tau,\xi), \end{aligned} \tag{1.3.13}$$

$$\begin{aligned} A_1 &:= \rho_\beta^{(1)}(t,x,y;\lambda,x',y') + \rho_\beta^{(1)}(\lambda,x',y';\tau,\xi,\eta) \\ &\geq \delta_1\rho_\beta^{(1)}(t,x,y;\tau,\xi,\eta), \end{aligned} \tag{1.3.14}$$

$$\begin{aligned} A_2 &:= \rho_\beta^{(2)}(t,x,y,z;\lambda,x',y',z') + \rho_\beta^{(2)}(\lambda,x',y',z';\tau,\xi,\eta,\zeta) \\ &\geq \delta_2\rho_\beta^{(2)}(t,x,y,z;\tau,\xi,\eta,\zeta), \end{aligned} \tag{1.3.15}$$

hold, with constants δ_1 and δ_2 depending only on β.

Proof. Let t,τ,x,ξ be fixed in an arbitrary way, and let $r := |x-x'|$, $a := |x-\xi|$. Using the fact that $(1/(1-\beta)) > 1$, and $|x'-\xi| = |(x-\xi)-(x-x')| \geq |a-r|$ for any $x' \in \mathbb{R}^n$, we have

$$A_0 \geq \rho_{0\beta}(t-\lambda, r) + \rho_{0\beta}(\lambda-\tau, |a-r|) =: f_\lambda(r).$$

It is sufficient to prove that for any $\lambda \in (\tau,t)$ and $r \geq 0$,

$$f_\lambda(r) \geq \rho_{0\beta}(t-\tau, a). \tag{1.3.16}$$

If $r \geq a$, then for $\lambda \in (\tau,t)$,

$$f_\lambda(r) \geq \rho_{0\beta}(t-\lambda, a) \geq \rho_{0\beta}(t-\tau, a). \tag{1.3.17}$$

Let us find $\min\limits_{r\in[0,a]} f_\lambda(r)$. Since

$$f_\lambda'(r) = \frac{1}{1-\beta}\left(\left(\frac{r}{t-\lambda}\right)^{\beta/(1-\beta)} - \left(\frac{a-r}{\lambda-\tau}\right)^{\beta/(1-\beta)}\right) = 0$$

for $r = r_0 := \dfrac{t-\lambda}{t-\tau}a$, we see that r_0 is the unique stationary point of the function f_λ on the interval $(0,a)$. For $\lambda \in (\tau,t)$,

$$\begin{aligned} f_\lambda(0) &= \rho_{0\beta}(\lambda-\tau, a) \geq \rho_{0\beta}(t-\tau, a), \\ f_\lambda(a) &= \rho_{0\beta}(t-\lambda, a) \geq \rho_{0\beta}(t-\tau, a), \\ f_\lambda(r_0) &= \rho_{0\beta}(t-\tau, a), \end{aligned}$$

so that $\min\limits_{r\in[0,a]} f_\lambda(r) = \rho_{0\beta}(t-\tau,a)$. Together with (1.3.17) this implies the inequality (1.3.16), so that (1.3.13) is proved.

Let us prove the inequality (1.3.14). Using the inequalities (1.3.9) with $p = 1/(1-\beta)$ and the fact that

$$t-\lambda \leq t-\tau \quad \text{and} \quad \lambda-\tau \leq t-\tau,$$

we have

$$\begin{aligned}
&\rho_{1\beta}(t-\lambda,|y-y'+(t-\lambda)x|)+\rho_{1\beta}(\lambda-\tau,|y'-\eta+(\lambda-\tau)x'|)\\
&\geq (t-\tau)^{-(1+\beta)/(1-\beta)}\Big(|y-y'+(t-\lambda)x|^{1/(1-\beta)}+|y'-\eta+(\lambda-\tau)x'|^{1/(1-\beta)}\Big)\\
&\geq 2^{-\beta/(1-\beta)}(t-\tau)^{-(1+\beta)/(1-\beta)}|y-\eta+(t-\tau)x-(\lambda-\tau)(x-x')|^{1/(1-\beta)}\\
&\geq 2^{-\beta/(1-\beta)}(t-\tau)^{-(1+\beta)/(1-\beta)}\Big(2^{-\beta/(1-\beta)}|y-\eta+(t-\tau)x|^{1/(1-\beta)}\\
&\quad -|(\lambda-\tau)(x-x')|^{1/(1-\beta)}\Big)\\
&=2^{-2\beta/(1-\beta)}\rho_{1\beta}(t-\tau,|y-\eta+(t-\tau)x|)\\
&\quad -2^{-\beta/(1-\beta)}|(\lambda-\tau)(t-\tau)^{-1-\beta}(x-x')|^{1/(1-\beta)}\\
&\geq 2^{-2\beta/(1-\beta)}\rho_{1\beta}(t-\tau,|y-\eta+(t-\tau)x|)-2^{-\beta/(1-\beta)}\rho_{0\beta}(t-\tau,|x-x'|)\\
&\geq 2^{-2\beta/(1-\beta)}\rho_{1\beta}(t-\tau,|y-\eta+(t-\tau)x|)-2^{-\beta/(1-\beta)}\rho_{0\beta}(t-\lambda,|x-x'|).
\end{aligned}$$

Using this inequality and (1.3.13) we find that

$$\begin{aligned}
A_1 &\geq \Big(1-2^{-\beta/(1-\beta)}\Big)A_0+2^{-2\beta/(1-\beta)}\rho_{1\beta}(t-\tau,|y-\eta+(t-\tau)x|)\\
&\geq \Big(1-2^{-\beta/(1-\beta)}\Big)\rho_\beta^{(0)}(t,x;\tau,\xi)+2^{-2\beta/(1-\beta)}\rho_{1\beta}(t-\tau,|y-\eta+(t-\tau)x|).
\end{aligned}$$

This implies the inequality (1.3.14) with $\delta_1:=\min\Big(1-2^{-\beta/(1-\beta)},2^{-2\beta/(1-\beta)}\Big)$.

In order to prove (1.3.15), we proceed in a similar way. First we use the inequalities (1.3.9) and (1.3.10) with $p=1/(1-\beta)$ to find that

$$\begin{aligned}
&\rho_{2\beta}(t-\lambda,|z-z'+(t-\lambda)y+2^{-1}(t-\lambda)^2x|)\\
&\quad+\rho_{2\beta}(\lambda-\tau,|z'-\zeta+(\lambda-\tau)y'+2^{-1}(\lambda-\tau)^2x'|)\\
&\geq (t-\tau)^{-(2+\beta)/(1-\beta)}\Big(|z-z'+(t-\lambda)y+2^{-1}(t-\lambda)^2x|^{1/(1-\beta)}\\
&\quad+|z'-\zeta+(\lambda-\tau)y'+2^{-1}(\lambda-\tau)^2x'|^{1/(1-\beta)}\Big)\\
&\geq 2^{-\beta/(1-\beta)}(t-\tau)^{-(2+\beta)/(1-\beta)}\Big|z-\zeta+(t-\tau)y+2^{-1}(t-\tau)^2x\\
&\quad-(\lambda-\tau)(y-y'+(t-\lambda)x)-2^{-1}(\lambda-\tau)^2(x-x')\Big|^{1/(1-\beta)}\\
&\geq 2^{-\beta/(1-\beta)}\Big(4^{-\beta/(1-\beta)}\rho_{2\beta}(t-\tau,|z-\zeta+(t-\tau)y+2^{-1}(t-\tau)^2x|)
\end{aligned}$$

$$- 2^{-\beta/(1-\beta)}\left|(\lambda-\tau)(t-\tau)^{-2-\beta}(y-y'+(t-\lambda)x)\right|^{1/(1-\beta)}$$
$$- 2^{-1/(1-\beta)}\left|(\lambda-\tau)^2(t-\tau)^{-2-\beta}(x-x')\right|^{1/(1-\beta)}\Big)$$
$$\geq 2^{-3\beta/(1-\beta)}\rho_{2\beta}(t-\tau,|z-\zeta+(t-\tau)y+2^{-1}(t-\tau)^2x|)$$
$$- 2^{-2\beta/(1-\beta)}\rho_{1\beta}(t-\lambda,|y-y'+(t-\tau)x|) - 2^{-(1+\beta)/(1-\beta)}\rho_{0\beta}(t-\lambda,|x-x'|).$$

Then, using the last inequality and (1.3.14), we obtain that

$$A_2 \geq \left(1-2^{-(1+\beta)/(1-\beta)}\right)A_0 + \left(1-2^{-2\beta/(1-\beta)}\right)\Big(\rho_{1\beta}(t-\tau,|y-y'+(t-\lambda)x|)$$
$$+\rho_{1\beta}(\lambda-\tau,|y'-\eta+(\lambda-\tau)x'|)\Big) + 2^{-3\beta/(1-\beta)}\rho_{2\beta}(t-\tau,|z-\zeta+(t-\tau)y$$
$$+2^{-1}(t-\tau)^2x|)$$
$$\geq \left(1-2^{-2\beta/(1-\beta)}\right)A_1 + 2^{-3\beta/(1-\beta)}\rho_{2\beta}(t-\tau,|z-\zeta+(t-\tau)y+2^{-1}(t-\tau)^2x|)$$
$$\geq \delta_2\rho_\beta^{(2)}(t,x,y,z;\tau,\xi,\eta,\zeta),$$

where $\delta_2 := \min\left((1-2^{-2\beta/(1-\beta)})\delta_1, 2^{-3\beta/(1-\beta)}\right)$. □

In order to formulate other properties of the functions (1.3.8), we consider the functions

$$k_{j\beta}(t,a) := c_0 a\left(c_0^{(1-\beta)/\beta} - a^{(1-\beta)/\beta}t^{1+j/\beta}\right)^{-\beta/(1-\beta)},$$
$$t\in[0,T], \quad j\in\{0,1,2\}, \tag{1.3.18}$$

similar to (1.1.32). Here c_0 is a positive number, a is nonnegative, and T is a positive number smaller than $(c_0/a)^{(1-\beta)/(j+\beta)}$. These functions are increasing, equal to a for $t=0$, and have the following properties:

$$k_{0\beta}(t-\tau,k_{0\beta}(\tau,a)) = k_{0\beta}(t,a),$$
$$k_{j\beta}(t-\tau,k_{j\beta}(\tau,a)) \leq k_{j\beta}(t,a), \quad j\in\{1,2\},$$
$$0\leq\tau\leq t\leq T \tag{1.3.19}$$

which are verified in a straightforward way.

Lemma 1.4. *Let $n\in\mathbb{N}$ and $\beta\in(0,1)$. For any $\{t,\tau\}\subset[0,T]$, $\tau<t$, and $\{x,\xi\}\subset\mathbb{R}^n$,*

$$- c_0\rho_{j\beta}(t-\tau,|x-\xi|) + k_{j\beta}(\tau,a)|\xi|^{1/(1-\beta)}$$
$$\leq k_{j\beta}(t,a)|x|^{1/(1-\beta)}, \quad j\in\{0,1,2\}. \tag{1.3.20}$$

Proof. Fix arbitrary t, x, j and denote $|\xi|$ by r. It is clear that

$$\begin{aligned} &- c_0\rho_{j\beta}(t-\tau, |x-\xi|) + k_{j\beta}(\tau, a)|\xi|^{1/(1-\beta)} \\ &\quad \le -c_0\rho_{j\beta}(t-\tau, ||x|-r|) + k_{j\beta}(\tau, a)r^{1/(1-\beta)} =: g_\tau(r). \end{aligned} \tag{1.3.21}$$

If $r \le |x|$ and $\tau < t$, we find immediately that

$$g_\tau(r) \le k_{j\beta}(t, a)|x|^{1/(1-\beta)}. \tag{1.3.22}$$

We used the fact that the functions $k_{j\beta}$ are increasing.

On the interval $(|x|, \infty)$ the function g_τ attains its maximal value for

$$r = r_1 := c_0^{(1-\beta)/\beta}\Big(c_0^{(1-\beta)/\beta} - (k_{j\beta}(\tau, a))^{(1-\beta)/\beta}(t-\tau)^{1+j/\beta}\Big)^{-1}|x|.$$

The maximum equals $g_\tau(r_1) = k_{j\beta}(t-\tau, k_{j\beta}(\tau, a))|x|^{1/(1-\beta)}$. Taking into account the properties (1.3.19) and the inequalities (1.3.21) and (1.3.22), we derive the required inequalities (1.3.20). □

1.3.3. Estimates of some integrals. In the lemmas below the numbers $n \in \mathbb{N}$, $\beta \in (0,1)$, $c \in (0,\infty)$, and $c_0 \in (0, c)$ are given, while the functions $\rho_{j\beta}$, $\rho_\beta^{(j)}$, and $k_{j\beta}$, $j \in \{0,1,2\}$, are defined by the formulas (1.3.8), (1.3.12), and (1.3.18) respectively.

Lemma 1.5. *Let*

$$\begin{aligned} I_c^{(0)} &:= \int_{\mathbb{R}^n} \exp\Big\{-c\Big(\rho_\beta^{(0)}(t, x; \lambda, x') + \rho_\beta^{(0)}(\lambda, x'; \tau, \xi)\Big)\Big\} \\ &\qquad \times \Big((t-\lambda)(\lambda-\tau)\Big)^{-n\beta} dx', \\ I_c^{(1)} &:= \int_{\mathbb{R}^{2n}} \exp\Big\{-c\Big(\rho_\beta^{(1)}(t, x, y; \lambda, x', y') + \rho_\beta^{(1)}(\lambda, x', y'; \tau, \xi, \eta)\Big)\Big\} \\ &\qquad \times \Big((t-\lambda)(\lambda-\tau)\Big)^{-n(2\beta+1)} dx'dy', \\ I_c^{(2)} &:= \int_{\mathbb{R}^{3n}} \exp\Big\{-c\Big(\rho_\beta^{(2)}(t, x, y, z; \lambda, x', y', z') + \rho_\beta^{(2)}(\lambda, x', y', z'; \tau, \xi, \eta, \zeta)\Big)\Big\} \\ &\qquad \times \Big((t-\lambda)(\lambda-\tau)\Big)^{-3n(\beta+1)} dx'dy'dz'. \end{aligned}$$

For an arbitrary $\varepsilon \in (0,1)$, there exist such constants $C_\varepsilon^{(j)}$, $j \in \{0,1,2\}$, that for any $\{t, \lambda, \tau\} \subset \mathbb{R}$, $\tau < \lambda < t$, and $\{x, y, z, \xi, \eta, \zeta\} \subset \mathbb{R}^n$,

$$I_c^{(0)} \le C_\varepsilon^{(0)}(t-\tau)^{-n\beta}\exp\Big\{-c(1-\varepsilon)\rho_\beta^{(0)}(t, x; \tau, \xi)\Big\}, \tag{1.3.23}$$

$$I_c^{(1)} \le C_\varepsilon^{(1)}(t-\tau)^{-n(2\beta+1)} \exp\Big\{-c(1-\varepsilon)\delta_1\rho_\beta^{(1)}(t,x,y;\tau,\xi,\eta)\Big\}, \tag{1.3.24}$$

$$I_c^{(2)} \le C_\varepsilon^{(2)}(t-\tau)^{-3n(\beta+1)} \exp\Big\{-c(1-\varepsilon)\delta_2\rho_\beta^{(2)}(t,x,y,z;\tau,\xi,\eta,\zeta)\Big\}. \tag{1.3.25}$$

Proof. Writing c as $c = c(1-\varepsilon) + c\varepsilon$ and using the inequalities (1.3.13)–(1.3.15), we find that

$$\begin{aligned}
I_c^{(0)} &\le \exp\Big\{-c(1-\varepsilon)\rho_\beta^{(0)}(t,x;\tau,\xi)\Big\} I_{c\varepsilon}^{(0)},\\
I_c^{(1)} &\le \exp\Big\{-c(1-\varepsilon)\delta_1\rho_\beta^{(1)}(t,x,y;\tau,\xi,\eta)\Big\} I_{c\varepsilon}^{(1)},\\
I_c^{(2)} &\le \exp\Big\{-c(1-\varepsilon)\delta_2\rho_\beta^{(2)}(t,x,y,z;\tau,\xi,\eta,\zeta)\Big\} I_{c\varepsilon}^{(2)}.
\end{aligned} \tag{1.3.26}$$

Let us give estimates for the integrals $I_{c\varepsilon}^{(j)}$, $j \in \{0,1,2\}$. We shall consider two possible cases:

1) $0 < \lambda - \tau \le (t-\tau)/2$;
2) $((t-\tau)/2) < \lambda - \tau < t - \tau$.

In the first case $t - \lambda \ge (t-\tau)/2$, hence

$$\begin{aligned}
I_{c\varepsilon}^{(0)} &\le 2^{n\beta}(t-\tau)^{-n\beta} \int_{\mathbb{R}^n} \exp\Big\{-c\varepsilon\rho_\beta^{(0)}(\lambda,x';\tau,\xi)\Big\}(\lambda-\tau)^{-n\beta}dx',\\
I_{c\varepsilon}^{(1)} &\le 2^{n(2\beta+1)}(t-\tau)^{-n(2\beta+1)}\\
&\quad\times \int_{\mathbb{R}^{2n}} \exp\Big\{-c\varepsilon\rho_\beta^{(1)}(\lambda,x',y';\tau,\xi,\eta)\Big\}(\lambda-\tau)^{-n(2\beta+1)}dx'dy',\\
I_{c\varepsilon}^{(2)} &\le 2^{3n(\beta+1)}(t-\tau)^{-3n(\beta+1)}\\
&\quad\times \int_{\mathbb{R}^{3n}} \exp\Big\{-c\varepsilon\rho_\beta^{(2)}(\lambda,x',y',z';\tau,\xi,\eta,\zeta)\Big\}(\lambda-\tau)^{-3n(\beta+1)}dx'dy'dz'.
\end{aligned}$$

Passing from the integration in x', y', z' to the new variables x'', y'', z'' by the formulas

$$\begin{aligned}
x' &= \xi + \varepsilon^{\beta-1}(\lambda-\tau)^\beta x'',\\
y' &= \eta - (\lambda-\tau)x' + \varepsilon^{\beta-1}(\lambda-\tau)^{1+\beta}y'',\\
z' &= \zeta - (\lambda-\tau)y' - \frac{1}{2}(\lambda-\tau)^2x' + \varepsilon^{\beta-1}(\lambda-\tau)^{2+\beta}z'',
\end{aligned}$$

we obtain that

$$I_{c\varepsilon}^{(j)} \le C_\varepsilon^{(j)}(t-\tau)^{-n(j+1)(\beta+j/2)}, \quad j \in \{0,1,2\}, \tag{1.3.27}$$

where

$$C_\varepsilon^{(j)} := 2^{n(j+1)(\beta+j/2)}\varepsilon^{n(j+1)(\beta-1)}\Big(\int\limits_{\mathbb{R}^n}\exp\Big\{-c|x''|^{1/(1-\beta)}\Big\}dx''\Big)^{j+1}.$$

The estimates (1.3.27) for the case 2 are proved in a similar way.

The inequalities (1.3.26) and (1.3.27) imply the estimates (1.3.23)–(1.3.25). □

Let us consider the case $\beta = 1/2$ separately. For this case we introduce the following notation:

$$\begin{aligned}
&\rho^{(0)}(t,x;\tau,\xi) := 4^{-1}\rho_{1/2}(t,x;\tau,\xi),\\
&\rho^{(1)}(t,x,y;\tau,\xi,\eta) := \rho^{(0)}(t,x;\tau,\xi) + 3\rho_{1(1/2)}(t-\tau,|y-\eta+2^{-1}(t-\tau)(x+\xi)|),\\
&\rho^{(2)}(t,x,y,z;\tau,\xi,\eta,\zeta) := \rho^{(1)}(t,x,y;\tau,\xi,\eta)\\
&\quad + 180\rho_{2(1/2)}(t-\tau,|z-\zeta+2^{-1}(t-\tau)(y+\eta)+12^{-1}(t-\tau)^2(x-\xi)|). \qquad (1.3.28)
\end{aligned}$$

Lemma 1.6. *There exist such constants $C^{(j)}$, $j \in \{0,1,2\}$, that for any $\{t,\lambda,\tau\} \subset \mathbb{R}$, $\tau < \lambda < t$, and $\{x,y,z,\xi,\eta,\zeta\} \subset \mathbb{R}^n$,*

$$\int\limits_{\mathbb{R}^n}\exp\Big\{-c\Big(\rho^{(0)}(t,x;\lambda,x')+\rho^{(0)}(\lambda,x';\tau,\xi)\Big)\Big\}\Big((t-\lambda)(\lambda-\tau)\Big)^{-n/2}dx'$$
$$=C^{(0)}(t-\tau)^{-n/2}\exp\Big\{-c\rho^{(0)}(t,x;\tau,\xi)\Big\}, \qquad (1.3.29)$$

$$\int\limits_{\mathbb{R}^{2n}}\exp\Big\{-c\Big(\rho^{(1)}(t,x,y;\lambda,x',y')+\rho^{(1)}(\lambda,x',y';\tau,\xi,\eta)\Big)\Big\}\Big((t-\lambda)(\lambda-\tau)\Big)^{-2n}dx'dy'$$
$$=C^{(1)}(t-\tau)^{-2n}\exp\Big\{-c\rho^{(1)}(t,x,y;\tau,\xi,\eta)\Big\}, \qquad (1.3.30)$$

$$\int\limits_{\mathbb{R}^{3n}}\exp\Big\{-c\Big(\rho^{(2)}(t,x,y,z;\lambda,x',y',z')+\rho^{(2)}(\lambda,x',y',z';\tau,\xi,\eta,\zeta)\Big)\Big\}$$
$$\times\Big((t-\lambda)(\lambda-\tau)\Big)^{-9n/2}dx'dy'dz'$$
$$=C^{(2)}(t-\tau)^{-9n/2}\exp\Big\{-c\rho^{(2)}(t,x,y,z;\tau,\xi,\eta,\zeta)\Big\}. \qquad (1.3.31)$$

Proof. Note (see (3.1.39), (3.1.40)) that the functions

$$(t-\tau)^{-n/2}\exp\Big\{-c\rho^{(0)}(t,x;\tau,\xi)\Big\},\quad (t-\tau)^{-2n}\exp\Big\{-c\rho^{(1)}(t,x,y;\tau,\xi,\eta)\Big\}$$

and

$$(t-\tau)^{-9n/2}\exp\Big\{-c\rho^{(2)}(t,x,y,z;\tau,\xi,\eta,\zeta)\Big\},\quad \tau<t,\quad \{x,y,z,\xi,\eta,\zeta\}\subset\mathbb{R}^n,$$

are, up to constant factors, FSCP for the equation from Example 1.1 and Example 1.10 respectively, for the case where $n_1=n_2=n$, $n_3=0$, and where $n_1=n_2=n_3=n$, if in those equations $a=1/c$. The equalities (1.3.29)–(1.3.31) follow from the convolution formula for FSCP valid for the above equations (see (2.1.112), (3.1.46)). □

Lemma 1.7. *There exists such a constant $C>0$ that for any $\{t,\tau\}\subset[0,T]$, $\tau<t$, and $x\in R^n$,*

$$\begin{aligned}J_j&:=\int\limits_{\mathbb{R}^n}\exp\Big\{-c\rho_{j\beta}(t-\tau,|x-\xi|)+k_{j\beta}(\tau,a)|\xi|^{1/(1-\beta)}\Big\}(t-\tau)^{-n(j+\beta)}d\xi\\&\le C\exp\Big\{k_{j\beta}(t,a)|x|^{1/(1-\beta)}\Big\},\quad j\in\{0,1,2\}.\end{aligned}\tag{1.3.32}$$

Proof. Using the inequality (1.3.20) and the change of the integration variable $\xi=x+(t-\tau)^{j+\beta}\eta$ we obtain

$$\begin{aligned}J_j&\le\exp\Big\{k_{j\beta}(t,a)|x|^{1/(1-\beta)}\Big\}\\&\quad\times\int\limits_{\mathbb{R}^n}\exp\Big\{-(c-c_0)\rho_{j\beta}(t-\tau,|x-\xi|)\Big\}(t-\tau)^{-n(j+\beta)}d\xi\\&=C\exp\Big\{k_{j\beta}(t,a)|x|^{1/(1-\beta)}\Big\},\end{aligned}$$

where

$$C:=\int\limits_{\mathbb{R}^n}\exp\Big\{-(c-c_0)|\eta|^{1/(1-\beta)}\Big\}d\eta.$$ □

1.3.4. Some Volterra integral equations of the second kind. Let $N\in\mathbb{N}$, $t_0\in[0,T)$ and $P^{\delta}_{[t_0,T]}:=\Big\{(t,x;\tau,\xi)\in(\Pi_{[t_0,T]}\times\Pi_{[t_0,T]})\,|\,t-\tau>\delta\Big\}$, $\delta>0$. Consider an integral equation of the form

$$u(t,x)=f(t,x)+\int\limits_{\tau}^{t}d\tau\int\limits_{\mathbb{R}^n}K(t,x;\tau,\xi)u(\tau,\xi)d\xi,\quad (t,x)\in\Pi_{[t_0,T]},\tag{1.3.33}$$

with a continuous kernel

$$K:P^0_{[t_0,T]}\to\mathbb{C}_{NN}.\tag{1.3.34}$$

It is known that, under appropriate conditions upon the kernel K, there exists the unique solution of the equation (1.3.33), for any admissible function f, given by the formula

$$u(t,x) = f(t,x) + \int_{t_0}^{t} d\tau \int_{\mathbb{R}^n} R(t,x;\tau,\xi) f(\tau,\xi) d\xi, \quad (t,x) \in \Pi_{[t_0,T]}. \tag{1.3.35}$$

Here

$$R(t,x;\tau,\xi) := \sum_{m=1}^{\infty} K_m(t,x;\tau,\xi), \quad (t,x;\tau,\xi) \in P^0_{[t_0,T]}, \tag{1.3.36}$$

where $K_1 := K$, while K_m, $m > 1$, are the iterated kernel determined from the recursive relation

$$K_m(t,x;\tau,\xi) = \int_{\tau}^{t} d\lambda \int_{\mathbb{R}^n} K(t,x;\lambda,y) K_{m-1}(\lambda,y;\tau,\xi) dy,$$
$$(t,x;\tau,\xi) \in P^0_{[t_0,T]}. \tag{1.3.37}$$

The function R is called the resolvent of the equation (1.3.33).

The lemmas below deal with some classes of the kernels K which emerge in the course of constructing FSCP's for equations considered in this book. For these kernels the series (1.3.36) is absolutely and uniformly convergent in $P^\delta_{[t_0,T]}$, for any $\delta \in (0, T - t_0)$. Hence there exists a resolvent R for the corresponding equation (1.3.33), and its solution is given by the formula (1.3.35).

Below $\rho_{j\beta}$, $\rho^{(j)}_\beta$, and $\rho^{(j)}$, $j \in \{0,1,2\}$, are the functions defined by (1.3.8), (1.3.12), and (1.3.28).

Lemma 1.8. *Let* $\{n, N\} \subset \mathbb{N}$, $\beta_j \in (0,1)$, $j \in \{1,\ldots,n\}$, $\beta := (\beta_1,\ldots,\beta_n)$, $M := \beta_1 + \cdots + \beta_n$, *and*

$$E^{(0)}_c(t,x;\tau,\xi) := \exp\Big\{-c \sum_{j=1}^{n} \rho^{(0)}_{\beta_j}(t,x_j;\tau,\xi_j)\Big\}, \ t > \tau, \ \{x,\xi\} \subset \mathbb{R}^n, \ c > 0.$$

If the kernel (1.3.34) *is continuous and satisfies the inequality*

$$|K(t,x;\tau,\xi)| \le C_1 (t-\tau)^{-M+\chi-1} E^{(0)}_{c_1}(t,x;\tau,\xi),$$
$$(t,x;\tau,\xi) \in P^0_{[t_0,T]}, \tag{1.3.38}$$

with some constants $C_1 > 0$, $c_1 > 0$, *and* $\chi \in (0,1)$, *then the resolvent (1.3.36) exists, is continuous, and has the bound*

$$|R(t,x;\tau,\xi)| \le C (t-\tau)^{-M+\chi-1} E^{(0)}_{c}(t,x;\tau,\xi),$$
$$(t,x;\tau,\xi) \in P^0_{[t_0,T]}, \tag{1.3.39}$$

where $C > 0$ *and* $c \in (0, c_1)$.

Proof. The consecutive estimation of the iterated kernels K_m will be carried out in two stages. At the first stage we shall employ Lemma 1.5. Using (1.3.37) and (1.3.38) for $(t,x;\tau,\xi) \in P^0_{[t_0,T]}$ we have

$$
\begin{aligned}
&|K_2(t,x;\tau,\xi)| \\
&\le C_1^2 \int_\tau^t \Big((t-\lambda)(\lambda-\tau)\Big)^{\chi-1} d\lambda \int_{\mathbb{R}^n} E^{(0)}_{c_1}(t,x;\lambda,y) E^{(0)}_{c_1}(\lambda,y;\tau,\xi)\Big((t-\lambda)(\lambda-\tau)\Big)^{-M} dy \\
&= C_1^2 \int_\tau^t \Big((t-\lambda)(\lambda-\tau)\Big)^{\chi-1} d\lambda \\
&\times \prod_{j=1}^n \int_{\mathbb{R}} \exp\Big\{-c_1\Big(\rho^{(0)}_{\beta_j}(t,x_j;\lambda,y_j)+\rho^{(0)}_{\beta_j}(\lambda,y_j;\tau,\xi_j)\Big)\Big\}\Big((t-\lambda)(\lambda-\tau)\Big)^{-\beta_j} dy_j.
\end{aligned}
$$

By the inequality (1.3.23) with $n=1$, $\beta=\beta_j$, and some $\varepsilon \in (0,1)$, and the identity

$$
\int_\tau^t \Big((t-\lambda)(\lambda-\tau)\Big)^{\chi-1} d\lambda = (t-\tau)^{2\chi-1} B(\chi,\chi), \tag{1.3.40}
$$

where B is Euler's Beta function, we find that

$$
|K_2(t,x;\tau,\xi)| \le C_{2\varepsilon}(t-\tau)^{-M+2\chi-1} E^{(0)}_{c_1(1-\varepsilon)}(t,x;\tau,\xi), \tag{1.3.41}
$$

where $C_{2\varepsilon} := C_1^2 C_\varepsilon^{(0)} B(\chi,\chi)$, $C_\varepsilon^{(0)}$ is the product of the constants from the estimates (1.3.23).

Similarly, the estimate (1.3.41) implies an estimate of the kernel K_3. Continuing the process we come to the following estimates for $m \ge 2$:

$$
|K_m(t,x;\tau,\xi)| \le C_{m\varepsilon}(t-\tau)^{-M+m\chi-1} E^{(0)}_{c_1(1-(m-1)\varepsilon)}(t,x;\tau,\xi). \tag{1.3.42}
$$

Let $m_0 := [(M+1)/\chi]+1$ where $[a]$ is the integer part of the number a. Then

$$
|K_{m_0}(t,x;\tau,\xi)| \le C_{m_0\varepsilon} E^{(0)}_{c_1(1-(m_0-1)\varepsilon)}(t,x;\tau,\xi).
$$

If we take $\varepsilon := \varepsilon_0/(m_0-1)$, $\varepsilon_0 \in (0,1)$ and set

$$
c := c_1(1-\varepsilon_0), \quad C_0 := \max_{m\in\{1,\dots,m_0\}} C_{m\varepsilon},
$$

where $C_{1\varepsilon} := C_1$, then the last estimate has the form

$$
|K_{m_0}(t,x;\tau,\xi)| \le C_0 E^{(0)}_c(t,x;\tau,\xi). \tag{1.3.43}
$$

The second stage of estimation is based on Lemma 1.3 and makes it possible to maintain the exponential factor in estimates of the kernels K_m for $m > m_0$.

The inequality (1.3.13) of Lemma 1.3 (the case $n = 1$, $\beta = \beta_j$, $j \in \{1, \dots, n\}$) implies the inequality

$$\begin{aligned} E_c^{(0)}(t,x;\lambda,y)E_c^{(0)}(\lambda,y;\tau,\xi) &= \exp\Big\{-c\sum_{j=1}^{n}\Big(\rho_{\beta_j}^{(0)}(t,x_j;\lambda,y_j) + \rho_{\beta_j}^{(0)}(\lambda,y_j;\tau,\xi_j)\Big)\Big\} \\ &\le \exp\Big\{-c\sum_{j=1}^{n}\rho_{\beta_j}^{(0)}(t,x_j;\tau,\xi_j)\Big\} \\ &= E_c^{(0)}(t,x;\tau,\xi), \\ \tau < \lambda < t, &\quad \{x,y,\xi\} \subset \mathbb{R}^n, \end{aligned} \tag{1.3.44}$$

which will be basic for subsequent estimates.

Below we shall use the equality

$$\begin{aligned} \int\limits_{\mathbb{R}^n} E_{c_1\varepsilon_0}^{(0)}(t,x;\lambda,y)(t-\lambda)^{-M}dy = \int\limits_{\mathbb{R}^n} \exp\Big\{-c_1\varepsilon_0\sum_{j=1}^{n}|\eta_j|^{1/(1-\beta_j)}\Big\}d\eta =: D, \\ \lambda < t, \quad \{x,y\} \subset \mathbb{R}^n, \end{aligned} \tag{1.3.45}$$

obtained by the change of variables $y_j = x_j + (t-\lambda)^{\beta_j}\eta_j$, $j \in \{1, \dots, n\}$.

Using (1.3.37), (1.3.38), (1.3.43)–(1.3.45) we have

$$\begin{aligned} &|K_{m_0+1}(t,x;\tau,\xi)| \\ &\quad \le C_0^2\int\limits_{\tau}^{t}(t-\lambda)^{-M+\chi-1}d\lambda\int\limits_{\mathbb{R}^n}E_{c_1}^{(0)}(t,x;\lambda,y)E_c^{(0)}(\lambda,y;\tau,\xi)dy \\ &\quad \le C_0^2\int\limits_{\tau}^{t}E_c^{(0)}(t,x;\tau,\xi)\int\limits_{\tau}^{t}(t-\lambda)^{\chi-1}d\lambda\int\limits_{\mathbb{R}^n}E_{c_1\varepsilon_0}^{(0)}(t,x;\lambda,y)(t-\lambda)^{-M}dy \\ &\quad = C_0^2DB(\chi,1)(t-\tau)^{\chi}E_c^{(0)}(t,x;\tau,\xi). \end{aligned}$$

Suppose that for $j > 1$,

$$|K_{m_0+j}(t,x;\tau,\xi)| \le C_0(C_0D)^j\prod_{s=0}^{j-1}B(\chi,s\chi+1)(t-\tau)^{j\chi}E_c^{(0)}(t,x;\tau,\xi). \tag{1.3.46}$$

Then, as before,

$$\begin{aligned} &|K_{m_0+j+1}(t,x;\tau,\xi)| \\ &\quad \le C_0(C_0D)^{j+1}\prod_{s=0}^{j-1}B(\chi,s\chi+1)E_c^{(0)}(t,x;\tau,\xi)\int\limits_{\tau}^{t}(t-\lambda)^{\chi-1}(\lambda-\tau)^{j\chi}d\lambda \end{aligned}$$

$$= C_0(C_0D)^{j+1}\prod_{s=0}^{j} B(\chi, s\chi+1)(t-\tau)^{(j+1)\chi} \times E_c^{(0)}(t,x;\tau,\xi).$$

The well-known relation for Euler's Beta and Gamma functions $B(a,b) = \Gamma(a)\Gamma(b)/\Gamma(a+b)$, $a>0$, $b>0$, shows that

$$\prod_{s=0}^{j-1} B(\chi, s\chi+1) = \prod_{s=0}^{j-1}\Big(\Gamma(\chi)\Gamma(s\chi+1)/\Gamma((s+1)\chi+1)\Big)$$
$$= (\Gamma(\chi))^j/\Gamma(j\chi+1). \tag{1.3.47}$$

It follows from (1.3.42), (1.3.46), and (1.3.47) that the series (1.3.35) is majorized by the series

$$C_0\left(\sum_{m=1}^{m_0}(t-\tau)^{-M+m\chi-1} + \sum_{j=1}^{\infty}\Big(C_0D\Gamma(\chi)(t-\tau)^{\chi}\Big)^j\Big/\Gamma(j\chi+1)\right)E_c^{(0)}(t,x;\tau,\xi).$$

Therefore the series (1.3.36) is absolutely and uniformly convergent in $P^{\delta}_{[t_0,T]}$ for any $\delta \in (0, T-t_0)$, and its sum satisfies (1.3.39). □

Let the natural numbers n_1, n_2, n_3 and the variables $x := (x_1, x_2, x_3) \in \mathbb{R}^n$, $x_j := (x_{j1}, \dots, x_{jn_j}) \in \mathbb{R}^{n_j}$, $j \in \{1,2,3\}$, be the same as in Section 1.1.1. We shall use also the notation: $x_j' := (x_{j1}, \dots, x_{jn_3})$, $x_j'' := (x_{j(n_3+1)}, \dots, x_{jn_2})$, $j \in \{1,2\}$, $x_1''' := (x_{1(n_2+1)}, \dots, x_{1n_1})$, $\hat{x}_1 := (x_1', x_1'')$ (the symbols ξ_j', ξ_j'', $\hat{\xi}_1$ have a similar meaning if $\xi := (\xi_1, \xi_2, \xi_3)$, $\xi_j \in \mathbb{R}^{n_j}$, $j \in \{1,2\}$);

$$E_c^{(1)}(t,x;\tau,\xi) := \exp\Big\{-c\Big(\rho_{0\beta}(t-\tau,|x_1-\xi_1|) + \rho_{1\beta}(t-\tau,|x_2-\xi_2+(t-\tau)\hat{x}_1|)$$
$$+\rho_{2\beta}(t-\tau,|x_3-\xi_3+(t-\tau)x_2'+2^{-1}(t-\tau)^2x_1'|)\Big)\Big\}$$
$$= \exp\Big\{-c\Big(\rho_\beta^{(2)}(t,x_1',x_2',x_3;\tau,\xi_1',\xi_2',\xi_3)$$
$$+\rho_\beta^{(1)}(t,x_1'',x_2'';\tau,\xi_1'',\xi_2'') + \rho_\beta^{(0)}(t,x_1''';\tau,\xi_1''')\Big)\Big\},$$
$$E_c^{(0)}(t,x_1;\tau,\xi_1) := \exp\Big\{-c\rho_\beta^{(0)}(t,x_1;\tau,\xi_1)\Big\}.$$

Lemma 1.9. *Let the numbers $\{n_1,n_2,n_3,n\} \subset \mathbb{N}$ be such that $n_3 \le n_2 \le n_1$, $n_1+n_2+n_3 = n$, let $\beta \in (0,1)$, and $M := n_1\beta + n_2(\beta+1) + n_3(\beta+2)$. If the kernel* (1.3.34) *is continuous and satisfies the inequality*

$$|K(t,x;\tau,\xi)| \le C_1(t-\tau)^{-M+\chi-1}E_{c_1}^{(1)}(t,x;\tau,\xi),$$
$$(t,x;\tau,\xi) \in P^0_{[t_0,T]}, \tag{1.3.48}$$

with some constants $C_1 > 0$, $c_1 > 0$, *and* $\chi \in (0,1)$, *then there exists the resolvent* (1.3.36) *which is a continuous function, such that*

$$|R(t,x;\tau,\xi)| \le C(t-\tau)^{-M+\chi-1} E_c^{(0)}(t,x_1;\tau,\xi_1)$$
$$\times \sum_{j=0}^{\infty} \Big(\hat{C}\Gamma(\chi)(t-\tau)^{\chi}\Big)^j \Big(\Gamma(j\chi+1)\Big)^{-1} E^{(1)}_{c\delta_0^j}(t,x;\tau,\xi),$$
$$(t,x;\tau,\xi) \in P^0_{[t_0,T]}. \tag{1.3.49}$$

Here C, $\hat{C}$, *and* c *are certain positive constants,* $c < c_1$, Γ *is Euler's Gamma function,* $\delta_0 := \min(\delta_1, \delta_2) < 1$, δ_1, *and* δ_2 *are the constants from Lemma* 1.3.

The *proof* is carried out, with the use of Lemmas 1.3 and 1.5, by the same scheme as the proof of Lemma 1.8.

The first stage employs the inequalities

$$I_m := \int_{\mathbb{R}^n} E^{(1)}_{c_1}(t,x;\lambda,y) E^{(1)}_{c_m}(\lambda,y;\tau,\xi)\Big((t-\lambda)(\lambda-\tau)\Big)^{-M} dy$$
$$\le C_{m\varepsilon}(t-\tau)^{-M} E^{(1)}_{c_{m+1}}(t,x;\tau,\xi),$$
$$c_m := c_1(1-(m-2)\varepsilon)\delta_0^{m-2}, \quad m \ge 2. \tag{1.3.50}$$

They are obtained consecutively if I_m is written in the form

$$I_m = \int_{\mathbb{R}^{3n_3}} \exp\Big\{-c_1\rho_\beta^{(2)}(t,x_1',x_2',x_3;\lambda,y_1',y_2',y_3)$$
$$- c_m\rho_\beta^{(2)}(\lambda,y_1',y_2',y_3;\tau,\xi_1',\xi_2',\xi_3)\Big\}\Big((t-\lambda)(\lambda-\tau)\Big)^{-n_3(3\beta+3)} dy_1' dy_2' dy_3$$
$$\times \int_{\mathbb{R}^{2(n_2-n_3)}} \exp\Big\{-c_1\rho_\beta^{(1)}(t,x_1'',x_2'';\lambda,y_1'',y_2'') - c_m\rho_\beta^{(1)}(\lambda,y_1'',y_2'';\tau,\xi_1'',\xi_2'')\Big\}$$
$$\times \Big((t-\lambda)(\lambda-\tau)\Big)^{-(n_2-n_3)(2\beta+1)} dy_1'' dy_2'' \int_{\mathbb{R}^{n_1-n_2}} \exp\Big\{-c_1\rho_\beta^{(0)}(t,x_1''';\lambda,y_1''')$$
$$- c_m\rho_\beta^{(0)}(\lambda,y_1''';\tau,\xi_1''')\Big\}\Big((t-\lambda)(\lambda-\tau)\Big)^{-(n_1-n_2)\beta} dy_1'''$$

and the inequalities (1.3.23)–(1.3.25) are used for the cases, in which $n_1 - n_2$, $n_2 - n_3$ and n_3 respectively are put instead of n.

Just as in Lemma 1.8, using (1.3.37), (1.3.48), and (1.3.50) we obtain the estimates

$$|K_m(t,x;\tau,\xi)| \le C_0(t-\tau)^{-M+m\chi-1} E^{(1)}_{c_{m+1}}(t,x;\tau,\xi) \tag{1.3.51}$$

for $m < m_0$, and

$$|K_{m_0}(t,x;\tau,\xi)| \le C_0 E_{2c}^{(1)}(t,x;\tau,\xi) \tag{1.3.52}$$

for $m = m_0$, where m_0 and ε are chosen as those in Lemma 1.8, while $2c := c_1(1-\varepsilon_0)\delta_0^{m_0-1}$, $\varepsilon_0 \in (0,1)$.

In order to estimate the kernels K_m with $m > m_0$, we use the inequalities

$$E_{2c}^{(1)}(t,x;\tau,\xi) \le E_c^{(0)}(t,x_1;\tau,\xi_1) E_c^{(1)}(t,x;\tau,\xi), \tag{1.3.53}$$

$$E_c^{(0)}(t,x_1;\lambda,y_1) E_c^{(0)}(\lambda,y_1;\tau,\xi_1) \le E_c^{(0)}(t,x_1;\tau,\xi_1), \tag{1.3.54}$$

$$E_c^{(1)}(t,x;\lambda,y) E_{c\delta_0^{j-1}}^{(1)}(\lambda,y;\tau,\xi) \le E_{c\delta_0^j}^{(1)}(t,x;\tau,\xi), \quad j \ge 1. \tag{1.3.55}$$

The inequality (1.3.53) is obvious, the inequality (1.3.54) is a consequence of (1.3.13) for $n = n_1$, and the inequalities (1.3.55) are proved by induction, with the use of (1.3.13)–(1.3.15) in which n is substituted by $n_1 - n_2$, $n_2 - n_3$, and n_3 respectively.

Just as in the proof of Lemma 1.8, the inequalities (1.3.52)–(1.3.55) are employed in the proof, for any $j \ge 1$, of the bounds

$$|K_{m_0+j}(t,x;\tau,\xi)| \le C_0 \Big(C_0 D \Gamma(\chi)(t-\tau)^{\chi} \Big)^j \Big(\Gamma(j\chi+1) \Big)^{-1} \times E_c^{(0)}(t,x_1;\tau,\xi_1) E_{c\delta_0^j}^{(1)}(t,x;\tau,\xi), \tag{1.3.56}$$

in which

$$D := \int_{\mathbb{R}^n} \exp\Big\{ -(c_1 - 2c)\Big(|\eta_1|^{1/(1-\beta)} + |\eta_2|^{1/(1-\beta)} + |\eta_3|^{1/(1-\beta)} \Big) \Big\} d\eta.$$

It follows from (1.3.48), (1.3.51), (1.3.52), and (1.3.56), that the series

$$C_0 \Bigg(\sum_{m=1}^{m_0} (t-\tau)^{-M+m\chi-1} E_{2c}^{(1)}(t,x;\tau,\xi) + E_c^{(0)}(t,x_1;\tau,\xi_1) \times \sum_{j=1}^{\infty} \Big(C_0 D\Gamma(\chi)(t-\tau)^{\chi} \Big)^j \Big(\Gamma(j\chi+1) \Big)^{-1} E_{c\delta_0^j}^{(1)}(t,x;\tau,\xi) \Bigg)$$

majorizes the series (1.3.36). Therefore the series (1.3.36) is absolutely and uniformly convergent on $P_{[t_0,T]}^{\delta}$ for any $\delta \in (0, T-t_0)$, and its sum has the bound (1.3.49). □

By Lemma 1.6, if $\beta = 1/2$, then all the iterated kernels have estimates of the same form. As a result, the most precise estimate is obtained. The next lemma describes this case. Set

$$E_c^{(2)}(t,x;\tau,\xi) := \exp\Big\{ -c\Big(\rho^{(2)}(t,x_1',x_2',x_3;\tau,\xi_1',\xi_2',\xi_3) + \rho^{(1)}(t,x_1'',x_2'';\tau,\xi_1'',\xi_2'') + \rho^{(0)}(t,x_1''';\tau,\xi_1''') \Big) \Big\}.$$

Lemma 1.10. *Let the kernel* (1.3.34) *be continuous and satisfy the inequality*

$$|K(t,x;\tau,\xi)| \leq C_1(t-\tau)^{-M+\chi-1}E_c^{(2)}(t,x;\tau,\xi),$$
$$(t,x;\tau,\xi) \in P^0_{[t_0,T]}, \tag{1.3.57}$$

where $C_1 > 0$, $c > 0$, $M := (n_1 + 3n_2 + 5n_3)/2$, $\chi \in (0,1)$. *Then the resolvent has the bound*

$$|R(t,x;\tau,\xi)| \leq C(t-\tau)^{-M+\chi-1}E_c^{(2)}(t,x;\tau,\xi),$$
$$(t,x;\tau,\xi) \in P^0_{[t_0,T]}. \tag{1.3.58}$$

Proof. Estimating the iterated kernels we use the equality

$$\int\limits_{\mathbb{R}^n} E_c^{(2)}(t,x;\lambda,y)E_c^{(2)}(\lambda,y;\tau,\xi)\Big((t-\lambda)(\lambda-\tau)\Big)^{-M} dy$$
$$= \int\limits_{\mathbb{R}^{3n_3}} \exp\Big\{-c\Big(\rho^{(2)}(t,x_1',x_2',x_3;\lambda,y_1',y_2',y_3)$$
$$+\rho^{(2)}(\lambda,y_1',y_2',y_3;\lambda,\xi_1',\xi_2',\xi_3)\Big)\Big\}\Big((t-\lambda)(\lambda-\tau)\Big)^{-9n_3/2} dy_1'dy_2'dy_3$$
$$\times \int\limits_{\mathbb{R}^{2(n_2-n_3)}} \exp\Big\{-c(\rho^{(1)}(t,x_1'',x_2'';\lambda,y_1'',y_2'')+\rho^{(1)}(\lambda,y_1'',y_2'';\lambda,\xi_1'',\xi_2''))\Big\}$$
$$\times\Big((t-\lambda)(\lambda-\tau)\Big)^{-2(n_2-n_3)} dy_1''dy_2'' \int\limits_{\mathbb{R}^{n_1-n_2}} \exp\Big\{-c\Big(\rho^{(0)}(t,x_1''';\lambda,y_1''')$$
$$+\rho^{(0)}(\lambda,y_1''';\tau,\xi_1''')\Big)\Big\}\Big((t-\lambda)(\lambda-\tau)\Big)^{-(n_1-n_2)/2} dy_1'''$$
$$= C_0(t-\tau)^{-M}E_c^{(2)}(t,x;\tau,\xi), \tag{1.3.59}$$

which is a consequence of (1.3.29)–(1.3.31).

Using the inequality (1.3.57) and the equalities (1.3.40), (1.3.59) we find that

$$|K_2(t,x;\tau,\xi)| \leq C_1^2 \int\limits_\tau^t \Big((t-\lambda)(\lambda-\tau)\Big)^{\chi-1} d\lambda \int\limits_{\mathbb{R}^n} E_c^{(2)}(t,x;\lambda,y)$$
$$\times E_c^{(2)}(\lambda,y;\tau,\xi)\Big((t-\lambda)(\lambda-\tau)\Big)^{-M} dy$$
$$= C_1^2C_0B(\chi,\chi)(t-\tau)^{-M+2\chi-1}E_c^{(2)}(t,x;\tau,\xi)$$
$$= C_1^2C_0(\Gamma(\chi))^2(\Gamma(2\chi))^{-1}(t-\tau)^{-M+2\chi-1}E_c^{(2)}(t,x;\tau,\xi).$$

The subsequent iterated kernels are estimated in a similar way. It is then proved by induction that for any $m \geq 2$ we have the estimates

$$|K_m(t,x;\tau,\xi)| \leq C_1^m C_0^{m-1}(\Gamma(\chi))^m(\Gamma(m\chi))^{-1}(t-\tau)^{-M+m\chi-1}E_c^{(2)}(t,x;\tau,\xi),$$

which imply the assertion of Lemma 1.10. □

1.3.5. Integrals and integral equations with polar singularities.

Lemma 1.11. *Let* $\gamma > 1$,

$$J(x,\zeta,t,\tau) := \int_\tau^t (t-\mu)^{-\alpha/\gamma}(\mu-\tau)^{-\beta/\gamma}\,d\mu \times \int_{\mathbb{R}^n} \Big((t-\mu)^{1/\gamma}+|x-\eta|\Big)^{-n-b_1}\Big((\mu-\tau)^{1/\gamma}+|\eta-\xi|\Big)^{-n-b_2}\,d\eta,$$

where $0 \leq \tau < t$, $b_1, b_2 > 0$, $\alpha + b_1 < \gamma$, *and* $\beta + b_2 < \gamma$. *Then*

$$\begin{aligned} J(x,\zeta,t,\tau) &\leq C\Big\{B\Big(1-\frac{\alpha+b_1}{\gamma},1-\frac{\beta}{\gamma}\Big)(t-\tau)^{-(\alpha+\beta+b_1-\gamma)/\gamma}\Big((t-\tau)^{1/\gamma}+|x-\xi|\Big)^{-n-b_2} \\ &\quad + B\Big(1-\frac{\alpha}{\gamma},1-\frac{\beta+b_2}{\gamma}\Big)(t-\tau)^{-(\alpha+\beta+b_2-\gamma)/\gamma}\Big((t-\tau)^{1/\gamma}+|x-\xi|\Big)^{-n-b_1}\Big\}, \end{aligned} \tag{1.3.60}$$

where C *is independent of* n, b_1, *and* b_2, *and* B *is the beta-function.*

Proof. Let as decompose the domain of integration $(\tau,t)\times\mathbb{R}^n$ as $\Pi_1 \cup \Pi_2$ where

$$\Pi_1 := (\tau,(t+\tau)/2)\times\mathbb{R}^n, \quad \Pi_2 := ((t+\tau)/2,t)\times\mathbb{R}^n.$$

Further, let

$$\begin{aligned} \Pi_{11} &:= \Big\{(\mu,\eta)\in\Pi_1 \big| (\mu-\tau)^{1/\gamma}+|\eta-\xi| \leq \frac{1}{2}\Big((t-\tau)^{1/\gamma}+|x-\xi|\Big)\Big\}, \\ \Pi_{12} &:= \Pi_1 \setminus \Pi_{11}, \\ \Pi_{21} &:= \Big\{(\mu,\eta)\in\Pi_2 \big| (t-\mu)^{1/\gamma}+|x-\eta| \leq \frac{1}{2}\Big((t-\tau)^{1/\gamma}+|x-\xi|\Big)\Big\}, \\ \Pi_{22} &:= \Pi_2 \setminus \Pi_{21}. \end{aligned}$$

Now

$$(\tau,t)\times\mathbb{R}^n = \bigcup_{j,k=1}^2 \Pi_{jk} \quad\text{and}\quad J(x,\zeta,t,\tau) = \sum_{j,k=1}^2 J_{jk},$$

where J_{jk} is the integral over Π_{jk}.

Let us use the easily verified inequality $(u-v)^\rho \geq u^\rho - v^\rho$, where $0 \leq v \leq u/2$ and $\rho \leq 1$. If $(\mu, \eta) \in \Pi_{11}$, then $\mu - \tau \leq (t-\tau)/2$, and

$$\begin{aligned}(t-\mu)^{1/\gamma} + |x-\eta| &= ((t-\tau)-(\mu-\tau))^{1/\gamma} + |(x-\xi)-(\eta-\xi)| \\ &\geq (t-\tau)^{1/\gamma} + |x-\xi| - \left((\mu-\tau)^{1/\gamma} + |\eta-\xi|\right) \\ &\geq \frac{1}{2}\left((t-\tau)^{1/\gamma} + |x-\xi|\right),\end{aligned}$$

which gives us that

$$\begin{aligned}J_{11} &\leq 2^{n+b_1}\left((t-\tau)^{1/\gamma} + |x-\xi|\right)^{-n-b_1} \int\limits_{\tau}^{(t+\tau)/2} (t-\mu)^{-\alpha/\gamma}(\mu-\tau)^{-\beta/\gamma}\, d\mu \\ &\quad \times \int\limits_{\mathbb{R}^n} \left((\mu-\tau)^{1/\gamma} + \eta|\right)^{-n-b_2} d\eta \\ &\leq C(n,b_1,b_2)\left((t-\tau)^{1/\gamma} + |x-\xi|\right)^{-n-b_1} \int\limits_{\tau}^{(t+\tau)/2} (t-\mu)^{-\alpha/\gamma}(\mu-\tau)^{-(\beta+b_2)/\gamma}\, d\mu.\end{aligned}$$

After the substitution $t - \mu = (t-\tau)\nu$ we get that

$$\begin{aligned}J_{11} &\leq C(n,b_1,b_2)\left((t-\tau)^{1/\gamma} + |x-\xi|\right)^{-n-b_1} \\ &\quad \times (t-\tau)^{-(\alpha+\beta+b_2-\gamma)/\gamma} \int\limits_{1/2}^{1} \nu^{-\alpha/\gamma}(1-\nu)^{-(\beta+b_2)/\gamma}\, d\nu \\ &\leq C(n,b_1,b_2)B\left(1-\frac{\alpha}{\gamma}, 1-\frac{\beta+b_2}{\gamma}\right) \\ &\quad \times (t-\tau)^{-(\alpha+\beta+b_2-\gamma)/\gamma}\left((t-\tau)^{1/\gamma} + |x-\xi|\right)^{-n-b_1}.\end{aligned}$$

Further,

$$\begin{aligned}J_{12} &\leq 2^{n+b_1}\left((t-\tau)^{1/\gamma} + |x-\xi|\right)^{-n-b_2} \int\limits_{\tau}^{(t+\tau)/2} (t-\mu)^{-\alpha/\gamma}(\mu-\tau)^{-\beta/\gamma}\, d\mu \\ &\quad \times \int\limits_{\mathbb{R}^n} \left((t-\mu)^{1/\gamma} + \eta|\right)^{-n-b_1} d\eta.\end{aligned}$$

It is proved analogously that

$$J_{12} \leq C(n,b_1,b_2)B\left(1-\frac{\alpha+b_1}{\gamma}, 1-\frac{\beta}{\gamma}\right)$$

$$\times (t-\tau)^{-(\alpha+\beta+b_1-\gamma)/\gamma}\left((t-\tau)^{1/\gamma}+|x-\xi|\right)^{-n-b_2}.$$

Estimating J_{21} and J_{22} similarly, we arrive at (1.3.60). □

Several subsequent lemmas involve the reference function $\rho_\beta^{(0)}$. To simplify the notation, in this subsection we drop the sub- and superscripts and write

$$\rho(t,x;\lambda,y)=\left(\frac{|x-y|}{(t-\lambda)^\beta}\right)^{\frac{1}{1-\beta}},\quad 0<\beta<1.$$

Lemma 1.12. *For any $a_1,a_2>0$, any $b_1,b_2>0$, such that $b_1+b_2<n$, and any $\mu>0$, $\varepsilon\in(0,1)$*

$$\int_\tau^t (t-\lambda)^{a_1-1}(\lambda-\tau)^{a_2-1}$$
$$\times\int_{\mathbb{R}^n}|x-y|^{b_1-n}|y-\xi|^{b_2-n}\exp\{-\mu\,(\rho(t,x;\lambda,y)+\rho(\lambda,y;\tau,\xi))\}\,dy\,d\lambda$$
$$\le C(\varepsilon,a_1,a_2,b_1,b_2)(t-\tau)^{a_1+a_2-1}|x-\xi|^{b_1+b_2-n}\exp\{-(1-\varepsilon)\mu\rho(t,x;\tau,\xi)\}. \tag{1.3.61}$$

Proof. Writing $\mathbb{R}^n=\bigcup_{k=1}^4 V_k$ where

$$V_1:=\left\{y\in\mathbb{R}^n\big||x-y|>(t-\lambda)^\beta,\ |y-\xi|>(\lambda-\tau)^\beta\right\},$$
$$V_2:=\left\{y\in\mathbb{R}^n\big||x-y|>(t-\lambda)^\beta,\ |y-\xi|\le(\lambda-\tau)^\beta\right\},$$
$$V_3:=\left\{y\in\mathbb{R}^n\big||x-y|\le(t-\lambda)^\beta,\ |y-\xi|>(\lambda-\tau)^\beta\right\},$$
$$V_4:=\left\{y\in\mathbb{R}^n\big||x-y|\le(t-\lambda)^\beta,\ |y-\xi|\le(\lambda-\tau)^\beta\right\},$$

we decompose the left-hand side in (1.3.61) into the sum of four integrals J_1, J_2, J_3, J_4.

If $y\in V_1$, then $|x-y|^{b_1-n}<(t-\lambda)^{\beta(b_1-n)}$, $|y-\xi|^{b_2-n}<(\lambda-\tau)^{\beta(b_2-n)}$, so that

$$J_1\le\int_\tau^t (t-\lambda)^{a_1+\beta b_1-1}(\lambda-\tau)^{a_2+\beta b_2-1}((t-\lambda)(\lambda-\tau))^{-n\beta}$$
$$\times\int_{\mathbb{R}^n}\exp\{-\mu(\rho(t,x;\lambda,y)+\rho(\lambda,y;\tau,\xi))\}\,dy\,d\lambda,$$

and by Lemma 1.5

$$J_1\le C(t-\tau)^{a_1+a_2+(b_1+b_2-n)\beta-1}\exp\{-(1-\frac{\varepsilon}{2})\mu\rho(t,x;\tau,\xi)\}. \tag{1.3.62}$$

Since

$$(t-\tau)^{(b_1+b_2-n)\beta}\exp\{-\frac{\varepsilon}{2}\mu\rho(t,x;\tau,\xi)\}$$
$$=(\rho(t,x;\tau,\xi))^{(1-\beta)(n-b_1-b_2)}\exp\{-\frac{\varepsilon}{2}\mu\rho(t,x;\tau,\xi)\}|x-\xi|^{b_1+b_2-n}$$
$$\le C|x-\xi|^{b_1+b_2-n},$$

the inequality (1.3.62) means that J_1 does not exceed the right-hand side of (1.3.61).

In order to estimate J_2, we consider two distinct cases. Suppose first that $|x-\xi|\ge 2(t-\tau)^\beta$. Then $|x-y|\ge(t-\tau)^\beta$, so that by Lemma 1.3,

$$J_2\le(t-\tau)^{\beta(b_1-n)}\exp\{-\mu\rho(t,x;\tau,\xi)\}$$
$$\times\int_\tau^t(t-\lambda)^{a_1-1}(\lambda-\tau)^{a_2-1}\,d\lambda\int_{\{y\in\mathbb{R}^n\,\big|\,|y-\xi|\le(\lambda-\tau)^\beta\}}|y-\xi|^{b_2-n}\,dy$$
$$=C(t-\tau)^{a_1+a_2-1}(t-\tau)^{\beta(b_1+b_2-n)}\exp\{-\mu\rho(t,x;\tau,\xi)\}$$
$$\le C(\varepsilon)(t-\tau)^{a_1+a_2-1}|x-\xi|^{b_1+b_2-n}\exp\{-(1-\varepsilon)\mu\rho(t,x;\tau,\xi)\}. \tag{1.3.63}$$

If $|x-\xi|<2(t-\tau)^\beta$, then $|x-y|<3(t-\tau)^\beta$,

$$J_2\le\exp\{-\mu\rho(t,x;\tau,\xi)\}\int_\tau^t(t-\lambda)^{a_1-1}(\lambda-\tau)^{a_2-1}\,d\lambda$$
$$\times\int_{\{y\in\mathbb{R}^n\,\big|\,|y-\xi|<3(t-\tau)^\beta\}}|x-y|^{b_1-n}|y-\xi|^{b_2-n}\,dy,$$

and by Lemma 2 from Chapter 1 of [83] we get

$$J_2\le C(t-\tau)^{a_1+a_2-1}|x-\xi|^{b_1+b_2-n}\exp\{-\mu\rho(t,x;\tau,\xi)\}.$$

Together with (1.3.63) this yields the required bound for J_2.

The estimate for J_3 is proved similarly to that of J_2.

Let us consider J_4. It follows from the definition of V_4 that $J_4=0$ if $|x-\xi|\ge 2(t-\tau)^\beta$. Therefore we may assume that $|x-\xi|<2(t-\tau)^\beta$ and note that

$$V_4\subset\left\{y\in\mathbb{R}^n\big||y-\xi|<2(t-\tau)^\beta\right\}.$$

Then

$$J_4\le\exp\{-\mu\rho(t,x;\tau,\xi)\}\int_\tau^t(t-\lambda)^{a_1-1}(\lambda-\tau)^{a_2-1}\,d\lambda$$

$$\times \int\limits_{\{y\in\mathbb{R}^n \,|\, |y-\xi|<2(t-\tau)^\beta\}} |x-y|^{b_1-n}|y-\xi|^{b_2-n}\,dy.$$

Using again Lemma 2 from Chapter 1 of [83] we find that

$$J_4 \leq C(t-\tau)^{a_1+a_2-1}|x-\xi|^{b_1+b_2-n}\exp\{-\mu\rho(t,x;\tau,\xi)\},$$

as desired. □

Remark. If $b_1+b_2>n$, then the assertion of Lemma 1.12 still holds if the factor $|x-\xi|^{b_1+b_2-n}$ is omitted from the right-hand side of (1.3.61). The proof is similar; we have only to use the appropriate estimate from the same lemma of [83].

Lemma 1.13. *If* $0\leq c_2<c_1<1$, $\gamma>0$, *then*

$$\int\limits_{\mathbb{R}^n} |x-y|^{-n+\gamma}\exp\{-\mu(c_1\rho(t,x;\lambda,y)+c_2\rho(\lambda,y;\tau,\xi))\}\,dy$$
$$\leq C\exp\{-\mu c_2\rho(t,x;\tau,\xi)\}(t-\lambda)^{\beta\gamma} \tag{1.3.64}$$

where the constant depends on c_2-c_1.

Proof. Let us write the integral in the left-hand side of (1.3.64) as the sum I_1+I_2 of two integrals corresponding to the decomposition $\mathbb{R}^n=W_1\cup W_2$ where $W_1:=\{y\in\mathbb{R}^n\,|\,|y-x|\leq(t-\lambda)^\beta\}$, $W_2:=\{y\in\mathbb{R}^n\,|\,|y-x|>(t-\lambda)^\beta\}$.

Using Lemma 1.3, we get

$$I_1\leq \exp\{-\mu c_2\rho(t,x;\tau,\xi)\}\int\limits_{|y-x|\leq(t-\lambda)^\beta}|x-y|^{-n+\gamma}\,dy$$
$$=C(t-\lambda)^{\beta\gamma}\exp\{-\mu c_2\rho(t,x;\tau,\xi)\}.$$

Next, again by Lemma 1.3,

$$I_2=\int\limits_{\{y\in\mathbb{R}^n\,|\,|y-x|>(t-\lambda)^\beta\}}|x-y|^{-n+\gamma}\exp\{-\mu[c_1\rho(t,x;\lambda,y)+c_2\rho(\lambda,y;\tau,\xi)]\}\,dy$$
$$\leq(t-\lambda)^{-n\beta+\gamma\beta}\exp\{-\mu c_2\rho(t,x;\tau,\xi)\}\int\limits_{\mathbb{R}^n}\exp\{-\mu(c_1-c_2)\rho(t,x;\lambda,y)\,dy\}$$
$$\leq(t-\lambda)^{\beta\gamma}\exp\{-\mu c_2\rho(t,x;\tau,\xi)\},$$

as desired. □

The next two lemmas give estimates of the iterated kernels

$$K_m(t,x;\tau,\xi)=\int\limits_\tau^t d\lambda\int\limits_{\mathbb{R}^n}K_1(t,x;\lambda,y)K_{m-1}(\lambda,y;\tau,\xi)\,dy,\quad m\geq 2, \tag{1.3.65}$$

where $K_1=K$ is a given kernel. We treat the cases $n\geq 2$ and $n=1$ separately.

Lemma 1.14. *Let $n \geq 2$. Suppose that $K(t,x;\tau,\xi)$ is a continuous function on*

$$P_n := \{(t,x;\tau,\xi)|\{x,\xi\} \subset \mathbb{R}^n, x \neq \xi, 0 \leq \tau < t \leq T\},$$

such that

$$|K(t,x;\tau,\xi)| \leq A_1(t-\tau)^{\nu_0\beta-1}|x-\xi|^{-n+\nu_1}\exp\{-\mu\rho(t,x;\tau,\xi)\} \tag{1.3.66}$$

where $A_1, \mu > 0$, $0 < \nu_0, \nu_1, \beta < 1$, and $n\nu_1^{-1}$ is not an integer. Then the series $R(t,x;\tau,\xi) = \sum_{m=1}^{\infty} K_m(t,x;\tau,\xi)$ is absolutely and uniformly convergent on $P_n \cap \{(t,x;\tau,\xi): t-\tau \geq \delta > 0, |x-\xi| \geq \delta > 0\}$, and

$$|R(t,x;\tau,\xi)| \leq C(t-\tau)^{\nu_0\beta-1}|x-\xi|^{-n+\nu_1}\exp\{-\mu^*\rho(t,x;\tau,\xi)\}, \tag{1.3.67}$$

with some $\mu^* \in (0,\mu)$.

Proof. The bounds for the kernels (1.3.65) are obtained by two stages. First we use Lemma 1.12. We find, for a small $\varepsilon > 0$, that

$$|K_2(t,x;\tau,\xi)| \leq C_2(\varepsilon)(t-\tau)^{2\nu_0\beta-1}|x-\xi|^{-n+2\nu_1}\exp\{-\mu(1-\varepsilon)\rho(t,x;\tau,\xi)\},$$

and so on, so that

$$|K_s(t,x;\tau,\xi)| \leq C_s(\varepsilon)(t-\tau)^{s\nu_0\beta-1}|x-\xi|^{-n+s\nu_1}\exp\{-\mu(1-(s-1)\varepsilon)\rho(t,x;\tau,\xi)\}, \tag{1.3.68}$$

whenever $-n + s\nu_1 < 0$. Let $s^* = \min\{s| -n + s\nu_1 > 0\}$. Then by the remark after Lemma 1.12,

$$|K_{s^*}(t,x;\tau,\xi)| \leq C_0(t-\tau)^{a_0-1}\exp\{-\mu^*\rho(t,x;\tau,\xi)\}, \tag{1.3.69}$$

where $a_0 := s^*\nu_0\beta$, $\mu^* := [1-(s^*-1)\varepsilon]\mu$.

For all the next iterations, we use (1.3.66), (1.3.69), and Lemma 1.13. Using the identity

$$\int_\tau^t (t-\lambda)^{a_1-1}(\lambda-\tau)^{a_2-1}\,d\lambda = B(a_1,a_2)(t-\tau)^{a_1+a_2-1}$$

we get by induction that

$$|K_{s^*+m}(t,x;\tau,\xi)| \leq C_0C_1^m\left\{\prod_{j=1}^m B((\nu_0+\nu_1)\beta, a_0+(\nu_0+\nu_1)(j-1))\right\} \times (t-\tau)^{m(\nu_0+\nu_1)\beta+a_0}\exp\{-\mu^*\rho(t,x;\tau,\xi)\}. \tag{1.3.70}$$

The product in the right-hand side of (1.3.70) equals $\dfrac{\Gamma(a_0)[\Gamma((\nu_0+\nu_1)\beta)]^{m-1}}{\Gamma(a_0+(m-1)(\nu_0+\nu_1)\beta)}$. This implies the convergence of the series for $\mathbb{R}$. The estimate (1.3.67) follows from (1.3.68) and (1.3.70). □

Let us consider the iterations (1.3.65) for $n=1$.

Lemma 1.15. *Let $n=1$. Suppose that the function $K(t,x;\tau,\xi)$ is continuous on*

$$P_1=\left\{(t,x;\tau,\xi)\big|\{x,\xi\}\subset\mathbb{R}, x\neq\xi, 0\le\tau<t\le T\right\},$$

and satisfies the inequality

$$|K(t,x;\tau,\xi)|\le A_1(t-\tau)^{\gamma\beta-\beta-1}\exp\{-\mu\rho(t,x;\tau,\xi)\}$$

where $A_1,\mu>0$, $0<\gamma,\beta<1$. Then the series $R(t,x;\tau,\xi)=\sum\limits_{m=1}^{\infty}K_m(t,x;\tau,\xi)$ converges absolutely and uniformly on $P_1\cap\{(t,x;\tau,\xi):\ t-\tau\ge\delta>0\}$, and

$$|R(t,x;\tau,\xi)|\le C(t-\tau)^{\gamma\beta-\beta-1}\exp\{-\mu^*\rho(t,x;\tau,\xi)\},\tag{1.3.71}$$

with some $\mu^*\in(0,\mu)$.

Proof. Using Lemma 1.5, we find for any $\varepsilon>0$ that

$$\begin{aligned}|K_2(t,x;\tau,\xi)|&\le A_1^2\int\limits_{\tau}^{t}((t-\lambda)(\lambda-\tau))^{\gamma\beta-\beta-1}\,d\lambda\\&\quad\times\int\limits_{-\infty}^{\infty}\exp\{\mu(\rho(t,x;\lambda,y)+\rho(\lambda,y;\tau,\xi))\,dy\\&\le A_2(\varepsilon)(t-\tau)^{2\gamma\beta-\beta-1}\exp\{-\mu(1-\varepsilon)\rho(t,x;\tau,\xi)\}.\end{aligned}$$

Repeating the procedure we obtain the inequality

$$|K_m(t,x;\tau,\xi)|\le A_m(\varepsilon)(t-\tau)^{m\gamma\beta-\beta-1}\exp\{-\mu(1-(m-1)\varepsilon)\rho(t,x;\tau,\xi)\}.$$

Let $m^*:=\left[\dfrac{1+\beta}{\beta\gamma}\right]+1$. Then

$$|K_{m^*}(t,x;\tau,\xi)|\le A_{m^*}(\varepsilon)\exp\{-\mu(1-m^*\varepsilon)\rho(t,x;\tau,\xi)\}.\tag{1.3.72}$$

Setting $\varepsilon:=\dfrac{1}{pm^*}$, $p>2$, $\mu^*:=\mu(1-p^{-1})$, $A_0:=\max\limits_{1\le m\le m^*}A_m(\varepsilon)$, we can rewrite (1.3.72) as

$$|K_{m^*}(t,x;\tau,\xi)|\le A_0\exp\{-\mu^*\rho(t,x;\tau,\xi)\}.\tag{1.3.73}$$

Now, for $m > m^*$, we proceed in a different way, in order to preserve the exponential factor μ^* in the estimates of further iterated kernels. By Lemma 1.3,

$$|K_{m^*+1}(t,x;\tau,\xi)| \le A_0^2 \int_\tau^t (t-\lambda)^{\gamma\beta-\beta-1}\, d\lambda \times \int_{-\infty}^{\infty} \exp\{-\mu(\rho(t,x;\lambda,y) + (1-p^{-1})\rho(\lambda,y;\tau,\xi))\}\, dy$$

$$\le A_0^2 F \exp\{-\mu^*\rho(t,x;\tau,\xi)\} \int_\tau^t (t-\lambda)^{\gamma\beta-1}\, d\lambda$$

$$= A_0^2 F B(\gamma\beta,1)(t-\tau)^{\gamma\beta} \exp\{-\mu^*\rho(t,x;\tau,\xi)\},$$

where $F := \int_0^\infty \exp\left\{-\frac{\mu}{p}|y|^{\frac{1}{1-\beta}}\right\} dy$.

Similarly, we prove by induction that

$$|K_{m^*+j}(t,x;\tau,\xi)| \le A_0(A_0F)^j(t-\tau)^{\gamma\beta j} \left\{\prod_{s=0}^{j-1} B(\gamma\beta, 1+s\gamma\beta)\right\} \exp\{-\mu^*\rho(t,x;\tau,\xi)\}.$$

Using the identity

$$\prod_{s=0}^{j-1} B(\gamma\beta, 1+s\gamma\beta) = \frac{(\Gamma(\gamma\beta))^j}{\Gamma(1+\gamma\beta j)}$$

we see that the series for R is majorized by the convergent series

$$\left\{\sum_{m=1}^{m^*} A_m(t-\tau)^{\gamma m\beta-\beta-1} + A_0\sum_{j=1}^{\infty} \frac{(\Gamma(\gamma\beta)A_0F)^j}{\Gamma(1+\gamma\beta j)}(t-\tau)^{\gamma\beta j}\right\} \exp\{-\mu^*\rho(t,x;\tau,\xi)\}.$$

This implies (1.3.71). □

Chapter 2

Parabolic Equations of a Quasi-Homogeneous Structure

2.1 Fundamental solution of the Cauchy problem for equations with bounded coefficients

2.1.1. Main assumptions. We start from the case of a $\overrightarrow{2b}$-parabolic equation of the form (1.1.10), of the first order with respect to the variable t, with bounded coefficients, that is the equation

$$L_N(t,x,\partial_t,\partial_x)u(t,x) := \left(I\partial_t - \sum_{\|k\|\le 2b} a_k(t,x)\partial_x^k\right)u(t,x) = f(t,x), \quad (t,x)\in\Pi_{(0,T]}, \tag{2.1.1}$$

in which $N\in\mathbb{N}$, the coefficients $a_k : \Pi_{[0,T]} \to \mathbb{C}_{NN}$, $\|k\|\le 2b$, are bounded functions, $f : \Pi_{(0,T]} \to \mathbb{C}_{N1}$ is a given function, $u : \Pi_{(0,T]} \to \mathbb{C}_{N1}$ is an unknown function.

The main assumptions regarding the coefficients of the equation (2.1.1) are as follows.

Since the variables $t, x_1, \dots, x_n$ in $\overrightarrow{2b}$-parabolic equations have different meanings, we shall use the following special distances between the points x and ξ, (t,x) and (τ,ξ):

$$p(x;\xi) := \left(\sum_{j=1}^n |x_j-\xi_j|^{2/m_j}\right)^{1/2},$$

$$p(t,x;\tau,\xi) := (|t-\tau|^{1/b} + (p(x;\xi))^2)^{1/2},$$

$$q(x;\xi) := \left(\sum_{j=1}^n |x_j-\xi_j|^{q_j}\right)^{1/q'}. \tag{2.1.2}$$

Here $m_j = b/b_j$, $j\in\{1,\dots,n\}$, b is the least common multiple of the numbers $b_1,\dots,b_n$, $q_j = 2b_j/(2b_j-1)$, $j\in\{1,\dots,n\}$, and $q' = \max_{j\in\{1,\dots,n\}} q_j$.

The functions p and q indeed satisfy the distance axioms. The axioms of identity and symmetry are evidently satisfied. The triangle axiom is checked using the Minkowski inequality

$$\left(\sum_{j=1}^{n}(x_j+y_j)^r\right)^{1/r} \leq \left(\sum_{j=1}^{n}x_j^r\right)^{1/r} + \left(\sum_{j=1}^{n}y_j^r\right)^{1/r}$$

and the inequality

$$(x_1+y_1)^s \leq x_1^s + y_1^s,$$

which are valid for any $x_j \geq 0$, $y_j \geq 0$, $r > 1$, and $s \in (0,1]$.

The Hölder property of functions will be understood with respect to the distance p.

Increments of functions will be denoted as

$$\Delta_t^\tau f(t,\cdot) := f(t,\cdot) - f(\tau,\cdot), \quad \Delta_x^\xi f(\cdot,x) := f(\cdot,x) - f(\cdot,\xi),$$
$$\Delta_{t,x}^{\tau,\xi} f(t,x,\cdot) := f(t,x,\cdot) - f(\tau,\xi,\cdot).$$

Let Ω and Q be sets of points $x \in \mathbb{R}^n$ and $(t,x) \in \mathbb{R}^{n+1}$ respectively, γ be a number from $(0,1]$, H be a positive constant, and $\{m,r\} \subset \mathbb{N}$.

Definition 2.1.

1) *A function $f : \Omega \to \mathbb{C}_{mr}$ is called Hölder continuous on Ω with the exponent γ and the constant H, if for any $\{x,\xi\} \subset \Omega$,*

$$|\Delta_x^\xi f(x)| \leq H(p(x;\xi))^\gamma.$$

2) *A function $f : Q \to \mathbb{C}_{mr}$ is called Hölder continuous in t,x on Q with the exponent γ and the constant H, if for arbitrary $(t,x), (\tau,\xi) \in Q$,*

$$|\Delta_{t,x}^{\tau,\xi} f(t,x)| \leq H(d(t,x;\tau,\xi))^\gamma.$$

3) *A function $f : Q \to \mathbb{C}_{mr}$ is called Hölder continuous in x uniformly with respect to t on Q with the exponent γ and the constant H, if for any (t,x), $(t,\xi) \in Q$,*

$$|\Delta_x^\xi f(t,x)| \leq H(p(x;\xi))^\gamma.$$

4) *A function $f : Q \to \mathbb{C}_{mr}$ is called locally Hölder continuous in t,x or in x, uniformly with respect to t, if it is Hölder continuous in a respective way on each compact subset of Q.*

5) *We shall call a function Hölder continuous in t,x or in x, if it has this property with some exponent γ and some constant H.*

If the coefficients $a_k, \|k\| \leq 2b$, of the equation (2.1.1) have the derivatives $\partial_x^k a_k$, then for such an equation there exists the Lagrange adjoint equation

$$L_N^*(t,x,\partial_t,\partial_x)v(t,x) := -\partial_t v(t,x) - \sum_{\|k\|\leq 2b}(-\partial_x)^k(\bar{a}_k'(t,x)v(t,x))$$
$$= g(t,x), \quad (t,x) \in \Pi_{[0,T)}, \tag{2.1.3}$$

where the bar over a_k means complex conjugation, and the dash means a transposed matrix.

Using the formulas for the expressions L_N and L_N^* from (2.1.1) and (2.1.3), we obtain an important "divergent" identity valid for all sufficiently smooth functions u and v with values in $\mathbb{C}_{N1}$:

$$\bar{v}' L_N u - \overline{(L_N^* v)}' u = \partial_t(\bar{v}' u) + \sum_{j=1}^{n} \partial_{x_j} B^j[v, u]. \tag{2.1.4}$$

Here

$$B^j[v, u] := - \sum_{0 < \|k\| \le 2b} \sum_{\nu_j = 0}^{k_j - 1} (-\partial_{x_1})^{k_1} \cdots (-\partial_{x_{j-1}})^{k_{j-1}} \times (-\partial_{x_j})^{\nu_j} (\bar{v}' a_k) \partial_{x_j}^{k_j - \nu_j} \partial_{x_{j+1}}^{k_{j+1}} \cdots \partial_{x_n}^{k_n} u, \quad j \in \{1, \dots, n\};$$

for $k_j = 0$ the sum is assumed equal to zero.

The assumptions regarding the coefficients of (2.1.1) are as follows.

(A_{11}) The equation (2.1.1) is uniformly $\overrightarrow{2b}$-parabolic on the set $\Pi_{[0,T]}$, with the $\overrightarrow{2b}$-parabolicity constant δ;

(A_{12}) the coefficients $a_k, \|k\| \le 2b$, are bounded, continuous in t (for $N > 1$ the continuity in t of a_k with $\|k\| = 2b$ is uniform with respect to $x \in \mathbb{R}^n$), and Hölder continuous in x uniformly with respect to t, with the exponent $\gamma \in (0, 1)$ and the constant H, on $\Pi_{[0,T]}$;

(A_{13}) the coefficients a_k, $\|k\| \le 2b$, are bounded and Hölder continuous in t, x with the exponent $\gamma \in (0, 1)$ and the constant H on $\Pi_{[0,T]}$;

(A_{14}) there exist the derivatives $\partial_x^k a_k$, $\|k\| \le 2b$, satisfying the conditions A_{12}.

2.1.2. The case of an equation whose coefficients do not depend on the spatial variables. Here we shall construct and study a FSCP for the equation (2.1.1) for the case of the coefficients independent of x.

For brevity we shall use the following notation:

$$A(t, \partial_x) := A_0(t, \partial_x) + A_1(t, \partial_x),$$
$$A_0(t, \partial_x) := \sum_{\|k\| = 2b} a_k(t) \partial_x^k,$$
$$A_1(t, \partial_x) := \sum_{\|k\| < 2b} a_k(t) \partial_x^k;$$
$$r(s) := \sum_{j=1}^{n} |s_j|^{2b_j},$$
$$s_t := (t^{-1/(2b_1)} s_1, \dots, t^{-1/(2b_n)} s_n),$$
$$s := (s_1, \dots, s_n) \in \mathbb{C}^n.$$

Let τ be an arbitrary fixed number from the interval $[0, T)$. Consider the Cauchy problem

$$(I\partial_t - A(t, \partial_x))u(t, x) = 0, \quad (t, x) \in \Pi_{(\tau,T]}, \tag{2.1.5}$$

$$u(t, x)|_{t=\tau} = \varphi(x), \quad x \in \mathbb{R}^n, \tag{2.1.6}$$

assuming that the coefficients of the equation (2.1.5) are continuous on $[0, T]$, and the uniform $\overrightarrow{2b}$-parabolicity condition holds with the constant δ, that is λ-roots of the equation

$$\det(I\lambda - A_0(t, i\sigma)) = 0 \tag{2.1.7}$$

satisfy the condition

$$\operatorname{Re} \lambda_j(t, \sigma) \leq -\delta r(\sigma), \quad j \in \{1, \ldots, N\}, \tag{2.1.8}$$

for any $t \in [0, T]$ and $\sigma \in \mathbb{R}^n$.

Suppose that for the function φ there exists the Fourier transform

$$\psi(\sigma) := (F_{x \to \sigma}[\varphi])(\sigma), \quad \sigma \in \mathbb{R}^n. \tag{2.1.9}$$

A solution of the problem (2.1.7), (2.1.8) is sought in the form

$$u(t, x) = (F^{-1}_{\sigma \to x}[v(t, \sigma)])(t, x), \quad (t, x) \in \Pi_{(\tau,T]}, \tag{2.1.10}$$

where v is an unknown function.

Substituting the expression (2.1.10) into (2.1.5), (2.1.6), and using (2.1.9) and properties of the Fourier transform we come to the problem

$$\left(I\frac{d}{dt} - A(t, i\sigma)\right)v(t, \sigma) = 0, \quad (t, \sigma) \in \Pi_{(\tau,T]}, \tag{2.1.11}$$

$$v(t, \sigma)|_{t=\tau} = \psi(\sigma), \quad \sigma \in \mathbb{R}^n. \tag{2.1.12}$$

for the function v. Its solution is given by the formula

$$v(t, \sigma) = V(t, \tau, \sigma)\psi(\sigma), \quad (t, \sigma) \in \Pi_{(\tau,T]}, \tag{2.1.13}$$

where V is a FSCP for the equation (2.1.11), that is a solution of this equation, such that $V(\tau, \tau, \sigma) = I$, $\sigma \in \mathbb{R}^n$.

Substituting (2.1.13) into (2.1.10) we obtain the formula

$$u(t, x) = \int_{\mathbb{R}^n} G(t, x; \tau, \xi)\varphi(\xi)d\xi, \quad (t, x) \in \Pi_{(\tau,T]}, \tag{2.1.14}$$

for the solution of the problem (2.1.5), (2.1.6), in which

$$G(t, x, ; \tau, \xi) := \hat{G}(t, \tau, x - \xi), \tag{2.1.15}$$

$$\hat{G}(t, \tau, x) := (F^{-1}_{\sigma \to x}[V(t, \tau, \sigma)])(t, \tau, x), \tag{2.1.16}$$

$$0 \leq \tau < t \leq T, \quad \{x, \xi\} \subset \mathbb{R}^n.$$

In order to substantiate the above procedures, proving that the formula (2.1.14) indeed determines a solution of the problem (2.1.5), (2.1.6), so that the function G is a FSCP, we have first to study properties of the function V.This will be based on the following lemmas.

Lemma 2.1. *Consider a matrix $A \in \mathbb{C}_{NN}$. Let $\lambda_1, \ldots, \lambda_N$ be its eigenvalues, and $\Lambda := \max_{j \in \{1,\ldots,N\}} \operatorname{Re} \lambda_j$. Then for any $t \geq 0$,*

$$|\exp\{tA\}| \leq \exp\{t\Lambda\} \sum_{j=0}^{N-1} (2t|A|)^j.$$

For the proof see Chapter 2, §6 in [88].

Lemma 2.2. *Let a function $f : \mathbb{C}^n \to \mathbb{R}$ be continuous and homogeneous in the following generalized sense: for any $s := (s_1, \ldots, s_n) \in \mathbb{C}^n$ and $a > 0$,*

$$f((a^{1/(2b_1)} s_1, \ldots, a^{1/(2b_n)} s_n)) = af(s). \tag{2.1.17}$$

Suppose that for any $\sigma \in \mathbb{R}^n$,

$$f(\sigma) \leq -\delta r(\sigma) \tag{2.1.18}$$

with some constant $\delta > 0$. Then for each $\delta_1 \in (0, \delta)$ there exists such a constant $c_1 > 0$, that for all $s = \sigma + i\gamma \in \mathbb{C}^n$

$$f(s) \leq -\delta_1 r(\sigma) + c_1 r(\gamma).$$

Proof. It is sufficient to show that the function

$$g(\sigma, \gamma) := (r(\gamma))^{-1}(f(\sigma + i\gamma) + \delta_1 r(\sigma)), \quad \{\sigma, \gamma\} \subset \mathbb{R}^n,$$

is bounded from above by a certain constant $c > 0$. It follows from the condition (2.1.17) with $a = (r(\gamma))^{-1}$ that

$$\begin{aligned} g(\sigma, \gamma) &= f(((r(\gamma))^{-1/(2b_1)}(\sigma_1 + i\gamma_1), \ldots \\ &\quad \ldots, r(\gamma))^{-1/(2b_n)}(\sigma_n + i\gamma_n))) + \delta_1 (r(\gamma))^{-1} r(\sigma) \\ &= f(\xi + i\eta) + \delta_1 r(\xi) \\ &=: g_0(\xi, \eta), \end{aligned}$$

where

$$\begin{gathered} \xi := (\xi_1, \ldots, \xi_n), \qquad \eta := (\eta_1, \ldots, \eta_n), \\ \xi_j := (r(\gamma))^{-1/(2b_j)} \sigma_j, \quad \eta_j := (r(\gamma))^{-1/(2b_j)} \gamma_j, \\ j \in \{1, \ldots, n\}. \end{gathered}$$

The function g_0, hence also the function g, is bounded from above. Indeed, if $r(\xi) > R$ where R is large enough, then due to (2.1.17) the number

$$g_0(\xi,\eta) = r(\xi)(f(((r(\xi))^{-1/(2b_1)}(\xi_1 + i\eta_1), \dots, (r(\xi))^{-1/(2b_n)}(\xi_n + i\eta_n))) + \delta_1),$$

for all η, such that $r(\eta) = 1$, is close to the number

$$r(\xi)(f(((r(\xi))^{-1/(2b_1)}\xi_1, \dots, (r(\xi))^{-1/(2b_n)}\xi_n)) + \delta_1),$$

which is negative by virtue of (2.1.18). If $r(\xi) \le R$, then the boundedness of the function g_0 is a consequence of its continuity. □

Lemma 2.3. *Let u, v be nonnegative continuous functions on the interval $[t_0, T]$, C be a positive constant. Then the inequality*

$$u(t) \le C + \int_{t_0}^{t} u(\tau)v(\tau)d\tau, \quad t \in [t_0, T], \tag{2.1.19}$$

implies the inequality

$$u(t) \le C \exp\left\{ \int_{t_0}^{t} v(\tau)d\tau \right\}, \quad t \in [t_0, T].$$

Proof. The inequality (2.1.19) can be written as

$$\left(C + \int_{t_0}^{t} u(\tau)v(\tau)d\tau \right)^{-1} u(t)v(t) \le v(t), \quad t \in [t_0, T].$$

Integrating it from t_0 to t, we find that

$$\ln\left(C + \int_{t_0}^{t} u(\tau)v(\tau)d\tau \right) - \ln C \le \int_{t_0}^{t} v(\tau)d\tau,$$

whence, due to (2.1.19), we get

$$u(t) \le C + \int_{t_0}^{t} u(\tau)v(\tau)d\tau \le C \exp\left\{ \int_{t_0}^{t} v(\tau)d\tau \right\}.$$ □

Let us turn to properties of the function V. For fixed t and τ, the function $V(t,\tau,\sigma)$, $\sigma \in \mathbb{R}^n$, can be extended onto $\mathbb{C}^n$ to an entire function $V(t,\tau,s)$, $s \in \mathbb{C}^n$, a solution of the Cauchy problem for the ordinary differential equation

$$\left(I\frac{d}{dt} - A(t, is) \right) V(t,\tau,s) = 0, \quad 0 \le \tau < t \le T, \quad s \in \mathbb{C}^n, \tag{2.1.20}$$

with polynomial in s coefficients. Let us give an estimate of this function.

Define a function

$$\Lambda(s) := \max_{t\in[0,T],j\in\{1,\ldots,N\}} \operatorname{Re}\lambda_j(t,s)$$

where λ_j, $j \subset \{1,\ldots,N\}$, are the roots of the equation (2.1.7), in which s is substituted for σ. This function is continuous and homogeneous in the generalized sense; see (2.1.17). By (2.1.8) it satisfies the inequality $\Lambda(\sigma) \le -\delta r(\sigma)$. Therefore by Lemma 2.2,

$$\Lambda(s) \le -\delta_1 r(\sigma) + c_1 r(\gamma), \quad s = \sigma + i\gamma \in \mathbb{C}^n. \tag{2.1.21}$$

Let t_0 be an arbitrary fixed point from $[0,T]$. Consider an auxiliary equation with constant coefficients

$$\left(I\frac{d}{dt} - A_0(t_0, is)\right)v = 0, \quad s \in \mathbb{C}^N.$$

Its FSCP is the function

$$V_0(t-\tau,s) := \exp\{A_0(t_0,is)(t-\tau)\}, \quad 0 \le \tau \le t \le T, \quad s \in \mathbb{C}^n.$$

We apply Lemma 2.1 to this function, taking into account the inequality (2.1.21) and the fact that

$$\begin{aligned}|A_0(t_0,is)| &\le C_0 \sum_{\|k\|=2b} |s_1|^{k_1}\ldots|s_n|^{k_n}\\ &\le C_0 \sum_{\|k\|=2b}\sum_{j=1}^{n}(|s_j|^{2b_j}k_j/(2b_j)) \le C_1 r(s).\end{aligned}$$

We have used the inequality

$$\prod_{j=1}^{n} a_j^{p_j} \le \sum_{j=1}^{n} a_j p_j, \quad a_j \ge 0, p_j \ge 0, \quad j \in \{1,\ldots,n\}, \quad \sum_{j=1}^{n} p_j = 1, \tag{2.1.22}$$

in which $a_j = |s_j|^{2b_j}$ and $p_j = k_j/(2b_j)$, $j \in \{1,\ldots,n\}$. Then we find that

$$\begin{aligned}|V_0(t-\tau,s)| &\le \exp\{(-\delta_1 r(\sigma) + c_1 r(\gamma))(t-\tau)\}\sum_{j=0}^{N-1}(2C_1(t-\tau)r(s))^j\\ &\le C_2\exp\{(-\delta_2 r(\sigma) + c_2 r(\gamma))(t-\tau)\}\end{aligned} \tag{2.1.23}$$

where $\delta_2 \in (0,\delta_1)$, $c_2 > c_1$.

Since the function V is a solution of the equation (2.1.20), writing the latter as

$$\left(I\frac{d}{dt} - A_0(t_0,is)\right)V = ((A_0(t,is) - A_0(t_0,is) + A_1(t,is))V$$

and using the function V_0 we obtain the equality

$$V(t,\tau,s) = V_0(t-t_0,s)V(t_0,\tau,s) + \int_{t_0}^{t} V_0(t-\theta,s)((A_0(\theta,is) - A_0(t_0,is) + A_1(\theta,is))V(\theta,\tau,s)d\theta,$$

$$0 \le \tau \le t_0 \le t \le T, \quad s \in \mathbb{C}^n. \tag{2.1.24}$$

Due to the uniform continuity of the coefficients a_k, $\|k\| = 2b$, for any $\eta > 0$ there exists such $\varepsilon > 0$ that for all $\{t_0, \theta\} \subset [\tau, T]$, $|t_0 - \theta| \le \varepsilon$, and $s \in \mathbb{C}^n$ we have

$$|A_0(t_0,is) - A_0(\theta,is)| \le \eta \sum_{\|k\|=2b} |s_1|^{k_1} \dots |s_n|^{k_n} \le C_3 \eta r(s). \tag{2.1.25}$$

Let us estimate $|A_1(\theta,is)|$. Using the boundedness of the coefficients a_k, $\|k\| < 2b$, and the inequality (2.1.22) with $a_j = |s_j|^{\|k\|/m_j}$, $p_j = m_j k_j/\|k\|$, $j \in \{1,\dots,n\}$, $\|k\| > 0$, we find that

$$\begin{aligned} |A_1(\theta,is)| &\le C_4 \sum_{\|k\|<2b} |s_1|^{k_1} \dots |s_n|^{k_n} \\ &\le C_4 \left(\sum_{0<\|k\|<2b} \sum_{j=1}^{n} (|s_j|^{\|k\|/m_j} m_j k_j/\|k\|) \right) + 1 \\ &\le C_5 \sum_{\|k\|<2b} (r(s))^{\|k\|/(2b)}. \end{aligned}$$

If $r(s) > R$, then

$$|A_1(\theta,is)| \le \eta r(s), \quad t_0 \le \theta \le T, \tag{2.1.26}$$

where the choice of R depends on η.

Let us find a bound for $|V(t,\tau,s)|$ as $t - t_0 \le \varepsilon$ and $r(s) > R$. Using the relation (2.1.4) and the inequalities (2.1.23), (2.1.25), and (2.1.26) we obtain that

$$|V(t,\tau,s)| \le C_2 \exp\{(-\delta_2 r(\sigma) + c_2 r(\gamma))(t-t_0)\}|V(t_0,\tau,s)| + C_2(C_3+1)\eta r(s) \int_{t_0}^{t} \exp\{(-\delta_2 r(\sigma) + c_2 r(\gamma))(t-\theta)\}|V(\theta,\tau,s)|d\theta.$$

Set

$$u(t) := |V(t,\tau,s)| \exp\{(\delta_2 r(\sigma) - c_2 r(\gamma))(t-t_0)\};$$

then the last inequality takes the form

$$u(t) \le C_2|V(t_0,\tau,s)| + C_2(C_3+1)\eta r(s) \int_{t_0}^{t} u(\theta)d\theta,$$

whence, after using Lemma 2.3,

$$u(t) \le C_2|V(t_0,\tau,s)|\exp\{C_2(C_3+1)\eta r(s)(t-t_0)\},$$

that is

$$|V(t,\tau,s)| \le C_2|V(t_0,\tau,s)|\exp\{(-\delta_2 r(\sigma)+c_2 r(\gamma)+C_2(C_3+1)\eta r(s))(t-t_0)\}.$$

By virtue of (1.3.9) we have

$$\begin{aligned} r(s) &= \sum_{j=1}^{n}|s_j|^{2b_j} = \sum_{j=1}^{n}(|\sigma_j|^2+|\gamma_j|^2)^{b_j} \\ &\le \sum_{j=1}^{n}2^{b_j-1}(|\sigma_j|^{2b_j}+|\gamma_j|^{2b_j}) \le 2^{b'-1}(r(\sigma)+r(\gamma)), \\ &\qquad b' := \max_{j\in\{1,\dots,n\}} b_j. \end{aligned}$$

Choose η in such a way that $\delta_2 - 2^{b'-1}C_2(C_3-1)\eta =: \delta_3 > 0$; then

$$\begin{gathered} |V(t,\tau,s)| \le C_2|V(t_0,\tau,s)|\exp\{(-\delta_3 r(\sigma)+c_3 r(\gamma))(t-t_0)\}, \\ t-t_0\le\varepsilon, \quad r(s)>R, \end{gathered} \tag{2.1.27}$$

where $c_3 := c_2 + 2^{b'-1}C_2(C_3+1)\eta$.

Taking $\tau, \tau+\varepsilon, \dots, \tau+m_0\varepsilon$, where m_0 is the largest integer smaller than $(t-\tau)/\varepsilon$, instead of t_0, and estimating consecutively $V(\tau+\varepsilon,\tau,s),\dots,V(\tau+m_0\varepsilon,\tau,s)$ using the inequality (2.1.27), we get

$$|V(t,\tau,s)| \le C_2^{m_0}\exp\{(-\delta_3 r(\sigma)+c_3 r(\gamma))(t-\tau)\}.$$

Now for arbitrary τ and t, $0\le\tau\le t\le T$, we have

$$|V(t,\tau,s)| \le C_2^{m}\exp\{(-\delta_3 r(\sigma)+c_3 r(\gamma))(t-\tau)\}, \tag{2.1.28}$$

where m is the largest integer smaller than T/ε.

The estimate (2.1.28) was proved under the assumption that $r(s)>R$. If $r(s)\le R$, then for any τ and t, $0\le\tau\le t\le T$, we obtain that

$$\begin{aligned} |V(t,\tau,s)| &\le C_6 \le C_6\exp\{(-\delta_3 r(\sigma)+c_3 r(\gamma))(t-\tau)+\delta_3 RT\} \\ &= C_7\exp\{(-\delta_3 r(\sigma)+c_3 r(\gamma))(t-\tau)\}. \end{aligned} \tag{2.1.29}$$

The estimates (2.1.28) and (2.1.29) imply the final estimate

$$\begin{gathered} |V(t,\tau,s)| \le C_8\exp\{(-\delta_3 r(\sigma)+c_3 r(\gamma))(t-\tau)\}, \\ 0\le\tau\le t\le T, \quad s\in\mathbb{C}^n. \end{gathered} \tag{2.1.30}$$

Taking $(t-\tau)^{-1/(2b_j)}s_j$, $j \in \{1,\dots,n\}$, instead of s_j, we can rewrite this estimate for $\tau < t$ in the following more convenient form:

$$|V(t,\tau,s_{t-\tau})| \le C_8 \exp\{-\delta_3 r(\sigma) + c_3 r(\gamma)\}, \quad 0 \le \tau < t \le T, \quad s \in \mathbb{C}^n. \tag{2.1.31}$$

Remark 1. *If $N = 1$, the estimate* (2.1.31) *can be proved without using the uniform continuity of the leading coefficients a_k, $\|k\| = 2b$, if we use the formula*

$$V(t,\tau,s) = \exp\left\{\int_\tau^t A(\theta, is)d\theta\right\}.$$

valid for this case.

It is easy to note that the constants C_8, δ_3, and c_3 from the estimate (2.1.31) depend only on

$$n, N, b_1, \dots, b_n, \delta, T, \max_{t\in[0,T], \|k\|\le 2b} |a_k(t)|, \tag{2.1.32}$$

and, if $N > 1$, also on the kind of continuity of the leading coefficients.

The above properties of the function V make it possible to obtain a complete analytic description of the function G and its derivatives.

By (2.1.15) and (2.1.16) we have

$$\begin{aligned} \partial_x^k G(t,x;\tau,\xi) &= (F^{-1}_{\sigma\to x}[(i\sigma)^k V(t,\tau,\sigma)])(t,\tau,x-\xi) \\ &= (t-\tau)^{-M-\|k\|/(2b)}(F^{-1}_{\sigma\to x}[(i\sigma_{t-\tau})^k V(t,\tau,\sigma_{t-\tau})])(t,\tau,(x-\xi)_{t-\tau}), \\ &\quad 0 \le \tau < t \le T, \quad \{x,\xi\} \subset \mathbb{R}^n, \quad k \in \mathbb{Z}^n_+, \end{aligned} \tag{2.1.33}$$

where $M := \sum_{j=1}^n (1/(2b_j))$. The functions

$$V^{(k)}(t,\tau,s_{t-\tau}) := (is_{t-\tau})^k V(t,\tau,s_{t-\tau}), \quad s \in \mathbb{C}^n, \quad k \in \mathbb{Z}^n_+,$$

for fixed t and τ, are entire, and the inequality (2.1.31) implies the estimates

$$\begin{aligned} &|V^{(k)}(t,\tau,s_{t-\tau})| \le C^{(k)} \exp\{-\delta_4 r(\sigma) + c_4 r(\gamma)\}, \\ &\quad 0 \le \tau < t \le T, \quad s \in \mathbb{C}^n, \quad k \in \mathbb{Z}^n_+, \end{aligned} \tag{2.1.34}$$

where $0 < \delta_4 < \delta_3$ and $c_4 > c_3$. Applying to these functions Lemma 1.1 with $p_j = q_j = 2b_j$, $j \in \{1,\dots,n\}$, taking into account the equalities (2.1.33) and the estimates (2.1.34), we can give a full description of properties of the function G. In particular, we can prove that G is a FSCP for the equation (2.1.5). We formulate all the results as the following main theorem.

Theorem 2.1. *Let the coefficients of the equation* (2.1.5) *be continuous on the interval* $[0,T]$, *and the uniform $\overrightarrow{2b}$-parabolicity condition* (2.1.8) *holds. Then:*

1) *There exists the unique FSCP G for the equation* (2.1.5) *determined by the formulas* (2.1.15) *and* (2.1.16).
2) *The function G and its derivatives can be extended onto* $\mathbb{C}^n$, *and for these extensions*

$$\partial_x^k G(t,z;\tau,\zeta) = (t-\tau)^{-M-\|k\|/(2b)}\Omega_k(t,\tau,(z-\zeta)_{t-\tau}),$$
$$0 \le \tau < t \le T, \quad \{z = x+iy, \zeta = \xi + i\eta\} \subset \mathbb{C}^n, \quad k \in \mathbb{Z}_+^n, \qquad (2.1.35)$$

where the functions $\Omega_k(t,\tau,z)$, $z = (z_1,\dots,z_n) \in \mathbb{C}^n$, $k \in \mathbb{Z}_+^n$, *are, for fixed* t *and* τ, *entire functions of* $z_1,\dots,z_n$, *with the growth orders* $q_1,\dots,q_n$ *respectively, and the same orders of decrease for real arguments. Here* $q_j = 2b_j/(2b_j-1)$, $j \in \{1,\dots,n\}$.
3) *G satisfies the estimates*

$$|\partial_x^k G(t,z;\tau,\zeta)| \le C_k(t-\tau)^{-M-\|k\|/(2b)}$$
$$\times \exp\left\{-c\sum_{j=1}^n (t-\tau)^{1-q_j}|x_j-\xi_j|^{q_j} + c'\sum_{j=1}^n (t-\tau)^{1-q_j}|y_j-\eta_j|^{q_j}\right\},$$
$$0 \le \tau < t \le T, \quad \{z,\zeta\} \subset \mathbb{C}^n, \quad k \in \mathbb{Z}_+^n, \qquad (2.1.36)$$

where C_k, c, *and* c' *are positive constants depending only on the numbers* (2.1.32), *and, if* $N > 1$, *also on the kind of continuity of the leading coefficients.*
4) *The formula*

$$I_0(t,\tau,x) := \int_{\mathbb{R}^n} G(t,x;\tau,\xi)d\xi = V(t,\tau,0), \quad 0 \le \tau < t \le T, \quad x \in \mathbb{R}^n, \qquad (2.1.37)$$

holds implying, in particular, that

$$I_0(t,\tau,x) \underset{t\to\tau}{\to} I \qquad (2.1.38)$$

uniformly with respect to $x \in \mathbb{R}^n$. *If in* (2.1.5) $a_0 = 0$, *then*

$$I_0(t,\tau,x) = I, \quad 0 \le \tau < t \le T, \quad x \in \mathbb{R}^n. \qquad (2.1.39)$$

Corollary 1. *Set*

$$E_c^{(0)}(t,x;\tau,\xi) := \exp\left\{-c\sum_{j=1}^n (t-\tau)^{1-q_j}|x_j-\xi_j|^{q_j}\right\}, \quad t > \tau, \quad \{x,\xi\} \subset \mathbb{R}^n, \qquad (2.1.40)$$

where c is a positive constant. The estimates (2.1.36) *imply the inequalities*

$$|\partial_x^k G(t,x;\tau,\xi)| \le C_k(t-\tau)^{-M-\|k\|/(2b)} E_c^{(0)}(t,x;\tau,\xi),$$
$$0 \le \tau < t \le T, \quad \{x,\xi\} \subset \mathbb{R}^n, \quad k \in \mathbb{Z}_+^n. \qquad (2.1.41)$$

Corollary 2. *The equalities* (2.1.35) *and* (2.1.37) *imply, for* $k \neq 0$, *the equalities*

$$\int_{\mathbb{R}^n} \partial_x^k G(t,x;\tau,\xi)d\xi = 0, \tag{2.1.42}$$

$$\int_{\mathbb{R}^n} \Omega_k(t,\tau,\xi)d\xi = 0, \tag{2.1.43}$$

$$0 \le \tau < t \le T, \quad x \in \mathbb{R}^n.$$

Proof of Theorem 2.1. The assertions 2) and 3) are direct consequences of Lemma 1.1 and the above properties of the function V. The equality (2.1.37) is proved with the use of (1.3.1), (2.1.15), and (2.1.16) as follows:

$$\begin{aligned} I_0(t,\tau,x) &= \int_{\mathbb{R}^n} \hat{G}(t,\tau,x-\xi)d\xi = \int_{\mathbb{R}^n} \hat{G}(t,\tau,x)dx \\ &= \int_{\mathbb{R}^n} \exp\{-i(0,x)\}F^{-1}_{\sigma\to x}[V(t,\tau,\sigma)]dx \\ &= F_{x\to 0}[F^{-1}_{\sigma\to x}[V(t,\tau,\sigma)]] = V(t,\tau,0). \end{aligned}$$

It remains to verify that the function G is indeed a FSCP for the equation (2.1.5).

Using the formulas (2.1.15) and (2.1.16), properties of the function V and the Fourier transforms, we show easily that the function $G(t,x;\tau,\xi)$, $(t,x) \in \Pi_{(\tau,T]}$, with fixed τ and ξ, is a solution of the equation (2.1.5).

Let the function $\varphi : \mathbb{R}^n \to \mathbb{C}_{N1}$ be continuous and bounded while $\tau \in [0,T)$ is fixed. Consider the integral (2.1.14). The estimates (2.1.41) ensure the uniform convergence in $\Pi_{[\tau+\varepsilon,T]}$, $\varepsilon > 0$, of the integral (2.1.14) and the ones obtained by formal differentiation under the integral's symbol. Since the function G is a solution of (2.1.5), the function (2.1.14) satisfies the equation (2.1.5) on $\Pi_{(\tau,T]}$. Let us prove the equality

$$\lim_{t\to\tau} u(t,x) = \varphi(x), \quad x \in \mathbb{R}^n, \tag{2.1.44}$$

which, together with the above results, will mean that G is a FSCP for the equation (2.1.5).

By (2.1.38), in order to prove (2.1.44), it suffices to show that

$$\lim_{t\to\tau} \int_{\mathbb{R}^n} G(t,x;\tau,\xi)\Delta_\xi^x \varphi(\xi)d\xi = 0. \tag{2.1.45}$$

The integral from (2.1.45) can be represented as the sum of the integral I_1 over the ball $B(x,\delta) := \{\xi \in \mathbb{R}^n \mid q(x;\xi) \le \delta\}$ and the integral I_2 over $\mathbb{R}^n \setminus B(x,\delta)$

where q is the distance defined in (2.1.2). Due to the uniform continuity of φ on $B(x,\delta)$, we have $|\Delta_\xi^x \varphi(\xi)| \le \omega(\delta)$, $\lim\limits_{\delta\to 0}\omega(\delta)$. Since the function φ is bounded, we have

$$|\Delta_\xi^x \varphi(\xi)| \le C_\varphi, \quad \xi\in\mathbb{R}^n\setminus B(x,\delta), \quad x\in\mathbb{R}^n.$$

Now, using the estimate (2.1.41), we find that

$$|I_1| \le C_0\omega(\delta)\int\limits_{\mathbb{R}^n}(t-\tau)^{-M}E_c^{(0)}(t,x;\tau,\xi)d\xi \le C_0 D\omega(\delta),$$

$$\begin{aligned}|I_2| &\le C_0C_\varphi \int\limits_{\mathbb{R}^n\setminus B(x,\delta)}(t-\tau)^{-M}E_c^{(0)}(t,x;\tau,\xi)d\xi\\ &\le C_0C_\varphi \exp\{-(c/2)(t-\tau)^{1-q''}\delta^{q'}\}\int\limits_{\mathbb{R}^n}(t-\tau)^{-M}E_{c/2}^{(0)}(t,x;\tau,\xi)d\xi\\ &= C_0C_\varphi D\exp\{-(c/2)(t-\tau)^{1-q''}\delta^{q'}\},\end{aligned}$$

if $t-\tau<1$. We have used the equality

$$\int\limits_{\mathbb{R}^n}(t-\tau)^{-M}E_{c_0}^{(0)}(t,x;\tau,\xi)d\xi = \int\limits_{\mathbb{R}^n}\exp\left\{-c_0\sum_{j=1}^n|y_j|^{q_j}\right\}dy =: D \tag{2.1.46}$$

for $c_0=c/2$, and the inequality

$$\begin{aligned}\sum_{j=1}^n(t-\tau)^{1-q_j}|x_j-\xi_j|^{q_j} &\ge (t-\tau)^{1-q''}\sum_{j=1}^n|x_j-\xi_j|^{q_j}\\ &= (t-\tau)^{1-q''}(q(x;\xi))^{q'} \ge (t-\tau)^{1-q''}\delta^{q'}, \qquad (2.1.47)\\ &t-\tau<1, \quad \xi\in\mathbb{R}^n\setminus B(x,\delta),\end{aligned}$$

where $q' = \max\limits_{j\in\{1,\dots,n\}} q_j$, $q'' = \min\limits_{j\in\{1,\dots,n\}} q_j$. The equality (2.1.46) is obtained after the change of variables $\xi_j = x_j + (t-\tau)^{1/(2b_j)}y_j$, $j\in\{1,\dots,n\}$.

In order to complete the proof of (2.1.45), we choose, for any $\varepsilon>0$, such δ that $C_0D\omega(\delta)<\varepsilon/2$, and then take $t-\tau$ so small that

$$C_0C_\varphi D\exp\{-(c/2)(t-\tau)^{1-q''}\delta^{q'}\}<\varepsilon/2. \qquad \square$$

Example 2.1. For the $\overrightarrow{2b}$-parabolic equation

$$(\partial_t - \partial_{x_1}^2 + \partial_{x_2}^4)u = 0,$$

in which $n=2$, $b_1=1$ and $b_2=2$, the FSCP has the form

$$\begin{aligned}G(t,x;\tau,\xi) &= 2^{-1}\pi^{-3/2}(t-\tau)^{-3/4}\exp\{-(4(t-\tau))^{-1}(x_1-\xi_1)^2\}\\ &\times\int\limits_0^\infty \exp\{-\alpha^4\}\cos((t-\tau)^{-1/4}(x_2-\xi_2)\alpha)d\alpha, \quad \tau<t, \{x,\xi\}\subset\mathbb{R}^2.\end{aligned}$$

Indeed, by the equalities (1.3.2), (2.1.15), and (2.1.16) we have

$$G(t,x;\tau,\xi) = (2\pi)^{-2} \int_{\mathbb{R}} \exp\{ix_1\sigma_1 - (t-\tau)\sigma_1^2\}d\sigma_1 \times \int_{\mathbb{R}} \exp\{ix_2\sigma_2 - (t-\tau)\sigma_2^4\}d\sigma_2.$$

The first integral equals

$$\pi^{1/2}(t-\tau)^{-1/2}\exp\{-(4(t-\tau))^{-1}(x_1-\xi_1)^2\},$$

while the second is equal, by virtue of the Euler formula, to

$$2(t-\tau)^{-1/4}\int_0^\infty \exp\{-\alpha^4\}\cos((t-\tau)^{-1/4}(x_2-\xi_2)\alpha)d\alpha.$$

2.1.3. Properties of the fundamental solution of the Cauchy problem for an equation with parameters. Let us consider some properties of the FSCP for an equation (2.1.5) with coefficients depending also on the parametric point $y \in \mathbb{R}^n$. These properties will imply necessary properties of the parametrix of the equation (2.1.1).

Consider the equation

$$(I\partial_t - A(t,y,\partial_x))u(t,x) = 0, \quad (t,x) \in \Pi_{(0,T]}, \tag{2.1.48}$$

where

$$A(t,y,\partial_x) := \sum_{\|k\|\le 2b} a_k(t,y)\partial_x^k.$$

We shall make the following assumptions.

(B_{11}) the equation (2.1.48) is uniformly, with respect to $(t,y) \in \Pi_{[0,T]}$, $\overrightarrow{2b}$-parabolic with a constant δ;

(B_{12}) the coefficients of the equation are bounded and continuous in t, and in the case $N > 1$ the continuity of the coefficients a_k, $\|k\| = 2b$, is uniform in $y \in \mathbb{R}^n$.

Under these conditions, just as in Section 2.1.2, for the equation (2.1.48) we construct a FSCP $G(\cdot,\cdot;\cdot,\cdot;y)$ and give its complete analytic description, which implies, in particular, the estimates

$$|\partial_x^k G(t,x;\tau,\xi;y)| \le C_k(t-\tau)^{-M-\|k\|/(2b)}E_c^{(0)}(t,x;\tau,\xi),$$
$$0 \le \tau < t \le T, \quad \{x,\xi,y\} \subset \mathbb{R}^n, \quad k \in \mathbb{Z}_+^n, \tag{(2.1.49)}$$

where the constants C_k and c depend only on

$$n, N, b_1, \dots, b_n, \delta, T, \max_{\|k\|\le 2b} \sup_{(t,y)\in\Pi_{[0,T]}} |a_k(t,y)|,$$

and for $N > 1$ also on the kind of continuity in t of the coefficients a_k, $\|k\| = 2b$.

Property 2.1. *Let the equation* (2.1.48) *and the equation*

$$\begin{aligned}(I\partial_t - B(t,y,\partial_x))u(t,x) &= 0, \quad (t,x) \in \Pi_{(0,T]}, \\ B(t,y,\partial_x) &:= \sum_{\|k\|\le 2b} b_k(t,y)\partial_x^k,\end{aligned} \tag{2.1.50}$$

satisfy the conditions B_{11} *and* B_{12}, *and also the inequalities*

$$|a_k(t,y) - b_k(t,y)| \le \varepsilon, \quad (t,y) \in \Pi_{[0,T]}, \quad \|k\| \le 2b.$$

Let G_A *and* G_B *be the FSCP for the equations* (2.1.48) *and* (2.1.50) *respectively. Then*

$$|\partial_x^k (G_A - G_B)(t,x;\tau,\xi;y)| \le C_k \varepsilon (t-\tau)^{-M-\|k\|/(2b)} E_c^{(0)}(t,x;\tau,\xi),$$
$$0 \le \tau < t \le T, \quad \{x,\xi,y\} \subset \mathbb{R}^n, \quad k \in \mathbb{Z}_+^n.$$

Proof. It is sufficient to obtain an estimate for the difference

$$(V_A - V_B)(t,\tau,s;y), \quad 0 \le \tau < t \le T, \quad s \in \mathbb{C}^n, \quad y \in \mathbb{R}^n,$$

where V_A and V_B are respectively the FSCP for the equations

$$\left(I\frac{d}{dt} - A(t,y,is)\right)v = 0, \quad \left(I\frac{d}{dt} - B(t,y,is)\right)v = 0.$$

Using the identity

$$\left(I\frac{d}{dt} - A(t,y,is)\right)V_B = (B(t,y,is) - A(t,y,is))V_B,$$

we come to the equality

$$(V_B - V_A)(t,\tau,s;y) = \int_\tau^t V_A(t,\theta,s;y)(B(\theta,y,is) - A(\theta,y,is))V_B(\theta,\tau,s;y)d\theta.$$

The last integral will be estimated with the use of the estimates for V_A and V_B similar to (2.1.30). We have

$$\begin{aligned}|(V_A - V_B)(t,\tau,s;y)| &\le C_8^2 \varepsilon \int_\tau^t \exp\{(-\delta_3 r(\sigma) + c_3 r(\gamma))(t-\theta)\} \\ &\quad\times \sum_{\|k\|\le 2b} |s_1|^{k_1}\dots|s_n|^{k_n} \exp\{(-\delta_3 r(\sigma) + c_3 r(\gamma))(\theta-\tau)\}d\theta \\ &\le C_9 \varepsilon \exp\{(-\delta_4 r(\sigma) + c_4 r(\gamma))(t-\tau)\}.\end{aligned}$$

This estimate and Lemma 1.1 imply the assertion of Property 2.1. □

The next property is a direct consequence of Property 2.1.

Property 2.2. *If the equation* (2.1.48) *satisfies the conditions* B_{11}, B_{12}, *and its coefficients are Hölder continuous in* y *uniformly with respect to* t *with an exponent* $\gamma \in (0, 1]$ *and a constant* H *on* $\Pi_{[0,T]}$, *then*

$$|\Delta_y^z \partial_x^k G(t, x; \tau, \xi; y)| \le C_k H(p(y; z))^\gamma (t - \tau)^{-M - \|k\|/(2b)} E_c^{(0)}(t, x; \tau, \xi),$$
$$0 \le \tau < t \le T, \quad \{x, \xi, y, z\} \subset \mathbb{R}^n, \quad k \in \mathbb{Z}_+^n.$$

Property 2.3. *Suppose that the equation* (2.1.48) *satisfies the conditions* B_{11}, B_{12}, *and its coefficients have continuous and bounded derivatives* $\partial_y^l a_k$, $\|k\| \le 2b$, $\|l\| \le r$, $r \in \mathbb{N}$. *Then there exist the derivatives* $\partial_x^k \partial_y^l G$, $k \in \mathbb{Z}_+^n$, $\|l\| \le r$, *satisfying the estimates*

$$|\partial_x^k \partial_y^l G(t, x; \tau, \xi; y)| \le C_{kl}(t - \tau)^{-M - \|k\|/(2b)} E_c^{(0)}(t, x; \tau, \xi),$$
$$0 \le \tau < t \le T, \quad \{x, \xi, y\} \subset \mathbb{R}^n, \quad k \in \mathbb{Z}_+^n, \quad \|l\| \le r. \tag{2.1.51}$$

If, in addition, the derivatives of a_k, $\|k\| \le 2b$, *are Hölder continuous in* y *uniformly with respect to* t *with an exponent* $\gamma \in (0, 1]$ *and a constant* H *on* $\Pi_{[0,T]}$, *then*

$$|\Delta_y^z \partial_x^k \partial_y^l G(t, x; \tau, \xi; y)| \le C_{kl} H(p(y; z))^\gamma (t - \tau)^{-M - \|k\|/(2b)} \times E_c^{(0)}(t, x; \tau, \xi),$$
$$0 \le \tau < t \le T, \quad \{x, \xi, y, z\} \subset \mathbb{R}^n, \quad k \in \mathbb{Z}_+^n, \quad \|l\| \le r. \tag{2.1.52}$$

Proof. The existence of the above derivatives of G follows from the existence of the derivatives $\partial_y^l V$ and properties of the Fourier transform. In order to prove (2.1.51) it is sufficient to verify the inequalities

$$|\partial_y^l V(t, \tau, s; y)| \le C_l \exp\{(-\delta_l r(\sigma) + c_l r(\gamma))(t - \tau)\}, \quad \|l\| \le r. \tag{2.1.53}$$

For $l = 0$ this is proved by the arguments from Section 2.1.2. For an arbitrary multi-index l, $\|l\| \le r$, the inequality (2.1.53) is proved by induction, with the use of the fact that the function $\hat{V} := \partial_y^m \partial_{y_j} V$, $\|m\| \le r$, $j \in \{1, \dots, n\}$, is a solution of the Cauchy problem

$$\left(I \frac{d}{dt} - A\right) \hat{V} = P(A, V), \quad \hat{V}|_{t=\tau} = 0,$$

where

$$P(A, V) := \sum_{p \le m} C_m^p \partial_y^p \partial_{y_j} A \partial_y^{m-p} V + \sum_{0 < p \le m} C_m^p \partial_y^p A \partial_y^{m-p} \partial_{y_j} V,$$

$C_m^p := \prod_{j=1}^n C_{m_j}^{p_j}$, $C_{m_j}^{p_j}$ is the binomial coefficient. Therefore

$$\hat{V}(t, \tau, s; y) = \int_\tau^t V(t, \theta, s; y) P(A(\theta, y, is), V(\theta, \tau, s; y)) d\theta.$$

This equality, together with estimates similar to those carried out in the proof of Property 2.1, and with the assumption of induction, implies the estimates (2.1.53).

The estimates (2.1.52) are consequences of Lemma 1.1 and the estimates for the increments $\Delta_y^z \partial_y^l V$. The latter are proved by induction, with the use of (2.1.53) and the above method. □

Property 2.4. *Uniformly with respect to* $x \in \mathbb{R}^n$,

$$\int\limits_{\mathbb{R}^n} G(t,x;\tau,\xi;y)d\xi = V(t,\tau,0;y) \underset{t\to\tau}{\to} I.$$

If $a_0 = 0$, *then*

$$\int\limits_{\mathbb{R}^n} G(t,x;\tau,\xi;y)d\xi = I.$$

For $k \neq 0$,

$$\int\limits_{\mathbb{R}^n} \partial_x^k G(t,x;\tau,\xi;y)d\xi = 0, \quad 0 \leq \tau < t \leq T, \quad \{x,y\} \subset \mathbb{R}^n.$$

This is proved just as the similar assertions from Theorem 2.1.

Corollary. *Suppose that the equation* (2.1.48) *satisfies the conditions from Property 2.2. Then the FSCP* $G(\cdot,\cdot;\cdot,\cdot;y)$ *and its derivatives are given by the formulas* (2.1.35), *in which* $\Omega_k(\cdot,\cdot,\cdot)$ *should be substituted by* $\Omega_k(\cdot,\cdot,\cdot;y)$. *The functions* Ω_k *satisfy the estimates*

$$|\Omega_k(t,\tau,x;y)| \leq C_k \exp\left\{-c\sum_{j=1}^n |x_j|^{q_j}\right\},$$

$$|\Delta_y^z \Omega_k(t,\tau,x;y)| \leq C_k (p(y;z))^\gamma \exp\left\{-c\sum_{j=1}^n |x_j|^{q_j}\right\},$$

$$0 \leq \tau < t \leq T, \quad \{x,y,z\} \subset \mathbb{R}^n,$$

and for $k \neq 0$,

$$\int\limits_{\mathbb{R}^n} \Omega_k(t,\tau,x;y)dx = 0, \quad 0 \leq \tau \leq t \leq T, \quad y \in \mathbb{R}^n.$$

Remark. *Similar properties hold for the FSCP* $G(\cdot,\cdot;\cdot,\cdot;\beta,y)$ *for the equation*

$$\left(I\partial_t - \sum_{\|k\|\leq 2b} a_k(\beta,y)\partial_x^k\right)u = 0, \tag{2.1.54}$$

which is uniformly, with respect to $(\beta, y) \in \Pi_{[0,T]}$, $\overrightarrow{2b}$*-parabolic, and its coefficients are bounded and Hölder continuous in* β, y *with the exponent* γ *on* $\Pi_{[0,T]}$. *In particular, in this case the functions* Ω_k, *which depend only on* x, β, *and* y, *satisfy the estimates*

$$|\Delta_{\beta,y}^{\zeta,z}\Omega_k(x;\beta,y)| \le C_k(p(\beta,y;\zeta,z))^\gamma \exp\left\{-c\sum_{j=1}^n |x_j|^{q_j}\right\},$$
$$x \in \mathbb{R}^n, \quad \{(\beta,y),(\zeta,z)\} \subset \Pi_{[0,T]}.$$

2.1.4. Parametrix and its properties. As it was said in Section 1.1.4, as a parametrix Z_0 for the equation (2.1.1) we shall take the function (1.1.43), where $G_0(\cdot,\cdot;\cdot,\cdot;y)$ is the FSCP for the equation (2.1.48) with only leading terms, that is the FSCP for the equation (1.1.39). Some properties of the parametrix needed in the sequel are as follows. In this section we assume that the conditions of Property 2.2 hold for the equation (1.1.39).

Property 2.5. *The following estimates are valid:*

$$|\partial_x^k Z_0(t,x;\tau,\xi)| \le C_k(t-\tau)^{-M-\|k\|/(2b)}E_c^{(0)}(t,x;\tau,\xi), \tag{2.1.55}$$

$$|\Delta_x^{x'}\partial_x^k Z_0(t,x;\tau,\xi)| \le C_k(p(x;x'))^{\gamma_0}(t-\tau)^{-M-(\|k\|+\gamma_0)/(2b)}E_c^{(0)}(t,x;\tau,\xi),$$
$$(p(x;x'))^{2b} \le t-\tau, \tag{2.1.56}$$

$$|\Delta_t^{t'}\partial_x^k Z_0(t,x;\tau,\xi)| \le C_k(t'-t)^{\gamma_0}(t-\tau)^{-M-\gamma_0-\|k\|/(2b)}E_c^{(0)}(t,x;\tau,\xi),$$
$$t'-t \le t-\tau, \quad k \in \mathbb{Z}_+^n; \tag{2.1.57}$$

$$\left|\int_{\mathbb{R}^n} \partial_x^k Z_0(t,x;\tau,y)dy\right| \le C_k(t-\tau)^{-(\|k\|-\gamma)/(2b)}, \tag{2.1.58}$$

$$\left|\Delta_x^{x'}\int_{\mathbb{R}^n} \partial_x^k Z_0(t,x;\tau,y)dy\right| \le C_k(p(x;x'))^{\gamma_0}(t-\tau)^{-(\|k\|-\gamma+\gamma_0)/(2b)},$$
$$(p(x;x'))^{2b} \le t-\tau, \tag{2.1.59}$$

$$\left|\Delta_t^{t'}\int_{\mathbb{R}^n} \partial_x^k Z_0(t,x;\tau,y)dy\right| \le C_k(t'-t)^{\gamma_0}(t-\tau)^{-\gamma_0-(\|k\|-\gamma)/(2b)},$$
$$t'-t \le t-\tau, \quad k \in \mathbb{Z}_+^n \setminus \{0\}, \tag{2.1.60}$$

where $0 \le \tau < t < t' \le T$, $\{x, x', \xi\} \subset \mathbb{R}^n$, γ_0 *is an arbitrary fixed number from* $(0,1]$.

Proof. The estimates (2.1.55) follow directly from the estimates (2.1.49). In order to obtain (2.1.56), one has to use the mean value theorem, the estimates (2.1.55), and the inequality

$$E_c^{(0)}(t,x'';\tau,\xi) \le C_1 E_{c_1}^{(0)}(t,x;\tau,\xi)$$

where x'' is a point on the line interval joining the points x and x', $c_1 \in (0, c)$. The last inequality is proved using the first inequality in (1.3.9) and the assumption $(p(x;x'))^{2b} \leq t-\tau$.

By the mean value theorem, the estimates (2.1.55), and the condition $t'-t \leq t-\tau$ we have

$$\begin{aligned}
|\Delta_t^{t'} \partial_x^k Z_0(t,x;\tau,\xi)| &= |(t'-t)\partial_{t''}\partial_x^k Z_0(t'',x;\tau,\xi)| \\
&= (t'-t)\left|\sum_{\|l\|=2b} a_l(t'',\xi)\partial_x^{k+l} Z_0(t'',x;\tau,\xi)\right| \\
&\leq C(t'-t)(t''-\tau)^{-M-1-\|k\|/(2b)} E_c^{(0)}(t'',x;\tau,\xi) \\
&\leq C_k(t'-t)^{\gamma_0}(t-\tau)^{-M-\gamma_0-\|k\|/(2b)} E_c^{(0)}(t,x;\tau,\xi), \\
&\qquad t'' \in (t,t').
\end{aligned}$$

Thus, the estimate (2.1.57) has been proved.

Due to Property 2.4 for G_0,

$$\int_{\mathbb{R}^n} \partial_x^k Z_0(t,x;\tau,y)dy = \int_{\mathbb{R}^n} \Delta_z^x \partial_x^k G_0(t,x;\tau,y;z)|_{z=y} dy.$$

From this representation, using Property 2.2, the inequality

$$\begin{aligned}
(p(x;y))^\gamma E_c^{(0)}(t,x;\tau,y) \leq C(t-\tau)^{\gamma/(2b)} E_{c_1}^{(0)}(t,x;\tau,y),& \\
\tau < t, \quad \{x,y\} \subset \mathbb{R}^n, \quad c_1 \in (0,c),& \qquad (2.1.61)
\end{aligned}$$

and the equality (2.1.46), we obtain the estimates (2.1.58).

In order to prove the estimates (2.1.59), we use the representation

$$\Delta_x^{x'} \int_{\mathbb{R}^n} \partial_x^k Z_0(t,x;\tau,y)dy = \sum_{j=1}^n \int_{x_j}^{x_j'} \partial_{\zeta_j}\left(\int_{\mathbb{R}^n} \partial_{\zeta^{(j)}}^k G_0(t,\zeta^{(j)};\tau,y;y)dy\right)d\zeta_j,$$

where $\zeta^{(j)} := (x_1,\dots,x_{j-1},\zeta_j,x_{j+1}',\dots,x_n')$. From Properties 2.2, 2.4, the inequality (2.1.61), and the equality (2.1.46) we find that

$$\begin{aligned}
\left|\Delta_x^{x'} \int_{\mathbb{R}^n} \partial_x^k Z_0(t,x;\tau,y)dy\right| &\leq \sum_{j=1}^n \left|\int_{x_j}^{x_j'}\left(\int_{\mathbb{R}^n}\left|\Delta_z^{\zeta^{(j)}} \partial_{\zeta_j}\partial_{\zeta^{(j)}}^k G_0(t,\zeta^{(j)};\tau,y;z)\right|_{z=y} dy\right)d\zeta_j\right| \\
&\leq C_k \sum_{j=1}^n \left|\int_{x_j}^{x_j'}\left(\int_{\mathbb{R}^n} (p(\zeta^{(j)};y))^\gamma (t-\tau)^{-M-(\|k\|+m_j)/(2b)}\right.\right.
\end{aligned}$$

$$\times E_c^{(0)}(t,\zeta^{(j)};\tau,y)dy\Bigg)d\zeta_j\Bigg|$$

$$\leq C_k \sum_{j=1}^{n} |x_j - x_j^{'}|(t-\tau)^{-(\|k\|-\gamma+m_j)/(2b)}.$$

Due to the inequality $(p(x;x'))^{2b} \leq t-\tau$, we have $|x_j - x_j^{'}|^{2b_j} \leq t-\tau$, $j \in \{1,\dots,n\}$, so that

$$\sum_{j=1}^{n} |x_j - x_j^{'}|(t-\tau)^{-m_j/(2b)} \leq \sum_{j=1}^{n} |x_j - x_j^{'}|^{\gamma_0/m_j}(t-\tau)^{-\gamma_0/(2b)}$$
$$\leq C(p(x;x'))^{\gamma_0}(t-\tau)^{-\gamma_0/(2b)}.$$

Hence the above estimates imply the required estimates (2.1.59).

The estimates (2.1.60) are proved as follows:

$$\left|\Delta_t^{t'} \int_{\mathbb{R}^n} \partial_x^k Z_0(t,x;\tau,y)dy\right| = \left|\int_t^{t'} \partial_\lambda \left(\int_{\mathbb{R}^n} \partial_x^k Z_0(\lambda,x;\tau,y)dy\right) d\lambda\right|$$
$$= \left|\int_t^{t'} \left(\int_{\mathbb{R}^n} \Delta_z^x \partial_\lambda \partial_x^k G_0(\lambda,x;\tau,y;z)\Big|_{z=y}\right) d\lambda\right|$$
$$\leq C_k \int_t^{t'} (\lambda-\tau)^{-1-(\|k\|-\gamma)/(2b)} d\lambda$$
$$\leq C_k (t-\tau)^{-1-(\|k\|-\gamma)/(2b)}(t'-t)$$
$$\leq C_k (t'-t)^{\gamma_0}(t-\tau)^{-\gamma_0-(\|k\|-\gamma)/(2b)}.$$

Here we used the definition of Z_0, Properties 2.2 and 2.4, the inequality (2.1.61), and the equality (2.1.46). □

Property 2.6. *Let $t_0 \in [0,T)$. Suppose that the function $f : \Pi_{[t_0,T]} \to \mathbb{C}_{N1}$ is bounded, continuous, and locally Hölder continuous in x, uniformly with respect to t, on $\Pi_{[t_0,T]}$. Then the function*

$$u(t,x;t_0) := \int_{t_0}^{t} d\tau \int_{\mathbb{R}^n} Z_0(t,x;\tau,\xi)f(\tau,\xi)d\xi, \quad (t,x) \in \Pi_{(t_0,T]}, \tag{2.1.62}$$

has the continuous derivatives $\partial_t u$ and $\partial_x^k u$, $\|k\| \leq 2b$. The derivatives $\partial_x^k u$, $\|k\| < 2b$, can be obtained by formal differentiation under the integral symbol, while the *derivatives $\partial_x^k u$, $\|k\| = 2b$, and $\partial_t u$ are given by the formulas:*

$$\partial_x^k u(t,x;t_0) = \int_{t_0}^{t} d\tau \int_{\mathbb{R}^n} \partial_x^k Z_0(t,x;\tau,\xi)\Delta_\xi^x f(\tau,\xi)d\xi \tag{2.1.63}$$

$$+ \int_{t_0}^{t} \Bigg(\int_{\mathbb{R}^n} \partial_x^k Z_0(t,x;\tau,\xi)d\xi \Bigg) f(\tau,x)d\tau, \quad (t,x) \in \Pi_{(t_0,T]}, \ \|k\| = 2b,$$

and

$$\partial_t u(t,x;t_0) = f(t,x) + \int_{t_0}^{t} d\tau \int_{\mathbb{R}^n} \partial_t Z_0(t,x;\tau,\xi)\Delta_\xi^x f(\tau,\xi)d\xi \tag{2.1.64}$$

$$+ \int_{t_0}^{t} \Bigg(\int_{\mathbb{R}^n} \partial_t Z_0(t,x;\tau,\xi)d\xi \Bigg) f(\tau,x)d\tau, \quad (t,x) \in \Pi_{(t_0,T]}.$$

Proof. For $t_0 \le \tau < t \le T$, $x \in \mathbb{R}^n$, and $\|k\| \le 2b$, we have

$$\partial_x^k \int_{\mathbb{R}^n} Z_0(t,x;\tau,\xi)f(\tau,\xi)d\xi = \int_{\mathbb{R}^n} \partial_x^k Z_0(t,x;\tau,\xi)f(\tau,\xi)d\xi$$

$$=: I^{(k)}(t,x;\tau). \tag{2.1.65}$$

In order to prove this, we have to verify that the integral $I^{(k)}$ converges, uniformly with respect to $x \in \mathbb{R}^n$, for arbitrary fixed t and τ. Making in the integral $I^{(k)}(t,x;\tau)$ the change of variables $\xi_j = x_j + (t-\tau)^{1/(2b_j)}y_j$, $j \in \{1,\dots,n\}$, we find that

$$I^{(k)}(t,x;\tau) = (t-\tau)^M \int_{\mathbb{R}^n} (\partial_x^k Z_0(t,x;\tau,\xi))f(\tau,\xi)\Big|_{\xi=\hat\xi(t-\tau,x,y)} dy,$$

$$\hat\xi(t,x,y) := (x_1 + t^{1/(2b_1)}y_1, \dots, x_n + t^{1/(2b_n)}y_n). \tag{2.1.66}$$

Due to the estimates (2.1.55) and the boundedness of the function f,

$$|\partial_x^k Z_0(t,x;\tau,\xi)f(\tau,\xi)|_{\xi=\hat\xi(t-\tau,x,y)}| \le C(t-\tau)^{-M-\|k\|/(2b)} \exp\Bigg\{-c\sum_{j=1}^{n}|y_j|^{q_j}\Bigg\},$$

$$\{x,y\} \subset \mathbb{R}^n.$$

Therefore the integral (2.1.66) converges, uniformly with respect to $x \in \mathbb{R}^n$, for fixed t, τ, and

$$|I^{(k)}(t,x;\tau)| \le C(t-\tau)^{-\|k\|/(2b)}, \quad t_0 \le \tau < t \le T, \quad x \in \mathbb{R}^n, \quad \|k\| \le 2b. \tag{2.1.67}$$

It follows from the equality (2.1.65) and the estimate (2.1.67) that for $\|k\|<2b$,

$$\partial_x^k u(t,x;t_0) = \int_{t_0}^{t} d\tau \int_{\mathbb{R}^n} \partial_x^k Z_0(t,x;\tau,\xi) f(\tau,\xi) d\xi, \quad (t,x) \in \Pi_{(t_0,T]},$$

since for $\|k\| < 2b$ the integral

$$\int_{t_0}^{t} (t-\tau)^{-\|k\|/(2b)} d\tau$$

is convergent for each $t \in (t_0, T]$.

If $\|k\| = 2b$, then the estimate (2.1.67) does not work. It can be improved for $x \in B_R := B(0,R)$ where R is an arbitrary fixed number. In order to do that, we write the representation

$$\begin{aligned} I^{(k)}(t,x;\tau) &= \int_{\mathbb{R}^n} \partial_x^k Z_0(t,x;\tau,\xi) \Delta_\xi^x f(\tau,\xi) d\xi + \int_{\mathbb{R}^n} \partial_x^k Z_0(t,x;\tau,\xi) d\xi f(\tau,x) \\ &=: I_1 + I_2. \end{aligned} \tag{2.1.68}$$

By the estimates (2.1.58), for $\|k\| = 2b$ we have

$$|I_2| \le C(t-\tau)^{-1+\gamma/(2b)}, \quad t_0 \le \tau < t \le T, \quad x \in \mathbb{R}^n. \tag{2.1.69}$$

Assuming that $x \in B_R$, we write I_1 as

$$I_1 = \int_{B_{2R}} \partial_x^k Z_0(t,x;\tau,\xi) \Delta_\xi^x f(\tau,\xi) d\xi + \int_{\mathbb{R}^n \setminus B_{2R}} \partial_x^k Z_0(t,x;\tau,\xi) \Delta_\xi^x f(\tau,\xi) d\xi =: I_1' + I_1''.$$

Using the estimates (2.1.55), the Hölder continuity of f, the inequality (2.1.61), and the equality (2.1.46) we obtain that

$$\begin{aligned} |I_1'| &\le C \int_{B_{2R}} (t-\tau)^{-M-1} (p(x;\xi))^\gamma E_c^{(0)}(t,x;\tau,\xi) d\xi \\ &\le C(t-\tau)^{-M-1+\gamma/(2b)} \int_{\mathbb{R}^n} E_{c_1}^{(0)}(t,x;\tau,\xi) d\xi = C(t-\tau)^{-1+\gamma/(2b)}, \\ &\qquad t_0 \le \tau < t \le T, \quad x \in B_R. \end{aligned} \tag{2.1.70}$$

We shall employ the inequality

$$\begin{aligned} &E_{c_0}^{(0)}(t,x;0,\xi) \le \exp\{-c_0 t^{-q_0} R^{q'}\}, \\ &t > 0, \quad x \in B_R, \quad \xi \in \mathbb{R}^n \setminus B_{2R}, \end{aligned} \tag{2.1.71}$$

where $c_0 > 0$, $q_0 = q'' - 1$, if $t \le 1$, and $q_0 = q' - 1$, if $t > 1$. It is a consequence of the estimates

$$\begin{aligned}\sum_{j=1}^{n} t^{1-q_j}|x_j - \xi_j|^{q_j} &\ge t^{-q_0}(q(x;\xi))^{q'}\\ &\ge t^{-q_0}(q(\xi;0) - q(x;0))^{q'}\\ &\ge t^{-q_0}(2R - R)^{q'}\\ &= t^{-q_0}R^{q'}.\end{aligned}$$

Using the estimates (2.1.55), the boundedness of f, the inequality (2.1.71), and the equality (2.1.46), we get

$$\begin{aligned}|I_1''| &\le C(t-\tau)^{-M-1} \int\limits_{\mathbb{R}^n\setminus B_{2R}} E_c^{(0)}(t,x;\tau,\xi)d\xi\\ &\le C(t-\tau)^{-M-1}\exp\{-(c/2)(t-\tau)^{-q_0}R^{q'}\}\int\limits_{\mathbb{R}^n} E_{c/2}^{(0)}(t,x;\tau,\xi)d\xi\\ &= C(t-\tau)^{-1}\exp\{-(c/2)(t-\tau)^{-q_0}R^{q'}\}\\ &\le C(t-\tau)^{-1+\gamma/(2b)},\\ &\qquad t_0 \le \tau < t \le T, \quad x \in B_R.\end{aligned} \tag{2.1.72}$$

From (2.1.68)–(2.1.70) and (2.1.72) we obtain the estimate

$$\begin{aligned}&|I^{(k)}(t,x;\tau)| \le C(t-\tau)^{-1+\gamma/(2b)},\\ &t_0 \le \tau < t \le T, \quad x \in B_R, \quad \|k\| = 2b.\end{aligned} \tag{2.1.73}$$

This estimate, together with the equality (2.1.68), implies the formula (2.1.63) for $(t,x) \in (t_0,T] \times B_R$. Since R is an arbitrary positive number, we come to the formula (2.1.63) for any $(t,x) \in \Pi_{(t_0,T]}$.

Let us prove the formula (2.1.64) for $(t,x) \in \Pi_{[t_1,T]}$ where $t_1 > t_0$. Consider the following family of functions depending on the parameter h:

$$\begin{aligned}u_h(t,x,;t_0) &:= \int\limits_{t_0}^{t-h} d\tau \int\limits_{\mathbb{R}^n} Z_0(t,x;\tau,\xi)f(\tau,\xi)d\xi,\\ &(t,x) \in \Pi_{[t_1,T]}, \quad h \in (0, t_1 - t_0).\end{aligned} \tag{2.1.74}$$

Since the integrand in (2.1.74) has no singularities,

$$\begin{aligned}\partial_t u_h(t,x;t_0) &= \int\limits_{\mathbb{R}^n} Z_0(t,x;t-h,\xi)f(t-h,\xi)d\xi + \int\limits_{t_0}^{t-h} d\tau \int\limits_{\mathbb{R}^n} \partial_t Z_0(t,x;\tau,\xi)f(\tau,\xi)d\xi\\ &=: I_h(t,x) + J_h(t,x), \qquad (t,x) \in \Pi_{[t_1,T]}.\end{aligned} \tag{2.1.75}$$

To substantiate the differentiation ∂_t under the integral in $\mathbb{R}^n$, it is sufficient to prove that the integral over $\mathbb{R}^n$ from $J_h(t,x)$ (denote it by $J^0(t,x;\tau)$) converges uniformly with respect to $t \in [t_2 - h/3, \min(t_2 + h/3, T)]$, where t_2 is an arbitrary fixed point from $[t_1, T]$, for any fixed $\tau \in [t_0, t_2 - 2h/3]$ and $x \in \mathbb{R}^n$. The last assertion is proved just as we proved the uniform convergence of the integral $I^{(k)}$. Here we have to use the equality

$$\partial_t Z_0(t,x;\tau,\xi) = \sum_{\|k\|=2b} a_k(t,\xi)\partial_x^k Z_0(t,x;\tau,\xi),$$
$$t_0 \leq \tau < t \leq T, \quad \{x,\xi\} \subset \mathbb{R}^n, \tag{2.1.76}$$

which, by (2.1.55), implies the estimate

$$\partial_t Z_0(t,x;\tau,\xi) \leq C(t-\tau)^{-M-1} E_c^{(0)}(t,x;\tau,\xi),$$
$$t_0 \leq \tau < t \leq T, \quad \{x,\xi\} \subset \mathbb{R}^n. \tag{2.1.77}$$

For the integral J^0 we have the estimate

$$|J^0(t,x;\tau)| \leq C(t-\tau)^{-1+\gamma/(2b)},$$
$$t_0 \leq \tau < t \leq T, \quad x \in B_R, \tag{2.1.78}$$

proved in the same way as (2.1.73). In this case

$$J^0(t,x;\tau) = \int_{\mathbb{R}^n} \partial_t Z_0(t,x;\tau,\xi)\Delta_\xi^x f(\tau,\xi)d\xi + \int_{\mathbb{R}^n} \partial_t Z_0(t,x;\tau,\xi)d\xi f(\tau,x)$$
$$=: J_1 + J_2.$$

The estimate (2.1.78) for J_1 is obtained just as the estimate for I_1, with the use of (2.1.77) instead of (2.1.54). In order to estimate J_2, we use the equality (2.1.76) and the estimates (2.1.58).

Note that:

1) $I_h(t,x) \underset{h\to 0}{\to} f(t,x)$, uniformly with respect to $(t,x) \in [t_1,T] \times B_R$;

2) $\displaystyle\int_0^{t-h} J^0(t,x;\tau)d\tau \underset{h\to 0}{\to} \int_0^t J^0(t,x;\tau)d\tau$

$$= \int_0^t d\tau \int_{\mathbb{R}^n} \partial_t Z_0(t,x;\tau,\xi)\Delta_\xi^x f(\tau,\xi)d\xi + \int_0^t \left(\int_{\mathbb{R}^n} \partial_t Z_0(t,x;\tau,\xi)d\xi \right) f(\tau,x)d\tau$$

uniformly with respect to $(t,x) \in [t_1,T] \times B_R$.

Using these facts, the equality (2.1.75), and the fact that $u_h(t,x;t_0) \underset{h\to 0}{\to} u(t,x;t_0)$ uniformly with respect to $(t,x) \in \Pi_{[t_1,T]}$, and also the arbitrariness of

R and t_1 in the intervals $(0,\infty)$ and $(0,T)$ respectively, we obtain the formula (2.1.64).

Let us prove 1) and 2). We have

$$\int\limits_{\mathbb{R}^n} Z_0(t,x;t-h,\xi)f(t-h,\xi)d\xi - f(t,x) = \int\limits_{B_{2R}} Z_0(t,x;t-h,\xi)\Delta^t_{t-h}f(t-h,\xi)d\xi$$

$$+ \int\limits_{\mathbb{R}^n\setminus B_{2R}} Z_0(t,x;t-h,\xi)f(t-h,\xi)d\xi + \int\limits_{B_{2R}} \Delta^x_z G_0(t,x;t-h,\xi;z)|_{z=\xi}f(t,\xi)d\xi$$

$$+ \left(\int\limits_{B_{2R}} G_0(t,x;t-h,\xi;x)f(t,\xi)d\xi - f(t,x)\right) =: \sum_{j=1}^{4} L_j.$$

Using the estimates (2.1.55), (2.1.61), and (2.1.71), the equality (2.1.46), Properties 2.2 and 2.4, as well as the continuity and boundedness of the function f, we get

$$|L_1| \le C \max_{(t,\xi)\in[t_1,T]\times B_{2R}} |\Delta^t_{t-h}f(t-h,\xi)|h^{-M}\int\limits_{\mathbb{R}^n} E^{(0)}_c(h,x;0,\xi)d\xi$$

$$= C \max_{(t,\xi)\in[t_1,T]\times B_{2R}} |\Delta^t_{t-h}f(t-h,\xi)| \underset{h\to 0}{\to} 0;$$

$$|L_2| \le C\exp\{-(c/2)h^{-q_0}R^{q'}\}h^{-M}\int\limits_{\mathbb{R}^n} E^{(0)}_{c/2}(h,x;0,\xi)d\xi$$

$$= C\exp\{-(c/2)h^{-q_0}R^{q'}\} \underset{h\to 0}{\to} 0;$$

$$|L_3| \le C\int\limits_{\mathbb{R}^n} (p(x;\xi))^\gamma h^{-M}E^{(0)}_c(h,x;0,\xi)d\xi \le Ch^{\gamma/(2b)} \underset{h\to 0}{\to} 0;$$

$$|L_4| = \left|\int\limits_{B_{2R}} G_0(t,x;t-h,\xi;x)\Delta^x_\xi f(t,\xi)d\xi\right.$$

$$\left. - \int\limits_{\mathbb{R}^n\setminus B_{2R}} G_0(t,x;t-h,\xi;x)d\xi f(t,x)\right| \underset{h\to 0}{\to} 0$$

uniformly with respect to $(t,x)\in[t_1,T]\times B_R$ (this is proved just as similar results for L_j, $j\in\{1,2,3\}$). Thus the statement 1) is proved.

The assertion 2) is a consequence of the estimate

$$\left|\int\limits_{t-h}^{t} J^0(t,x;\tau)d\tau\right| \le C\int\limits_{t-h}^{t}(t-\tau)^{-1+\gamma/(2b)}d\tau = Ch^{\gamma/(2b)},$$

based on the estimate (2.1.78). □

Remark 1. The investigation of differential properties of the integrals

$$u_y(t,x;t_0) := \int_{t_0}^{t} d\tau \int_{\mathbb{R}^n} G(t,x;\tau,\xi;y) f(\tau,\xi)d\xi,$$

$$u_{\beta,y}(t,x;t_0) := \int_{t_0}^{t} d\tau \int_{\mathbb{R}^n} G(t,x;\tau,\xi;\beta,y) f(\tau,\xi)d\xi,$$

$$(t,x) \in \Pi_{(t_0,T]},$$

where $G(\cdot,\cdot;\cdot,\cdot;y)$ and $G(\cdot,\cdot;\cdot,\cdot;\beta,y)$ are the FSCPs for the equations (2.1.48) and (2.1.54) respectively, is carried out as in Property 2.6. Here the arguments are simplified, and the formulas for the leading derivatives have the form

$$\partial_x^k u_y(t,x;t_0) = \int_{t_0}^{t} d\tau \int_{\mathbb{R}^n} \partial_x^k G(t,x;\tau,\xi;y) \Delta_\xi^x f(\tau,\xi)d\xi,$$

$$\partial_t u_y(t,x;t_0) = f(t,x) + \int_{t_0}^{t} d\tau \int_{\mathbb{R}^n} \partial_t G(t,x;\tau,\xi;y) \Delta_\xi^x f(\tau,\xi)d\xi,$$

$$\partial_x^k u_{\beta,y}(t,x;t_0) = \int_{t_0}^{t} d\tau \int_{\mathbb{R}^n} \partial_x^k G(t,x;\tau,\xi;\beta,y) \Delta_\xi^x f(\tau,\xi)d\xi,$$

$$\partial_t u_{\beta,y}(t,x;t_0) = f(t,x) + \int_{t_0}^{t} d\tau \int_{\mathbb{R}^n} \partial_t G(t,x;\tau,\xi;\beta,y) \Delta_\xi^x f(\tau,\xi)d\xi,$$

$$(t,x) \in \Pi_{(t_0,T]}, \quad \|k\| = 2b.$$

Remark 2. The function

$$u(t,x;t_0,y) := \int_{t_0}^{t} d\tau \int_{\mathbb{R}^n} Z_0(t,x;\tau,\xi) f(\tau,\xi;t_0,y)d\xi,$$

$$(t,x) \in \Pi_{(t_0,T]}, \quad y \in \mathbb{R}^n, \tag{2.1.79}$$

satisfies a property similar to Property 2.6, if the function $f(\cdot,\cdot;t_0,\cdot) : \Pi_{(t_0,T]} \times \mathbb{R}^n \to \mathbb{C}_{N1}$ is continuous and satisfies the following conditions:

$$|f(t,x;t_0,y)| \le C(t-t_0)^{-M-p/(2b)} E_c^{(0)}(t,x;t_0,y),$$

$$|\Delta_x^\xi f(t,x;t_0,y)| \le C(p(x;\xi))^{\gamma_0}(t-t_0)^{-M-(p-\gamma_0)/(2b)}$$

$$\times (E_c^{(0)}(t,x;t_0,y) + E_c^{(0)}(t,\xi;t_0,y)),$$

$$t_0 < t \le T, \quad \{x,y,\xi\} \subset \mathbb{R}^n,$$

where $C > 0$, $c > 0$, $\gamma_0 \in (0,1]$ and $p < 2b - \gamma_0$.

Property 2.7. *If a function f satisfies the above conditions, then the function* (2.1.79) *has the continuous derivatives $\partial_t u$ and $\partial_x^k u$, $\|k\| \le 2b$. The derivatives $\partial_x^k u$, $\|k\| < 2b$, can be calculated by direct differentiation under the integrals, while the leading derivatives satisfy the formulas* (2.1.63) *and* (2.1.64) *with $f(\tau,\xi)$ and $f(\tau,x)$ replaced by $f(\tau,\xi;t_0,y)$ and $f(\tau,x;t_0,y)$ respectively.*

The *proof* is similar to the one for Property 2.6 and uses the inequality

$$\int_{\mathbb{R}^n} E_c^{(0)}(t,x;\tau,\xi)E_c^{(0)}(\tau,\xi;t_0,y)((t-\tau)(\tau-t_0))^{-M}d\xi$$
$$\le C_\varepsilon(t-t_0)^{-M}E_{c(1-\varepsilon)}^{(0)}(t,x;t_0,y),$$
$$t_0<\tau<t,\quad \{x,\xi,y\}\subset\mathbb{R}^n,\quad \varepsilon\in(0,1), \tag{2.1.80}$$

implied by the inequality (1.3.23) with $n=1$ and $\beta = 1/(2b_j)$, $j\in\{1,\dots,n\}$.

Remark 3. If the coefficients of the equation (2.1.1) are Hölder continuous not only in x but also in t, then, as mentioned in Section 1.4.4, a parametrix can be chosen as

$$Z_1(t,x;\tau,\xi) := G_0(t,x;\tau,\xi;\tau,\xi),\quad 0\le\tau<t\le T,\quad \{x,\xi\}\subset\mathbb{R}^n, \tag{2.1.81}$$

where $G_0(\cdot,\cdot;\cdot,\cdot;\beta,y)$ is the FSCP for the equation (1.1.41) without lower order terms. By the remark from Section 2.1.3, Properties 2.5–2.7 hold for the function Z_1.

2.1.5. Fundamental solution of the Cauchy problem for equations whose coefficients depend on all the variables. Let us construct and study the FSCP for the equation (2.1.1) for the general case.

Theorem 2.2. *Assume the conditions A_{11} and A_{12}. Then there exists the FSCP Z for the equation* (2.1.1) *satisfying the estimates*

$$|\partial_x^k Z(t,x;\tau,\xi)| \le C(t-\tau)^{-M-\|k\|/(2b)}E_c^{(0)}(t,x;\tau,\xi), \tag{2.1.82}$$
$$|\Delta_x^{x'}\partial_x^k Z(t,x;\tau,\xi)| \le C(p(x;x'))^\gamma(t-\tau)^{-M-(\|k\|+\gamma)/(2b)}$$
$$\times (E_c^{(0)}(t,x;\tau,\xi)+E_c^{(0)}(t,x';\tau,\xi)), \tag{2.1.83}$$
$$0\le\tau<t\le T,\quad \{x,x',\xi\}\subset\mathbb{R}^n,\quad \|k\|\le 2b,$$

where the positive constants C and c depend only on

$$n,\ N,\ b_1,\dots,b_n,\ T,\ \max_{\|k\|\le 2b}\sup_{(t,x)\in\Pi}|a_k(t,x)|,$$

the constants δ,γ, and H from the conditions A_{11} and A_{12}, and, for $N>1$, also on the kind of the continuity in t of the coefficients a_k, $\|k\|=2b$.

Proof. As it was mentioned in Section 1.1.4, the FSCP for the equation (2.1.1) will be sought in the form (1.1.45), that is

$$Z(t,x;\tau,\xi) = Z_0(t,x;\tau,\xi)+W(t,x;\tau,\xi),$$
$$0\le\tau<t\le T,\quad \{x,\xi\}\subset\mathbb{R}^n, \tag{2.1.84}$$

where

$$W(t,x;\tau,\xi) := \int_{\tau}^{t} d\lambda \int_{\mathbb{R}^n} Z_0(t,x;\lambda,y)Q(\lambda,y;\tau,\xi)dy, \tag{2.1.85}$$

Z_0 is the parametrix described in Section 2.1.4. The function $Q(\cdot,\cdot;\tau,\xi) : \Pi_{(\tau,T]} \to \mathbb{C}_{NN}$ must be chosen for the function $Z(\cdot,\cdot;\tau,\xi) : \Pi_{(\tau,T]} \to \mathbb{C}_{NN}$ to be a solution of the equation (2.1.1) with $f = 0$, for any fixed point $(\tau,\xi) \in \Pi_{[0,T)}$. Let us assume that the sought function Q is continuous and satisfies the estimates

$$|Q(t,x;\tau,\xi)| \leq C(t-\tau)^{-M-1+\gamma/(2b)} E_c^{(0)}(t,x;\tau,\xi), \tag{2.1.86}$$

$$\begin{aligned}|\Delta_x^{x'} Q(t,x;\tau,\xi)| \leq{}& C(p(x;x'))^{\gamma_1}(t-\tau)^{-M-1+\gamma_2/(2b)} \\ &\times (E_c^{(0)}(t,x;\tau,\xi) + E_c^{(0)}(t,x;\tau,\xi)), \qquad (2.1.87)\\ &0 \leq \tau < t \leq T, \quad \{x,x',\xi\} \subset \mathbb{R}^n,\end{aligned}$$

where $\gamma_1 \in (0,\gamma)$, $\gamma_2 := \gamma - \gamma_1$, γ is the constant from the condition A_{12}.

Applying the differential expression

$$I\partial_t - \sum_{\|k\|\leq 2b} a_k(t,x)\partial_x^k$$

to the function (2.1.84) and using the a priori assumptions regarding Q and Property 2.7, we obtain the integral equation (1.1.47) for Q, in which

$$K(t,x;\tau,\xi) := \left(\sum_{\|k\|=2b} \Delta_x^{\xi} a_k(t,x)\partial_x^k + \sum_{\|k\|<2b} a_k(t,x)\partial_x^k \right) Z_0(t,x;\tau,\xi).$$

In order to find a bound for K, we use the estimates (2.1.55), the condition A_{12}, and the inequality (2.1.61). We get

$$\begin{aligned}|K(t,x;\tau,\xi)| &\leq C\left((p(x;\xi))^{\gamma}(t-\tau)^{-M-1} + \sum_{\|k\|<2b} (t-\tau)^{-M-\|k\|/(2b)} \right) E_c^{(0)}(t,x;\tau,\xi) \\ &\leq C(t-\tau)^{-M-1+\gamma/(2b)} E_{c_1}^{(0)}(t,x;\tau,\xi), \\ &\quad 0\leq\tau<t\leq T, \quad \{x,\xi\}\subset\mathbb{R}^n. \end{aligned} \tag{2.1.88}$$

Thus the kernel K satisfies the conditions of Lemma 1.8 with $\beta_j = 1/(2b_j)$, $j \in \{1,\ldots,n\}$ and $\chi = \gamma/(2b)$. By this lemma there exists a resolvent R of the equation (1.1.47) satisfying the estimate (1.3.39) with $\chi = \gamma/(2b)$. Since (1.1.47) is an inhomogeneous equation with the known function K, the solution Q coincides with the resolvent R. Therefore the function Q satisfies the estimate (2.1.86).

Let us prove (2.1.87). If $(p(x;x'))^{2b} > (t-\tau)/2$, then the estimate (2.1.87) is a direct consequence of (2.1.86). Then it is sufficient to consider the case where $(p(x;x'))^{2b} \leq (t-\tau)/2$.

First we estimate the difference $\Delta_x^{x'} K$. Using the condition A_{12}, the inequalities (2.1.55), (2.1.56), (2.1.61), and the fact that $(p(x';\xi))^\gamma \leq (p(x;x'))^\gamma + (p(x;\xi))^\gamma$, $\{x,x',\xi\} \subset \mathbb{R}^n$, we get

$$\begin{aligned}
&|\Delta_x^{x'} K(t,x;\tau,\xi)| \\
&\leq \sum_{\|k\|=2b} \left(|\Delta_x^{x'} a_k(t,x)||\partial_x^k Z_0(t,x;\tau,\xi)| + |\Delta_{x'}^{\xi} a_k(t,x')||\Delta_x^{x'} \partial_x^k Z_0(t,x;\tau,\xi)| \right) \\
&\quad + \sum_{\|k\|<2b} (|\Delta_x^{x'} a_k(t,x)||\partial_x^k Z_0(t,x;\tau,\xi)| + |a_k(t,x')||\Delta_x^{x'} \partial_x^k Z_0(t,x;\tau,\xi)|) \\
&\leq C(t-\tau)^{-M-1}(p(x;x'))^\gamma \Bigg(1 + (p(x';\xi))^\gamma (t-\tau)^{-\gamma/(2b)} \\
&\quad + \sum_{\|k\|<2b} (t-\tau)^{1-\|k\|/(2b)}(1+(t-\tau)^{-\gamma/(2b)}) \Bigg) E_c^{(0)}(t,x;\tau,\xi) \\
&\leq C(p(x;x'))^\gamma (t-\tau)^{-M-1} E_{c_1}^{(0)}(t,x;\tau,\xi), \\
&\qquad\qquad 0 \leq \tau < t \leq T, \quad \{x,x',\xi\} \subset \mathbb{R}^n, \\
&\qquad\qquad (p(x;x'))^{2b} \leq t-\tau, \quad c_1 \in (0,c).
\end{aligned} \tag{2.1.89}$$

In particular, this implies the estimate

$$\begin{aligned}
&|\Delta_x^{x'} K(t,x;\tau,\xi)| \leq C(p(x;x'))^{\gamma_1}(t-\tau)^{-M-1+\gamma_2/(2b)} E_{c_1}^{(0)}(t,x;\tau,\xi), \\
&0 \leq \tau < t \leq T, \quad \{x,x',\xi\} \subset \mathbb{R}^n, \quad (p(x;x'))^{2b} \leq t-\tau.
\end{aligned} \tag{2.1.90}$$

Next we consider $\Delta_x^{x'} Q$. By (1.1.47), we write

$$\begin{aligned}
\Delta_x^{x'} Q(t,x;\tau,\xi) &= \Delta_x^{x'} K(t,x;\tau,\xi) \\
&\quad + \int_\tau^\eta d\lambda \int_{\mathbb{R}^n} \Delta_x^{x'} K(t,x;\lambda,y) Q(\lambda,y;\tau,\xi) dy \\
&\quad + \int_\eta^t d\lambda \int_{\mathbb{R}^n} K(t,x;\lambda,y) Q(\lambda,y;\tau,\xi)\, dy \\
&\quad - \int_\eta^t d\lambda \int_{\mathbb{R}^n} K(t,x';\lambda,y) Q(\lambda,y;\tau,\xi) dy =: \sum_{j=1}^4 I_j,
\end{aligned} \tag{2.1.91}$$

where $\eta := t - (p(x;x'))^{2b}$. The term I_1 has been estimated; see (2.1.90). Since $t - \lambda \geq (p(x;x'))^{2b}$ in I_2 , using the estimates (2.1.80), (2.1.86), and (2.1.90) we obtain that

$$|I_2| \le C(p(x;x'))^{\gamma_1} \int_\tau^t (t-\lambda)^{-1+\gamma_2/(2b)}(\lambda-\tau)^{-1+\gamma/(2b)}d\lambda$$
$$\times \int_{\mathbb{R}^n} E_{c_1}^{(0)}(t,x;\lambda,y)E_c^{(0)}(\lambda,y;\tau,\xi)((t-\lambda)(\lambda-\tau))^{-M}dy$$
$$\le C(p(x;x'))^{\gamma_1}(t-\tau)^{-M-1+(\gamma\gamma_2)/(2b)}E_{c_2}^{(0)}(t,x;\tau,\xi), \tag{2.1.92}$$

where $c_2 \in (0, c_1)$.

The integrals I_3 and I_4 are treated in a similar way. For example, let us consider the first one. Due to (2.1.80), (2.1.86), and (2.1.88) we get

$$|I_3| \le C\int_\eta^t ((t-\lambda)(\lambda-\tau))^{-1+\gamma/(2b)}d\lambda \int_{\mathbb{R}^n} E_{c_1}^{(0)}(t,x;\lambda,y)$$
$$\times E_c^{(0)}(\lambda,y;\tau,\xi)((t-\lambda)(\lambda-\tau))^{-M}dy \le C(p(x;x'))^{\gamma}$$
$$\times (t-\tau)^{-M-1+\gamma/(2b)}E_{c_2}^{(0)}(t,x;\tau,\xi), \tag{2.1.93}$$

since for any $\lambda \in [\eta, t]$ $\lambda-\tau \ge \eta-\tau = t-\tau-(p(x;x'))^{2b} \ge (t-\tau)/2$ and

$$\int_\eta^t ((t-\lambda)(\lambda-\tau))^{-1+\gamma/(2b)}d\lambda \le ((t-\tau)/2)^{-1+\gamma/(2b)}\int_\eta^t (t-\lambda)^{-1+\gamma/(2b)}d\lambda$$
$$= (2b/\gamma)((t-\tau)/2)^{-1+\gamma/(2b)}(t-\eta)^{\gamma/(2b)}$$
$$= (2b/\gamma)((t-\tau)/2)^{-1+\gamma/(2b)}(p(x;x'))^{\gamma}.$$

From (2.1.90)–(2.1.93) we obtain the estimate (2.1.87) for the case $(p(x;x'))^{2b} \le (t-\tau)/2$.

Note that the function Q is in fact Hölder continuous in x with the exponent γ, that is

$$|\Delta_x^{x'}Q(t,x;\tau,\xi)| \le C(p(x;x'))^{\gamma}(t-\tau)^{-M-1}$$
$$\times (E_c^{(0)}(t,x;\tau,\xi) + E_c^{(0)}(t,x';\tau,\xi)),$$
$$0 \le \tau < t \le T, \quad \{x,x',\xi\} \subset \mathbb{R}^n. \tag{2.1.94}$$

Indeed, if $(p(x;x'))^{2b} > (t-\tau)/4$, then this inequality is a direct consequence of the estimate (2.1.86). If $(p(x;x'))^{2b} \le (t-\tau)/4$, we use the equality (1.1.44) and find that

$$
\begin{aligned}
|\Delta_x^{x'} Q(t,x;\tau,\xi)| \le\ & |\Delta_x^{x'} K(t,x;\tau,\xi)| \\
& + \int_\tau^{t_1} d\lambda \int_{\mathbb{R}^n} |\Delta_x^{x'} K(t,x;\lambda,y)||Q(\lambda,y;\tau,\xi)| dy \\
& + \int_{t_1}^{\eta} d\lambda \int_{\mathbb{R}^n} |\Delta_x^{x'} K(t,x;\lambda,y)||\Delta_y^{x} Q(\lambda,y;\tau,\xi)| dy \\
& + \int_{t_1}^{\eta} \left| \int_{\mathbb{R}^n} \Delta_x^{x'} K(t,x;\lambda,y) dy \right| |Q(\lambda,x;\tau,\xi)| d\lambda \\
& + \int_{\eta}^{t} d\lambda \int_{\mathbb{R}^n} |K(t,x;\lambda,y) dy||Q(\lambda,y;\tau,\xi)| dy \\
& + \int_{\eta}^{t} d\lambda \int_{\mathbb{R}^n} |K(t,x';\lambda,y)||Q(\lambda,y;\tau,\xi)| dy \\
=:\ & \sum_{j=1}^{6} J_j,
\end{aligned}
\tag{2.1.95}
$$

where $t_1 := t - (t-\tau)/2$, and η is the same as in (2.1.91).

Let us give estimates for J_j, $j \in \{1, \dots, 6\}$. It follows from (2.1.89) that

$$
\begin{gathered}
J_1 \le C(p(x;x'))^\gamma (t-\tau)^{-M-1} E_c^{(0)}(t,x;\tau,\xi), \\
0 \le \tau < t \le T, \quad \{x, x', \xi\} \subset \mathbb{R}^n, \quad (p(x;x'))^{2b} \le t - \tau.
\end{gathered}
\tag{2.1.96}
$$

Using (2.1.61), (2.1.80), (2.1.86), (2.1.87), and (2.1.96) we get

$$
\begin{aligned}
J_2 \le\ & C(p(x;x'))^\gamma \int_\tau^{t_1} (t-\lambda)^{-1} (\lambda-\tau)^{-1+\gamma/(2b)} d\lambda \\
& \times \int_{\mathbb{R}^n} E_c^{(0)}(t,x;\lambda,y) E_c^{(0)}(\lambda,y;\tau,\xi)((t-\lambda)(\lambda-\tau))^{-M} dy \\
\le\ & C(p(x;x'))^\gamma (t-\tau)^{-M-1} \int_\tau^{t_1} (\lambda-\tau)^{-1+\gamma/(2b)} d\lambda E_c^{(0)}(t,x;\tau,\xi) \\
=\ & C(p(x;x'))^\gamma (t-\tau)^{-M-1+\gamma/(2b)} E_c^{(0)}(t,x;\tau,\xi),
\end{aligned}
$$

$$J_3 \leq C(p(x;x'))^\gamma \int_{t_1}^{\eta} (t-\lambda)^{-1}(\lambda-\tau)^{-1+\gamma_2/(2b)} \int_{\mathbb{R}^n} (p(x;y))^{\gamma_1/(2b)}$$
$$\times E_c^{(0)}(t,x;\lambda,y)(E_c^{(0)}(\lambda,y;\tau,\xi) + E_c^{(0)}(\lambda,x;\tau,\xi))$$
$$\times ((t-\lambda)(\lambda-\tau))^{-M} dy \leq C(p(x;x'))^\gamma (t-\tau)^{-M} \int_{t_1}^{\eta} (t-\lambda)^{-1+\gamma_1/(2b)}$$
$$\times (\lambda-\tau)^{-1+\gamma_2/(2b)} d\lambda E_c^{(0)}(t,x;\tau,\xi)$$
$$\leq C(p(x;x'))^\gamma (t-\tau)^{-M-1+\gamma/(2b)} E_c^{(0)}(t,x;\tau,\xi),$$
$$J_5 \leq C \int_{\eta}^{t} ((t-\lambda)(\lambda-\tau))^{-1+\gamma/(2b)} d\lambda \int_{\mathbb{R}^n} E_c^{(0)}(t,x;\lambda,y)$$
$$\times E_c^{(0)}(\lambda,y;\tau,\xi)((t-\lambda)(\lambda-\tau))^{-M} dy \leq C(t-\tau)^{-M-1+\gamma/(2b)}$$
$$\times \int_{\eta}^{t} (t-\lambda)^{-1+\gamma/(2b)} d\lambda E_c^{(0)}(t,x;\tau,\xi) = C(p(x;x'))^\gamma (t-\tau)^{-M-1+\gamma/(2b)}$$
$$\times E_c^{(0)}(t,x;\tau,\xi).$$

The estimate

$$J_6 \leq C(p(x;x'))^\gamma (t-\tau)^{-M-1+\gamma/(2b)} E_c^{(0)}(t,x;\tau,\xi)$$

is proved in a similar way.

It remains to obtain an estimate for J_4. First we consider the integral

$$J := \int_{\mathbb{R}^n} \Delta_x^{x'} K(t,x;\lambda,y) dy.$$

We have

$$J = \sum_{\|k\|\leq 2b} \Delta_x^{x'} a_k(t,x) \int_{\mathbb{R}^n} \partial_x^k Z_0(t,x;\lambda,y)dy + \sum_{\|k\|<2b} a_k(t,x')\Delta_x^{x'} \int_{\mathbb{R}^n} \partial_x^k Z_0(t,x;\lambda,y)dy$$
$$+ \sum_{\|k\|=2b} \int_{\mathbb{R}^n} \Delta_{x'}^{y} a_k(t,x') \Delta_x^{x'} \partial_x^k Z_0(t,x;\lambda,y)dy.$$

Using the estimates (2.1.55), (2.1.58), (2.1.59) with $\gamma_0 = \gamma$, (2.1.56) with $\gamma_0 > \gamma$, the condition A_{12}, and the inequality (2.1.61), we find that

$$|J| \leq C(p(x;x'))^\gamma (t-\lambda)^{-1+\gamma/(2b)} + C(p(x;x'))^\gamma (t-\lambda)^{-1+1/(2b)}$$
$$+ C(p(x;x'))^{\gamma_0} \int_{\mathbb{R}^n} (p(x';y))^\lambda (t-\lambda)^{-M-1-\gamma_0/(2b)} E_c^{(0)}(t,x;\lambda,y)dy$$
$$\leq C(p(x;x'))^\gamma (t-\lambda)^{-1+\gamma/(2b)} + C(p(x;x'))^{\gamma_0} (t-\lambda)^{-1+(\gamma-\gamma_0)/(2b)}.$$

From this estimate and (2.1.86) we get

$$J_4 \le C \int_{t_1}^{\eta} ((p(x;x'))^{\gamma}(t-\lambda)^{-1+\gamma/(2b)} + (p(x;x')^{\gamma_0} \times (t-\lambda)^{-1+(\gamma-\gamma_0)/(2b)}(\lambda-\tau)^{-M-1+\gamma/(2b)} E_c^{(0)}(\lambda,x;\tau,\xi)d\lambda.$$

Using the inequalities

$$t-\tau \ge \lambda-\tau \ge t_1-\tau = (t-\tau)/2, \quad \lambda\in[t_1,\eta],$$

$$\begin{aligned}\int_{t_1}^{\eta}(t-\lambda)^{-1+\gamma/(2b)}d\lambda &= (2b/\gamma)((t-t_1)^{\gamma/(2b)} - (t-\eta)^{\gamma/(2b)})\\ &\le (2b/\gamma)(t-t_1)^{\gamma/(2b)}\\ &= (2b/\gamma)((t-\tau)/2)^{\gamma/(2b)},\end{aligned}$$

$$\begin{aligned}\int_{t_1}^{\eta}(t-\lambda)^{-1+(\gamma-\gamma_0)/(2b)}d\lambda &= (2b/(\gamma-\gamma_0))((t-\eta)^{(\gamma-\gamma_0)/(2b)} - (t-t_1)^{(\gamma-\gamma_0)/(2b)})\\ &\le (2b/(\gamma-\gamma_0))(t-\eta)^{(\gamma-\gamma_0)/(2b)}\\ &= (2b/(\gamma-\gamma_0))(p(x;x'))^{\gamma-\gamma_0},\end{aligned}$$

we obtain that

$$J_4 \le C(p(x;x'))^{\gamma}(t-\tau)^{-M-1+\gamma/(2b)} E_c^{(0)}(t,x;\tau,\xi).$$

It follows from (2.1.95) and the above estimates of J_j, $j\in\{1,\ldots,6\}$, that (2.1.94) holds for the case $(p(x;x'))^{2b} \le (t-\tau)/4$.

Let us prove now the estimates (2.1.82) for Z. Since the first summand in (2.1.84) satisfies (2.1.55), we have to obtain only the estimates for the second summand (2.1.85) and its derivatives..

We shall use the estimates (2.1.55), (2.1.58), (2.1.61), (2.1.80), (2.1.86), and (2.1.94). If $\|k\| < 2b$, then by Property 2.7 we have

$$\begin{aligned}|\partial_x^k W(t,x;\tau,\xi)| &\le C\int_{\tau}^{t}(t-\lambda)^{-\|k\|/(2b)}(\lambda-\tau)^{-1+\gamma/(2b)}d\lambda\\ &\quad\times\int_{\mathbb{R}^n} E_c^{(0)}(t,x;\lambda,y)E_c^{(0)}(\lambda,y;\tau,\xi)((t-\lambda)(\lambda-\tau))^{-M}dy\\ &\le C(t-\tau)^{-M-(\|k\|-\gamma)/(2b)} E_c^{(0)}(t,x;\tau,\xi).\end{aligned}$$

In the case $\|k\| = 2b$, by the same property we write the equality

$$\partial_x^k W(t,x;\tau,\xi) = \int\limits_{\tau}^{t_1} d\lambda \int\limits_{\mathbb{R}^n} \partial_x^k Z_0(t,x;\lambda,y) Q(\lambda,y;\tau,\xi)dy$$
$$+ \int\limits_{t_1}^{t} d\lambda \int\limits_{\mathbb{R}^n} \partial_x^k Z_0(t,x;\lambda,y)\Delta_y^x Q(\lambda,y;\tau,\xi)dy$$
$$+ \int\limits_{t_1}^{t} \Bigg(\int\limits_{\mathbb{R}^n} \partial_x^k Z_0(t,x;\lambda,y)dy \Bigg) Q(\lambda,x;\tau,\xi)d\lambda =: \sum_{j=1}^{3} K_j \quad (2.1.97)$$

where the number t_1 is the same as in (2.1.95). The integrals K_j are estimated as follows:

$$|K_1| \le C \int\limits_{\tau}^{t_1} (t-\lambda)^{-1}(\lambda-\tau)^{-1+\gamma/(2b)} d\lambda \int\limits_{\mathbb{R}^n} E_c^{(0)}(t,x;\lambda,y)$$
$$\times E_c^{(0)}(\lambda,y;\tau,\xi)((t-\lambda)(\lambda-\tau))^{-M} dy$$
$$\le C(t-\tau)^{-M-1} \int\limits_{\tau}^{t} (\lambda-\tau)^{-1+\gamma/(2b)} d\lambda E_c^{(0)}(t,x;\tau,\xi)$$
$$= C(t-\tau)^{-M-1+\gamma/(2b)} E_c^{(0)}(t,x;\tau,\xi),$$
$$|K_2| \le C \int\limits_{t_1}^{t} d\lambda \int\limits_{\mathbb{R}^n} (t-\lambda)^{-M-1} E_c^{(0)}(t,x;\lambda,y)(p(x;y))^{\gamma}(\lambda-\tau)^{-M-1}$$
$$\times (E_c^{(0)}(\lambda,y;\tau,\xi) + E_c^{(0)}(\lambda,x;\tau,\xi))dy \le C(t-\tau)^{-M-1}$$
$$\times \int\limits_{t_1}^{t} (t-\lambda)^{-1+\gamma/(2b)} d\lambda E_c^{(0)}(t,x;\tau,\xi)$$
$$= C(t-\tau)^{-M-1+\gamma/(2b)} E_c^{(0)}(t,x;\tau,\xi),$$
$$|K_3| \le C \int\limits_{t_1}^{t} (t-\lambda)^{-1+\gamma/(2b)}(\lambda-\tau)^{-M-1+\gamma/(2b)} E_c^{(0)}(\lambda,x;\tau,\xi)d\lambda$$
$$\le C(t-\tau)^{-M-1+\gamma/(2b)} E_c^{(0)}(t,x;\tau,\xi) \int\limits_{t_1}^{t} (t-\lambda)^{-1+\gamma/(2b)} d\lambda$$
$$= C(t-\tau)^{-M-1+2\gamma/(2b)} E_c^{(0)}(t,x;\tau,\xi).$$

Hence for W we have the estimates

$$|\partial_x^k W(t,x;\tau,\xi)| \le C(t-\tau)^{-M-(\|k\|-\gamma)/(2b)} E_c^{(0)}(t,x;\tau,\xi),$$
$$0 \le \tau < t \le T, \quad \{x,\xi\} \subset \mathbb{R}^n, \quad \|k\| \le 2b. \tag{2.1.98}$$

Let us prove (2.1.83). By (2.1.56), the first summand from (2.1.84) satisfies the estimates (2.1.83). It remains to consider the differences of derivatives from W. We assume that $(p(x;x'))^{2b} \le (t-\tau)/2$. If $(p(x;x'))^{2b} > (t-\tau)/2$, then the required estimate of the difference $\Delta_x^{x'} \partial_x^k W$ is a direct consequence of (2.1.98).

Using the formula (2.1.97) we write

$$\begin{aligned}
\Delta_x^{x'} \partial_x^k W(t,x;\tau,\xi) = & \int_\tau^{t_1} d\lambda \int_{\mathbb{R}^n} \Delta_x^{x'} \partial_x^k Z_0(t,x;\lambda,y) Q(\lambda,y;\tau,\xi) dy \\
& + \int_{t_1}^{\eta} d\lambda \int_{\mathbb{R}^n} \Delta_x^{x'} \partial_x^k Z_0(t,x;\lambda,y) \Delta_y^x Q(\lambda,y;\tau,\xi) dy \\
& + \int_{\eta}^{t} d\lambda \int_{\mathbb{R}^n} \partial_x^k Z_0(t,x;\lambda,y) \Delta_y^x Q(\lambda,y;\tau,\xi) dy \\
& + \int_{\eta}^{t} d\lambda \int_{\mathbb{R}^n} \partial_{x'}^k Z_0(t,x';\lambda,y) \Delta_{x'}^y Q(\lambda,x';\tau,\xi) dy \\
& + \int_{t_1}^{\eta} \Delta_x^{x'} \left(\int_{\mathbb{R}^n} \partial_x^k Z_0(t,x;\lambda,y) dy \right) Q(\lambda,x;\tau,\xi) d\lambda \\
& + \int_{\eta}^{t} \left(\int_{\mathbb{R}^n} \partial_x^k Z_0(t,x;\lambda,y) dy \right) Q(\lambda,x;\tau,\xi) d\lambda \\
& + \int_{t}^{\eta} \left(\int_{\mathbb{R}^n} \partial_{x'}^k Z_0(t,x';\lambda,y) dy \right) Q(\lambda,x';\tau,\xi) d\lambda =: \sum_{j=1}^{7} L_j,
\end{aligned}$$

where the numbers t_1 and η are the same as in (2.1.95).

The summand L_1 is estimated with the use of the inequality (2.1.56) with $\gamma_0 = \gamma$ and (2.1.86), the estimate is

$$|L_1| \le C(p(x;x'))^{\gamma} (t-\tau)^{-M-\|k\|/(2b)} E_c^{(0)}(t,x;\tau,\xi).$$

In order to find an estimate of L_2, we use (2.1.94) and (2.1.56) with $\gamma_0 = \gamma$ for $\|k\| < 2b$, and $\gamma_0 > \gamma$ for $\|k\| = 2b$.

We have

$$|L_2| \le C(p(x;x'))^{\gamma_0} \int_{t_1}^{\eta} d\lambda \int_{\mathbb{R}^n} (t-\lambda)^{-M-(\|k\|+\gamma_0)/(2b)}$$
$$\times (\lambda-\tau)^{-M-1}(p(x;y))^{\gamma} E_c^{(0)}(t,x;\lambda,y)(E_c^{(0)}(\lambda,y;\tau,\xi)+E_c^{(0)}(\lambda,x;\tau,\xi))dy$$
$$\le C(p(x;x'))^{\gamma_0}(t-\tau)^{-M-1}E_c^{(0)}(t,x;\tau,\xi)\int_{t_1}^{\eta}(t-\lambda)^{-(\|k\|+\gamma_0-\gamma)/(2b)}d\lambda.$$

Since

$$\int_{t_1}^{\eta}(t-\lambda)^{-(\|k\|+\gamma_0-\gamma)/(2b)}d\lambda \le \begin{cases} C(t-t_1)^{(1-\|k\|)/(2b)}, & \|k\|<2b,\\ C(t-\eta)^{(\gamma-\gamma_0)/(2b)}, & \|k\|=2b,\end{cases}$$

we get

$$|L_2| \le C(p(x;x'))^{\gamma}(t-\tau)^{-M-\|k\|/(2b)}E_c^{(0)}(t,x;\tau,\xi).$$

The integrals L_3 and L_4 are treated similarly. For example, consider L_3. Using the estimates (2.1.55) and (2.1.94) we find that

$$|L_3| \le C\int_{\eta}^{t} d\lambda \int_{\mathbb{R}^n}(t-\lambda)^{-M-\|k\|/(2b)}(\lambda-\tau)^{-M-1}(p(x;y))^{\gamma}$$
$$\times E_c^{(0)}(t,x;\lambda,y)(E_c^{(0)}(\lambda,y;\tau,\xi)+E_c^{(0)}(\lambda,y;\tau,\xi))dy$$
$$\le C(t-\tau)^{-M-1}E_c^{(0)}(t,x;\tau,\xi)\int_{\eta}^{t}(t-\lambda)^{-(\|k\|-\gamma)/(2b)}d\lambda$$
$$= C(t-\tau)^{-M-1}(t-\eta)^{1-(\|k\|-\gamma)/(2b)}E_c^{(0)}(t,x;\tau,\xi)$$
$$\le C(p(x;x'))^{\gamma}(t-\tau)^{-M-\|k\|/(2b)}E_c^{(0)}(t,x;\tau,\xi).$$

By the estimates (2.1.86) and (2.1.59) with $\gamma_0=\gamma$ for $\|k\|\in(0,2b)$, and $\gamma_0>\gamma$ for $\|k\|=2b$, arguing as for L_2, we obtain for the case $\|k\|\in(0,2b]$ that

$$|L_5| \le C(p(x;x'))^{\gamma_0}(t-\tau)^{-M-1+\gamma/(2b)}E_c^{(0)}(t,x;\tau,\xi)\int_{t_1}^{\eta}(t-\lambda)^{-(\|k\|+\gamma_0-\gamma)/(2b)}d\lambda$$
$$\le C(p(x;x'))^{\gamma}(t-\tau)^{-M-(\|k\|-\gamma)/(2b)}E_c^{(0)}(t,x;\tau,\xi).$$

If $\|k\|=0$, then

$$|L_5| \le C(p(x;x'))^{\gamma}(t-\tau)^{-M}E_c^{(0)}(t,x;\tau,\xi),$$

since in this case we obtain from (2.1.56) that

$$\left|\Delta_x^{x'} \int_{\mathbb{R}^n} Z_0(t,x;\lambda,y)dy\right| \le C(p(x;x'))^\gamma \int_{\mathbb{R}^n} (t-\lambda)^{-M-\gamma/(2b)} E_c^{(0)}(t,x;\lambda,y)dy$$
$$\le C(p(x;x'))^\gamma (t-\lambda)^{-\gamma/(2b)},$$
$$0 \le \lambda < t \le T, \quad \{x,x'\} \subset \mathbb{R}^n, \quad (p(x;x'))^{2b} \le (t-\lambda).$$

It remains to consider the integrals L_6 and L_7. Since the procedures are quite similar, we shall deal only with L_6. By (2.1.58) and (2.1.86) we find for $\|k\| \in (0,2b]$ that

$$|L_6| \le C \int_\eta^t (t-\lambda)^{-(\|k\|-\gamma)/(2b)} (\lambda-\tau)^{-M-1+\gamma/(2b)} E_c^{(0)}(\lambda,x;\tau,\xi)d\lambda$$
$$\le C(t-\tau)^{-M-1+\gamma/(2b)} E_c^{(0)}(t,x;\tau,\xi) \int_\eta^t (t-\lambda)^{-(\|k\|-\gamma)/(2b)} d\lambda$$
$$= C(t-\tau)^{-M-1+\gamma/(2b)} E_c^{(0)}(t,x;\tau,\xi)(t-\eta)^{1-(\|k\|-\gamma)/(2b)}$$
$$\le C(p(x;x'))^\gamma (t-\tau)^{-M-(\|k\|-\gamma)/(2b)} E_c^{(0)}(t,x;\tau,\xi).$$

If $\|k\| = 0$, then by (2.1.55),

$$|L_6| \le C(t-\tau)^{-M-1+\gamma/(2b)} E_c^{(0)}(t,x;\tau,\xi) \int_\eta^t d\lambda$$
$$= C(t-\tau)^{-M-1+\gamma/(2b)} E_c^{(0)}(t,x;\tau,\xi)(p(x;x'))^{2b}$$
$$\le C(p(x;x'))^\gamma (t-\tau)^{-M} E_c^{(0)}(t,x;\tau,\xi).$$

Therefore

$$|\Delta_x^{x'} \partial_x^k W(t,x;\tau,\xi)| \le C(p(x;x'))^\gamma (t-\tau)^{-M-\|k\|/(2b)}$$
$$\times (E_c^{(0)}(t,x;\tau,\xi) + E_c^{(0)}(t,x';\tau,\xi)),$$
$$0 \le \tau < t \le T, \quad \{x,x',\xi\} \subset \mathbb{R}^n, \quad \|k\| \le 2b. \tag{2.1.99}$$

The estimates (2.1.83) follow from the inequalities (2.1.56), (2.1.99), and the formula (2.1.84). □

Remark 1. *Using the equality*

$$\left(I\partial_t - \sum_{\|k\| \le 2b} a_k(t,x)\partial_x\right) Z(t,x;\tau,\xi) = 0,$$
$$0 \le \tau < t \le T, \quad \{x,\xi\} \subset \mathbb{R}^n, \tag{2.1.100}$$

the estimates (2.1.82), (2.1.83), *and the condition* A_{12}, *we obtain also the following estimates:*

$$|\partial_t Z(t,x,;\tau,\xi)| \le C(t-\tau)^{-M-1} E_c^{(0)}(t,x;\tau,\xi),$$
$$|\Delta_x^{x'} \partial_t Z(t,x;\tau,\xi)| \le C(p(x;x'))^\gamma (t-\tau)^{-M-1-\gamma/(2b)} \times (E_c^{(0)}(t,x;\tau,\xi) + E_c^{(0)}(t,x';\tau,\xi)),$$
$$0 \le \tau < t \le T, \quad \{x,x',\xi\} \subset \mathbb{R}^n. \tag{2.1.101}$$

Remark 2. *Let* $N > 1$, *and let the equation* (2.1.1) *be a uniformly strongly* $\overrightarrow{2b}$-*parabolic system on* $\Pi_{[0,T]}$, *that is there exists such a constant* $\delta > 0$ *that for any* $(t,x) \in \Pi_{[0,T]}$, $\sigma \in \mathbb{R}^n$ *and* $a \in \mathbb{C}_{N1}$

$$\operatorname{Re}(A_0(t,x,i\sigma)a,a) \le -\delta r(\sigma)|a|^2, \tag{2.1.102}$$

where

$$A_0(t,x;i\sigma) := \sum_{\|k\|=2b} a_k(t,x)(is)^k,$$

$(\cdot,\cdot)$ *is the inner product in* $\mathbb{C}_{N1}$, *while the function* r *was defined in Section* 2.1.2. *Then the FSCP can be constructed without the assumption that the continuity of leading coefficients in* t *is uniform with respect to* x *(under all other assumptions of Theorem* 2.2).

Indeed, consider the system dual to (1.1.39) in the sense of the Fourier transform:

$$\left(I\frac{d}{dt} - A_0(t,y,is)\right)v = 0.$$

Let V_j be the j-th column of the normal fundamental matrix of solutions $V(t,\tau,s;y)$ for this system. Then

$$\begin{aligned}\operatorname{Re}(A_0(t,y,is)V_j,V_j) &= \frac{1}{2}((A_0(t,y,is)V_j,V_j) + \overline{(A_0(t,y,is)V_j,V_j)})\\ &= \frac{1}{2}\left(\left(\frac{d}{dt}V_j,V_j\right) + \left(V_j,\frac{d}{dt}V_j\right)\right)\\ &= \frac{1}{2}\frac{d}{dt}|V_j|^2.\end{aligned}$$

It follows from (2.1.102) and Lemma 2.2 that for any $s = \sigma + i\gamma \in \mathbb{C}^n$,

$$\operatorname{Re}(A_0(t,y,is)V_j,V_j) \le (-\delta_1 r(\sigma) + c_1 r(\gamma))|V_j|^2,$$
$$\delta_1 \in (0,\delta), \quad c_1 > 0.$$

Hence

$$\frac{d}{dt}|V_j|^2 \le (-2\delta_1 r(\sigma) + 2c_1 r(\gamma))|V_j|^2,$$

so that

$$|V_j(t,\tau,s;y)|^2 \leq \exp\{(-2\delta_1 r(\sigma)+2c_1 r(\gamma))(t-\tau)\},$$

whence

$$|V(t,\tau,s;y)| \leq C\exp\{(-\delta_1 r(\sigma)+c_1 r(\gamma))(t-\tau)\},$$
$$0\leq\tau<t\leq T, \quad s\in\mathbb{R}^n. \tag{2.1.103}$$

Thus, the main estimate (2.1.103) has been obtained without the assumption of the uniform, with respect to x, continuity in t of the coefficients. Since that assumption was needed only to obtain the estimate (2.1.103), the FSCP for strongly $\overrightarrow{2b}$-parabolic systems can be constructed without it.

Remark 3. *If, in addition to the conditions of Theorem* 2.2, *the coefficients of the equation* (2.1.1) *are Hölder continuous in t, that is instead of the condition A_{12} the condition A_{13} is satisfied, then the FSCP Z can be constructed also in the form*

$$Z(t,x;\tau,\xi) = Z_1(t,x;\tau,\xi)+W_1(t,x;\tau,\xi),$$
$$W_1(t,x;\tau,\xi) := \int_\tau^t d\lambda \int_{\mathbb{R}^n} Z_1(t,x;\lambda,y)Q_1(\lambda,y;\tau,\xi)dy,$$
$$0\leq\tau<t\leq T, \quad \{x,\xi\}\subset\mathbb{R}^n, \tag{2.1.104}$$

where Z_1 is a parametrix defined by the formula (2.1.81), and, by Remark 3 *from Section* 2.1.4, *possessing Properties* 2.5–2.7. *The function Q_1 satisfies the inequalities* (2.1.86), (2.1.87), *and* (2.1.94). *Therefore the estimates* (2.1.98) *and* (2.1.99) *hold for the function W_1.*

Note that in the present case the derivatives of Z are Hölder continuous not only in x, but also in t, that is, in addition to the estimates (2.1.82), (2.1.83), and (2.1.101) we have the estimates

$$|\Delta_t^{t'}\partial_t^{k_0}\partial_x^k Z(t,x;\tau,\xi)| \leq C(t'-t)^{1-k_0-(\|k\|-\gamma)/(2b)}((t-\tau)^{-M-1-\gamma/(2b)}$$
$$\times E_c^{(0)}(t,x;\tau,\xi)+(t'-\tau)^{-M-1-\gamma/(2b)}E_c^{(0)}(t',x;\tau,\xi)),$$
$$0\leq\tau<t<t'\leq T, \quad \{x,\xi\}\subset\mathbb{R}^n, \quad 0<2bk_0+\|k\|\leq 2b. \tag{2.1.105}$$

It is sufficient to prove them for the case $t'-t\leq t-\tau$, since if $t'-t>t-\tau$, then the estimates (2.1.105) are direct consequences of (2.1.82) and (2.1.101).

Using (2.1.103) and Property 2.7, we write

$$\Delta_t^{t'}\partial_x^k Z(t,x;\tau,\xi) = \Delta_t^{t'}\partial_x^k Z_1(t,x;\tau,\xi)$$
$$+\int_\tau^{t_1} d\lambda \int_{\mathbb{R}^n} \Delta_t^{t'}\partial_x^k Z_1(t,x;\lambda,y)Q_1(\lambda,y;\tau,\xi)dy$$

$$
\begin{aligned}
&+\int_{t_1}^{t} d\lambda \int_{\mathbb{R}^n} \Delta_t^{t'} \partial_x^k Z_1(t,x;\lambda,y) \Delta_y^x Q_1(\lambda,y;\tau,\xi)dy \\
&+\int_{t_1}^{t} \Big(\Delta_t^{t'} \int_{\mathbb{R}^n} \partial_x^k Z_1(t,x;\lambda,y)dy \Big) Q_1(\lambda,x;\tau,\xi)d\lambda \\
&-\int_{t}^{t'} d\lambda \int_{\mathbb{R}^n} \partial_x^k Z_1(t',x;\lambda,y) \Delta_y^x Q_1(\lambda,y;\tau,\xi)dy \\
&-\int_{t}^{t'} \Big(\int_{\mathbb{R}^n} \partial_x^k Z_1(t',x;\lambda,y)dy \Big) Q_1(\lambda,x;\tau,\xi)d\lambda,
\end{aligned}
$$

$$
t_1 := (t+\tau)/2, \quad 0 < \|k\| \le 2b.
$$

The first summand satisfies the estimate (2.1.57). Using it, as well as the estimate (2.1.60), we find bounds for other summands, just as it was done for the summands L_j from the proof of Theorem 2.2. The estimate (2.1.105) for $k_0 = 1$ is obtained with the use of the equality (2.1.100), the estimates (2.1.105) with $k_0 = 0$, and the condition A_{13}.

Note that the function W_1 satisfies the estimates

$$
\begin{aligned}
|\Delta_t^{t'} \partial_x^k W_1(t,x;\tau,\xi)| \le{}& C(t'-t)^{1-(\|k\|-\gamma)/(2b)} ((t-\tau)^{-M-1} \\
&\times E_c^{(0)}(t,x;\tau,\xi) + (t'-\tau)^{-M-1} E_c^{(0)}(t',x;\tau,\xi)), \\
&0 \le \tau < t < t' \le T, \quad \{x,\xi\} \subset \mathbb{R}^n, \quad 0 < \|k\| \le 2b.
\end{aligned}
$$

Corollary. *The results described in Remark 3, together with Property 2.5 for the parametrix Z_1, imply the following estimates:*

$$
\begin{aligned}
|\Delta_{t,x}^{t',x'} \partial_x^k Z(t,x;\tau,\xi)| \le{}& C(p(t,x;t',x'))^\gamma ((t-\tau)^{-M-(\|k\|+\gamma)/(2b)} E_c^{(0)}(t,x;\tau,\xi) \\
&+ (t'-\tau)^{-M-(\|k\|+\gamma)/(2b)} E_c^{(0)}(t',x';\tau,\xi)), \\
&\|k\| \le 2b, \qquad (2.1.106)
\end{aligned}
$$

$$
\Big| \int_{\mathbb{R}^n} \partial_x^k Z(t,x;\tau,y)dy \Big| \le C(t-\tau)^{-(\|k\|-\gamma/(2b))}, \quad 0 < \|k\| \le 2b, \qquad (2.1.107)
$$

$$
\Big| \Delta_{t,x}^{t',x'} \int_{\mathbb{R}^n} \partial_x^k Z(t,x;\tau,y)dy \Big| \le C(p(t,x;t',x'))^\gamma (t-\tau)^{-\|k\|/(2b)}, \quad 0 < \|k\| \le 2b, \qquad (2.1.108)
$$

in which $0 \le \tau < t < t' \le T$, and $\{x,x',\xi\} \subset \mathbb{R}^n$.

Remark 4. *Looking carefully into the proof of Theorem 2.2 and the derivation of the results of Remark 3, we notice that the estimates (2.1.105) with* $k_0 = 0$ *remain valid if only the conditions* A_{11} *and* A_{12} *are satisfied. Thus the estimates (2.1.106)–(2.1.108) hold in this situation too.*

Remark 5. *If, in addition to the conditions of Theorem 2.2, the condition* A_{14} *is satisfied, then for the equation adjoint to (2.1.1),*

$$L_N^*(\tau, \xi, \partial_\tau, \partial_\xi)v(\tau, \xi) = 0, \quad (\tau, \xi) \in \Pi_{[0,T)}, \tag{2.1.109}$$

where L_N^* *was defined in (2.1.3), there exists the FSCP defined as a function* $Z^*(\tau, \xi; t, x)$, $0 \le \tau < t \le T$, $\{\xi, x\} \subset \mathbb{R}^n$, *such that the formula*

$$v(\tau, \xi) = \int_{\mathbb{R}^n} Z^*(\tau, \xi; t, x)\varphi(x)dx, \quad (\tau, \xi) \in \Pi_{[0,t)},$$

determines a solution of the equation (2.1.109) on $\Pi_{[0,t)}$, *satisfying the condition*

$$v(\tau, \xi)|_{\tau=t} = \varphi(\xi), \quad \xi \in \mathbb{R}^n,$$

for any $t \in (0, T]$ *and an arbitrary continuous bounded function* φ. *This follows from Theorem 2.2, since the equation (2.1.109) is inversely* $\overrightarrow{2b}$*-parabolic, that is becomes* $\overrightarrow{2b}$*-parabolic with respect to the new independent variable* $-\tau$.

Consider some properties of the FSCP. Their proofs will be based on the following Green-Ostrogradsky formula obtained by integrating the equality (2.1.4):

$$\int_{t_1}^{t_2} d\theta \int_{B_R} (\bar{v}' L_N u - \overline{(L_N^* v)}' u)(\theta, y)dy \tag{2.1.110}$$

$$= \int_{B_R} (\bar{v}'u)(t_2, y)dy - \int_{B_R} (\bar{v}'u)(t_1, y)dy + \int_{t_1}^{t_2} d\theta \int_{\Gamma_R} \sum_{j=1}^{n} B^j[v, u](\theta, y)\mu_j dS_y,$$

where $t_1 < t_2$, $(\mu_1, \ldots, \mu_n)$ is a unit vector of the outer normal to the boundary Γ_R of the ball B_R.

Property 2.8. *Assume the conditions* A_{11}, A_{12}, *and* A_{14}. *Then the FSCPs* Z *and* Z^* *for the equations (2.1.1) and (2.1.109) respectively satisfy the equalities*

$$Z^*(\tau, \xi; t, x) = \bar{Z}'(t, x; \tau, \xi), \tag{2.1.111}$$

$$Z(t, x; \tau, \xi) = \int_{\mathbb{R}^n} Z(t, x; \lambda, y)Z(\lambda, y; \tau, \xi)dy \tag{2.1.112}$$

where $0 \le \tau < \lambda < t \le T$, $\{x, \xi\} \subset \mathbb{R}^n$.

The equality (2.1.111) means the *property of normality* of the FSCP, while the equality (2.1.112) is called the *convolution formula.*

Proof. The formula (2.1.111) is obtained if we put in the formula (2.1.110) $t_1 = \tau + \varepsilon$, $t_2 = \tau - \varepsilon$, where ε is a sufficiently small positive number, $u(\theta, y) = Z(\theta, y; \tau, \xi)$, $v(\theta, y) = Z^*(\theta, y; t, x)$, and pass to the limit first as $R \to \infty$, and then as $\varepsilon \to 0$. Here we use the estimates (2.1.82) and similar estimates for Z^*. The equality (2.1.112) is proved similarly, only instead of t_1 in the formula (2.1.110) we take λ and use the formula (2.1.111). □

2.1.6. The fundamental solution and the Green matrix of the Cauchy problem for an equation of an arbitrary order in the time variable. Results similar to those of Sections 2.1.1–2.1.5 are valid also for a general $\overrightarrow{2b}$-parabolic equation

$$L_N(t, x, \partial_t, \partial_x)u(t,x) = f(t,x), \quad (t,x) \in \Pi_{(0,T]}, \tag{2.1.113}$$

where the elements L_{Njl} of the matrix differential expression L_N are of the form

$$L_{Njl}(t, x, \partial_t, \partial_x) := \delta_{jl}\partial_t^{r_j} - \sum_{\substack{2bk_0 + \|k\| \le 2br_l \\ (k_0 < r_l)}} a^{jl}_{k_0 k}(t,x)\partial_t^{k_0}\partial_x^k,$$
$$\{j, l\} \subset \{1, \dots, N\}.$$

The next theorem deals with the existence and estimates of the FSCP for such an equation.

Theorem 2.3. *Let the equation* (2.1.113) *be uniformly $\overrightarrow{2b}$-parabolic in the layer $\Pi_{[0,T]}$, the coefficients $a^{jl}_{k_0 k}$ be bounded, continuous in t (for $N > 1$ the continuity in t of $a^{jl}_{k_0 k}$ with $2bk_0 + \|k\| = 2br_l$ is assumed to be uniform in $x \in \mathbb{R}^n$), and Hölder continuous in x uniformly with respect to t with the exponent $\gamma \in (0,1)$ on $\Pi_{[0,T]}$. Then there exists a FSCP, a matrix Z whose elements Z_{jl}, $\{j,l\} \subset \{1, \dots, N\}$, satisfy the estimates*

$$|\partial_t^{k_0}\partial_x^k Z_{jl}(t, x; \tau, \xi)| \le C(t-\tau)^{r_j - 1 - M - k_0 - \|k\|/(2b)} E_c^{(0)}(t, x; \tau, \xi),$$
$$|\Delta_x^{x'} \partial_t^{k_0}\partial_x^k Z_{jl}(t, x; \tau, \xi)| \le C(p(x;x'))^\gamma (t-\tau)^{r_j - 1 - M - k_0 - (\|k\| + \gamma)/(2b)}$$
$$\times (E_c^{(0)}(t, x; \tau, \xi) + E_c^{(0)}(t, x'; \tau, \xi)),$$
$$0 \le \tau < t \le T, \quad \{x, x', \xi\} \subset \mathbb{R}^n, \quad 2bk_0 + \|k\| \le 2br_j,$$

where C and c are positive constants depending on the same parameters as the constants of Theorem 2.2.

Proof. The scheme of the proof is the same as in Theorem 2.2. First we consider the case, in which the coefficients do not depend on x. In this case we obtain a theorem, similar to Theorem 2.1, about a complete analytic description of the

FSCP defined by the formulas (2.1.15) and (2.1.16), in which V is a solution of the following Cauchy problem for an ordinary differential equation:

$$L_N\left(t, \frac{d}{dt}, i\sigma\right)V = 0,$$

$$\left.\frac{d^{s-1}}{dt^{s-1}}V_{jl}\right|_{t=\tau} = \delta_{sr_j}\delta_{jl}, \quad s \in \{1, \dots, r_j\}, \quad \{j, l\} \subset \{1, \dots, N\},$$

where V_{jl} are elements of the matrix V. The function V is studied by the techniques used in Section 2.1.2 for a first order equation in t, and in [45] for a general equation, parabolic in the sense of Petrovsky.

After that we investigate properties of the FSCP for the equation

$$L_N(t, y, \partial_t, \partial_x)u = 0,$$

whose coefficients depend on t and the parameter $y \in \mathbb{R}^n$. That is done just as in Section 2.1.3. This investigation is basic for studying properties of the parametrix and implementing the Levi method. We take as a parametrix the function (1.1.43) where $G_0(\cdot, \cdot; \cdot, \cdot; y)$ is the FSCP for the equation $L_N^0(t, y, \partial_t, \partial_x)u = 0$, in which L_N^0 is a matrix differential expression with the elements

$$L_{Njl}^0(t, y, \partial_t, \partial_x) := \delta_{jl}\partial_t^{r_j} - \sum_{\substack{2bk_0+\|k\|=2br_l \\ (k_0<r_l)}} a_{k_0k}^{jl}(t, y)\partial_t^{k_0}\partial_x^k, \quad \{j, l\} \subset \{1, \dots, N\}.$$

□

Let us consider the Cauchy problem for the equation (2.1.113) with the initial conditions (1.1.20). Proceeding from the definition and properties of the FSCP Z in the same way as for the case of general Petrovsky parabolic equations [45, 52], we can prove that for any sufficiently smooth functions f_l and $\varphi_l^{r_l}$, $l \in \{1, \dots, N\}$, with compact supports, the components of the solution $u := (u_1, \dots, u_N)'$ of the problem (1.1.20), (2.1.113) with $f := (f_1, \dots, f_N)'$ and $\varphi_l^s = 0$, $s \in \{1, \dots, r_l - 1\}$, $l \in \{1, \dots, N\}$, can be represented as

$$u_j(t, x) = \sum_{l=1}^N \left(\int_0^t d\tau \int_{\mathbb{R}^n} Z_{jl}(t, x; \tau, \xi) f_l(\tau, \xi) d\xi + \int_{\mathbb{R}^n} Z_{jl}(t, x; 0, \xi)\varphi_l^{r_l}(\xi) d\xi \right),$$

$$(t, x) \in \Pi_{(0,T]}, \quad j \in \{1, \dots, N\}. \qquad (2.1.114)$$

Solutions of the Cauchy problem with general initial conditions are represented with the use of the Green matrix of the Cauchy problem.

Definition 2.2. *The Green matrix of the Cauchy problem* (1.1.20), (2.1.113) *is the matrix*

$$G := (G_0, G_1, \dots, G_N), \quad G_0 := (G_{0jl})_{j=1,l=1}^{N,N}, \quad G_l := (G_{ljs})_{j=1,s=1}^{N,r_l},$$

such that the components of the solution u of this problem, for arbitrary sufficiently smooth functions f_l and φ_l^s, $s \in \{1, \dots, r_l\}$, $l \in \{1, \dots, N\}$, with compact supports, can be represented as

$$u_j(t,x) = \sum_{l=1}^{N} \Bigg(\int_0^t d\tau \int_{\mathbb{R}^n} G_{0jl}(t,x;\tau,\xi) f_l(\tau,\xi) d\xi + \sum_{s=0}^{r_l} \int_{\mathbb{R}^n} G_{ljs}(t,x;\xi) \varphi_l^s(\xi) d\xi \Bigg),$$

$$(t,x) \in \Pi_{(0,T]}, \quad j \in \{1, \dots, N\}. \tag{2.1.115}$$

Under some additional (compared to Theorem 2.3) assumptions on the coefficients of the equation (2.1.113) we obtain the following result on the existence and structure of the Green matrix.

Theorem 2.4. *Suppose that the conditions of Theorem 2.3 are satisfied and, in addition, the coefficients of the equation (2.1.113) have, on the layer $\Pi_{[0,T]}$, the bounded, continuous in t, and Hölder continuous in x, uniformly with respect to t, derivatives $\partial_t^{k_0} \partial_x^k a_{k_0 k}^{jl}$, $2bk_0 + \|k\| \le 2br_l$, $\{j,l\} \subset \{1, \dots, N\}$. Then there exists the Green matrix of the Cauchy problem for the equation (2.1.113), whose elements are expressed via elements of the FSCP Z and coefficients of the equation as follows:*

$$G_{0jl}(t,x;\tau,\xi) = Z_{jl}(t,x;\tau,\xi),$$

$$G_{ljs}(t,x;\xi) = \sum_{p=1}^{N} \Bigg(\delta_{lp} (-\partial_\beta)^{r_l - s} Z_{jp}(t,x;\beta,\xi)$$

$$- \theta(r_l - s - 1) \sum_{\substack{2bk_0 + \|k\| \le 2b(r_l - s) \\ (k_0 < r_l - s)}} (-\partial_\beta)^{k_0} (-\partial_\xi)^k \big(Z_{jp}(t,x;\beta,\xi) a_{(k_0+s)k}^{pl}(\beta,\xi) \big) \Bigg) \Bigg|_{\beta=0},$$

$$0 \le \tau < t \le T, \quad \{x,\xi\} \subset \mathbb{R}^n, \quad s \in \{1, \dots, r_l\}, \quad \{j,l\} \subset \{1, \dots, N\}, \tag{2.1.116}$$

where θ is the Heavyside function.

Proof. We give only a scheme of the proof. For the detailed proof see [109]. We reduce the problem (1.1.20), (2.1.113) to the problem with the zero initial data. In order to do that, consider the functions

$$u_{jh}(t,x) := \zeta(t/h) u_j(t,x), \quad (t,x) \in \Pi_{(0,T]}, \quad j \in \{1, \dots, N\},$$

where h is a number from $(0,T)$, ζ is a sufficiently smooth function on $[0,\infty)$, such that $\zeta = 0$ on $[0,1/2]$, and $\zeta = 1$ on $[1,\infty)$. Since the functions u_j, $j \in \{1, \dots, N\}$, are components of the solution of the problem (1.1.20), (2.1.113), for u_{jh}, $j \in \{1, \dots, N\}$, we obtain the following problem with the zero initial data:

$$\sum_{l=1}^{N} L_{Njl}(t,x,\partial_t,\partial_x) u_{lh}(t,x) = f_{jh}(t,x), \quad (t,x) \in \Pi_{(0,T]},$$

$$\partial_t^{s-1} u_{jh}(t,x)|_{t=0} = 0, \quad x \in \mathbb{R}^n, \quad s \in \{1, \dots, r_j\}, \quad j \in \{1, \dots, N\},$$

where

$$f_{jh}(t,x) := \zeta(t/h)f_j(t,x) + \sum_{l=1}^{N}(L_{Njl}(t,x,\partial_t,\partial_x)(\zeta(t/h)u_l(t,x)) \\ - \zeta(t/h)L_{Njl}(t,x,\partial_t,\partial_x)u_l(t,x)).$$

By (2.1.114), we have the equalities

$$u_{jh}(t,x) = \sum_{l=1}^{N}\int_0^t d\tau \int_{\mathbb{R}^n} Z_{jl}(t,x;\tau,\xi)f_{lh}(\tau,\xi)d\xi, \quad (t,x)\in\Pi_{(0,T]}, \quad j\in\{1,\ldots,N\}.$$

for components of the solution of this problem. If we fix an arbitrary point (t,x) inside the layer $\Pi_{(0,T]}$, assume that $h\in(0,t)$, and pass to the limit as $h\to 0$, we obtain the formulas (2.1.115), in which the kernels of the integrals are given by the equalities (2.1.116). □

2.2 Cauchy problem for equations with bounded coefficients

2.2.1. Function spaces. We begin with the definition of function spaces for the Cauchy problem for the equation (2.1.1) with bounded coefficients. As we noted in Section 1.1.3, for such an equation it is natural to consider the spaces of functions which increase rapidly as $|x|\to\infty$, with the growth type determined by the vector-function $\vec{k}(t,\vec{a}) := (k_1(t,a_1),\ldots,k_n(t,a_n))$, $t\in[0,T]$ where the functions k_j, $j\in\{1,\ldots,n\}$, are defined by the formulas (1.1.34). There we assume $c_0,a_1,\ldots,a_n$ to be fixed numbers, such that $c_0\in(0,c)$, $a_j\geq 0$, $j\in\{1,\ldots,n\}$, $T<\min\limits_{j\in\{1,\ldots,n\}}(c_0/a_j)^{2b_j-1}/2$, where c is the smallest of the constants c in the estimates (2.1.82), (2.1.83), and (2.1.106); $\vec{a} := (a_1,\ldots,a_n)$.

Let us introduce the following weight functions:

$$\Psi_\nu(t,x) := \exp\left\{\nu\sum_{j=1}^{n}k_j(t,a_j)|x_j|^{q_j}\right\}, \quad (t,x)\in\Pi_{[0,T]}, \quad \nu\in\{-1,1\}. \tag{2.2.1}$$

The behaviour of functions from the spaces defined below, as $|x|\to\infty$, will be described by the function Ψ_1.

Note that the functions k_j, $j\in\{1,\ldots,n\}$, are the functions defined by the formula (1.3.18), with 0, $1/(2b_j)$, and a_j substituted for j, β, and a respectively. Therefore, by the definition, the equality from (1.3.19), and Lemma 1.4, these functions possess the following properties:

$$\vec{k}(0,\vec{a}) = \vec{a}, \tag{2.2.2}$$

$$k_j(t-\tau,k_j(\tau,a_j)) = k_j(t,a_j), \quad j\in\{1,\ldots,n\}, \tag{2.2.3}$$

$$E_{c_0}^{(0)}(t,x;\tau,\xi)\Psi_1(\tau,\xi)\leq\Psi_1(t,x), \tag{2.2.4}$$

$$0\leq\tau<t\leq T, \quad \{x,\xi\}\subset\mathbb{R}^n.$$

Recall that the function $E_{c_0}^{(0)}$ was defined in (2.1.40).

Below we assume that p is a number from the interval $[1,\infty)$ or $p=\infty$, λ is a number from (0,1), m is a number from $\mathbb{N}\cup\{0\}$, and $\vec{a}$ is the vector from (2.2.2).

We use the following spaces of functions:

$$\varphi : \mathbb{R}^n \to \mathbb{C}_{N1}, \tag{2.2.5}$$

$$\mu : \mathcal{B}_n \to \mathbb{C}_{N1}, \tag{2.2.6}$$

$$u : \Pi_{[0,T]} \to \mathbb{C}_{N1}, \tag{2.2.7}$$

$$u : \Pi_{(0,T]} \to \mathbb{C}_{N1}, \tag{2.2.8}$$

where $\mathcal{B}_n$ is the σ-algebra of Borel sets in $\mathbb{R}^n$:

$L_p^{\vec{a}}$ and $L_1^{-\vec{k}(T,\vec{a})}$, the spaces of the Lebesgue measurable functions (2.2.5) with finite norms

$$\|\varphi\|_p^{\vec{a}} := \|\varphi(\cdot)\Psi_{-1}(0,\cdot)\|_{L_p(\mathbb{R}^n)} \quad \text{and} \quad \|\varphi\|_1^{-\vec{k}(T,\vec{a})} := \|\varphi(\cdot)\Psi_1(T,\cdot)\|_{L_p(\mathbb{R}^n)},$$

respectively, where $L_p(\mathbb{R}^n)$ is the Lebesgue L_p-space of functions (2.2.5);

$M^{\vec{a}}$, the space of σ-additive set functions (2.2.6) (generalized Borel measures on $\mathbb{R}^n$) satisfying the condition

$$\|\mu\|^{\vec{a}} := \int_{\mathbb{R}^n} \Psi_{-1}(0,x)d|\mu|(x) < \infty,$$

where $|\mu|$ is the full variation of μ;

$C_0^{-\vec{k}(T,\vec{a})}$, the space of continuous functions (2.2.5), with the property

$$|\varphi(x)|\Psi_1(T,x) \underset{|x|\to\infty}{\longrightarrow} 0$$

(for $\varphi \in C_0^{-\vec{k}(T,\vec{a})}$ we set $\|\varphi\|_0^{-\vec{k}(T,\vec{a})} := \sup\limits_{x\in\mathbb{R}^n}(|\varphi(x)|\Psi_1(T,x)))$;

$C_{0,0}^{\vec{a}}$ and $C_{0,\lambda}^{\vec{a}}$, the space of continuous functions (2.2.5), for which the norms $\|\varphi\|_{0,0}^{\vec{a}}$ and $\|\varphi\|_{0,\lambda}^{\vec{a}} := \|\varphi\|_{0,0}^{\vec{a}} + \langle\varphi\rangle_\lambda^{\vec{a}}$ respectively are finite. Here

$$\|\varphi\|_{0,0}^{\vec{a}} := \sup_{x\in\mathbb{R}^n}(|\varphi(x)|\Psi_{-1}(0,x)), \langle\varphi\rangle_\lambda^{\vec{a}}$$

$$:= \sup_{\substack{\{x,x'\}\subset\mathbb{R}^n \\ x\neq x'}} \left(\frac{|\Delta_x^{x'}\varphi(x)|}{(p(x;x'))^\lambda(\Psi_1(0,x)+\Psi_1(0,x'))} \right);$$

$C_{m,0}^{\vec{a}}$ and $C_{m,\lambda}^{\vec{a}}$, the space of functions (2.2.5) with the continuous derivatives $\partial_x^l\varphi$, $\|l\| \le m$, and the finite respective norms

$$\|\varphi\|_{m,0}^{\vec{a}} := \sum_{\|l\|\le m} \|\partial_x^l\varphi\|_{0,0}^{\vec{a}} \quad \text{and} \quad \|\varphi\|_{m,\lambda}^{\vec{a}} := \sum_{\|l\|\le m} \|\partial_x^l\varphi\|_{0,\lambda}^{\vec{a}};$$

$C^{\vec{k}(\cdot,\vec{a})}_{0,0,0}$, $C^{\vec{k}(\cdot,\vec{a})}_{0,0,\lambda}$, and $C^{\vec{k}(\cdot,a)}_{0,\lambda/(2b),\lambda}$, the spaces of continuous functions (2.2.7) with the finite respective norms

$$\|u\|^{\vec{k}(\cdot,\vec{a})}_{0,0,0} := \sup_{(t,x)\in\Pi_{[0,T]}} \left(|u(t,x)|\Psi_{-1}(t,x)\right), \quad \|u\|^{\vec{k}(\cdot,\vec{a})}_{0,0,\lambda} := \|u\|^{\vec{k}(\cdot,\vec{a})}_{0,0,0} + \langle u\rangle^{\vec{k}(\cdot,\vec{a})}_{\lambda},$$

and

$$\|u\|^{\vec{k}(\cdot,\vec{a})}_{0,\lambda/(2b),\lambda} := \|u\|^{\vec{k}(\cdot,\vec{a})}_{0,0,0} + \langle\!\langle u\rangle\!\rangle^{\vec{k}(\cdot,\vec{a})}_{\lambda},$$

where

$$\langle u\rangle^{\vec{k}(\cdot,\vec{a})}_{\lambda} := \sup_{\substack{\{(t,x),(t,x')\}\subset\Pi_{[0,T]}\\ x\neq x'}} \left(\frac{|\Delta^{x'}_{x} u(t,x)|}{(p(x;x'))^{\lambda}(\Psi_1(t,x)+\Psi_1(t,x'))}\right),$$

$$\langle\!\langle u\rangle\!\rangle^{\vec{k}(\cdot,\vec{a})}_{\lambda} := \sup_{\substack{\{(t,x),(t',x')\}\subset\Pi_{[0,T]}\\ (t,x)\neq(t',x')}} \left(\frac{|\Delta^{t',x'}_{t,x} u(t,x)|}{(p(t,x;t',x'))^{\lambda}(\Psi_1(t,x)+\Psi_1(t',x'))}\right);$$

$C^{\vec{k}(\cdot,\vec{a})}_{m,0,0}$, $C^{\vec{k}(\cdot,\vec{a})}_{m,0,\lambda}$, and $C^{\vec{k}(\cdot,\vec{a})}_{m,\lambda/(2b),\lambda}$, the spaces of functions (2.2.7) with the continuous derivatives $\partial^l_x u$, $\|l\|\le m$, and the finite respective norms

$$\|u\|^{\vec{k}(\cdot,\vec{a})}_{m,0,0} := \sum_{\|l\|\le m} \|\partial^l_x u\|^{\vec{k}(\cdot,\vec{a})}_{0,0,0}, \quad \|u\|^{\vec{k}(\cdot,\vec{a})}_{m,0,\lambda} := \sum_{\|l\|\le m} \|\partial^l_x u\|^{\vec{k}(\cdot,\vec{a})}_{0,0,\lambda},$$

$$\text{and} \quad \|u\|^{\vec{k}(\cdot,\vec{a})}_{m,\lambda/(2b),\lambda} := \sum_{\|l\|\le m} \|\partial^l_x u\|^{\vec{k}(\cdot,\vec{a})}_{0,\lambda/(2b),\lambda};$$

$\hat{C}^{\vec{k}(\cdot,\vec{a})}_{0,0}$ and $\hat{C}^{\vec{k}(\cdot,\vec{a})}_{0,\lambda}$, the spaces of continuous functions (2.2.8), for which the numbers $\|u(t,\cdot)\|^{\vec{k}(t,\vec{a})}_{0,0}$, $\|u(t,\cdot)\|^{\vec{k}(t,\vec{a})}_{0,\lambda}$ for any $t\in(0,T]$, respectively, and the norms

$$\widehat{\|u\|}^{\vec{k}(\cdot,\vec{a})}_{0,0} := \int_0^T \|u(t,\cdot)\|^{\vec{k}(t,\vec{a})}_{0,0}\,dt, \quad \widehat{\|u\|}^{\vec{k}(\cdot,\vec{a})}_{0,\lambda} := \int_0^T \|u(t,\cdot)\|^{\vec{k}(t,\vec{a})}_{0,\lambda}\,dt,$$

where

$$\|u(t,\cdot)\|^{\vec{k}(t,\vec{a})}_{0,\lambda} := \|u(t,\cdot)\|^{\vec{k}(t,\vec{a})}_{0,0} + \langle u(t,\cdot)\rangle^{\vec{k}(t,\vec{a})}_{\lambda},$$

$$\|u(t,\cdot)\|^{\vec{k}(t,\vec{a})}_{0,0} := \sup_{x\in\mathbb{R}^n} \left(|u(t,x)|\Psi_{-1}(t,x)\right),$$

$$\langle u(t,\cdot)\rangle^{\vec{k}(t,\vec{a})}_{\lambda} := \sup_{\substack{\{x,x'\}\subset\mathbb{R}^n\\ x\neq x'}} \left(\frac{|\Delta^{x'}_{x} u(t,x)|}{(p(x;x'))^{\lambda}(\Psi_1(t,x)+\Psi_1(t,x'))}\right);$$

are finite.

$\hat{C}^{\vec{k}(\cdot,\vec{a})}_{m,0}$, the space of functions (2.2.8) belonging, together with the derivatives $\partial^l_x u$, $\|l\| \le m$, to the space $\hat{C}^{\vec{k}(\cdot,\vec{a})}_{0,0}$ thus having the finite norm

$$\widehat{\|u\|}^{\vec{k}(\cdot,\vec{a})}_{m,0} := \sum_{\|l\|\le m} \widehat{\|\partial^l_x u\|}^{\vec{k}(\cdot,\vec{a})}_{0,0};$$

$H^{\gamma,\lambda}$ and $\hat{H}^{\gamma,\lambda}$, $0 < \lambda \le \gamma < 1$, the spaces of functions (2.2.7) and (2.2.8) respectively, possessing the derivatives $\partial^l_x u$, $\|l\| \le 2b$, which belong to the spaces $C^{\vec{k}(\cdot,\vec{a})}_{0,\gamma/(2b),\gamma}$ and $\hat{C}^{\vec{k}(\cdot,\vec{a})}_{0,\gamma}$ respectively, if $\|l\| < 2b$, and to the spaces $C^{\vec{k}(\cdot,\vec{a})}_{0,\lambda/(2b),\lambda}$ and $\hat{C}^{\vec{k}(\cdot,\vec{a})}_{0,\lambda}$, if $\|l\| = 2b$. The norms are defined by the equalities

$$\|u\|_{H^{\gamma,\lambda}} := \sum_{\|l\|<2b} \|\partial^l_x u\|^{\vec{k}(\cdot,\vec{a})}_{0,\gamma/(2b),\gamma} + \sum_{\|l\|=2b} \|\partial^l_x u\|^{\vec{k}(\cdot,\vec{a})}_{0,\lambda/(2b),\lambda},$$

$$\|u\|_{\hat{H}^{\gamma,\lambda}} := \sum_{\|l\|<2b} \widehat{\|\partial^l_x u\|}^{\vec{k}(\cdot,\vec{a})}_{0,\gamma} + \sum_{\|l\|=2b} \widehat{\|\partial^l_x u\|}^{\vec{k}(\cdot,\vec{a})}_{0,\lambda};$$

note that $H^{\lambda,\lambda} = C^{\vec{k}(\cdot,\vec{a})}_{2b,\lambda/(2b),\lambda}$.

In the above expressions for norms $p(x;x')$ and $p(t,x;t',x')$ denote the distances defined by the formulas (2.1.2).

Apart from the above norms, we shall use also the following norms for the Lebesgue measurable in x functions (2.2.8):

$$\|u(t,\cdot)\|^{\vec{k}(t,\vec{a})}_p := \|u(t,\cdot)\Psi_{-1}(t,\cdot)\|_{L_p(\mathbb{R}^n)}, \quad t \in (0,T], \quad p \in [1,\infty].$$

2.2.2. Properties of the Poisson integrals. Consider some properties of the Poisson integrals generated by the FSCP for the equation (2.1.1) according to (1.1.24) and (1.1.25). We assume the conditions of Theorem 2.2, so that for the equation (2.1.1) there exists the FSCP Z, with the properties described in Section 2.1.5.

Let us study the Poisson integral

$$u(t,x) := (P\varphi)(t,x) := \int_{\mathbb{R}^n} Z(t,x;0,\xi)\varphi(\xi)d\xi, \quad (t,x) \in \Pi_{(0,T]}, \tag{2.2.9}$$

of a function φ, and the Poisson integral

$$u_0(t,x) := (P\mu)(t,x) := \int_{\mathbb{R}^n} Z(t,x;0,\xi)d\mu(\xi), \quad (t,x) \in \Pi_{(0,T]}. \tag{2.2.10}$$

of a generalized measure μ.

The proofs of the lemmas below employ systematically the estimates (2.1.82), for Z, the inequality (2.2.4), and the equality (2.1.46). Therefore we shall not mention their use each time.

Lemma 2.4. *If $\varphi \in C^{\vec{a}}_{0,0}$, then the function (2.2.9) has the following properties:*

1) *there exists such a constant $C > 0$ that for any $t \in (0,T]$ and $l \in \mathbb{Z}^n_+$, $\|l\| \le 2b$,*

$$\|\partial^l_x u(t,\cdot)\|^{\vec{k}(t,\vec{a})}_{0,0} \le C\|\varphi\|^{\vec{a}}_{0,0} t^{-\|l\|/(2b)}; \tag{2.2.11}$$

2) *for any compact set $K \subset \mathbb{R}^n$,*

$$u(t,x) \underset{t\to 0}{\longrightarrow} \varphi(x) \quad \text{uniformly} \quad \text{with respect to} \quad x \in K. \tag{2.2.12}$$

Proof. 1) For $(t,x) \in \Pi_{(0,T]}$, we have

$$\begin{aligned}|\partial^l_x u(t,x)| &\le Ct^{-M-\|l\|/(2b)} \\ &\quad\times \int_{\mathbb{R}^n} E^{(0)}_{c-c_0}(t,x;0,\xi)(E^{(0)}_{c_0}(t,x;0,\xi)\Psi_1(0,\xi))(|\varphi(\xi)|\Psi_{-1}(0,\xi))d\xi \\ &\le C\|\varphi\|^{\vec{a}}_{0,0}t^{-\|l\|/(2b)}\Psi_1(t,x)\int_{\mathbb{R}^n} t^{-M}E^{(0)}_{c-c_0}(t,x;0,\xi)d\xi \\ &= C\|\varphi\|^{\vec{a}}_{0,0}t^{-\|l\|/(2b)}\Psi_1(t,x), \quad \|l\| \le 2b,\end{aligned}$$

which implies the estimates (2.2.11).

2) First we prove that, uniformly with respect to $x \in \mathbb{R}^n$,

$$\int_{\mathbb{R}^n} Z(t,x;0,\xi)d\xi \underset{t\to 0}{\longrightarrow} I. \tag{2.2.13}$$

By (1.1.41) and (2.1.84) we have

$$\begin{aligned}\int_{\mathbb{R}^n} Z(t,x;0,\xi)d\xi &= \int_{\mathbb{R}^n} G_0(t,x;0,\xi;x)d\xi + \int_{\mathbb{R}^n} \Delta^x_y G_0(t,x;0,\xi;y)\Big|_{y=\xi} d\xi \\ &\quad + \int_{\mathbb{R}^n} W(t,x;0,\xi)d\xi =: \sum_{j=1}^{3} I_j(t,x).\end{aligned} \tag{2.2.14}$$

Using Properties 2.2 and 2.4, the inequalities (2.1.61) and (2.1.98) we find that

$$I_1(t,x) \underset{t\to 0}{\longrightarrow} I \quad \text{uniformly w.r.t.} \quad x \in \mathbb{R}^n; \tag{2.2.15}$$

$$\begin{aligned}|I_2(t,x)| &\le C\int_{\mathbb{R}^n} t^{-M}(p(x;\xi))^{\gamma}E^{(0)}_c(t,x;0,\xi)d\xi \\ &\le Ct^{\gamma/(2b)}\int_{\mathbb{R}^n} t^{-M}E^{(0)}_{c_1}(t,x;0,\xi)d\xi = Ct^{\gamma/(2b)} \underset{t\to 0}{\longrightarrow} 0, \\ &\qquad x \in \mathbb{R}^n, \quad c_1 \in (0,c);\end{aligned} \tag{2.2.16}$$

$$|I_3(t,x)| \le Ct^{\gamma/(2b)} \int_{\mathbb{R}^n} t^{-M} E_c^{(0)}(t,x;0,\xi)d\xi = Ct^{\gamma/(2b)} \underset{t\to 0}{\longrightarrow} 0,$$
$$x \in \mathbb{R}^n. \tag{2.2.17}$$

(2.2.14)–(2.2.17) imply (2.2.13).

Due to (2.2.13), in order to prove (2.2.12), it suffices to show that the function

$$v(t,x) := \int_{\mathbb{R}^n} Z(t,x;0,\xi)\Delta_\xi^x \varphi(\xi)d\xi, \quad (t,x) \in \Pi_{(0,T]},$$

is such that

$$v(t,x) \underset{t\to 0}{\longrightarrow} 0 \quad \text{uniformly} \quad \text{with respect to} \quad x \in K,$$

that is for any $\varepsilon > 0$ there exists such a $\delta \in (0,T)$ that for any $t \in (0,\delta)$ and $x \in K$

$$|v(t,x)| < \varepsilon. \tag{2.2.18}$$

Let K' be such a compact set in $\mathbb{R}^n$ that $K \subset K'$, and $\varepsilon > 0$ be a given number. Due to the uniform continuity of the function φ on K', for any $\varepsilon_1 > 0$ there exists such $\eta > 0$ that for all $x \in K$ and $\xi \in B(x,\eta)$,

$$|\Delta_\xi^x \varphi(\xi)| < \varepsilon_1. \tag{2.2.19}$$

Represent $v(t,x)$ as

$$v(t,x) = \int_{B(x,\eta)} Z(t,x;0,\xi)\Delta_\xi^x\varphi(\xi)d\xi + \int_{\mathbb{R}^n \setminus B(x,\eta)} Z(t,x;0,\xi)\Delta_\xi^x\varphi(\xi)d\xi$$

and use the inequalities (2.1.47) and (2.2.19). Then

$$\begin{aligned}
|v(t,x)| &\le C\varepsilon_1 \int_{\mathbb{R}^n} t^{-M}E_c^{(0)}(t,x;0,\xi)d\xi + C \int_{\mathbb{R}^n\setminus B(x,\eta)} t^{-M}(E_{c-c_0}^{(0)}(t,x;0,\xi) \\
&\quad \times (E_{c_0}^{(0)}(t,x;0,\xi)\Psi_1(0,\xi))(|\varphi(\xi)|\Psi_{-1}(0,\xi)) \\
&\quad + E_c^{(0)}(t,x;0,\xi)\Psi_1(0,x)(|\varphi(x)|\Psi_{-1}(0,x))) \\
&\le C_1\varepsilon_1 + C_0\|\varphi\|_{0,0}^{\vec{a}}\Psi_1(t,x)\exp\{-c_1t^{1-q''}\eta^{q'}\} \\
&\le C_1\varepsilon_1 + C_0\|\varphi\|_{0,0}^{\vec{a}}\Psi_0\exp\{-c_1t^{1-q''}\eta^{q'}\},
\end{aligned}$$

where $c_1 := (c-c_0)/2$, $\Psi_0 := \max_{x\in K}\Psi_1(T,x)$.

Let the number $\varepsilon_1 > 0$ be such that $C_1\varepsilon_1 < \varepsilon/2$, and let η correspond to it in accordance with (2.2.19). Choose a small number $\delta > 0$ in such a way that for all $t \in (0,\delta)$,

$$C_0\|\varphi\|_{0,0}^{\vec{a}}\Psi_0\exp\{-c_1t^{1-q''}\eta^{q'}\} < \varepsilon/2.$$

Then we obtain the required inequality (2.2.18). □

Lemma 2.5. *Let $\varphi \in L_p^{\vec{a}}$, $p \in [1,\infty]$. Then the function (2.2.9) has the following properties:*

1) *there exists such a constant $C > 0$ that for any $t \in (0,T]$ and $l \in \mathbb{Z}_+^n$, $\|l\| \le 2b$,*
$$\|\partial_x^l u(t,\cdot)\|_p^{\vec{k}(t,\vec{a})} \le C\|\varphi\|_p^{\vec{a}} t^{-\|l\|/(2b)}; \tag{2.2.20}$$
2) *if $p \in [1,\infty)$, then*
$$\lim_{t\to 0} \|u(t,\cdot) - \varphi(\cdot)\|_p^{\vec{k}(t,\vec{a})} = 0, \tag{2.2.21}$$
while for $p = \infty$ $u(t,\cdot) \xrightarrow[t\to 0]{w} \varphi(\cdot)$, that is for any $\psi \in L_1^{-\vec{k}(T,\vec{a})}$,
$$\lim_{t\to 0} \int\limits_{\mathbb{R}^n} \psi'(x)\bar{u}(t,x)dx = \int\limits_{\mathbb{R}^n} \psi'(x)\bar{\varphi}(x)dx, \tag{2.2.22}$$
where, as usual, the dash means taking the conjugate matrix while the bar denotes the complex conjugation.

Proof. 1) Below l is a multi-index with $\|l\| \le 2b$. Consider first the case $p = \infty$. For $(t,x) \in \Pi_{(0,T]}$, we have

$$\begin{aligned}
&|\partial_x^l u(t,x)| \\
&\le Ct^{-M-\|l\|/(2b)} \int\limits_{\mathbb{R}^n} E_{c-c_0}^{(0)}(t,x;0,\xi)(E_{c_0}^{(0)}(t,x;0,\xi)\Psi_1(0,\xi))(|\varphi(\xi)|\Psi_{-1}(0,\xi))d\xi \\
&\le C\|\varphi\|_\infty^{\vec{a}} t^{-\|l\|/(2b)}\Psi_1(t,x) \int\limits_{\mathbb{R}^n} t^{-M} E_{c-c_0}^{(0)}(t,x;0,\xi)d\xi = C\|\varphi\|_\infty^{\vec{a}} t^{-\|l\|/(2b)}\Psi_1(t,x),
\end{aligned}$$

which implies the estimates (2.2.20) for $p = \infty$.

If $p \in (1,\infty)$, then using the Hölder inequality we obtain that

$$\begin{aligned}
&|\partial_x^l u(t,x)| \\
&\le Ct^{-M-\|l\|/(2b)} \int\limits_{\mathbb{R}^n} (|\varphi(\xi)|\Psi_{-1}(0,\xi))(E_{c_0}^{(0)}(t,x;0,\xi)\Psi_1(0,\xi))E_{c-c_0}^{(0)}(t,x;0,\xi)d\xi \\
&\le Ct^{-\|l\|/(2b)}\Psi_1(t,x)\left(\int\limits_{\mathbb{R}^n} (|\varphi(\xi)|\Psi_{-1}(0,\xi))^p E_{(c-c_0)p/2}^{(0)}(t,x;0,\xi)t^{-M}d\xi\right)^{1/p} \\
&\quad\times \left(\int\limits_{\mathbb{R}^n} E_{(c-c_0)p'/2}^{(0)}(t,x;0,\xi)t^{-M}d\xi\right)^{1/p'} \\
&= Ct^{-\|l\|/(2b)}\Psi_1(t,x)\left(\int\limits_{\mathbb{R}^n} (|\varphi(\xi)|\Psi_{-1}(0,\xi))^p E_{(c-c_0)p/2}^{(0)}(t,x;0,\xi)t^{-M}d\xi\right)^{1/p},
\end{aligned}$$

$(t,x) \in \Pi_{(0,T]}$. Here $p' = p/(p-1)$.

It follows that

$$\begin{aligned}
&\|\partial_x^l u(t,\cdot)\|_p^{\vec{k}(t,\vec{a})} \\
&\leq Ct^{-\|l\|/(2b)}\left(\int\limits_{\mathbb{R}^n}\left(\int\limits_{\mathbb{R}^n}(|\varphi(\xi)|\Psi_{-1}(0,\xi))^p E^{(0)}_{(c-c_0)p/2}(t,x;0,\xi)t^{-M}d\xi\right)dx\right)^{1/p} \\
&= Ct^{-\|l\|/(2b)}\left(\int\limits_{\mathbb{R}^n}(|\varphi(\xi)|\Psi_{-1}(0,\xi))^p\left(\int\limits_{\mathbb{R}^n}E^{(0)}_{(c-c_0)/2}(t,x;0,\xi)t^{-M}dx\right)d\xi\right)^{1/p} \\
&= C\|\varphi\|_p^{\vec{a}}t^{-\|l\|/(2b)}, \quad t\in(0,T].
\end{aligned}$$

Similarly, if $p=1$, then

$$\begin{aligned}
&|\partial_x^l u(t,x)| \\
&\leq Ct^{-M-\|l\|/(2b)}\int\limits_{\mathbb{R}^n}(|\varphi(\xi)|\Psi_{-1}(0,\xi))(E^{(0)}_{c_0}(t,x;0,\xi)\Psi_1(0,\xi))E^{(0)}_{c-c_0}(t,x;0,\xi)d\xi \\
&\leq Ct^{-\|l\|/(2b)}\Psi_1(t,x)\int\limits_{\mathbb{R}^n}(|\varphi(\xi)|\Psi_{-1}(0,\xi))E^{(0)}_{c-c_0}(t,x;0,\xi)d\xi, \quad (t,x)\in\Pi_{(0,T]},
\end{aligned}$$

whence

$$\begin{aligned}
\|\partial_x^l u(t,\cdot)\|_1^{\vec{k}(t,\vec{a})} &\leq Ct^{-\|l\|/(2b)}\int\limits_{\mathbb{R}^n}(|\varphi(\xi)|\Psi_{-1}(0,\xi))\left(\int\limits_{\mathbb{R}^n}E^{(0)}_{c-c_0}(t,x;0,\xi)t^{-M}dx\right)d\xi \\
&= C\|\varphi\|_1^{\vec{a}}t^{-\|l\|/(2b)}, \quad t\in(0,T].
\end{aligned}$$

2) Let $p\in[1,\infty)$. We have to prove that for any $\varepsilon>0$ there exists such $\delta\in(0,T)$ that for all $t\in(0,\delta)$,

$$\|(P\varphi)(t,\cdot)-\varphi(\cdot)\|_p^{\vec{k}(t,\vec{a})}<\varepsilon. \tag{2.2.23}$$

For $R>0$, consider the function $\varphi^{(R)}$ defined by the equalities

$$\varphi^{(R)}(x):=\begin{cases}\varphi(x), & x\in B_R,\\ 0, & x\in\mathbb{R}^n\setminus B_R,\end{cases} \tag{2.2.24}$$

where $B_R:=B(0,R)$. We have

$$\begin{aligned}
\|(P\varphi)(t,\cdot)-\varphi(\cdot)\|_p^{\vec{k}(t,\vec{a})} &\leq \|(P(\varphi-\varphi^{(R)}))(t,\cdot)\|_p^{\vec{k}(t,\vec{a})} \\
&\quad+\|(P\varphi^{(R)})(t,\cdot)-\varphi^{(R)}(\cdot)\|_p^{\vec{k}(t,\vec{a})}+\|\varphi-\varphi^{(R)}\|_p^{\vec{k}(t,\vec{a})}, \\
&\quad t\in(0,T].
\end{aligned} \tag{2.2.25}$$

By (2.2.20),

$$\| \left(P(\varphi - \varphi^{(R)})\right)(t,\cdot)\|_p^{\vec{k}(t,\vec{a})} \leq C\|\varphi - \varphi^{(R)}\|_p^{\vec{a}}, \quad t \in (0,T]. \tag{2.2.26}$$

The inequalities (2.2.25) and (2.2.26) imply the inequality

$$\|(P\varphi)(t,\cdot) - \varphi(\cdot)\|_p^{\vec{k}(t,\vec{a})} \leq (C+1)\|\varphi - \varphi^{(R)}\|_p^{\vec{a}} + \|(P\varphi^{(R)})(t,\cdot) - \varphi^{(R)}(\cdot)\|_p^{\vec{k}(t,\vec{a})},$$
$$t \in (0,T].$$

Let $\varepsilon > 0$ be a given number. Choose $R > 0$ in such a way that

$$\|\varphi - \varphi^{(R)}\|_p^{\vec{a}} = \left(\int\limits_{\mathbb{R}^n \setminus B_R} (|\varphi(x)|\Psi_{-1}(0,x))^p dx\right)^{1/p} < \varepsilon/(2(C+1)).$$

Since

$$\|(P\varphi^{(R)})(t,\cdot) - \varphi^{(R)}(\cdot)\|_p^{\vec{k}(t,\vec{a})} \leq \|(P\varphi^{(R)})(t,\cdot) - \varphi^{(R)}(\cdot)\|_{L_p(\mathbb{R}^n)} =: J^{1/p},$$

(2.2.23) will be proved if we prove the existence of such a $\delta \in (0,T)$, that for any $t \in (0,\delta)$,

$$J^{1/p} < \varepsilon/2. \tag{2.2.27}$$

Represent J as $J = J_1 + J_2$ where

$$J_1 := \int\limits_{\mathbb{R}^n \setminus B_{2R}} \left| \int\limits_{B_R} Z(t,x;0,\xi)\varphi^{(R)}(\xi)d\xi\right|^p dx,$$

$$J_2 := \int\limits_{B_{2R}} \left| \int\limits_{B_R} Z(t,x;0,\xi)\varphi^{(R)}(\xi)d\xi d\xi - \varphi^{(R)}(x)\right|^p dx.$$

Note that for any $c_1 > 0$, $t \in (0,1]$, $x \in \mathbb{R}^n \setminus B_{2R}$ and $\xi \in B_R$,

$$E_{c_1}^{(0)}(t,x;0,\xi) \leq \exp\{-c_1 R^{q'} t^{1-q''}\}. \tag{2.2.28}$$

This inequality follows from the definition (2.1.40) of the function $E_{c_1}^{(0)}$, and the inequalities

$$\sum_{j=1}^n t^{1-q_j}|x_j - \xi_j|^{q_j} \geq t^{1-q''}\sum_{j=1}^n |x_j - \xi_j|^{q_j} = t^{1-q''}(q(x;\xi))^{q'}$$
$$\geq t^{1-q''}(q(x;0) - q(\xi;0))^{q'} \geq t^{1-q''}(2R - R)^{q'} = R^{q'}t^{1-q''}.$$

We used the definition (2.1.2) of the distance q and its properties.

If $p = 1$, using the inequality (2.2.28) with $c_1 = c - c_0$, we obtain for $t \leq 1$ that

$$
\begin{aligned}
J_1 &\leq C \int\limits_{\mathbb{R}^n \setminus B_{2R}} \left(\int\limits_{B_R} |\varphi^{(R)}(\xi)| E_c^{(0)}(t,x;0,\xi) t^{-M} d\xi \right) dx \\
&= C \int\limits_{B_R} |\varphi^{(R)}(\xi)| \left(\int\limits_{\mathbb{R}^n \setminus B_{2R}} E_c^{(0)}(t,x;0,\xi) t^{-M} dx \right) d\xi \\
&\leq C \exp\{-(c-c_0) R^{q'} t^{1-q''}\} \int\limits_{\mathbb{R}^n} |\varphi^{(R)}(\xi)| \left(\int\limits_{\mathbb{R}^n} E_{c_0}^{(0)}(t,x;0,\xi) t^{-M} dx \right) d\xi \\
&= C \|\varphi^{(R)}\|_{L_1(\mathbb{R}^n)} \exp\{-(c-c_0) R^{q'} t^{1-q''}\}. \qquad (2.2.29)
\end{aligned}
$$

If $p > 1$, we use the Hölder inequality and the inequality (2.2.28) with $c_1 = (c - c_0)/2$. For $t \in (0, 1]$ and $x \in \mathbb{R}^n \setminus B_{2R}$, we find that

$$
\begin{aligned}
&\left| \int\limits_{B_R} Z(t,x;0,\xi) \varphi^{(R)}(\xi) d\xi \right| \\
&\leq C \int\limits_{B_R} E_c^{(0)}(t,x;0,\xi) |\varphi^{(R)}(\xi)| t^{-M} d\xi \\
&\leq C \left(\int\limits_{B_R} |\varphi^{(R)}(\xi)|^p E_{(c-c_0)p/2}^{(0)}(t,x;0,\xi) t^{-M} d\xi \right)^{1/p} \\
&\quad \times \left(\int\limits_{B_R} E_{(c+c_0)p'/2}^{(0)}(t,x;0,\xi) t^{-M} d\xi \right)^{1/p'} \\
&\leq C \left(\int\limits_{B_R} |\varphi^{(R)}(\xi)|^p E_{(c-c_0)p/2}^{(0)}(t,x;0,\xi) t^{-M} d\xi \right)^{1/p} \\
&\quad \times \exp\{-(c-c_0) R^{q'} t^{1-q''}/2\} \left(\int\limits_{\mathbb{R}^n} E_{c_0 p'}^{(0)}(t,x;0,\xi) t^{-M} d\xi \right)^{1/p'} \\
&= C \left(\int\limits_{B_R} |\varphi^{(R)}(\xi)|^p E_{(c-c_0)p/2}^{(0)}(t,x;0,\xi) t^{-M} d\xi \right)^{1/p} \exp\{-(c-c_0) R^{q'} t^{1-q''}/2\},
\end{aligned}
$$

whence

$$
J_1 \leq C \int\limits_{\mathbb{R}^n \setminus B_{2R}} \left(\int\limits_{B_R} |\varphi^{(R)}(\xi)|^p E_{(c-c_0)p/2}^{(0)}(t,x;0,\xi) t^{-M} d\xi \right) dx
$$

$$\times \exp\{-(c-c_0)pR^{q'}t^{1-q''}/2\}$$
$$\leq C\|\varphi^{(R)}\|^p_{L_p(\mathbb{R}^n)} \exp\{-(c-c_0)pR^{q'}t^{1-q''}/2\}, \quad t \in (0,1]. \tag{2.2.30}$$

Let us estimate J_2. Let $\varphi_h^{(R)}$ be the averaging for the function $\varphi^{(R)}$ [214]. It follows from properties of the averaging that

$$\|\varphi^{(R)} - \varphi_h^{(R)}\|_{L_p(\mathbb{R}^n)} \underset{h\to 0}{\longrightarrow} 0. \tag{2.2.31}$$

Since $\varphi_h^{(R)}$ is an infinitely differentiable function with a compact support, we find, due to (2.2.12), that for a fixed $h > 0$, uniformly with respect to $x \in B_{2R}$,

$$\left| \int_{\mathbb{R}^n} Z(t,x;0,\xi)\varphi_h^{(R)}(\xi)d\xi - \varphi_h^{(R)}(x) \right| \underset{t\to 0}{\longrightarrow} 0. \tag{2.2.32}$$

We have

$$J_2^{1/p} \leq \left(\int_{B_{2R}} \left| \int_{\mathbb{R}^n} Z(t,x;0,\xi)(\varphi^{(R)}(\xi) - \varphi_h^{(R)}(\xi))d\xi \right|^p dx \right)^{1/p}$$
$$+ \left(\int_{B_{2R}} \left| \int_{\mathbb{R}^n} Z(t,x;0,\xi)\varphi_h^{(R)}(\xi)d\xi - \varphi_h^{(R)}(x) \right|^p dx \right)^{1/p}$$
$$+ \left(\int_{B_{2R}} |\varphi_h^{(R)}(x) - \varphi^{(R)}(x)|^p dx \right)^{1/p}.$$

Repeating for the first summand the estimates similar to the ones carried out in the proof of the estimates (2.2.20), and using the relations (2.2.31) and (2.2.32), we come to the existence of such a constant $\delta_2 > 0$ that for any $t \in (0, \delta_2)$,

$$J_2 < (1/2)(\varepsilon/2)^p. \tag{2.2.33}$$

The inequalities (2.2.29) and (2.2.30) imply the existence of such a $\delta_1 > 0$ that for all $t \in (0, \delta_1)$,

$$J_1 < (1/2)(\varepsilon/2)^p.$$

Together with (2.2.33) this shows that for all $t \in (0,\delta)$, $\delta := \min(\delta_1, \delta_2)$, the inequality (2.2.27) holds. Hence, we have proved (2.2.21).

Let us prove (2.2.22). Note first of all that the integrals from (2.2.22) make sense for any $\varphi \in L_\infty^{\vec{a}}$, $\psi \in L_1^{-\vec{k}(T,\vec{a})}$, and $t \in (0,T]$, since by (2.2.20), $\|u(t,\cdot)\|_\infty^{\vec{k}(t,\vec{a})} < \infty$ for any $t \in (0,T]$, if $\varphi \in L_\infty^{\vec{a}}$. Indeed, since $\Psi_1(0,x) \leq \Psi(t,x) \leq \Psi_1(T,x)$, $t \in (0,T]$, we have

$$\left| \int_{\mathbb{R}^n} \psi'(x)\bar{u}(t,x)dx \right| \leq \int_{\mathbb{R}^n} (|\psi(x)|\Psi_1(T,x))(|u(t,x)|\Psi_{-1}(t,x))dx$$

$$\leq \|u(t,\cdot)\|_\infty^{\vec{k}(t,\vec{a})} \|\psi\|_1^{-\vec{k}(T,\vec{a})} < \infty,$$

$$\left|\int_{\mathbb{R}^n} \psi'(x)\bar{\varphi}(x)dx\right| \leq \int_{\mathbb{R}^n} (|\psi(x)|\Psi_1(T,x))(|\varphi(x)|\Psi_{-1}(0,x))dx$$

$$\leq \|\varphi\|_\infty^{\vec{a}} \|\psi\|_1^{-\vec{k}(T,\vec{a})} < \infty.$$

Due to (2.2.9), in order to prove (2.2.22), it suffices to show that

$$\int_{\mathbb{R}^n} (v(t,\xi) - \psi(\xi))'\bar{\varphi}(\xi)d\xi \underset{t\to 0}{\longrightarrow} 0$$

where

$$v(t,\xi) := \int_{\mathbb{R}^n} \bar{Z}'(t,x;0,\xi)\psi(x)dx, \quad (t,\xi) \in \Pi_{(0,T]}. \tag{2.2.34}$$

Since $\varphi \in L_\infty^{\vec{a}}$, we find that

$$\left|\int_{\mathbb{R}^n} (v(t,\xi) - \psi(\xi))'\bar{\varphi}(\xi)d\xi\right| \leq \|\varphi\|_\infty^{\vec{a}} \int_{\mathbb{R}^n} |v(t,\xi) - \psi(\xi)|\Psi_1(0,\xi)d\xi,$$

and (2.2.22) will be proved if we prove that

$$\int_{\mathbb{R}^n} |v(t,\xi) - \psi(\xi)|\Psi_1(0,\xi)d\xi \underset{t\to 0}{\longrightarrow} 0. \tag{2.2.35}$$

Consider the functions

$$l_j(t,a_j) := c_0 k_j(T,a_j)(c_0^{2b_j-1} + (k_j(T,a_j))^{2b_j-1}t)^{1-q_j}, \quad j \in \{1,\dots,n\},$$
$$\vec{l}(t,\vec{a}) := (l_1(t,a_j),\dots,l_n(t,a_n)), \quad t \geq 0, \tag{2.2.36}$$

and also, by analogy with (2.2.1), the weight functions

$$\hat{\Psi}_\nu(t,\xi) := \exp\left\{\nu \sum_{j=1}^n l_j(t,a_j)|\xi_j|^{q_j}\right\}, \quad t > 0, \quad \xi \in \mathbb{R}^n, \quad \nu \in \{-1,1\}. \tag{2.2.37}$$

Since $a_j < k_j(T,a_j)$, $j \in \{1,\dots,n\}$, there exists such a $\gamma > 0$ that for all $t \in [0,\gamma)$,

$$a_j \leq l_j(t,a_j) \leq k_j(T,a_j), \quad j \in \{1,\dots,n\},$$

hence also the inequalities

$$\Psi_1(0,\xi) \leq \hat{\Psi}_1(t,\xi) \leq \Psi_1(T,\xi), \quad \xi \in \mathbb{R}^n. \tag{2.2.38}$$

Therefore, to prove (2.2.35) it is sufficient to show that for any $\varepsilon > 0$ there exists such a $\delta \in (0,\gamma)$ that for all $t \in (0,\delta)$,

$$\|v(t,\cdot) - \psi(\cdot)\|_1^{-\vec{l}(t,\vec{a})} := \|v(t,\cdot) - \psi(\cdot)\hat{\Psi}_1(t,\cdot)\|_{L_1(\mathbb{R}^n)} < \varepsilon. \tag{2.2.39}$$

The *proof* of (2.2.39) is similar to the one for the inequality (2.2.23). Just as it was done there, we introduce, for $R > 0$, a function $\psi^{(R)}$ by the equalities (2.2.24), with ψ substituted for φ. For $t \in (0,\gamma)$, we obtain that

$$\|v(t,\cdot) - \psi(\cdot)\|_1^{-\vec{l}(t,\vec{a})} \le \left\| \int_{\mathbb{R}^n} \bar{Z}'(t,x;0,\xi)(\psi - \psi^{(R)})(x)dx \right\|_1^{-\vec{l}(t,\vec{a})}$$

$$+ \left\| \int_{\mathbb{R}^n} \bar{Z}'(t,x;0,\xi)\psi^{(R)}(x)dx - \psi^{(R)}(\xi) \right\|_1^{-\vec{l}(t,\vec{a})} + \|\psi - \psi^{(R)}\|_1^{-\vec{l}(t,\vec{a})} =: \sum_{j=1}^{3} K_j. \tag{2.2.40}$$

Let us give an estimate for K_1. We have

$$\left| \int_{\mathbb{R}^n} \bar{Z}'(t,x;0,\xi)(\psi - \psi^{(R)})(x)dx \right|$$

$$\le C \int_{\mathbb{R}^n} E^{(0)}_{c-c_0}(t,x;0,\xi)(E^{(0)}_{c_0}(t,x;0,\xi)\Psi_{-1}(T,x))(|(\psi - \psi^{(R)})(x)|\Psi_1(T,x))t^{-M}dx$$

$$\le C\hat{\Psi}_{-1}(t,\xi) \int_{\mathbb{R}^n} E^{(0)}_{c-c_0}(t,x;0,\xi)(|(\psi - \psi^{(R)})(x)|\Psi_1(T,x))t^{-M}dx, \tag{2.2.41}$$

due to the inequality

$$E^{(0)}_{c_0}(t,x;0,\xi)\Psi_{-1}(T,x) \le \hat{\Psi}_{-1}(t,\xi), \quad t > 0, \quad \{x,\xi\} \subset \mathbb{R}^n, \tag{2.2.42}$$

proved just as (2.2.4). It follows from (2.2.40) that

$$K_1 \le C\|\psi - \psi^{(R)}\|_1^{-\vec{k}(T,\vec{a})}, \quad t \in (0,\gamma). \tag{2.2.43}$$

By the inequalities (2.2.38), $K_3 \le \|\psi - \psi^{(R)}\|_1^{-\vec{k}(T,\vec{a})}$, $t \in (0,\gamma)$, so that

$$K_1 + K_3 \le (C+1)\|\psi - \psi^{(R)}\|_1^{-\vec{k}(T,\vec{a})}, \quad t \in (0,\gamma).$$

Since

$$\|\psi - \psi^{(R)}\|_1^{-\vec{k}(T,\vec{a})} = \int_{\mathbb{R}^n \setminus B_R} |\psi(x)|\Psi_1(T,x)dx \underset{R\to\infty}{\longrightarrow} 0,$$

we get

$$K_1 + K_3 \underset{R\to\infty}{\longrightarrow} 0, \quad t \in (0,\gamma). \tag{2.2.44}$$

Consider the expression K_2. We rewrite it as

$$\begin{aligned} K_2 = & \int\limits_{\mathbb{R}^n\setminus B_{2R}} \left| \int\limits_{\mathbb{R}^n} \bar{Z}'(t,x;0,\xi)\psi^{(R)}(x)dx \right| \hat{\Psi}_1(t,\xi)d\xi \\ & + \int\limits_{B_{2R}} \left| \int\limits_{\mathbb{R}^n} \bar{Z}'(t,x;0,\xi)\psi^{(R)}(x)dx - \psi^{(R)}(\xi) \right| \hat{\Psi}_1(t,\xi)d\xi \\ =: & K_1' + K_2''. \end{aligned}$$

The inequality

$$\left\| \int\limits_{\mathbb{R}^n} \bar{Z}'(t,x;0,\xi)\psi^{(R)}(x)dx \right\|_1^{-\vec{l}(t,\vec{a})} \leq C\|\psi^{(R)}\|_1^{-\vec{k}(T,\vec{a})},$$

which is proved just as (2.2.43),implies

$$K_2' \underset{R\to\infty}{\longrightarrow} 0, \quad t \in (0,\gamma). \tag{2.2.45}$$

For K_2'' we have

$$K_2'' \leq \int\limits_{B_{2R}} \left| \int\limits_{\mathbb{R}^n} \bar{Z}'(t,x;0,\xi)\psi^{(R)}(x)dx - \psi^{(R)}(\xi) \right| d\xi.$$

The arguments similar to those for J_2 show that $K_2'' \underset{t\to 0}{\longrightarrow} 0$, $R > 0$. Together with (2.2.29), (2.2.44), (2.2.45) this implies the inequality (2.2.39). □

Lemma 2.6. *If $\mu \in M^{\vec{a}}$, then the function* (2.2.10) *has the following properties:*

1) *there exists such a constant $C > 0$ that for any $t \in (0,T]$ and $l \in \mathbb{Z}_+^n$, $\|l\| \leq 2b$,*

$$\|\partial_x^l u(t,\cdot)\|_1^{\vec{k}(t,\vec{a})} \leq C\|\mu\|^{\vec{a}} t^{-\|l\|/(2b)}; \tag{2.2.46}$$

2) $u(t,\cdot) \underset{t\to 0}{\overset{w}{\longrightarrow}} \mu$, *that is, for any $\psi \in C_0^{-\vec{k}(T,\vec{a})}$,*

$$\lim_{t\to 0} \int\limits_{\mathbb{R}^n} \psi'(x)\bar{u}(t,x)dx = \int\limits_{R^n} \psi'(x)d\bar{\mu}(x). \tag{2.2.47}$$

Proof. 1) For $l \in \mathbb{Z}_+^n$, $\|l\| \le 2b$, and $(t,x) \in \Pi_{(0,T]}$ we have

$$|\partial_x^l u(t,x)| \le Ct^{-M-\|l\|/(2b)}$$
$$\times \int\limits_{\mathbb{R}^n} E^{(0)}_{c\quad c_0}(t,x;0,\xi)(E^{(0)}_{c_0}(t,x;0,\xi)\Psi_1(0,\xi))\Psi_{-1}(0,\xi)d|\mu|(\xi)$$
$$\le Ct^{-\|l\|/(2b)}\Psi_1(t,x)\int\limits_{\mathbb{R}^n} t^{-M}E^{(0)}_{c-c_0}(t,x;0,\xi)\Psi_{-1}(0,\xi)d|\mu|(\xi),$$

whence

$$\|\partial_x^l u(t,\cdot)\|_1^{\vec{k}(t,\vec{a})} \le Ct^{-\|l\|/(2b)}\int\limits_{\mathbb{R}^n}\left(\int\limits_{\mathbb{R}^n} E^{(0)}_{c-c_0}(t,x;0,\xi)t^{-M}dx\right)\Psi_{-1}(0,\xi)d|\mu|(\xi)$$
$$= C\|\mu\|^{\vec{a}}t^{-\|l\|/(2b)}, \quad t \in (0,T], \quad \|l\| \le 2b.$$

2) The integrals from (2.2.47) make sense for any $\psi \in C_0^{-\vec{k}(T,\vec{a})}$, $\mu \in M^{\vec{a}}$, and $t \in (0,T]$. Indeed, using (2.2.46) we find that

$$\left|\int\limits_{\mathbb{R}^n} \psi'(x)\bar{u}(t,x)dx\right| \le \int\limits_{\mathbb{R}^n}(|\psi(x)|\Psi_1(T,x))(|u(t,x)|\Psi_{-1}(t,x))dx$$
$$\le \|\psi\|_0^{-\vec{k}(t,\vec{a})}\|u(t,\cdot)\|_1^{\vec{k}(t,\vec{a})}$$
$$\le C\|\psi\|_0^{-\vec{k}(T,\vec{a})}\|\mu\|^{\vec{a}} < \infty,$$
$$\left|\int\limits_{\mathbb{R}^n}\psi'(x)d\bar{\mu}(x)\right| \le \int\limits_{\mathbb{R}^n}(|\psi(x)|\Psi_1(T,x))\Psi_{-1}(0,x)d|\mu|(x) \le \|\psi\|_0^{-\vec{k}(T,\vec{a})}\|\mu\|^{\vec{a}} < \infty.$$

By virtue of the formula (2.2.10) we obtain that

$$\left|\int\limits_{\mathbb{R}^n}\psi'(x)\bar{u}(t,x)dx - \int\limits_{\mathbb{R}^n}\psi'(x)d\bar{\mu}(x)\right| \le \|v(t,\cdot)-\psi(\cdot)\|_\infty^{-\vec{a}}\|\mu\|^{\vec{a}},$$

where v is the function (2.2.34). Therefore it suffices to prove that

$$\|v(t,\cdot)-\psi(\cdot)\|_\infty^{-\vec{l}(t,\vec{a})} := \|(v(t,\cdot)-\psi(\cdot))\hat{\Psi}_1(t,\cdot)\|_{L_\infty(\mathbb{R}^n)} \underset{t\to 0}{\longrightarrow} 0, \qquad (2.2.48)$$

where the functions $\vec{l}$ and $\hat{\Psi}_1$ were defined in (2.2.36) and (2.2.37).

Suppose that $R > 0$, θ_R is a function from the space $C^\infty(\mathbb{R}^n)$, such that $0 \le \theta_R \le 1$ on $\mathbb{R}^n$, $\theta_R = 1$ on $B_{R/2}$, and $\theta_R = 0$ on $\mathbb{R}^n \setminus B_R$. Set $\psi_R := \theta_R\psi$. For $t \in (0,\gamma)$ we have

$$\|v(t,\cdot)-\psi(\cdot)\|_\infty^{-\vec{l}(t,\vec{a})} \le \left\|\int\limits_{\mathbb{R}^n}\bar{Z}'(t,x;0,\xi)(\psi-\psi_R)(x)dx\right\|_\infty^{-\vec{l}(t,\vec{a})}$$

$$+\left\|\int_{\mathbb{R}^n} \bar{Z}'(t,x;0,\xi)\psi_R(x)dx - \psi_R(\xi)\right\|_\infty^{-\vec{l}(t,\vec{a})} + \|\psi_R - \psi\|_\infty^{-\vec{l}(t,\vec{a})} =: \sum_{j=1}^{3} L_j. \tag{2.2.49}$$

Just as in the proof of the inequality (2.2.41), we obtain that

$$\left\|\int_{\mathbb{R}^n} \bar{Z}'(t,x;0,\xi)(\psi - \psi_R)(x)dx\right\|$$
$$\leq C\|\psi - \psi_R\|_\infty^{-\vec{k}(T,a)}\hat{\Psi}_{-1}(t,\xi)\int_{\mathbb{R}^n} E^{(0)}_{c-c_0}(t,x;0,\xi)t^{-M}dx,$$

which implies the inequality $L_1 \leq C\|\psi - \psi_R\|_\infty^{-\vec{k}(T,\vec{a})}$, $t \in (0,\gamma)$. Due to the inequalities (2.2.38), $L_3 \leq \|\psi - \psi_R\|_\infty^{-\vec{k}(T,\vec{a})}$, so that $L_1 + L_3 \leq (C+1)\|\psi - \psi_R\|_\infty^{-\vec{k}(T,\vec{a})}$, $t \in (0,\gamma)$. Since

$$\|\psi - \psi_R\|_\infty^{-\vec{k}(T,\vec{a})} \leq \sup_{x\in\mathbb{R}^n\setminus B_{R/2}} (|\psi(x)|\Psi_1(T,x)) \underset{R\to\infty}{\longrightarrow} 0,$$

we get

$$L_1 + L_3 \underset{R\to\infty}{\longrightarrow} 0, \quad t \in (0,\gamma). \tag{2.2.50}$$

Next,

$$L_2 \leq \sup_{\xi\in\mathbb{R}^n\setminus B_{2R}} \left(\left|\int_{\mathbb{R}^n} \bar{Z}'(t,x;0,\xi)\psi_R(x)dx\right|\hat{\Psi}_1(t,\xi)\right)$$
$$+ C_0 \sup_{\xi\in B_{2R}} \left|\int_{\mathbb{R}^n} \bar{Z}'(t,x;0,\xi)\psi_R(x)dx - \psi_R(\xi)\right| =: L_2' + L_2'', \tag{2.2.51}$$

where $C_0 := \exp\{\max_{j\in\{1,\dots,n\}} k_j(T,a_j)(2R)^{q'}\}$, because, due to (2.2.38),

$$\hat{\Psi}_1(t,\xi) \leq \Psi_1(T,\xi) \leq \exp\{\max_{j\in\{1,\dots,n\}} k_j(T,a_j)(q(\xi;0))^{q'}\}$$
$$\leq C_0, \quad t \in (0,\gamma), \quad \xi \in B_{2R}.$$

Using the inequalities (2.2.28) and (2.2.42) we find that

$$L_2' \leq \sup_{\xi\in\mathbb{R}^n\setminus B_{2R}} \left(C\int_{B_R} E^{(0)}_{c-c_0}(t,x;0,\xi)(E^{(0)}_{c_0}(t,x;0,\xi)\Psi_{-1}(T,x))\right.$$
$$\left.\times (|\psi_R(x)|\Psi_1(T,x))t^{-M}dx\hat{\Psi}_1(t,\xi)\right)$$

$$\leq C\|\psi_R\|_\infty^{-\vec{k}(T,\vec{a})} \exp\{-(c-c_0)R^{q'}t^{1-q''}/2\} \int_{\mathbb{R}^n} E^{(0)}_{(c-c_0)/2}(t,x;0,\xi)t^{-M}dx$$

$$\leq C\|\psi\|_\infty^{-\vec{k}(T,\vec{a})} \exp\{-(c-c_0)R^{q'}t^{1-q''}/2\} \underset{t\to 0}{\longrightarrow} 0, \quad R>0. \tag{2.2.52}$$

Since ψ_R is a bounded and continuous function, we prove, as in the proof of (2.2.12) that

$$L_2'' \underset{t\to 0}{\longrightarrow} 0, \quad R > 0. \tag{2.2.53}$$

The required relation (2.2.48) follows from (2.2.49)–(2.2.53).

Remark. The functions (2.2.9) and (2.2.10), for any fixed φ and μ from the spaces $C^{\vec{a}}_{0,0}$, $L^{\vec{a}}_p$, $p \in [1,\infty]$, and $M^{\vec{a}}$ respectively, are solutions of the equation (2.1.1) with $f = 0$. This is proved using properties of the FSCP, just as in the proof of Theorem 2.1.

2.2.3. Lemmas about integrals similar to derivatives of volume potentials. The proofs of theorems about the correct solvability of the Cauchy problem for the equation (2.1.1) will be based on properties of volume potentials generated by the corresponding FSCPs. Before turning to these properties, we give some lemmas about integrals similar to derivatives of volume potentials, which are also of independent interest.

We shall consider integrals of the form

$$u(t,x) := \int_0^t d\tau \int_{\mathbb{R}^n} K(t,x;\tau,\xi)f(\tau,\xi)d\xi, \quad (t,x) \in \Pi_{(0,T]}. \tag{2.2.54}$$

First we consider the kernel

$$K(t,x;\tau,\xi) := K_0(t-\tau, x-\xi), \quad 0 \leq \tau < t \leq T, \quad \{x,\xi\} \subset \mathbb{R}^n, \tag{2.2.55}$$

with the properties of derivatives of the FSCP G for the equation (2.1.5) with constant coefficients.

Let us describe properties of the function K_0. We assume the representation

$$K_0(t,x) = t^{-M-\nu}\Omega(x_t), \quad t>0, \quad x \in \mathbb{R}^n, \tag{2.2.56}$$

where, as before, $M = \sum_{j=1}^n (1/(2b_j))$, $x_t = (t^{-1/(2b_1)}x_1, \ldots, t^{-1/(2b_n)}x_n)$, $\nu \in (0,1]$, the function $\Omega : \mathbb{R}^n \to \mathbb{C}_{NN}$ is continuous and satisfies the following conditions (with certain constants $C > 0$, $c > 0$, and $\gamma_0 \in (0,1]$):

$$\int_{\mathbb{R}^n} \Omega(x)dx = 0, \tag{2.2.57}$$

and for any $\{x, x'\} \subset \mathbb{R}^n$,

$$|\Omega(x)| \leq C \exp\left\{-c\sum_{j=1}^{n} |x_j|^{q_j}\right\}, \tag{2.2.58}$$

$$|\Delta_x^{x'}\Omega(x)| \leq C(p(x;x'))^{\gamma_0}\left(\exp\left\{-c\sum_{j=1}^{n}|x_j|^{q_j}\right\} + \exp\left\{-c\sum_{j=1}^{n}|x_j'|^{q_j}\right\}\right). \tag{2.2.59}$$

For $\nu = 1$, the integral (2.2.54) with the above kernel is understood as the limit

$$\lim_{h\to 0}\int_0^{t-h} d\tau \int_{\mathbb{R}^n} K_0(t-\tau, x-\xi) f(\tau,\xi)d\xi,$$

existing for an appropriate f due to (2.2.57).

We describe properties of the integral (2.2.54) stating that the function u belongs to an appropriate space, depending on a space to which f belongs. In particular, we use the spaces $C_{0,0,0}^{\vec{k}(\cdot,\vec{a})}$, $C_{0,0,\lambda}^{\vec{k}(\cdot,\vec{a})}$, $C_{0,\lambda/(2b),\lambda}^{\vec{k}(\cdot,\vec{a})}$, and $\hat{C}_{0,\lambda}^{\vec{k}(\cdot,\vec{a})}$, defined in Section 2.2.1. Note that the constant c_0 in the expression for the function $\vec{k}(\cdot,\vec{a})$ will be a fixed positive constant, smaller than the constant c from the estimates (2.2.58) and (2.2.59).

We have the following estimates for the function K_0:

$$|K_0(t-\tau, x-\xi)| \leq C(t-\tau)^{-M-\nu}E_c^{(0)}(t,x;\tau,\xi), \tag{2.2.60}$$

$$\begin{aligned} |\Delta K_0| &\leq C_1 p_0^{\gamma_0}(t-\tau)^{-\gamma_0/(2b)}((t-\tau)^{-M-\nu}E_{c_1}^{(0)}(t,x;\tau,\xi) \\ &\quad + (t'-\tau)^{-M-\nu}E_{c_1}^{(0)}(t',x';\tau,\xi)), \\ &c_1 \in (c_0, c), \quad \tau < t \leq t', \quad \{x,x'\} \subset \mathbb{R}^n, \end{aligned} \tag{2.2.61}$$

where

$$\Delta K_0 := \Delta_{t,x}^{t',x'} K_0(t-\tau, x-\xi), \quad p_0 := p(t,x;t',x'). \tag{2.2.62}$$

Indeed, the equality (2.2.56), the estimate (2.2.58), and the definition (2.1.40) of the function $E_c^{(0)}$ imply the estimate (2.2.60) directly; the latter implies the estimate (2.2.61) for the case where $p_0^{2b} \geq t-\tau$.

Suppose that $p_0^{2b} < t-\tau$. By (2.2.56) and (2.2.58), we have

$$\begin{aligned} |\Delta K_0| &\leq \Delta_t^{t'}(t-\tau)^{-M-\nu}|\Omega((x-\xi)_{t-\tau})| + (t'-\tau)^{-M-\nu}|\Delta\Omega| \\ &\leq C(t'-t)(t-\tau)^{-M-\nu-1}E_c^{(0)}(t,x;\tau,\xi) + (t'-\tau)^{-M-\nu}|\Delta\Omega| \\ &\leq Cp_0^{\gamma_0}(t-\tau)^{-M-\nu-\gamma_0/(2b)}E_c^{(0)}(t,x;\tau,\xi) + (t'-\tau)^{-M-\nu}|\Delta\Omega|. \end{aligned} \tag{2.2.63}$$

Let us give an estimate for $\Delta\Omega$. Using (2.2.59) we find that

$$\begin{aligned}
&|\Delta\Omega|(E_c^{(0)}(t,x;\tau,\xi)+E_c^{(0)}(t',x';\tau,\xi))^{-1}\\
&\quad\le C(p((x-\xi)_{t-\tau};(x'-\xi)_{t'-\tau}))^{\gamma_0}\\
&\quad\le C((t-\tau)^{-\gamma_0/(2b)}(p(x;x'))^{\gamma_0}\\
&\qquad+(t'-t)^{\gamma_0/(2b)}(t-\tau)^{-\gamma_0/(2b)}(t'-\tau)^{-\gamma_0/(2b)}(p(x';\xi))^{\gamma_0})\\
&\quad\le Cp_0^{\gamma_0}(t-\tau)^{-\gamma_0/(2b)}(1+(t'-\tau)^{-\gamma_0/(2b)}(p(x';\xi))^{\gamma_0}).
\end{aligned}$$

Together with the inequality (2.1.61), this gives the required estimate of $\Delta\Omega$. The latter estimate and the estimate (2.2.63) imply (2.2.61) for this case.

Lemma 2.7. *The integral* (2.2.54), *with the kernel* K *satisfying* (2.2.55)–(2.2.59), *has the following properties:*

1) *if* $1-\nu-\gamma_0/(2b)>0$ *and* $f\in C_{0,0,0}^{\vec{k}(\cdot,\vec{a})}$, *then* $u\in C_{0,\gamma_0/(2b),\gamma_0}^{\vec{k}(\cdot,\vec{a})}$, *and*

$$\|u\|_{0,\gamma_0/(2b),\gamma_0}^{\vec{k}(\cdot,\vec{a})}\le C\|f\|_{0,0,0}^{\vec{k}(\cdot,\vec{a})}; \tag{2.2.64}$$

2) *if* $f\in C_{0,0,\lambda}^{\vec{k}(\cdot,\vec{a})}$, $\lambda\in(0,1)$, *then for* $1-\nu-(\gamma_0-\lambda)/(2b)>0$ $u\in C_{0,\gamma_0/(2b),\gamma_0}^{\vec{k}(\cdot,\vec{a})}$, *and*

$$\|u\|_{0,\gamma_0/(2b),\gamma_0}^{\vec{k}(\cdot,\vec{a})}\le C\,\langle\!\langle f\rangle\!\rangle_{\lambda}^{\vec{k}(\cdot,\vec{a})}, \tag{2.2.65}$$

while for $1-\nu-(\gamma_0-\lambda)/(2b)<0$ $u\in C_{0,\lambda/(2b),\lambda}^{\vec{k}(\cdot,\vec{a})}$, *and*

$$\|u\|_{0,\lambda/(2b),\lambda}^{\vec{k}(\cdot,\vec{a})}\le C\,\langle\!\langle f\rangle\!\rangle_{\lambda}^{\vec{k}(\cdot,\vec{a})}. \tag{2.2.66}$$

Proof. We shall use the inequalities (2.2.4), (2.2.60), (2.2.61), and the equality (2.1.46).

1) We have for $(t,x)\in\Pi_{(0,T]}$ that

$$\begin{aligned}
|u(t,x)|&\le C\int_0^t(t-\tau)^{-\nu}d\tau\int_{\mathbb{R}^n}(t-\tau)^{-M}E_{c-c_0}^{(0)}(t,x;\tau,\xi)\\
&\quad\times(E_{c_0}^{(0)}(t,x;\tau,\xi)\Psi_1(\tau,\xi))(|f(\tau,\xi)|\Psi_{-1}(\tau,\xi))d\xi\\
&\le C\|f\|_{0,0,0}^{\vec{k}(\cdot,\vec{a})}\Psi_1(t,x)\int_0^t(t-\tau)^{-\nu}d\tau=C\|f\|_{0,0,0}^{\vec{k}(\cdot,\vec{a})}\Psi_1(t,x)t^{1-\nu},
\end{aligned} \tag{2.2.67}$$

whence

$$\|u\|_{0,0,0}^{\vec{k}(\cdot,\vec{a})}\le C\|f\|_{0,0,0}^{\vec{k}(\cdot,\vec{a})}. \tag{2.2.68}$$

Let (t,x), (t',x') be arbitrary points of the layer $\Pi_{(0,T]}$, with $t\le t'$, and let p_0 be the distance (2.2.62) between them. Consider the difference $\Delta_{t,x}^{t',x'}u$.

If $p_0^{2b} \ge t$, then due to (2.2.67) and the fact that $1-\nu-\gamma_0/(2b) > 0$, we have

$$|\Delta_{t,x}^{t',x'} u(t,x)| \le C p_0^{\gamma_0} t^{1-\nu-\gamma_0/(2b)} \|f\|_{0,0,0}^{\vec{k}(\cdot,\vec{a})} (\Psi_1(t,x) + \Psi_1(t',x')). \tag{2.2.69}$$

For the case where $p_0^{2b} < t$ we get

$$\begin{aligned}
&|\Delta_{t,x}^{t',x'} u(t,x)| \\
&\le \int_0^t d\tau \int_{\mathbb{R}^n} |\Delta_{t,x}^{t',x'} K_0(t-\tau, x-\xi)||f(\tau,\xi)|d\xi \\
&\quad + \int_t^{t'} d\tau \int_{\mathbb{R}^n} |K_0(t'-\tau, x'-\xi)||f(\tau,\xi)|d\xi \\
&\le C\|f\|_{0,0,0}^{\vec{k}(\cdot,\vec{a})} (\Psi_1(t,x) \\
&\quad + \Psi_1(t',x'))\Bigg(p_0^{\gamma_0} \int_0^t (t-\tau)^{-\gamma_0/(2b)} d\tau \Bigg(\int_{\mathbb{R}^n} (t-\tau)^{-M-\nu} E_{c_1-c_0}^{(0)}(t,x;\tau,\xi)d\xi \\
&\quad + \int_{\mathbb{R}^n} (t'-\tau)^{-M-\nu} E_{c_1-c_0}^{(0)}(t',x';\tau,\xi)d\xi \Bigg) \\
&\quad + \int_t^{t'} d\tau \int_{\mathbb{R}^n} (t'-\tau)^{-M-\nu} E_{c-c_0}^{(0)}(t',x';\tau,\xi)d\xi \Bigg) \\
&\le C\|f\|_{0,0,0}^{\vec{k}(\cdot,\vec{a})} (\Psi_1(t,x) + \Psi_1(t',x')) \Bigg(p_0^{\gamma_0} \int_0^t (t-\tau)^{-\nu-\gamma_0/(2b)} d\tau + \int_t^{t'} (t'-\tau)^{-\nu} d\tau \Bigg) \\
&\le C\|f\|_{0,0,0}^{\vec{k}(\cdot,\vec{a})} p_0^{\gamma_0} \times (\Psi_1(t,x) + \Psi_1(t',x'))(t^{1-\nu-\gamma_0/(2b)} + (t'-t)^{1-\nu-\gamma_0/(2b)}) \\
&\le C\|f\|_{0,0,0}^{\vec{k}(\cdot,\vec{a})} p_0^{\gamma_0} (\Psi_1(t,x) + \Psi_1(t',x')).
\end{aligned} \tag{2.2.70}$$

It follows from (2.2.69) and (2.2.70) that

$$\langle\!\langle u \rangle\!\rangle_{\gamma_0}^{\vec{k}(\cdot,\vec{a})} \le C\|f\|_{0,0,0}^{\vec{k}(\cdot,\vec{a})},$$

which, together with (2.2.68), implies the estimate (2.2.64).

2) By virtue of the condition (2.2.57) we can rewrite the integral (2.2.54) as

$$u(t,x) = \int_0^t d\tau \int_{\mathbb{R}^n} K_0(t-\tau, x-\xi) \Delta_\xi^x f(\tau,\xi) d\xi \tag{2.2.71}$$

and carry out the estimates similar to those for the case 1). As a result, using the inequality (2.1.61), we obtain that

$$|u(t,x)| \le C \langle\!\langle f \rangle\!\rangle_{\lambda}^{\vec{k}(\cdot,\vec{a})} \Psi_1(t,x) t^{1-\nu+\lambda/(2b)}, \quad (t,x) \in \Pi_{(0,T]}, \tag{2.2.72}$$

whence

$$\|u\|_{0,0,0}^{\vec{k}(\cdot,\vec{a})} \le C \langle\!\langle f \rangle\!\rangle_{\lambda}^{\vec{k}(\cdot,\vec{a})}. \tag{2.2.73}$$

Consider the difference $\Delta_{t,x}^{t',x'} u$. If $p^{2b} \ge t$, then by (2.2.72) we have

$$\begin{aligned} |\Delta_{t,x}^{t',x'} u(t,x)| \le\, & C \langle\!\langle f \rangle\!\rangle_{\lambda}^{\vec{k}(\cdot,\vec{a})} (\Psi_1(t,x) + \Psi_1(t',x')) \\ & \times \begin{cases} p_0^{\gamma_0} t^{1-\nu-(\gamma_0-\lambda)/(2b)}, & 1-\nu-(\gamma_0-\lambda)/(2b) > 0, \\ p_0^{\lambda} t^{1-\nu}, & 1-\nu-(\gamma_0-\lambda)/(2b) < 0, \end{cases} \end{aligned}$$

whence

$$|\Delta_{t,x}^{t',x'} u(t,x)| \le C \langle\!\langle f \rangle\!\rangle_{\lambda}^{\vec{k}(\cdot,\vec{a})} p_0^{\alpha} (\Psi_1(t,x) + \Psi_1(t',x')), \tag{2.2.74}$$

where $\alpha = \gamma_0$ for $1-\nu-(\gamma_0-\lambda)/(2b) > 0$ and $\alpha = \lambda$ for $1-\nu-(\gamma_0-\lambda)/(2b) < 0$.

It remains to study the case $p_0^{2b} < t$. Let $t_1 := t - p_0^{2b}$. Using the equality (2.2.57), we write

$$\begin{aligned} \Delta_{t,x}^{t',x'} u(t,x) = & \int_0^{t_1} d\tau \int_{\mathbb{R}^n} \Delta_{t,x}^{t',x'} K_0(t-\tau, x-\xi) \Delta_\xi^x f(\tau,\xi) d\xi \\ & + \int_{t_1}^{t} d\tau \int_{\mathbb{R}^n} K_0(t-\tau, x-\xi) \Delta_\xi^x f(\tau,\xi) d\xi \\ & - \int_{t_1}^{t'} d\tau \int_{\mathbb{R}^n} K_0(t'-\tau, x'-\xi) \Delta_\xi^{x'} f(\tau,\xi) d\xi =: \sum_{j=1}^{3} I_j. \end{aligned}$$

Let us give an estimate for I_1. Since $(p(x;\xi))^\lambda \le (p(x';\xi))^\lambda + (p(x;x'))^\lambda$, $p_0^\lambda = (t-t_1)^{\lambda/(2b)} \le (t-\tau)^{\lambda/(2b)}$, and $(t'-\tau)^{\lambda/(2b)} \le (t'-t)^{\lambda/(2b)} + (t-\tau)^{\lambda/(2b)} \le p_0^\lambda + (t-\tau)^{\lambda/(2b)} \le 2(t-\tau)^{\lambda/(2b)}$, $\tau \in [0,t_1]$, we have

$$\begin{aligned} |I_1| \le\, & C \langle\!\langle f \rangle\!\rangle_{\lambda}^{\vec{k}(\cdot,\vec{a})} p_0^{\gamma_0} (\Psi_1(t,x) + \Psi_1(t',x')) \int_0^{t_1} (t-\tau)^{-\gamma_0/(2b)} \\ & \times ((t-\tau)^{-\nu+\lambda/(2b)} + (t'-\tau)^{-\nu+\lambda/(2b)} + p_0^\lambda (t'-\tau)^{-\nu}) d\tau \\ \le\, & C \langle\!\langle f \rangle\!\rangle_{\lambda}^{\vec{k}(\cdot,\vec{a})} p_0^{\gamma_0} (\Psi_1(t,x) + \Psi_1(t',x')) \int_0^{t_1} (t-\tau)^{-\nu-(\gamma_0-\lambda)/(2b)} d\tau. \end{aligned}$$

If $1-\nu-(\gamma_0-\lambda)/(2b)>0$, then

$$\int_0^{t_1}(t-\tau)^{-\nu-(\gamma_0-\lambda)/(2b)}d\tau \le Ct^{1-\nu-(\gamma_0-\lambda)/(2b)},$$

while for $1-\nu-(\gamma_0-\lambda)/(2b)<0$ we have

$$\int_0^{t_1}(t-\tau)^{-\nu-(\gamma_0-\lambda)/(2b)}d\tau \le Cp_0^{(1-\nu)/(2b)-\gamma_0+\lambda} \le Ct^{1-\nu}p_0^{-\gamma_0+\lambda},$$

so that we come to the estimate

$$|I_1| \le C\,\langle\!\langle f\rangle\!\rangle_\lambda^{\vec{k}(\cdot,\vec{a})}\, p_0^\alpha(\Psi_1(t,x)+\Psi_1(t',x')).$$

Next, for I_2 we get

$$\begin{aligned}|I_2| &\le C\,\langle\!\langle f\rangle\!\rangle_\lambda^{\vec{k}(\cdot,\vec{a})}\,\Psi_1(t,x)\int_{t_1}^{t}(t-\tau)^{-\nu+\lambda/(2b)}d\tau\\ &= C\,\langle\!\langle f\rangle\!\rangle_\lambda^{\vec{k}(\cdot,\vec{a})}\,\Psi_1(t,x)p_0^{2b(1-\nu)+\lambda}\\ &\le C\,\langle\!\langle f\rangle\!\rangle_\lambda^{\vec{k}(\cdot,\vec{a})}\,p_0^\alpha\Psi_1(t,x),\end{aligned}$$

while for I_3, in a similar way,

$$|I_3| \le C\,\langle\!\langle f\rangle\!\rangle_\lambda^{\vec{k}(\cdot,\vec{a})}\,p_0^\alpha\Psi_1(t',x').$$

The estimates (2.2.73), (2.2.74), and the estimates for I_j, $j\in\{1,2,3\}$, imply (2.2.65) and (2.2.66). □

Lemma 2.8. *Suppose that the kernel of the integral* (2.2.54) *is a function*

$$K(t,x;\tau,\xi) := K_0(t-\tau,x-\xi;t,x) := (t-\tau)^{-M-\nu}\Omega((x-\xi)_{t-\tau};t,x),$$
$$0\le\tau<t\le T,\quad \{x,\xi\}\subset\mathbb{R}^n,$$

where $\nu\in(0,1]$, *and* $\Omega(x;\beta,y)$, *as a function of* x, *satisfies the conditions* (2.2.57)–(2.2.59) *uniformly with respect to* $(\beta,y)\in\Pi_{[0,T]}$, *and also, with some constants* $C>0$, $c>0$, $\lambda\in(0,1)$, *the condition*

$$|\Delta_{\beta,y}^{\beta',y'}\Omega(x;\beta,y)| \le C(p(\beta,y;\beta',y'))^\lambda \exp\Big\{-c\sum_{j=1}^n|x_j|^{q_j}\Big\},$$
$$x\in\mathbb{R}^n,\quad \{(\beta,y),(\beta',y')\}\subset\Pi_{[0,T]}.$$

Then the assertion 2) *of Lemma 2.7 is valid.*

The proof is similar to that of Lemma 2.7, except the emergence of the additional term

$$\int_0^{t_1} d\tau \int_{\mathbb{R}^n} \Delta_{\beta,y}^{t',x'} K_0(t \quad \tau, x \quad \xi; \beta, y)\Big|_{(\beta,y)=(t,x)} \Delta_\xi^x f(\tau,\xi)d\xi,$$

in the course of estimating $\Delta_{t,x}^{t',x'} u$. It is treated as in (2.2.72).

Lemma 2.9. *The assertion* 1) *of Lemma* 2.7 *is valid for the integral* (2.2.54) *if its kernel satisfies, with some constants* $C > 0$, $c > 0$, $\nu \in (0,1]$, *and* $\gamma_0 \in (0,1]$, *the inequalities*

$$\begin{aligned} |K(t,x;\tau,\xi)| &\le C(t-\tau)^{-M-\nu} E_c^{(0)}(t,x;\tau,\xi), \\ |\Delta_{t,x}^{t',x'} K(t,x;\tau,\xi)| &\le C(p(t,x;t',x'))^{\gamma_0}((t-\tau)^{-M-\nu-\gamma_0/(2b)} \\ &\quad \times E_c^{(0)}(t,x;\tau,\xi) + (t'-\tau)^{-M-\nu-\gamma_0/(2b)} E_c^{(0)}(t',x';\tau,\xi)), \\ &0 \le \tau < t \le t' \le T, \quad \{x,x',\xi\} \subset \mathbb{R}^n. \end{aligned} \tag{2.2.75}$$

If, in addition, K *satisfies the conditions*

$$\begin{aligned} \left| \int_{\mathbb{R}^n} K(t,x;\tau,\xi)d\xi \right| &\le C(t-\tau)^{-\nu+\lambda/(2b)}, \\ \left| \Delta_{t,x}^{t',x'} \int_{\mathbb{R}^n} K(t,x;\tau,\xi)d\xi \right| &\le C(p(t,x;t',x'))^{\gamma_0}(t-\tau)^{-\nu-(\gamma_0-\lambda)/(2b)}, \\ 0 \le \tau < t \le t' \le T, &\quad \{x,x',\xi\} \subset \mathbb{R}^n, \end{aligned} \tag{2.2.76}$$

with some $\lambda \in (0,1)$, *then the assertion similar to the assertion* 2) *of Lemma* 2.7, *with the estimates*

$$\|u\|_{0,\gamma_0/(2b),\gamma_0}^{\vec{k}(\cdot,\vec{a})} \le C\|f\|_{0,0,\lambda}^{\vec{k}(\cdot,\vec{a})} \quad \text{and} \quad \|u\|_{0,\lambda/(2b),\lambda}^{\vec{k}(\cdot,\vec{a})} \le C\|f\|_{0,0,\lambda}^{\vec{k}(\cdot,\vec{a})},$$

instead of (2.2.65) *and* (2.2.66), *holds for the integral* (2.2.54).

The proof of Lemma 2.9 is based on the representation

$$u(t,x) = \int_0^t d\tau \int_{\mathbb{R}^n} K(t,x;\tau,\xi)\Delta_\xi^x f(\tau,\xi)d\xi + \int_0^t \left(\int_{\mathbb{R}^n} K(t,x;\tau,\xi)d\xi \right) f(\tau,x)d\tau.$$

A bound for the first summand is obtained like the bound for the corresponding integral from Lemma 2.7. The inequalities (2.2.75) yield the required estimates of the second summand.

Lemma 2.10. *Let u be the function defined by the formula (2.2.54) with a kernel K satisfying the conditions (2.2.55)–(2.2.59). If $1-\nu-\gamma_0/(2b)>0$ and $f\in\hat{C}_{0,0}^{\vec{k}(\cdot,\vec{a})}$, then $u\in\hat{C}_{0,\gamma_0}^{\vec{k}(\cdot,\vec{a})}$, and*

$$\widehat{\|u\|}_{0,\gamma_0}^{\vec{k}(\cdot,\vec{a})}\le C\widehat{\|f\|}_{0,0}^{\vec{k}(\cdot,\vec{a})}. \tag{2.2.77}$$

If $f\in\hat{C}_{0,\lambda}^{\vec{k}(\cdot,\vec{a})}$ and $1-\nu-(\gamma_0-\lambda)/(2b)>0$, then $u\in\hat{C}_{0,\gamma_0}^{\vec{k}(\cdot,\vec{a})}$ and

$$\widehat{\|u\|}_{0,\gamma_0}^{\vec{k}(\cdot,\vec{a})}\le C\int_0^T\langle f(t,\cdot)\rangle_\lambda^{\vec{k}(t,\vec{a})}dt. \tag{2.2.78}$$

Similar assertions are valid for the integral (2.2.54) with a kernel satisfying the conditions (2.2.75) and (2.2.76) with $t'=t$, if the estimate

$$\widehat{\|u\|}_{0,\gamma_0}^{\vec{k}(\cdot,\vec{a})}\le C\widehat{\|f\|}_{0,\lambda}^{\vec{k}(\cdot,\vec{a})}.$$

is substituted for the estimate (2.2.78).

Proof. We shall dwell only on the proof of the lemma for the case where K satisfies the conditions (2.2.55)–(2.2.59). Using the inequalities (2.2.4), (2.2.60), and the equality (2.1.46) with $1-\nu-\gamma_0/(2b)>0$, we find that

$$|u(t,x)|\le C\Psi_1(t,x)\int_0^t(t-\tau)^{-\nu}\|f(\tau,\cdot)\|_{0,0}^{\vec{k}(\tau,\vec{a})}d\tau,$$

$$|\Delta_x^{x'}u(t,x)|\le C(p(x;x'))^{\gamma_0}(\Psi_1(t,x)+\Psi_1(t,x'))$$
$$\times\int_0^t(t-\tau)^{-\nu-\gamma_0/(2b)}\|f(\tau,\cdot)\|_{0,0}^{\vec{k}(\tau,\vec{a})}d\tau,$$

which implies the inequalities for $t\in(0,T]$:

$$\|u(t,\cdot)\|_{0,0}^{\vec{k}(t,\vec{a})}\le C\int_0^t(t-\tau)^{-\nu}\|f(\tau,\cdot)\|_{0,0}^{\vec{k}(\tau,\vec{a})}d\tau,$$

$$\langle u(t,\cdot)\rangle_{\gamma_0}^{\vec{k}(t,\vec{a})}\le C\int_0^t(t-\tau)^{-\nu-\gamma_0/(2b)}\|f(\tau,\cdot)\|_{0,0}^{\vec{k}(\tau,\vec{a})}d\tau.$$

Integrating these inequalities in $t\in(0,T]$ and changing the order of integration we obtain

$$\widehat{\|u\|}_{0,0}^{\vec{k}(\cdot,\vec{a})}\le C\widehat{\|f\|}_{0,0}^{\vec{k}(\cdot,\vec{a})},\qquad\int_0^T\langle u(t,\cdot)\rangle_{\gamma_0}^{\vec{k}(t,\vec{a})}dt\le C\widehat{\|f\|}_{0,0}^{\vec{k}(\cdot,\vec{a})}$$

respectively. These inequalities imply the estimate (2.2.77).

Suppose that $1-\nu-(\gamma_0-\lambda)/(2b)>0$. Using the representation (2.2.71) as above, we come first to the inequality

$$|u(t,x)| \le C\Psi_1(t,x)\int_0^t (t-\tau)^{-\nu+\lambda/(2b)}\langle f(\tau,\cdot)\rangle_\lambda^{\vec{k}(\tau,\vec{a})}d\tau, \tag{2.2.79}$$

and then to the estimate

$$\widehat{\|u\|}_{0,0}^{\vec{k}(\cdot,\vec{a})} \le C\int_0^T \langle f(\tau,\cdot)\rangle_\lambda^{\vec{k}(\tau,\vec{a})}d\tau. \tag{2.2.80}$$

In order to obtain an estimate for $\Delta_x^{x'}u(t,x)$, we distinguish between two cases, $p(x;x')\ge t^{1/(2b)}$ and $p(x;x')<t^{1/(2b)}$.

In the first case, using the inequality (2.2.79) we get

$$\begin{aligned}|\Delta_x^{x'}u(t,x)| &\le C(p(x;x'))^{\gamma_0}(\Psi_1(t,x)+\Psi_1(t,x'))\\ &\quad\times\int_0^t (t-\tau)^{-\nu+\lambda/(2b)}t^{-\gamma_0/(2b)}\langle f(\tau,\cdot)\rangle_\lambda^{\vec{k}(\tau,\vec{a})}d\tau\\ &\le C(p(x;x'))^{\gamma_0}(\Psi_1(t,x)+\Psi_1(t,x'))\\ &\quad\times\int_0^t (t-\tau)^{-\nu-(\gamma_0-\lambda)/(2b)}\langle f(\tau,\cdot)\rangle_\lambda^{\vec{k}(\tau,\vec{a})}d\tau.\end{aligned} \tag{2.2.81}$$

For the second case, we use the representation

$$\begin{aligned}\Delta_x^{x'}u(t,x) &= \int_0^{t_1}d\tau\int_{\mathbb{R}^n}\Delta_x^{x'}K_0(t-\tau,x-\xi)\Delta_\xi^x f(\tau,\xi)d\xi\\ &\quad+\int_{t_1}^t d\tau\int_{\mathbb{R}^n}K_0(t-\tau,x-\xi)\Delta_\xi^x f(\tau,\xi)d\xi\\ &\quad-\int_{t_1}^t d\tau\int_{\mathbb{R}^n}K_0(t-\tau,x'-\xi)\Delta_\xi^{x'}f(\tau,\xi)d\xi,\end{aligned}$$

where $t_1:=t-(p(x;x'))^{2b}$. The integrals are estimated as above, which results in the inequality

$$\begin{aligned}|\Delta_x^{x'}u(t,x)| &\le C(p(x;x'))^{\gamma_0}(\Psi_1(t,x)+\Psi_1(t,x'))\\ &\quad\times\int_0^t (t-\tau)^{-\nu-(\gamma_0-\lambda)/(2b)}\langle f(\tau,\cdot)\rangle_\lambda^{\vec{k}(\tau,\vec{a})}d\tau.\end{aligned} \tag{2.2.82}$$

The inequalities (2.2.80)–(2.2.82) imply the estimate (2.2.78).

Remark. Since the conditions (2.2.55) and (2.2.76) were not used for $1 - \nu - \gamma_0/(2b) > 0$, the corresponding statements of Lemmas 2.7–2.10 remain valid also for $\nu = 0$.

2.2.4. Properties of volume potentials. Let us consider some properties of volume potentials generated by the FSCP Z for the equation (2.1.1), that is the integrals

$$u(t,x) := \int_0^t d\tau \int_{\mathbb{R}^n} Z(t,x;\tau,\xi) f(\tau,\xi) d\xi, \quad (t,x) \in \Pi_{(0,T]}. \tag{2.2.83}$$

We assume that the coefficients of the equation (2.1.1) satisfy the conditions A_{11} and A_{12}, or A_{11} and A_{13}. The norm and spaces used below were defined in Section 2.2.1.

The conditions used for the density f of the potential (2.2.83) are as follows.

(D_{11}) The function $f : \Pi_{(0,T]} \to \mathbb{C}_{N1}$ is continuous, locally Hölder continuous in x uniformly with respect to t, and for any $t \in (0,T]$ the expressions

$$\|f(t,\cdot)\|_{0,0}^{\vec{k}(t,\vec{a})}, \qquad F_0(t) := \int_0^t (t-\tau)^{-1+1/(2b)} \|f(\tau,\cdot)\|_{0,0}^{\vec{k}(\tau,\vec{a})} d\tau$$

are finite.

(D_{12p}) The function $f : \Pi_{(0,T]} \to \mathbb{C}_{N1}$ is continuous, locally Hölder continuous in x uniformly with respect to t, and for any $t \in (0,T]$ the expressions

$$\|f(t,\cdot)\|_p^{\vec{k}(t,\vec{a})}, \quad F_p(t) := \int_0^t (t-\tau)^{-1+1/(2b)} \|f(\tau,\cdot)\|_p^{\vec{k}(\tau,\vec{a})} d\tau,$$

where $p \in [1,\infty]$, are finite.

(D_{13}) $f \in C_{0,0,\gamma}^{\vec{k}(\cdot,\vec{a})}$;

(D_{14}) $f \in \hat{C}_{0,\gamma}^{\vec{k}(\cdot,\vec{a})}$.

Lemma 2.11. *Suppose that f satisfies one of the conditions D_{11}, D_{12p}, D_{13}, or D_{14}. Then the function u defined by the formula (2.2.83) has continuous derivatives $\partial_x^k u$, $\|k\| \le 2b$, $\partial_t u$, where the derivatives $\partial_x^k u$, $\|k\| < 2b$, can be obtained by the formal differentiation under the integrals while the derivatives $\partial_x^k u$, $\|k\| = 2b$, and $\partial_t u$ are calculated by the formulas*

$$\partial_x^k u(t,x) = \int_0^t d\tau \int_{\mathbb{R}^n} \partial_x^k Z(t,x;\tau,\xi) \Delta_\xi^x f(\tau,\xi) d\xi$$

$$+\int_0^t \left(\int_{\mathbb{R}^n} \partial_x^k Z(t,x;\tau,\xi)d\xi \right) f(\tau,x)d\tau, \quad \|k\| = 2b,$$

$$\partial_t u(t,x) = f(t,x) + \int_0^t d\tau \int_{\mathbb{R}^n} \partial_t Z(t,x;\tau,\xi)\Delta_\xi^x f(\tau,\xi)d\xi$$

$$+\int_0^t \left(\int_{\mathbb{R}^n} \partial_t Z(t,x;\tau,\xi)d\xi \right) f(\tau,x)d\tau, \quad (t,x) \in \Pi_{(0,T]}. \tag{2.2.84}$$

The proof is similar to the proof of Property 2.6 for the parametrix.

Lemma 2.12. *If the condition D_{11} or D_{12p} is satisfied, then the function* (2.2.83) *has, for any $t \in (0,T]$ and $l \in \mathbb{Z}_+^n$, $\|l\| < 2b$, the estimate*

$$\|\partial_x^l u(t,\cdot)\|_{0,0}^{\vec{k}(t,\vec{a})} \le CF_0(t) \tag{2.2.85}$$

or

$$\|\partial_x^l u(t,\cdot)\|_p^{\vec{k}(t,\vec{a})} \le CF_p(t). \tag{2.2.86}$$

respectively.

Proof. We use the inequalities

$$\left\| \int_{\mathbb{R}^n} \partial_x^l Z(t,\cdot;\tau,\xi) f(\tau,\xi)d\xi \right\|_{0,0}^{\vec{k}(t,\vec{a})} \le C(t-\tau)^{-\|l\|/(2b)} \|f(\tau,\cdot)\|_{0,0}^{\vec{k}(\tau,\vec{a})}, \tag{2.2.87}$$

$$\left\| \int_{\mathbb{R}^n} \partial_x^l Z(t,\cdot;\tau,\xi) f(\tau,\xi)d\xi \right\|_p^{\vec{k}(t,\vec{a})} \le C(t-\tau)^{-\|l\|/(2b)} \|f(\tau,\cdot)\|_p^{\vec{k}(\tau,\vec{a})}, \tag{2.2.88}$$

which are valid for $0 \le \tau < t \le T$ and can be proved with the use of (2.2.3) and (2.2.4), just as the inequalities (2.2.11) and (2.2.20).

Using (2.2.87) we have

$$\|\partial_x^l u(t,\cdot)\|_{0,0}^{\vec{k}(t,\vec{a})} \le \int_0^t \left\| \int_{\mathbb{R}^n} \partial_x^l Z(t,\cdot;\tau,\xi) f(\tau,\xi)d\xi \right\|_{0,0}^{\vec{k}(t,\vec{a})} d\tau$$

$$\le C \int_0^t (t-\tau)^{-\|l\|/(2b)} \|f(\tau,\cdot)\|_{0,0}^{\vec{k}(\tau,\vec{a})} d\tau \le CF_0(t), \quad t \in (0,T].$$

Similarly, from the Minkowski inequality and (2.2.88) we get

$$\|\partial_x^l u(t,\cdot)\|_p^{\vec{k}(t,\vec{a})} \leq \int_0^t \left\| \int_{\mathbb{R}^n} \partial_x^l Z(t,\cdot;\tau,\xi) f(\tau,\xi)d\xi \right\|_p^{\vec{k}(t,\vec{a})} d\tau$$
$$\leq C\int_0^t (t-\tau)^{-\|l\|/(2b)} \|f(\tau,\cdot)\|_p^{\vec{k}(\tau,\vec{a})} d\tau$$
$$\leq CF_p(t), \qquad t\in(0,T]. \qquad \square$$

Lemma 2.13. *Let the conditions* A_{11}*,* A_{13}*, and* D_{13} *be satisfied. Then the function* (2.2.83) *belongs to the space* $H^{\gamma,\gamma'}$ *with any* $\gamma'\in(0,\gamma)$*, and*

$$\|u\|_{H^{\gamma,\gamma'}} \leq C\|f\|_{0,0,\gamma}^{\vec{k}(\cdot,\vec{a})}.$$

Proof. By Lemma 2.11, the integral (2.2.83) and its derivatives are integrals of the form (2.2.54) with $K(t,x;\tau,\xi) = \partial_x^l Z(t,x;\tau,\xi)$, $\|l\|\leq 2b$. Now the assertion of Lemma 2.13 is a consequence of Lemma 2.9 and the remark from Section 2.2.3. Indeed, the conditions of the latter Lemma are satisfied for $\partial_x^l Z$ – the inequalities (2.1.82), (2.1.106)–(2.1.108) imply the inequalities (2.2.75) and (2.2.76) with $\gamma_0 = \lambda=\gamma$ for $\nu=\|l\|/(2b)<1$, and $\gamma_0=\gamma'<\gamma$, $\lambda=\gamma$ for $\nu=\|l\|/(2b)=1$. $\square$

Lemma 2.14. *If the conditions* A_{11}*,* A_{12}*, and* D_{14}*, are satisfied, then the function* u *from* (2.2.83) *belongs to the space* $\hat{H}^{\gamma,\gamma'}$ *with any* $\gamma'\in(0,\gamma)$*, and*

$$\|u\|_{\hat{H}^{\gamma,\gamma'}} \leq C\widehat{\|f\|_{0,\gamma}^{\vec{k}(\cdot,\vec{a})}}.$$

The proof is based on Lemma 2.10 and the fact that the inequalities (2.1.106)–(2.1.108) with $t'=t$ hold if the conditions A_{11} and A_{12} are satisfied.

Lemma 2.15. *Let* $G(\cdot,\cdot;\cdot,\cdot;\beta,y)$ *be the FSCP for the equation* (2.1.54) *satisfying the conditions stated in the remark from Section* 2.1.3. *If the condition* D_{13} *is satisfied, then the integral*

$$v(t,x;\beta,y) := \int_0^t d\tau \int_{\mathbb{R}^n} G(t,x;\tau,\xi;\beta,y)f(\tau,\xi)d\xi, \quad (t,x)\in\Pi_{(0,T]},$$

has the following property: the functions $v_l(t,x) := \partial_x^l G(t,x;\tau,\xi;\beta,y)|_{(\beta,y)=(t,x)}$*,* $(t,x)\in\Pi_{(0,T]}$*,* $\|l\|\leq 2b$*, belong to the space* $C_{0,\gamma/(2b),\gamma}^{\vec{k}(\cdot,\vec{a})}$*, and*

$$\sum_{\|l\|\leq 2b} \|v_l\|_{0,\gamma/(2b),\gamma}^{\vec{k}(\cdot,\vec{a})} \leq C\|f\|_{0,0,\gamma}^{\vec{k}(\cdot,\vec{a})}.$$

Proof. We use Lemma 2.8, with $K(t,x;\tau,\xi) = \partial_x^l G(t,x;\tau,\xi;\beta,y)|_{(\beta,y)=(t,x)}$. According to the remark from Section 2.1.3,

$$\partial_x^l G(t,x;\tau,\xi;\beta,y)|_{(\beta,y)=(t,x)} = (t-\tau)^{-M-\|l\|/(2b)}\Omega_l((x-\xi)_{t-\tau};t,x),$$
$$0 < \|l\| \le 2b,$$

where Ω_l satisfy the conditions of Lemma 2.8 with an arbitrary $\gamma_0 \in (0,1]$ and $\lambda = \gamma$. Applying that lemma with $\gamma_0 = \lambda = \gamma$ for $\|l\| < 2b$,and $\gamma_0 > \gamma$, $\lambda = \gamma$ for $\|l\| = 2b$, and taking into account the remark from Section 2.2.3, we come to the required result. □

2.2.5. Integral representations of solutions of the Cauchy problem. Here we shall deal with representations of solutions of the Cauchy problem for the equation (2.1.1) by Poisson integrals and volume potentials. In this section we assume the conditions A_{11}, A_{12}, and A_{14}.

Lemma 2.16. *Suppose that the function u belongs to the space $C_{0,0,0}^{\vec{k}(\cdot,\vec{a})}$, and that u is a solution on $\Pi_{(0,T]}$ of the equation (2.1.1) with $f \in \hat{C}_{0,0}^{\vec{k}(\cdot,\vec{a})}$. Then*

$$u(t,x) = \int_{\mathbb{R}^n} Z(t,x;0,\xi)u(0,\xi)d\xi$$
$$+ \int_0^t d\tau \int_{\mathbb{R}^n} Z(t,x;\tau,\xi)f(\tau,\xi)d\xi, \quad (t,x) \in \Pi_{(0,T]}. \tag{2.2.89}$$

Proof. We shall use the Green-Ostrogradsky formula (2.1.110). Denote $V_R := (0,T] \times B_R$, ζ a smooth enough function on $[0,\infty)$, such that $\zeta = 1$ on $[0,1/2]$, $\zeta = 0$ on $[3/4,\infty)$, and $\zeta' \le 0$, $\zeta_R(x) := \zeta(q(x;0)/R)$; (t,x) an arbitrary fixed point from $V_{R_0/4}$ where R_0 is an arbitrary positive number. Substitute in the formula (2.1.110) h, $t-\varepsilon$, τ, ξ, $u(\tau,\xi)$, and $v(\tau,\xi) = Z^*(\tau,\xi;t,x)\zeta_R(\xi)$ for t_1, t_2, θ, y, $u(\theta,y)$, and $v(\theta,y)$ respectively. Here $R \ge R_0$, $0 < h < t/2$, $0 < \varepsilon < t/2$, and u is a function satisfying the conditions of the lemma. Using properties of the function ζ_R, the equality (2.1.111), and the fact that $L_N u = f$, we obtain that

$$\int_{\mathbb{R}^n} Z(t,x;t-\varepsilon,\xi)\zeta_R(\xi)u(t-\varepsilon,\xi)d\xi$$
$$= \int_{\mathbb{R}^n} Z(t,x;h,\xi)\zeta_R(\xi)u(h,\xi)d\xi + \int_h^{t-\varepsilon} d\tau \int_{\mathbb{R}^n} Z(t,x;\tau,\xi)\zeta_R(\xi)f(\tau,\xi)d\xi$$
$$- \int_h^{t-\varepsilon} d\tau \int_{B_{3R/4}\setminus B_{R/2}} \overline{(L_N^*(Z^*(\tau,\xi;t,x)\zeta_R(\xi)))}' u(\tau,\xi)d\xi,$$

which results, after passing to the limit as $\varepsilon \to 0$, in the equality

$$\begin{aligned}u(t,x) &= \int\limits_{\mathbb{R}^n} Z(t,x;h,\xi)\zeta_R(\xi)u(h,\xi)d\xi + \int\limits_h^t d\tau \int\limits_{\mathbb{R}^n} Z(t,x;\tau,\xi)\zeta_R(\xi)f(\tau,\xi)d\xi \\ &\quad - \int\limits_h^t d\tau \int\limits_{B_{3R/4}\setminus B_{R/2}} (\overline{L_N^*(Z^*(\tau,\xi;t,x)\zeta_R(\xi))}' u(\tau,\xi))d\xi \\ &=: \sum_{j=1}^{3} I_j^{(R)}. \end{aligned} \tag{2.2.90}$$

Pass in (2.2.90) to the limit as $R \to \infty$. The integral $I_1^{(R)}$ tends to

$$I_1 := \int\limits_{\mathbb{R}^n} Z(t,x;h,\xi)u(h,\xi)d\xi.$$

Indeed, using the inequalities (2.1.82), (2.2.4), and the equalities (2.1.46) we get

$$\begin{aligned}|I_1 - I_1^{(R)}| &= \left| \int\limits_{\mathbb{R}^n} Z(t,x;h,\xi)(1-\zeta_R(\xi))u(h,\xi)d\xi \right| \\ &\le C \int\limits_{\mathbb{R}^n\setminus B_{R/2}} (t-h)^{-M} E_{c-c_0}^{(0)}(t,x;h,\xi)(E_{c_0}^{(0)}(t,x;h,\xi)\Psi_1(h,\xi)) \\ &\quad \times (|u(h,\xi)|\Psi_{-1}(h,\xi))d\xi \\ &\le C\|u(h,\cdot)\|_{0,0}^{\vec{k}(h,\vec{a})}\Psi_1(t,x) \\ &\quad \times \exp\{-c_1(R/4)^{q'}(t-h)^r\} \int\limits_{\mathbb{R}^n} (t-h)^{-M} E_{c_1}^{(0)}(t,x;h,\xi)d\xi \underset{R\to\infty}{\longrightarrow} 0,\end{aligned}$$

since for $x \in B_{R_0/4}$, $\xi \in \mathbb{R}^n \setminus B_{R/2}$, and $R \ge R_0$

$$E_{c_1}^{(0)}(t,x;h,\xi) \le \exp\{-c_1(R/4)^{q'}(t-h)^r\},$$

where $c_1 = (c-c_0)/2$, $r = 1-q''$ for $t-h \le 1$, and $r = 1-q'$, if $t-h > 1$. The proof of this inequality is the same as the one for (2.2.28).

Similarly we show that

$$I_2^{(R)} \underset{R\to\infty}{\longrightarrow} \int\limits_h^t d\tau \int\limits_{\mathbb{R}^n} Z(t,x;\tau,\xi)f(\tau,\xi)d\xi.$$

Let us prove now, that $I_3^{(R)} \underset{R\to\infty}{\longrightarrow} 0$. Since $L_N^* Z^* = 0$, the expression

$$L_N^*(Z^*(\tau,\xi;t,x)\zeta_R(\xi))$$

is a sum of products of expressions like $\partial_\xi^k \zeta_R(\xi)$ and $\partial_\xi^l(\bar{a}'_k(\tau,\xi)Z^*(\tau,\xi;t,x))$, $0 < \|k\| \le 2b$, $\|l\| < 2b$, with constant coefficients. Using the equality (2.1.111), the estimates (2.1.82), we find that for $R \ge 1$,

$$|L_N^*(Z^*(\tau,\xi;t,x)\zeta_R(\xi))| \le C(t-\tau)^{-M-1+1/(2b)} E_c^{(0)}(t,x;\tau,\xi),$$
$$\xi \in B_{3R/4} \setminus B_{R/2}.$$

Applying this estimate just as in the course of estimating $I_1 - I_1^{(R)}$, we have

$$|I_3^{(R)}| \le C\|u\|_{0,0,0}^{\vec{k}(\cdot,\vec{a})} \Psi_1(t,x)\exp\{-c_1(R/4)^{q'}(t-h)^r\}$$
$$\times \int_h^t (t-\tau)^{-1+1/(2b)} d\tau \int_{\mathbb{R}^n} (t-\tau)^{-M} E_{c_1}^{(0)}(t,x;\tau,\xi)d\xi \underset{R\to\infty}{\longrightarrow} 0.$$

Hence, after passing to the limit in (2.2.90), as $R \to \infty$, we obtain the equality

$$u(t,x) = \int_{\mathbb{R}^n} Z(t,x;h,\xi)u(h,\xi)d\xi + \int_h^t d\tau \int_{\mathbb{R}^n} Z(t,x;\tau,\xi)f(\tau,\xi)d\xi. \tag{2.2.91}$$

Now, if we pass to the limit in (2.2.91), as $h \to 0$, we come to the formula (2.2.89). □

Lemma 2.17. *Suppose that the solution u and the right-hand side f of the equation (2.1.1) satisfy the following conditions with some $p \in [1,\infty]$:*

1) *there exists such a constant $C > 0$ that for any $t \in (0,T]$ the inequality $\|u(t,\cdot)\|_p^{\vec{k}(t,\vec{a})} \le C$ holds;*
2) *u satisfies the initial condition, in the sense of the relations (2.2.21), if $p \in [1,\infty)$, or (2.2.22), if $p = \infty$, in which $\varphi \in L_p^{\vec{a}}$;*
3) *the function $f : \Pi_{(0,T]} \to \mathbb{C}_{N1}$ is continuous, and the integral*

$$\int_0^T \|f(t,\cdot)\|_p^{\vec{k}(t,\vec{a})} dt$$

is finite.

Then

$$u(t,x) = \int_{\mathbb{R}^n} Z(t,x;0,\xi)\varphi(\xi)d\xi + \int_0^t d\tau \int_{\mathbb{R}^n} Z(t,x;\tau,\xi)f(\tau,\xi)d\xi,$$
$$(t,x) \in \Pi_{(0,T]}. \tag{2.2.92}$$

Proof. As in Lemma 2.16, we prove, for a given solution, the formula (2.2.91). The representation (2.2.92) will be obtained if we pass to the limit in this formula, as $h \to 0$. In order to do that, it suffices to prove the relation

$$\lim_{h\to 0} \int_{\mathbb{R}^n} Z(t,x;h,\xi)u(h,\xi)d\xi = \int_{\mathbb{R}^n} Z(t,x;0,\xi)\varphi(\xi)d\xi \tag{2.2.93}$$

for any fixed point $(t,x) \in \Pi_{(0,T]}$.

We use the inequality

$$|\Delta_h^0 Z(t,x;h,\xi)| \le Cht^{-M-1}E_c^{(0)}(t,x;0,\xi) \tag{2.2.94}$$

valid for any $t \in (0,T]$, $\{x,\xi\} \subset \mathbb{R}^n$ and $h \in (0,t/2)$. This inequality is proved, with the use of the normality property of Z (2.1.111) and estimates of the FSCP Z^* for the adjoint equation (2.1.109), as follows:

$$\begin{aligned}|\Delta_h^0 Z(t,x;h,\xi)| &= \left|\int_0^h \partial_\tau Z(t,x;\tau,\xi)d\tau\right| \\ &\le C\int_0^h (t-\tau)^{-M-1}E_c^{(0)}(t,x;\tau,\xi)d\tau \\ &\le Cht^{-M-1}E_c^{(0)}(t,x;0,\xi),\end{aligned}$$

since $t \ge t-\tau \ge t-h \ge t/2$ for $\tau \in (0,h)$ and $h \in (0,t/2)$.

Using the inequalities (2.1.82) and (2.2.94), and also the inequality

$$\begin{aligned}E_c^{(0)}(t,x;0,\xi)\Psi_1(h,\xi) &= E_{c_0}^{(0)}(t+h,x;h,\xi)\Psi_1(h,\xi) \\ &\le \Psi_1(t+h,x),\end{aligned} \tag{2.2.95}$$

which is valid due to (2.2.4), we have

$$\begin{aligned}|\Delta_h| :=& \left|\int_{\mathbb{R}^n} Z(t,x;h,\xi)u(h,\xi)d\xi - \int_{\mathbb{R}^n} Z(t,x;0,\xi)\varphi(\xi)d\xi\right| \\ \le& \int_{\mathbb{R}^n} |\Delta_h^0 Z(t,x;h,\xi)||u(h,\xi)|d\xi + \int_{\mathbb{R}^n} |Z(t,x;0,\xi)||u(h,\xi)-\varphi(\xi)|d\xi \\ \le& Cht^{-M-1}\int_{\mathbb{R}^n} E_{c-c_0}^{(0)}(t,x;0,\xi)(E_{c_0}^{(0)}(t,x;0,\xi)\Psi_1(h,\xi)) \\ &\times (|u(h,\xi)|\Psi_{-1}(h,\xi))d\xi + Ct^{-M}\int_{\mathbb{R}^n} E_{c-c_0}^{(0)}(t,x;0,\xi)(E_{c_0}^{(0)}(t,x;0,\xi) \\ &\times \Psi_1(h,\xi))(|u(h,\xi)-\varphi(\xi)|\Psi_{-1}(h,\xi))d\xi \\ &+ C(ht^{-M-1}J_1^{(h)} + t^{-M}J_2^{(h)})\Psi_1(t+h,x)\end{aligned} \tag{2.2.96}$$

where

$$J_1^{(h)} := \int_{\mathbb{R}^n} E^{(0)}_{c-c_0}(t,x;0,\xi)(|u(h,\xi)|\Psi_{-1}(h,\xi))d\xi,$$

$$J_2^{(h)} := \int_{\mathbb{R}^n} E^{(0)}_{c-c_0}(t,x;0,\xi)(|u(h,\xi) - \varphi(\xi)|\Psi_{-1}(h,\xi))d\xi.$$

For $p = 1$ we have

$$J_1^{(h)} \le \|u(h,\cdot)\|_1^{\vec{k}(h,\vec{a})}, \qquad J_2^{(h)} \le \|u(h,\cdot) - \varphi(\cdot)\|_1^{\vec{k}(h,\vec{a})}, \tag{2.2.97}$$

whereas for $p \in (1,\infty)$ we get by the Hölder inequality that

$$\begin{aligned} J_1^{(h)} &\le \left(\int_{\mathbb{R}^n} E^{(0)}_{(c-c_0)p'}(t,x;0,\xi)t^{-M}d\xi\right)^{1/p'} t^{M/p'}\|u(h,\cdot)\|_p^{\vec{k}(h,\vec{a})} \\ &= Ct^{M/p'}\|u(h,\cdot)\|_p^{\vec{k}(h,\vec{a})}, \\ J_2^{(h)} &\le Ct^{M/p'}\|u(h,\cdot) - \varphi(\cdot)\|_p^{\vec{k}(h,\vec{a})}. \end{aligned} \tag{2.2.98}$$

By our assumptions for $p \in [1,\infty)$, the relation (2.2.93) follows from the inequalities (2.2.96)–(2.2.98).

This relation is valid also for $p = \infty$. Indeed, write the difference Δ_h as

$$\begin{aligned} \Delta_h &= \int_{\mathbb{R}^n} \Delta_h^0 Z(t,x;h,\xi)u(h,\xi)d\xi + \left(\int_{\mathbb{R}^n} Z(t,x;0,\xi)u(h,\xi)d\xi \right. \\ &\quad \left. - \int_{\mathbb{R}^n} Z(t,x;0,\xi)\varphi(\xi)d\xi\right) \\ &=: K_1^{(h)} + K_2^{(h)}. \end{aligned} \tag{2.2.99}$$

Just as in (2.2.96), we have

$$|K_1^{(h)}| \le Cht^{-M-1}\Psi_1(t+h,x)J_1^{(h)}, \tag{2.2.100}$$

and by the condition 1) of the lemma,

$$J_1^{(h)} \le Ct^M\|u(h,\cdot)\|_\infty^{\vec{k}(h,\vec{a})} \le Ct^M, \quad h > 0,$$

so that

$$\lim_{h\to 0} K_1^{(h)} = 0. \tag{2.2.101}$$

Since the function $\psi(\xi) := \bar{Z}'(t,x;0,\xi)$, $\xi \in \mathbb{R}^n$, due to the estimate (2.1.82) and the inequality (2.2.95) with $h = T$, is such that

$$|\psi(\xi)|\Psi_1(T,\xi) \le Ct^{-M}E^{(0)}_{c-c_0}(t,x;0,\xi)\Psi_1(t+T,x), \tag{2.2.102}$$

hence $\psi \in L_1^{-\vec{k}(T,\vec{a})}$ (in fact, each column of the matrix belongs to this space, if $N > 1$), it follows from (2.2.22) that

$$\lim_{h\to 0} K_2^{(h)} = 0.$$

Together with (2.2.99), (2.2.101), this implies the relation (2.2.93) for $p = \infty$. □

Lemma 2.18. *If the solution u and the right-hand side f of the equation* (2.1.1) *satisfy the conditions* 1) *and* 3) *of Lemma* 2.17 *with $p = 1$, and also the relation* (2.2.47)*, with $\mu \in M^{\vec{a}}$, holds, then*

$$u(t,x) = \int_{\mathbb{R}^n} Z(t,x;0,\xi)d\mu(\xi) + \int_0^t d\tau \int_{\mathbb{R}^n} Z(t,x;\tau,\xi)f(\tau,\xi)d\xi,$$
$$(t,x) \in \Pi_{(0,T]}. \tag{2.2.103}$$

Proof. Just as in Lemma 2.17, the solution can be represented by the formula (2.2.91). If we pass to the limit there, as $h \to 0$, we obtain (2.2.103). In order to substantiate that, it is sufficient to prove the relation

$$\lim_{h\to 0} \int_{\mathbb{R}^n} Z(t,x;h,\xi)u(h,\xi)d\xi = \int_{\mathbb{R}^n} Z(t,x;0,\xi)d\mu(\xi). \tag{2.2.104}$$

Consider the difference

$$\int_{\mathbb{R}^n} Z(t,x;h,\xi)u(h,\xi)d\xi - \int_{\mathbb{R}^n} Z(t,x;0,\xi)d\mu(\xi) = K_1^{(h)} + \tilde{K}_2^{(h)}, \tag{2.2.105}$$

where $K_1^{(h)}$ is from (2.2.99), and

$$\tilde{K}_2^{(h)} := \int_{\mathbb{R}^n} Z(t,x;0,\xi)u(h,\xi)d\xi - \int_{\mathbb{R}^n} Z(t,x;0,\xi)d\mu(\xi).$$

For $K_1^{(h)}$ we have the inequality (2.2.100), in which $J_1^{(h)}$ satisfies the estimate (2.2.97). By the condition 1) of Lemma 2.17 we get

$$\lim_{h\to 0} K_1^{(h)} = 0. \tag{2.2.106}$$

It follows from the inequality (2.2.102) that the function ψ from Lemma 2.17 (or each column of the matrix ψ, if $N > 1$) belongs to the space $C_0^{-\vec{k}(T,\vec{a})}$. Therefore, due to (2.2.47),

$$\lim_{h\to 0} \tilde{K}_2^{(h)} = 0. \tag{2.2.107}$$

Now (2.2.105)–(2.2.107) imply the relation (2.2.104). □

Remark. Lemmas 2.16–2.18 obviously imply results regarding the uniqueness of a solution of the Cauchy problem for an equation (2.1.1) possessing the Lagrange adjoint equation. Using them, in Section 2.2.6 we shall establish the uniqueness of a solution of the Cauchy problem under the only assumption of the Hölder continuity of the coefficient. Of course, the uniqueness classes will be more narrow.

2.2.6. Correct solvability of the Cauchy problem. Below we give theorems on the correct solvability of the Cauchy problem for the equation (2.1.1), which are based on results from Sections 2.2.2–2.2.5.

Theorem 2.5. *Suppose that the conditions A_{11}, A_{12}, A_{14} for the coefficients of the equation* (2.1.1), *and the condition D_{11} for f hold, and $\varphi \in C^{\vec{a}}_{0,0}$. Then the formula* (2.2.92) *determines the unique solution of the equation* (2.1.1), *which satisfies the following conditions:*

1) *there exists such a constant $C > 0$ that for any $t \in (0,T]$ and $l \in \mathbb{Z}^n_+$, $\|l\| < 2b$,*
$$\|\partial^l_x u(t,\cdot)\|^{\vec{k}(t,\vec{a})}_{0,0} \le C(\|\varphi\|^{\vec{a}}_{0,0} t^{-\|l\|/(2b)} + F_0(t)),$$
where the function F_0 appears in the condition D_{11};
2) *for any compact set $K \subset \mathbb{R}^n$ $u(t,x) \underset{t\to 0}{\longrightarrow} \varphi(x)$ uniformly with respect to $x \in K$.*

Proof. The assertions that the Poisson integral (2.2.9) and the volume potential (2.2.83) are solutions of the homogeneous and inhomogeneous equations respectively, are consequences of the remark from Section 2.2.2, Lemma 2.11, and the fact that the FSCP $Z(\cdot,\cdot;\tau,\xi)$ is a solution of the homogeneous equation. The statement 1) of the theorem follows from Lemmas 2.4 and 2.12. The statement 2) for the Poisson integral was proved in Lemma 2.4. It follows from the estimate (2.2.85) for the volume potential u, that $u(t,x) \underset{t\to 0}{\longrightarrow} 0$ uniformly with respect to $x \in K$. The uniqueness of a solution from the class under consideration follows from Lemma 2.16. □

Theorem 2.6. *If the conditions A_{11}, A_{12}, A_{14} for the coefficients of the equation* (2.1.1), *and the condition D_{12p} for f hold, and $\varphi \in L^{\vec{a}}_p$ with some $p \in [1,\infty]$, then the formula* (2.2.92) *determines the unique solution of the equation* (2.1.1) *satisfying the following conditions:*

1) *there exists such a constant $C > 0$ that for any $t \in (0,T]$ and $l \in \mathbb{Z}^n_+$, $\|l\| < 2b$,*
$$\|\partial^l_x u(t,\cdot)\|^{\vec{k}(t,\vec{a})}_p \le C(\|\varphi\|^{\vec{a}}_p t^{-\|l\|/(2b)} + F_p(t)),$$
where the function F_p appears in the condition D_{12p};
2) *if $p \in [1,\infty)$, then*
$$\lim_{t\to 0} \|u(t,\cdot) - \varphi(\cdot)\|^{\vec{k}(t,\vec{a})}_p = 0,$$
whereas for $p = \infty$,
$$u(t,\cdot) \underset{t\to 0}{\overset{w}{\longrightarrow}} \varphi(\cdot),$$
that is the relation (2.2.22) *holds.*

Proof. The assertions of Theorem 2.6 follow from Lemmas 2.5, 2.11, 2.12 and 2.17, just as the corresponding assertions of Theorem 2.3 are the consequences of Lemmas 2.4, 2.11, 2.12, and 2.16. □

Theorem 2.7. *Suppose that the conditions A_{11}, A_{12}, A_{14} for the coefficients of the equation* (2.1.1) *and the condition D_{121} for f hold, and the generalized measure μ belongs to $M^{\vec{a}}$. Then the formula* (2.2.103) *determines the unique solution of the equation* (2.1.1) *satisfying the following properties:*

1) *there exists such a constant $C > 0$ that for any $t \in (0, T]$ and $l \in \mathbb{Z}_+^n$, $\|l\| < 2b$,*
$$\|\partial_x^l u(t,\cdot)\|_1^{\vec{k}(t,\vec{a})} \le C(\|\mu\|^{\vec{a}} t^{-\|l\|/(2b)} + F_1(t));$$
2) $u(t,\cdot) \xrightarrow[t\to 0]{w} \mu$, *that is the relation* (2.2.47) *holds.*

Proof. The assertions of the theorem are proved with the use of Lemmas 2.6, 2.11, and 2.18 similarly to the proofs of the preceding theorems. □

The following theorems will deal with the Cauchy problem for the equation (2.1.1) with the initial condition

$$u(t,x)|_{t=0} = \varphi(x), \quad x \in \mathbb{R}^n, \tag{2.2.108}$$

to be satisfied at least as in Theorem 2.3.

Theorem 2.8. *If the coefficients of the equation* (2.1.1) *satisfy the conditions A_{11} and A_{13}, $\varphi \in C_{2b,\gamma}^{\vec{a}}$ and $f \in C_{0,0,\gamma}^{\vec{k}(\cdot,\vec{a})}$, then the problem* (2.1.1), (2.2.108) *has the unique solution $u \in C_{2b,\gamma/(2b),\gamma}^{\vec{k}(\cdot,\vec{a})}$, satisfying the estimate*

$$\|u\|_{2b,\gamma/(2b),\gamma}^{\vec{k}(\cdot,\vec{a})} \le C(\|\varphi\|_{2b,\gamma}^{\vec{a}} + \|f\|_{0,0,\gamma}^{\vec{k}(\cdot,\vec{a})}). \tag{2.2.109}$$

Proof. By the change of variables $u = \varphi + v$ the problem (2.1.1), (2.2.108) is reduced to the problem

$$\begin{aligned} L_N(t,x,\partial_t,\partial_x)v(t,x) = g(t,x), \quad (t,x) \in \Pi_{(0,T]}, \\ v(t,x)|_{t=0} = 0, \quad x \in \mathbb{R}^n, \end{aligned} \tag{2.2.110}$$

where
$$g(t,x) := f(t,x) + \sum_{\|k\|\le 2b} a_k(t,x)\partial_x^k \varphi(x).$$

Due to our assumptions, $g \in C_{0,0,\gamma}^{\vec{k}(\cdot,\vec{a})}$. Therefore, by Lemmas 2.11 and 2.13 the function

$$v(t,x) := \int_0^t d\tau \int_{\mathbb{R}^n} Z(t,x;\tau,\xi) g(\tau,\xi) d\xi, \quad (t,x) \in \Pi_{(0,T]},$$

is a solution of the problem (2.2.110) from the space $H^{\gamma,\gamma'}$ with any $\gamma' < \gamma$, and

$$\|v\|_{H^{\gamma,\gamma'}} \le C\|g\|^{\vec{k}(\cdot,\vec{a})}_{0,0,\gamma}.$$

Hence, the function $u := \varphi + v$ is a solution of the problem (2.1.1), (2.2.108), it belongs to the space $H^{\gamma,\gamma'}$, and

$$\|u\|_{H^{\gamma,\gamma'}} \le C(\|\varphi\|^{\vec{a}}_{2b,\gamma} + \|f\|^{\vec{k}(\cdot,\vec{a})}_{0,0,\gamma}). \tag{2.2.111}$$

To complete the proof, we have to show that each solution of the Cauchy problem (2.1.1), (2.2.108) belonging to the space $H^{\gamma,\gamma'}$ and satisfying the inequality (2.2.111), belongs to the space $H^{\gamma,\gamma} = C^{\vec{k}(\cdot,\vec{a})}_{2b,\gamma/(2b),\gamma}$, and satisfies the estimate (2.2.109). In order to do that, it suffices to prove that for such a solution the derivatives $\partial^l_x u$, $\|l\| = 2b$, belong to the space $C^{\vec{k}(\cdot,\vec{a})}_{0,\gamma/(2b),\gamma}$, and

$$\|\partial^l_x u\|^{\vec{k}(\cdot,\vec{a})}_{0,\gamma/(2b),\gamma} \le C(\|\varphi\|^{\vec{a}}_{2b,\gamma} + \|f\|^{\vec{k}(\cdot,\vec{a})}_{0,0,\gamma}). \tag{2.2.112}$$

We substitute a solution u with the above properties into the equation (2.1.1), and write the resulting identity as

$$\partial_t v(t,x) - \sum_{\|k\|=2b} a_k(\beta,y)\partial^k_x v(t,x) = f_1(t,x;\beta,y) + f_2(t,x;\beta,y),$$
$$(t,x) \in \Pi_{(0,T]}, \tag{2.2.113}$$

where (β,y) is an arbitrary fixed point of the layer $\Pi_{[0,T]}$,

$$v := u - \varphi,$$
$$f_1(t,x;\beta,y) := f(t,x) + \sum_{\|k\|<2b} a_k(t,x)\partial^k_x u(t,x) + \sum_{\|k\|=2b} a_k(\beta,y)\partial^k_x\varphi(x),$$
$$f_2(t,x;\beta,y) := \sum_{\|k\|=2b} \Delta^{\beta,y}_{t,x} a_k(t,x)\partial^k_x u(t,x).$$

Since the coefficients of (2.2.113) do not depend on x, so that this equation possesses an adjoint one, and $v|_{t=0} = 0$, we have by Lemma 2.16 that

$$u(t,x) = \varphi(x) + \int_0^t d\tau \int_{\mathbb{R}^n} G(t,x;\tau,\xi;\beta,y) f_1(\tau,\xi;\beta,y)d\xi$$
$$+ \int_0^t d\tau \int_{\mathbb{R}^n} G(t,x;\tau,\xi;\beta,y) f_2(\tau,\xi;\beta,y)d\xi, \quad (t,x) \in \Pi_{(0,T]}, \tag{2.2.114}$$

where $G(\cdot,\cdot;\cdot,\cdot;\beta,y)$ is the FSCP for the equation $\partial_t v = \sum_{\|k\|=2b} a_k(\beta,y)\partial^k_x v$.

According to Remark 1 to Property 2.6 from Section 2.1.4, the first integral from (2.2.114) has the derivatives of the form ∂_x^l, $\|l\| = 2b$, at every point $(t,x) \in \Pi_{(0,T]}$, for any fixed point (β, y). It is easy to see, repeating the proof of Property 2.6, that

$$\partial_x^l \int_0^t d\tau \int_{\mathbb{R}^n} G(t,x;\tau,\xi;\beta,y) f_2(\tau,\xi;\beta,y)d\xi \Bigg|_{(t,x)=(\beta,y)}$$
$$= \int_0^t d\tau \int_{\mathbb{R}^n} \partial_x^l G(\beta,x;\tau,\xi;\beta,y)\Big|_{x=y} f_2(\tau,\xi;\beta,y)d\xi$$

so that we obtain a representation

$$\partial_x^l u(t,x) = \partial_x^l \varphi(x) + \partial_x^l \int_0^t d\tau \int_{\mathbb{R}^n} G(t,x;\tau,\xi;\beta,y) f_1(\tau,\xi;\beta,y)d\xi \Bigg|_{(\beta,y)=(t,x)}$$
$$+ \sum_{\|k\|=2b} \int_0^t d\tau \int_{\mathbb{R}^n} \partial_x^l G(t,x;\tau,\xi;t,y)\Big|_{y=x} \Delta_{\tau,\xi}^{t,x} a_k(\tau,\xi) \partial_\xi^k u(\tau,\xi)d\xi,$$
$$(t,x) \in \Pi_{(0,T]}, \quad \|l\| = 2b.$$

We apply Lemma 2.15 to the second summand, and Lemma 2.9, with $K(t,x;\tau,\xi) = \partial_x^l G(t,x;\tau,\xi;t,y)|_{y=x} \Delta_{\tau,\xi}^{t,x} a_k(\tau,\xi)$, $f(t,x) = \partial_x^k u(t,x)$, $\nu = 1 - \gamma/(2b)$, $\gamma_0 = \gamma$, $\lambda = \gamma'$, to the integrals from the third summand. Using the assumptions regarding the solution u, we find that $\partial_x^l u \in C_{0,\gamma/(2b),\gamma'}^{\vec{k}(\cdot,\vec{a})}$, and the estimate (2.2.112) holds. □

Theorem 2.9. *Suppose that the coefficients of the equation* (2.1.1) *satisfy the conditions* A_{11} *and* A_{12}, $\varphi \in C_{0,\gamma}^{\vec{a}}$, *and* $f \in \hat{C}_{0,\gamma}^{\vec{k}(\cdot,\vec{a})}$. *Then the formula* (2.2.92) *determines the unique solution of the Cauchy problem* (2.1.1), (2.2.108), *belonging to the space* $\hat{H}^{\gamma,\gamma'}$ *with any* $\gamma' \in (0,\gamma)$, *and satisfying the estimate*

$$\|u\|_{\hat{H}^{\gamma,\gamma'}} \le C(\|\varphi\|_{0,\gamma}^{\vec{a}} + \widehat{\|f\|}_{0,\gamma}^{\vec{k}(\cdot,\vec{a})}). \tag{2.2.115}$$

Proof. Denote the first integral from (2.2.92) by u_1, and the second integral by u_2. According to the remark from Section 2.2.2, Lemmas 2.4 and 2.11, the function (2.2.92) is a solution of the problem (2.1.1), (2.2.108). By Lemma 2.14, $u_2 \in H^{\gamma,\gamma'}$ and

$$\|u_2\|_{\hat{H}^{\gamma,\gamma'}} \le C\widehat{\|f\|}_{0,\gamma}^{\vec{k}(\cdot,\vec{a})}. \tag{2.2.116}$$

We write the derivatives of u_1 as

$$\partial_x^l u_1(t,x) = \int\limits_{\mathbb{R}^n} \partial_x^l Z(t,x;0,\xi)\Delta_\xi^x \varphi(\xi)d\xi + \left(\int\limits_{\mathbb{R}^n} \partial_x^l Z(t,x;0,\xi)d\xi\right)\varphi(x),$$
$$(t,x) \in \Pi_{(0,T]}.$$

Using the estimates (2.1.106)–(2.1.108) and the method of proof of Lemmas 2.4 and 2.10, we get

$$\|\partial_x^l u_1(t,\cdot)\|_{0,\gamma}^{\vec{k}(t,\vec{a})} \le C\|\varphi\|_{0,\gamma}^{\vec{a}} t^{-\|l\|/(2b)}, \quad \|l\| < 2b,$$
$$\|\partial_x^l u_1(t,\cdot)\|_{0,\gamma'}^{\vec{k}(t,\vec{a})} \le C\|\varphi\|_{0,\gamma}^{\vec{a}} t^{-1+\gamma-\gamma'}, \quad \|l\| = 2b,$$

which implies the estimate $\|u_1\|_{\hat{H}^{\gamma,\gamma'}} \le C\|\varphi\|_{0,\gamma}^{\vec{a}}$, and, together with (2.2.116), also the estimate (2.2.115).

The uniqueness of the solution is a consequence of the next theorem. □

Theorem 2.10. *If the coefficients of the equation* (2.1.1) *satisfy the conditions* A_{11} *and* A_{12}, *then any solution of the problem* (2.1.1), (2.2.108) *with* $f = 0$ *and* $\varphi = 0$ *belonging to the space* $\hat{C}_{2b,0}^{\vec{k}(\cdot,\vec{a})}$, *equals zero identically.*

Proof. Let y be a fixed point in $\mathbb{R}^n$. Substituting the solution u into the equation (2.1.1) we obtain the equality

$$\partial_t u(t,x) - \sum_{\|k\|\le 2b} a_k(t,y)\partial_x^k u(t,x) = f(t,x;y), \quad (t,x) \in \Pi_{(0,T]},$$

where $f(t,x;y) := \sum_{\|k\|\le 2b} \Delta_x^y a_k(t,x)\partial_x^k u(t,x)$. By Lemma 2.16, we have the representation

$$u(t,x) = \int\limits_0^t d\tau \int\limits_{\mathbb{R}^n} G(t,x;\tau,\xi;y)f(\tau,\xi;y)d\xi, \quad (t,x) \in \Pi_{(0,T]},$$

where $G(\cdot,\cdot;\cdot,\cdot;y)$ is the FSCP for the equation (2.1.48).

By virtue of the Remark 1 to Property 2.6 from Section 2.1.4, lower derivatives of u satisfy the formulas

$$\partial_x^l u(t,x) = \int\limits_0^t d\tau \int\limits_{\mathbb{R}^n} \partial_x^l G(t,x;\tau,\xi;y)f(\tau,\xi;y)d\xi,$$
$$(t,x) \in \Pi_{(0,T]}, \quad \|l\| < 2b, \tag{2.2.117}$$

for any fixed point $y \in \mathbb{R}^n$. We cannot use the formulas for the leading derivatives from the above remark, since they require the Hölder continuity in x of the function f; meanwhile f contains the derivatives of u, whose Hölder continuity is not assumed. However, as in the proof of Theorem 2.8, it is easy to see that for $x = y$ there exist the leading derivatives of the integral from (2.2.117), which are determined by the formulas

$$\partial_x^l u(t,x)|_{x=y} = \int_0^t d\tau \int_{\mathbb{R}^n} \partial_x^l G(t,x;\tau,\xi;y)|_{x=y} f(\tau,\xi;y)d\xi, \quad \|l\| = 2b. \quad (2.2.118)$$

Since the equalities (2.2.117) and (2.2.118) are valid for any $y \in \mathbb{R}^n$, substituting there x for y we obtain the relations

$$\partial_x^l u(t,x) = \sum_{\|k\| \le 2b} \int_0^t d\tau \int_{\mathbb{R}^n} \partial_x^l G(t,x;\tau,\xi;y)\Big|_{y=x} \Delta_\xi^x a_k(\tau,\xi) \partial_\xi^k u(\tau,\xi) d\xi,$$
$$(t,x) \in \Pi_{(0,T]},$$

$\|l\| \le 2b$. Using these relations and performing the usual estimates carried out, for example, in the proof of Lemma 2.10, we come to the inequality

$$W(t) \le C \int_0^t (t-\tau)^{-1+\gamma/(2b)} W(\tau) d\tau$$

where

$$W(t) := \sum_{\|l\| \le 2b} \|\partial_x^l u(t,\cdot)\|_{0,0}^{\vec{k}(t,\vec{a})}.$$

Integrating we obtain that

$$\begin{aligned}\int_0^t W(\beta)d\beta &\le C \int_0^t \left(\int_0^\beta (\beta-\tau)^{-1+\gamma/(2b)} W(\tau)d\tau \right) d\beta \\ &= C \int_0^t \left(\int_\tau^t (\beta-\tau)^{-1+\gamma/(2b)} d\beta \right) W(\tau)d\tau \\ &= C_0 \int_0^t (t-\tau)^{\gamma/(2b)} W(\tau)d\tau \\ &\le C_0 t^{\gamma/(2b)} \int_0^t W(\tau)d\tau.\end{aligned}$$

It follows from this inequality that $\int\limits_0^t W(\tau)d\tau = 0$ for all $t \in (0, t_0]$, if t_0 satisfies the condition $C_0 t_0^{\gamma/(2b)} < 1$. Repeating, if necessary, these arguments we find that $\int\limits_0^t W(\tau)d\tau = 0$ for all $t \in (0, T]$. Hence $u = 0$ on $\Pi_{[0,T]}$. □

2.2.7. Integral representations and the sets of initial values for solutions defined on an open layer. The next theorem is, in a certain sense, converse to Theorems 2.6 and 2.7. Its analog, for the case of a homogeneous equation parabolic in the sense of Petrovsky, is the second part of Theorem 1.1.

Theorem 2.11. *Let the coefficients of the equation* (2.1.1) *satisfy the conditions* A_{11}, A_{12}, *and* A_{14}, *while its right-hand side* f *satisfies the condition* D_{12p} *with some* $p \in [1, \infty]$. *Let* u *be a solution of the equation* (2.1.1) *defined on the layer* $\Pi_{(0,T]}$ *and satisfying the condition*

$$\|u(t, \cdot)\|_p^{\vec{k}(t,\vec{a})} \leq C, \quad t \in (0, T], \tag{2.2.119}$$

with some constant $C > 0$. *Then, for* $p \in (1, \infty]$, *there exists the unique function* $\varphi \in L_p^{\vec{a}}$, *or, for* $p = 1$, *the unique generalized measure* $\mu \in M^{\vec{a}}$, *such that the solution* u *admits the representation* (2.2.92) *and* (2.2.103) *respectively.*

Proof. Set

$$v(t, x) := u(t, x) - \int\limits_0^t d\tau \int\limits_{\mathbb{R}^n} Z(t, x; \tau, \xi) f(\tau, \xi) d\xi, \quad (t, x) \in \Pi_{(0,T]}.$$

It follows from the condition (2.2.119) and Lemma 2.12 that

$$\|v(t, \cdot)\|_p^{\vec{k}(t,\vec{a})} \leq C, \quad t \in (0, T], \tag{2.2.120}$$

and v is a solution on $\Pi_{(0,T]}$ of the homogeneous equation (2.1.1).

Thus, in order to prove the theorem, it suffices to prove the following: let v be a solution of the homogeneous equation (2.1.1) satisfying the condition (2.2.120); then, for $p \in (1, \infty]$, there exists the unique function $\varphi \in L_p^{\vec{a}}$, and, for $p = 1$, there exists the unique generalized measure $\mu \in M^{\vec{a}}$, such that the solution v is represented as the Poisson integral of the function φ,

$$v(t, x) = \int\limits_{\mathbb{R}^n} Z(t, x; 0, \xi) \varphi(\xi) d\xi, \quad (t, x) \in \Pi_{(0,T]}, \tag{2.2.121}$$

and the generalized measure μ,

$$v(t, x) = \int\limits_{\mathbb{R}^n} Z(t, x; 0, \xi) d\mu(\xi), \quad (t, x) \in \Pi_{(0,T]}. \tag{2.2.122}$$

respectively.

Suppose first that $p \in (1, \infty]$. It follows from the condition (2.2.120) that the sequence of functions

$$\{v(1/\nu, x)\Psi_{-1}(1/\nu, x), x \in \mathbb{R}^n : \nu \geq 1\} \tag{2.2.123}$$

is bounded in the space $L_p(\mathbb{R}^n)$. This space is isometric to the conjugate of the space $L_{p'}(\mathbb{R}^n)$, $p' = p/(p-1)$. By the theorem of compactness of a bounded subset of a conjugate space, the sequence (2.2.123) is weakly compact in $L_p(\mathbb{R}^n)$. Therefore it has a subsequence

$$\{v(1/\nu(r), x)\Psi_{-1}(1/\nu(r), x), x \in \mathbb{R}^n : r \geq 1\} \tag{2.2.124}$$

weakly converging to some function $\chi \in L_p(\mathbb{R}^n)$, that is

$$\lim_{r\to\infty} \int\limits_{\mathbb{R}^n} \bar{\psi}'(\xi)\Psi_{-1}(1/\nu(r), \xi)v(1/\nu(r), \xi)d\xi = \int\limits_{\mathbb{R}^n} \bar{\psi}'(\xi)\chi(\xi)d\xi \tag{2.2.125}$$

for each function $\psi \in L_{p'}(\mathbb{R}^n)$.

Set $\varphi(\xi) := \chi(\xi)\Psi_1(0, \xi)$, $\xi \in \mathbb{R}^n$. Then $\varphi \in L_p^{\vec{a}}$, and the relation (2.2.125) can be written as

$$\lim_{r\to\infty} \int\limits_{\mathbb{R}^n} \bar{\psi}'(\xi)\Psi_{-1}(1/\nu(r), \xi)v(1/\nu(r), \xi)d\xi$$
$$= \int\limits_{\mathbb{R}^n} \bar{\psi}'(\xi)\Psi_{-1}(0, \xi)\varphi(\xi)d\xi, \quad \psi \in L_{p'}(\mathbb{R}^n). \tag{2.2.126}$$

Take a fixed point $(t, x) \in \Pi_{(0,T]}$ and consider the function

$$\psi(\xi) := Z^*(0, \xi; t, x)\Psi_1(0, \xi), \quad \xi \in \mathbb{R}^n. \tag{2.2.127}$$

It follows from the estimate

$$|\psi(\xi)| \leq Ct^{-M}\Psi_1(t, x)E^{(0)}_{c-c_0}(t, x; 0, \xi), \quad \xi \in \mathbb{R}^n, \tag{2.2.128}$$

which is obtained using the equality (2.1.111), the estimate (2.1.82), and the inequality (2.2.4), that $\psi \in L_{p'}(\mathbb{R}^n)$. Therefore by (2.1.111) and (2.2.126) we have

$$\lim_{r\to\infty} \int\limits_{\mathbb{R}^n} Z(t, x; 0, \xi)\Psi_1(0, \xi)\Psi_{-1}(1/\nu(r), \xi)v(1/\nu(r), \xi)d\xi = \int\limits_{\mathbb{R}^n} Z(t, x; 0, \xi)\varphi(\xi)d\xi. \tag{2.2.129}$$

We may assume that $1/\nu(r) \leq t/2$, $r \geq 1$. Just as we proved the formula (2.2.91), we can show that

$$v(t, x) = \int\limits_{\mathbb{R}^n} Z(t, x; 1/\nu(r), \xi)v(1/\nu(r), \xi)d\xi. \tag{2.2.130}$$

Due to this equality,

$$v(t,x) - \int_{\mathbb{R}^n} Z(t,x;0,\xi)\varphi(\xi)d\xi = \int_{\mathbb{R}^n} \Delta_\tau^0 Z(t,x;\tau,\xi)\Big|_{\tau=1/\nu(r)} v(1/\nu(r),\xi)d\xi$$
$$+ \int_{\mathbb{R}^n} Z(t,x;0,\xi)(1-\Psi_1(0,\xi)\Psi_{-1}(1/\nu(r),\xi))v(1/\nu(r),\xi)d\xi$$
$$+ \Bigg(\int_{\mathbb{R}^n} Z(t,x;0,\xi)\Psi_1(0,\xi)\Psi_{-1}(1/\nu(r),\xi)v(1/\nu(r),\xi)d\xi$$
$$- \int_{\mathbb{R}^n} Z(t,x;0,\xi)\varphi(\xi)d\xi\Bigg) =: \sum_{j=1}^{3} I_j^{(r)}, \qquad r \geq 1. \tag{2.2.131}$$

To prove the representation (2.2.121), it is sufficient to show that for $j \in \{1,2,3\}$,

$$\lim_{r\to\infty} I_j^{(r)} = 0. \tag{2.2.132}$$

The relation (2.2.129) implies (2.2.132) for $j = 3$. Let us prove (2.2.132) for $j = 2$. Using the Hölder inequality and the estimate (2.2.120) we find that

$$|I_2^{(r)}| \leq C\Bigg(\int_{\mathbb{R}^n} Y_r(\xi)d\xi\Bigg)^{1/p'} \tag{2.2.133}$$

where $Y_r(\xi) := |Z(t,x;0,\xi)|^{p'}|\Psi_1(1/\nu(r),\xi) - \Psi_1(0,\xi)|^{p'}$, $\xi \in \mathbb{R}^n$, $r \geq 1$.

Let us study properties of the functions Y_r, $r \geq 1$. Using the inequalities (2.1.82), (2.2.4), and (2.2.95) we obtain that

$$\begin{aligned}(Y_r(\xi))^{1/p'} &\leq Ct^{-M}(\Psi_1(t+1/\nu(r),x) + \Psi_1(t,x))E^{(0)}_{c-c_0}(t,x;0,\xi)\\ &\leq Ct^{-M}(\Psi_1(t+1/\nu(1),x) + \Psi_1(t,x))E^{(0)}_{c-c_0}(t,x;0,\xi),\\ &\qquad \xi \in \mathbb{R}^n, \quad r \geq 1,\end{aligned}$$

which implies the existence of an integrable majorant for the sequence $\{Y_r : r \geq 1\}$. Since for each $\xi \in \mathbb{R}^n$ $\lim\limits_{r\to\infty} Y_r(\xi) = 0$, by the Lebesgue majorated convergence theorem,

$$\lim_{r\to\infty} \int_{\mathbb{R}^n} Y_r(\xi)d\xi = 0.$$

This, together with (2.2.133), implies (2.2.132) for $j = 2$.

Finally, we prove the relation (2.2.132) for $j = 1$. Using the inequalities (2.2.94), (2.2.95), and (2.2.120), we have

$$\begin{aligned}|I_1^{(r)}| &\le C(1/\nu(r))t^{-M-1}\Psi_1(t+1/\nu(r),x)\\ &\quad\times \int\limits_{\mathbb{R}^n} E^{(0)}_{c-c_0}(t,x;0,\xi)(|v(1/\nu(r),\xi)|\Psi_{-1}(1/\nu(r),\xi))d\xi\\ &\le C(1/\nu(r))t^{-M-1}\Psi_1(t+1/\nu(r),x)\|v(1/\nu(r),\cdot)\|_p^{\vec{k}(1/\nu(r),\vec{a})}\\ &\quad\times\left(\int\limits_{\mathbb{R}^n} E^{(0)}_{(c-c_0)p'}(t,x;0,\xi)d\xi\right)^{1/p'} \underset{r\to\infty}{\longrightarrow} 0.\end{aligned}$$

Suppose that $p=1$. It follows from the condition (2.2.120) that the sequence (2.2.123) is bounded in $L_1(\mathbb{R}^n)$. This space is not a conjugate of any Banach space but it can be imbedded into the space $M(\mathbb{R}^n)$ of all the generalized measures $\lambda : \mathcal{B}_n \to \mathbb{C}_{N1}$ with the finite full variation $|\lambda|(\mathbb{R}^n)$. If we introduce a norm of λ by the formula $\|\lambda\| := |\lambda|(\mathbb{R}^n)$, then $M(\mathbb{R}^n)$ becomes a Banach space. This space is isometric to the conjugate of the space $C_0(\mathbb{R}^n)$ of continuous functions $\psi : \mathbb{R}^n \to \mathbb{C}_{N1}$ tending to zero at infinity. The boundedness in $L_1(\mathbb{R}^n)$ of the sequence (2.2.123) implies the boundedness of the corresponding sequence of elements of the space $M(\mathbb{R}^n)$, hence the weak compactness of the latter sequence. Thus there exists a subsequence (2.2.124) and a generalized measure $\lambda \in M(\mathbb{R}^n)$, such that

$$\lim_{r\to\infty}\int\limits_{\mathbb{R}^n}\bar{\psi}'(\xi)\Psi_{-1}(1/\nu(r),\xi)v(1/\nu(r),\xi)d\xi = \int\limits_{\mathbb{R}^n}\bar{\psi}'d\lambda(\xi) \tag{2.2.134}$$

for any function $\psi \in C_0(\mathbb{R}^n)$.

For bounded sets $A \in \mathcal{B}_n$ we define $\mu(A) := \int\limits_A \Psi_1(0,x)d\lambda(x)$; then

$$\int\limits_A \Psi_{-1}(0,x)d\mu(x) = \lambda(A).$$

If A is an unbounded set from $\mathcal{B}_n$, we consider a monotone non-decreasing sequence of bounded sets $A_k \in \mathcal{B}_n$, $k \ge 1$, such that $\bigcup\limits_{k=1}^{\infty} A_k = A$, and put

$$\int\limits_A \Psi_{-1}(0,x)d\mu(x) := \lim_{k\to\infty}\int\limits_{A_k}\Psi_{-1}(0,x)d\mu(x) = \lim_{k\to\infty}\lambda(A_k) = \lambda(A).$$

The function μ, just defined, belongs to the space $M^{\vec{a}}$, and the equality (2.2.134) is written as

$$\lim_{r\to\infty}\int_{\mathbb{R}^n}\bar{\psi}'(\xi)\Psi_{-1}(1/\nu(r),\xi)v(1/\nu(r),\xi)d\xi=\int_{\mathbb{R}^n}\bar{\psi}'(\xi)\Psi_{-1}(0,\xi)d\mu(\xi),\quad \psi\in C_0(\mathbb{R}^n). \tag{2.2.135}$$

It follows from the estimate (2.2.128) that the function (2.2.127) (if $N > 1$, each column of this matrix) belongs to $C_0(\mathbb{R}^n)$, for any fixed point $(t, x) \in \Pi_{(0,T]}$. Therefore, by the equalities (2.1.111) and (2.2.135) we obtain that

$$\lim_{r\to\infty}\int_{\mathbb{R}^n} Z(t,x;0,\xi)\Psi_1(0,\xi)\Psi_{-1}(1/\nu(r),\xi)v(1/\nu(r),\xi)d\xi = \int_{\mathbb{R}^n} Z(t,x;0,\xi)d\mu(\xi). \tag{2.2.136}$$

The subsequent reasoning is the same as for the case $p > 1$: using (2.2.130) we write the equality

$$v(t,x) - \int_{\mathbb{R}^n} Z(t,x;0,\xi)d\mu(\xi) = I_1^{(r)} + I_2^{(r)} + \tilde{I}_3^{(r)},\quad r\geq 1 \tag{2.2.137}$$

where $I_j^{(r)}$, $j \in \{1, 2\}$, are the same as in (2.2.131), while

$$\tilde{I}_3^{(r)} := \int_{\mathbb{R}^n} Z(t,x;0,\xi)\Psi_1(0,\xi)\Psi_{-1}(1/\nu(r),\xi)v(1/\nu(r),\xi)d\xi - \int_{\mathbb{R}^n} Z(t,x;0,\xi)d\mu(\xi);$$

next we prove the relation (2.2.132) for $j \in \{1, 2\}$. Since, by (2.2.136), $\lim_{r\to\infty} \tilde{I}_3^{(r)} = 0$, the representation (2.2.122) follows from (2.2.137).

The uniqueness of the function φ and the generalized measure μ from the representations (2.2.92) and (2.2.103) follows from Theorems 2.6 and 2.7. □

The next result follows from Theorems 2.6, 2.7, and 2.11.

Corollary. Under the assumptions regarding the coefficients and the right-hand side f of the equation (2.1.1) from Theorems 2.6, 2.7, and 2.11, the following assertions are valid:

1) the spaces $L_p^{\vec{a}}$ and $M^{\vec{a}}$ are the sets of initial values of solutions of the equation (2.1.1) if and only if the solutions satisfy the condition (2.2.119) for $p \in (1, \infty]$ and $p = 1$ respectively;
2) a solution of the equation (2.1.1) can be represented in the form (2.2.92) or (2.2.103) with $\varphi \in L_p^{\vec{a}}$ and $\mu \in M^{\vec{a}}$, if and only if the condition (2.2.119) is satisfied.

2.3 Equations with growing coefficients

2.3.1. Assumptions and auxiliary results. We shall consider the equation (2.1.1) from the class $\mathbf{E}_{12}$. For such an equation, the coefficients of its junior terms grow, as $|x| \to \infty$. As we mentioned in Section 1.1.1, the main notion for equations from the class $\mathbf{E}_{12}$ is the notion of dissipation.

We make the following assumptions regarding the coefficients of the equation (2.1.1):

(A_{15}) The equation (2.1.1) is dissipative $\overrightarrow{2b}$-parabolic on $\Pi_{[0,T]}$, with the dissipation characteristic D;

(A_{16}) The coefficients a_k, $\|k\| \le 2b$, have continuous derivatives $\partial_x^l a_k$, $\|l\| \le 2b$, satisfying the estimates

$$|\partial_x^l a_k(t,x)| \le C(D(x))^{2b-\|k\|+\|l\|(1-\varepsilon)}, \quad (t,x) \in \Pi_{[0,T]},$$

where $C > 0$, $\varepsilon \in (0,1)$; the functions $b_k(t,x) := a_k(t,x)(D(x))^{\|k\|-2b}$, $(t,x) \in \Pi_{[0,T]}$, $\|k\| \le 2b$, are continuous in t, uniformly with respect to $x \in \mathbb{R}^n$;

(A_{17}) The derivatives $\partial_x^l a_k$, $\|k\| \le 2b$, $\|l\| \le 2b$, are locally Hölder continuous in x, uniformly with respect to t, with the exponent γ.

There exists another set of assumptions which does not require a high smoothness of the coefficients, but a special restriction is imposed upon the dissipation characteristic D:

(B_{13}) The condition A_{15} holds;

(B_{14}) The coefficients a_k, $\|k\| \le 2b$, satisfy the inequality

$$|\Delta_x^y a_k(t,x)| \le C(p(x;y))^\gamma((D(x))^{2b-\|k\|} + (D(y))^{2b-\|k\|}),$$
$$\{(t,x),(t,y)\} \subset \Pi_{[0,T]},$$

in which $C > 0$ and $\gamma \in (0,1)$; the functions b_k, $\|k\| \le 2b$, are continuous in t uniformly with respect to $x \in \mathbb{R}^n$;

(B_{15}) the dissipation characteristic satisfies the condition $D(x) \le CD(y)$ for any $\{x,y\} \subset \mathbb{R}^n$, such that $q(x;y) \le 1$, and the condition

$$D(x) \le C\exp\left\{\varepsilon \sum_{j=1}^{n} |x_j - y_j|(D(y))^{m_j}\right\},$$

where $C > 0$ while ε is a sufficiently small positive number (to be chosen in specific situations), if $q(x;y) > 1$.

Let $g : \mathbb{R}^n \to [1,\infty)$ be a function connected with the dissipation characteristic D via one of the following sets of conditions:

(A_{18}) $g(x) \underset{|x|\to\infty}{\to} \infty$; there exist the derivatives $\partial_x^k g$, $0 < \|k\| \le 4b$, locally Hölder continuous with the exponent γ from the condition A_{17}, which are connected with the dissipation characteristic via the condition

$$|\partial_x^k g(x)| \le C\eta(D(x))^{\|k\|(1-\varepsilon)}, \quad x \in \mathbb{R}^n,$$

where $C > 0$, $\varepsilon \in (0,1)$, η is a sufficiently small positive number, to be chosen in specific situations;

(B_{16}) $g(x) \underset{|x|\to\infty}{\to} \infty$; there exist the derivatives $\partial_x^k g$, $0 < \|k\| \le 2b$, satisfying the inequalities

$$|\partial_x^k g(x)| \le C\eta(D(x))^{\|k\|},$$
$$|\Delta_x^y \partial_x^k g(x)| \le C\eta(p(x;y))^\gamma((D(x))^{\|k\|} + (D(y))^{\|k\|}),$$
$$\{x,y\} \subset \mathbb{R}^n,$$

where $C > 0$, λ is the number from the condition B_{14}, η is a sufficiently small positive number, to be chosen in specific situations.

Below we construct a FSCP for a dissipative equation (2.1.1) for the cases where the coefficients are sufficiently smooth (the conditions A_{15}–A_{17} are satisfied), or where they are Hölder continuous (the conditions B_{13}–B_{15} are satisfied). Levi's method will be used in both cases but, naturally enough, the parametrices will be chosen differently.

In order to define the parametrix, we consider an auxiliary equation

$$\left(I\partial_t - \sum_{\|k\|+k_{n+1}=2b} b_k(t,x)\partial_x^k(-i\partial_{x_{n+1}})^{k_{n+1}}\right)\hat{v}(t,x,x_{n+1}) = 0,$$
$$(t,x,x_{n+1}) \in \Pi_{(0,T]} \times \mathbb{R}, \tag{2.3.1}$$

where y is a fixed point from the space $\mathbb{R}^n$. If the condition A_{15} is satisfied, and the coefficients b_k, $\|k\| \le 2b$, are continuous in t uniformly with respect to $y \in \mathbb{R}^n$, then, due to the results from Section 2.1.3, there exists a FSCP $\hat{G}(t,x;\tau,\xi;x_{n+1} - \xi_{n+1};y)$, $0 \le \tau < t \le T$, $\{x,\xi,y\} \subset \mathbb{R}^n$, $\{x_{n+1},\xi_{n+1}\} \subset \mathbb{R}^n$, for the equation (2.3.1), satisfying the estimates

$$|\partial_x^k \partial_{x_{n+1}}^{k_{n+1}} \hat{G}(t,x;\tau,\xi;x_{n+1} - \xi_{n+1};y)|$$
$$\le C_{kk_{n+1}}(t-\tau)^{-M-(\|k\|+k_{n+1}+1)/(2b)}$$
$$\times E_c^{(0)}(t,x;\tau,\xi)\exp\{-c(t-\tau)^{1-q}|x_{n+1} - \xi_{n+1}|^q\},$$
$$k \in \mathbb{Z}_+^n, \quad k_{n+1} \in \mathbb{Z}_+^1,$$

where $C_{kk_{n+1}}$ and c are positive constants, $q := 2b/(2b-1)$. The derivative $\partial_x^k \hat{G}(t,x;\tau,\xi;z_{n+1};y)$, $z_{n+1} := x_{n+1} + iy_{n+1} \in \mathbb{C}$, as a function in $(t-\tau)^{-1/(2b)}z_{n+1}$

with fixed t, τ, x, ξ, and y, is entire and

$$|\partial_x^k \hat{G}(t,x;\tau,\xi;z_{n+1};y)| \le C_k(t-\tau)^{-M-(\|k\|+1)/(2b)} E_c^{(0)}(t,x;\tau,\xi) \times \exp\{-c(t-\tau)^{1-q}|x_{n+1}|^q + c'(t-\tau)^{1-q}|y_{n+1}|^q\}, \quad 0 \le \tau < t \le T, \quad \{x,\xi\} \subset \mathbb{R}^n, \quad z_{n+1} \in \mathbb{C}, \tag{2.3.2}$$

where $C_k > 0$, $c > 0$, and $c' > 0$.

Let $G(t,x;\tau,\xi;\eta;y) := F_{z_{n+1}\to\eta}[\hat{G}(t,x;\tau,\xi;z_{n+1};y)]$, where $F_{z_{n+1}\to\eta}$ is the Fourier transform in the variable z_{n+1}. Then the function $G(t,x;\tau,\xi;\eta;y)$, $0 \le \tau < t \le T$, $\{x,\xi\} \subset \mathbb{R}^n$, is a FSCP for the equation

$$\left(I\partial_t - \sum_{\|k\|+k_{n+1}=2b} b_k(t,y)\eta^{k_{n+1}}\partial_x^k\right)v(t,x) = 0, \quad (t,x) \in \Pi_{(0,T]},$$

for arbitrary fixed points $\eta \in \mathbb{R}$ and $y \in \mathbb{R}^n$. By Lemma 1.1, the estimates (2.3.2) imply the estimates

$$|\partial_x^k G(t,x;\tau,\xi;\eta;y)| \le C_k(t-\tau)^{-M-\|k\|/(2b)} E_c^{(0)}(t,x;\tau,\xi) \exp\{-c(t-\tau)\eta^{2b}\}, \quad 0 \le \tau < t \le T, \quad \{x,\xi,y\} \subset \mathbb{R}^n, \quad \eta \in \mathbb{R},$$

which, in turn imply the estimates

$$|\partial_x^k G_0(t,x;\tau,\xi;y)| \le C_k(t-\tau)^{-M-\|k\|/(2b)} E_c^{(0)}(t,x;\tau,\xi) \exp\{-c(t-\tau)(D(y))^{2b}\}, \quad 0 \le \tau < t \le T, \quad \{x,\xi,y\} \subset \mathbb{R}^n, \quad k \in \mathbb{Z}_+^n, \tag{2.3.3}$$

where we denoted

$$G_0(t,x;\tau,\xi;y) := G(t,x;\tau,\xi;D(y);y).$$

Note that the function $G_0(t,x;\tau,\xi;y)$, $0 \le \tau < t \le T$, $\{x,\xi\} \subset \mathbb{R}^n$, is a FSCP for the equation

$$L_N(t,y,\partial_t,\partial_x)v(t,x) = 0, \quad (t,x) \in \Pi_{(0,T]}, \tag{2.3.4}$$

where L_N was defined in (2.1.1), y is an arbitrary fixed point in $\mathbb{R}^n$.

The following properties of the function G_0 will be important in the sequel.

Property 2.9. *Suppose that the conditions A_{15} and A_{16} are satisfied. Then there exist the derivatives $\partial_x^k \partial_y^l G_0$, $\{k,l\} \subset \mathbb{Z}_+^n$, $\|l\| \le 2b$, for which*

$$|\partial_x^k \partial_y^l G_0(t,x;\tau,\xi;y)| \le C_{kl}(t-\tau)^{-M-(\|k\|+\|l\|(1-\varepsilon)/(2b))} \times E_c^{(0)}(t,x;\tau,\xi) \exp\{-c(t-\tau)(D(y))^{2b}\}, \quad 0 \le \tau < t \le T, \quad \{x,\xi,y\} \subset \mathbb{R}^n. \tag{2.3.5}$$

If, in addition, the condition A_{17} is satisfied too, then for any $R > 0$ there exists such a constant $C_{kl} > 0$ that for all $\{y, z\} \subset B_R$,

$$|\Delta_y^z \partial_x^k \partial_y^l G_0(t,x;\tau,\xi;y)| \le C_{kl}(p(y;z))^\gamma (t-\tau)^{-M-(\|k\|+\|l\|(1-\varepsilon))/(2b)} E_c^{(0)}(t,x;\tau,\xi),$$
$$0 \le \tau < t \le T, \quad \{x,\xi\} \subset \mathbb{R}^n. \tag{2.3.6}$$

In the estimates (2.3.5) *and* (2.3.6), ε *and* γ *are the numbers from the conditions A_{16} and A_{17}.*

The proof of this property is similar to that of Property 2.3. The emergence of the factor $(t-\tau)^{-\|l\|(1-\varepsilon)/(2b)}$ in the estimates (2.3.5) and (2.3.6) is caused by the fact that, due to the condition A_{16}, derivatives of the coefficients of the equation (2.3.4) are estimated via powers of $D(y)$.

The following property is proved similarly.

Property 2.10. *If the conditions B_{13} and B_{14} are satisfied, then*

$$|\Delta_y^z \partial_x^k G_0(t,x;\tau,\xi;y)|$$
$$\le C_k (p(y;z))^\gamma (t-\tau)^{-M-\|k\|/(2b)} E_c^{(0)}(t,x;\tau,\xi)$$
$$\times (\exp\{-c(t-\tau)(D(y))^{2b}\} + \exp\{-c(t-\tau)(D(z))^{2b}\})((D(y))^{2b} + (D(z))^{2b}),$$
$$0 \le \tau < t \le T, \quad \{x,\xi,y,z\} \subset \mathbb{R}^n.$$

Property 2.11. *All the assertions from Property 2.4 are valid for the function G_0.*

Let us define the functions

$$\hat{Z}_0(t,x;\tau,\xi) := G_0(t,x;\tau,\xi;x), \tag{2.3.7}$$
$$Z_0(t,x;\tau,\xi) := G_0(t,x;\tau,\xi;\xi), \tag{2.3.8}$$
$$0 \le \tau < t \le T, \quad \{x,\xi\} \subset \mathbb{R}^n,$$

which will be taken for parametrices, provided the conditions A_{15}–A_{17} and B_{13}–B_{15} respectively are satisfied.

The equality

$$\partial_x^k \hat{Z}_0(t,x;\tau,\xi) = \sum_{l \le k} C_k^l \partial_x^{k-l} \partial_y^l G_0(t,x;\tau,\xi;y)|_{y=x},$$

and the estimates (2.3.5) imply the estimates

$$|\partial_x^k \hat{Z}_0(t,x;\tau,\xi)| \le C(t-\tau)^{-M-\|k\|/(2b)} E_c^{(0)}(t,x;\tau,\xi) \exp\{-c(t-\tau)(D(x))^{2b}\},$$
$$0 \le \tau < t \le T, \quad \{x,\xi\} \subset \mathbb{R}^n, \quad \|k\| \le 2b, \tag{2.3.9}$$

while the estimates (2.3.3) give the inequalities

$$|\partial_x^k Z_0(t,x;\tau,\xi)| \le C_k(t-\tau)^{-M-\|k\|/(2b)} E_c^{(0)}(t,x;\tau,\xi) \exp\{-c(t-\tau)(D(\xi))^{2b}\},$$
$$0 \le \tau < t \le T, \quad \{x,\xi\} \subset \mathbb{R}^n, \quad \|k\| \in \mathbb{Z}_+^n. \tag{2.3.10}$$

The next statements describe differential properties of the integrals

$$\hat{u}(t,x;\tau,\xi) := \int_{\tau}^{t} d\lambda \int_{\mathbb{R}^n} \hat{Z}_0(t,x;\lambda,y)Q(\lambda,y;\tau,\xi)dy, \tag{2.3.11}$$

$$u(t,x;\tau,\xi) := \int_{\tau}^{t} d\lambda \int_{\mathbb{R}^n} Z_0(t,x;\lambda,y)Q(\lambda,y;\tau,\xi)dy, \tag{2.3.12}$$

$$0 \le \tau < t \le T, \quad \{x,\xi\} \subset \mathbb{R}^n,$$

where Q is a continuous function with values from $\mathbb{C}_{NN}$ satisfying certain conditions.

Property 2.12. *Suppose that the conditions A_{15} and A_{16} are fulfilled, while the function Q satisfies the estimate*

$$|Q(t,x;\tau,\xi)| \le C(t-\tau)^{-M-1+\chi/(2b)}E_c^{(0)}(t,x;\tau,\xi), \tag{2.3.13}$$
$$0 \le \tau < t \le T, \quad \{x,\xi\} \subset \mathbb{R}^n,$$

with some constants $C > 0$, $c > 0$, $\chi \in (0,1)$, and, for any $R > 0$, the estimate

$$|\Delta_x^y Q(t,x;\tau,\xi)| \le C(p(x;y))^{\gamma_1}(t-\tau)^{-M-1+\gamma_2/(2b)}(E_c^{(0)}(t,x;\tau,\xi)+E_c^{(0)}(t,y;\tau,\xi)),$$
$$0 \le \tau < t \le T, \quad \{x,y\} \subset B_R, \quad \xi \in \mathbb{R}^n, \tag{2.3.14}$$

with constants $C > 0$, $c > 0$, $\gamma_1 \in (0,\chi)$ and $\gamma_2 := \chi - \gamma_1$, which may depend on R. Then the function (2.3.11) *possesses the continuous derivatives involved in the equation* (2.3.4). *The derivatives $\partial_x^k \hat{u}$, $\|k\| < 2b$, are obtained by the formal differentiations under the symbols of the integrals while the leading derivatives can be found by the formulas*

$$\partial_x^k \hat{u}(t,x;\tau,\xi) = \int_{\tau}^{t_0} d\lambda \int_{\mathbb{R}^n} \partial_x^k \hat{Z}_0(t,x;\lambda,y)Q(\lambda,y;\tau,\xi)dy$$
$$+ \int_{t_0}^{t} d\lambda \int_{\mathbb{R}^n} \partial_x^k \hat{Z}_0(t,x;\lambda,y)\Delta_y^x Q(\lambda,y;\tau,\xi)dy$$
$$+ \int_{t_0}^{t} \left(\int_{\mathbb{R}^n} \partial_x^k \hat{Z}_0(t,x;\lambda,y)dy \right) Q(\lambda,x;\tau,\xi)d\lambda,$$
$$\|k\| = 2b, \tag{2.3.15}$$

$$\partial_t \hat{u}(t,x;\tau,\xi) = Q(t,x;\tau,\xi) + \int_{\tau}^{t_0} d\lambda \int_{\mathbb{R}^n} \partial_t \hat{Z}_0(t,x;\lambda,y) Q(\lambda,y;\tau,\xi)dy$$
$$+ \int_{t_0}^{t} d\lambda \int_{\mathbb{R}^n} \partial_t \hat{Z}_0(t,x;\lambda,y) \Delta_y^x Q(\lambda,y;\tau,\xi)dy$$
$$+ \int_{t_0}^{t} \left(\int_{\mathbb{R}^n} \partial_t \hat{Z}_0(t,x;\lambda,y)dy \right) Q(\lambda,y;\tau,\xi)d\lambda,$$
$$0 \le \tau < t \le T, \quad \{x,\xi\} \subset \mathbb{R}^n, \tag{2.3.16}$$

where $t_0 := (t+\tau)/2$.

Property 2.13. *If the conditions* B_{13}–B_{15} *are fulfilled, and the function* Q *satisfies the estimates*

$$|Q(t,x;\tau,\xi)| \le C_l((t-\tau)^{-M-1+\gamma/(2b)} \exp\{-c(t-\tau)(D(\xi))^{2b}\}$$
$$+ (t-\tau)^{\gamma/(2b)}(D(\xi))^{-l})E_c^{(0)}(t,x;\tau,\xi), \tag{2.3.17}$$

$$|\Delta_x^y Q(t,x;\tau,\xi)| \le C_l(p(x;y))^{\gamma_1}((t-\tau)^{-M-1+\gamma_2/(2b)} \exp\{-c(t-\tau)(D(\xi))^{2b}\}$$
$$+ (t-\tau)^{\gamma_2/(2b)}(D(\xi))^{-l})(E_c^{(0)}(t,x;\tau,\xi) + E_c^{(0)}(t,y;\tau,\xi)), \tag{2.3.18}$$
$$0 \le \tau < t \le T, \quad \{x,y,\xi\} \subset \mathbb{R}^n,$$

with some constants $C_l > 0$, $c > 0$, $\gamma_1 \in (0,\gamma)$, *and* $\gamma_2 := \gamma - \gamma_1$, *where* l *is an arbitrary fixed positive number, then the function* (2.3.12) *has the continuous derivatives involved in the equation* (2.3.4). *The derivatives* $\partial_x^k u$, $\|k\| < 2b$, *are obtained by the formal differentiations under the symbols of the integrals while the leading derivatives are computed by the formulas which differ from* (2.3.15) *and* (2.3.16) *only by the substitution of* u *and* Z_0 *for* $\hat{u}$ *and* $\hat{Z}_0$ *respectively.*

Properties 2.12 and 2.13 are proved by the method used in the proof of Properties 2.6 and 2.7. We show only how the integrals

$$I^{(k)} := \int_{\mathbb{R}^n} \partial_x^k Z_0(t,x;\lambda,y) Q(\lambda,y;\tau,\xi)dy,$$
$$0 \le \tau < t_0 \le \lambda < t \le T,$$
$$\{x,\xi\} \subset \mathbb{R}^n, \quad \|k\| = 2b.$$

are estimated.

By Property 2.11 and the formula (2.3.8) we may write

$$I^{(k)} = \int\limits_{\mathbb{R}^n} \partial_x^k Z_0(t,x;\lambda,y)\Delta_y^x Q(\lambda,y;\tau,\xi)dy$$
$$+ \left(\int\limits_{\mathbb{R}^n} \Delta_z^x \partial_x^k G_0(t,x;\lambda,y;z)|_{z=y} dy \right) Q(\lambda,x;\tau,\xi)$$
$$=: I_1 + I_2 Q(\lambda,x;\tau,\xi). \tag{2.3.19}$$

Using (2.1.46), (2.1.61), (2.1.80), (2.3.10), (2.3.18), and the fact that

$$(t-\tau)/2 \le \lambda - \tau \le t-\tau \qquad \text{and} \qquad \exp\{-c(t-\lambda)(D(y))^{2b}\} \le 1,$$

we have

$$|I_1| \le C_l \int\limits_{\mathbb{R}^n} (t-\lambda)^{-M-1+\gamma_1/(2b)} E_c^{(0)}(t,x;\lambda,y)$$
$$\times ((\lambda-\tau)^{-M-1+\gamma_2/(2b)} \exp\{-c(\lambda-\tau)(D(\xi))^{2b}\}$$
$$+ (\lambda-\tau)^{\gamma_2/(2b)}(D(\xi))^{-l}(E_c^{(0)}(\lambda,y;\tau,\xi) + E_c^{(0)}(\lambda,x;\tau,\xi))dy$$
$$\le C_l(t-\lambda)^{-1+\gamma_1/(2b)}((t-\tau)^{-M-1+\gamma_2/(2b)} \exp\{-c_1(t-\tau)(D(\xi))^{2b}\}$$
$$+ (t-\tau)^{\gamma_2/(2b)}(D(\xi))^{-l})E_{c_1}^{(0)}(t,x;\tau,\xi),$$
$$0 \le \tau < t_0 \le \lambda < t \le T, \quad \{x,\xi\} \subset \mathbb{R}^n. \tag{2.3.20}$$

By (2.1.46), (2.1.61), and Property 2.10,

$$|I_2| \le C(t-\lambda)^{-1+\gamma/(2b)} \int\limits_{\mathbb{R}^n} (t-\lambda)^{-M} E_{c_1}^{(0)}(t,x;\lambda,y)((D(x))^{2b} + (D(y))^{2b})dy$$
$$= C(t-\lambda)^{-1+\gamma/(2b)} \left((D(x))^{2b} + \int\limits_{B(x,1)} (t-\lambda)^{-M} E_{c_1}^{(0)}(t,x;\lambda,y)(D(y))^{2b} dy \right.$$
$$\left. + \int\limits_{\mathbb{R}^n \setminus B(x,1)} (t-\lambda)^{-M} E_{c_1}^{(0)}(t,x;\lambda,y)(D(y))^{2b} dy \right).$$

It follows from the condition B_{15} that $D(y) \le CD(x)$ if $y \in B(x,1)$. For $y \in \mathbb{R}^n \setminus B(x,1)$ we use the Young inequality and find that

$$(D(y))^{2b} \le C\exp\left\{ 2b\varepsilon \sum_{j=1}^n ((t-\lambda)^{-1/(2b_j)}|x_j - y_j|)((t-\lambda)^{1/(2b_j)}(D(x))^{m_j}) \right\}$$
$$\le CE_{-2b\varepsilon/q''}^{(0)}(t,x;\lambda,y)\exp\{2bM\varepsilon(t-\lambda)(D(x))^{2b}\}. \tag{2.3.21}$$

Therefore choosing ε in such a way that $c_1 - 2b\varepsilon/q'' > 0$, we get

$$|I_2| \le C(t-\lambda)^{-1+\gamma/(2b)}((D(x))^{2b} + \exp\{2bM\varepsilon(t-\lambda)(D(x))^{2b}\}). \tag{2.3.22}$$

The estimates (2.3.13) and (2.3.22) imply the estimate

$$\begin{aligned}|I_2Q(\lambda,x;\tau,\xi)| \le C_l(t-\lambda)^{-1+\gamma/(2b)}&((t-\tau)^{-M-1+\gamma/(2b)}\exp\{-c_1(t-\tau)(D(\xi))^{2b}\}\\ &+(t-\tau)^{\gamma/(2b)}(D(\xi))^{-l}E_c^{(0)}(t,x;\tau,\xi)((D(x))^{2b}\\ &+\exp\{2bM\varepsilon(t-\tau)(D(x))^{2b}\}),\\ &\quad 0\le\tau<t_0\le\lambda<t\le T,\quad \{x,\xi\}\subset\mathbb{R}^n,\end{aligned} \tag{2.3.23}$$

while (2.3.19), (2.3.20), and (2.3.23) lead to

$$\begin{aligned}|I^{(k)}| \le C_l(t-\lambda)^{-1+\gamma_1/(2b)}&((t-\tau)^{-M-1+\gamma_2/(2b)}\exp\{-c_1(t-\tau)(D(\xi))^{2b}\\ &+(t-\tau)^{\gamma_2/(2b)}(D(\xi))^{-l})E_{c_1}^{(0)}(t,x;\tau,\xi)((D(x))^{2b}\\ &+\exp\{2bM\varepsilon(t-\tau)(D(x))^{2b}\}),\\ &\quad 0\le\tau<t_0\le\lambda<t\le T,\quad \{x,\xi\}\subset\mathbb{R}^n.\end{aligned}$$

2.3.2. Fundamental solution of the Cauchy problem for dissipative equations. The next two theorems deal with the FSCPs for dissipative equations (2.1.1) with sufficiently smooth coefficients and with Hölder continuous coefficients.

Theorem 2.12. *Suppose that the conditions A_{15}–A_{17} are satisfied. Then there exists a FSCP Z for the equation* (2.1.1), *such that*

$$|\partial_x^k Z(t,x;\tau,\xi)| \le C(t-\tau)^{-M-\|k\|/(2b)}E_c^{(0)}(t,x;\tau,\xi) \tag{2.3.24}$$

and

$$\begin{aligned}&|\partial_x^k Z(t,x;\tau,\xi)|\\ &\le C\sum_{j=0}^{\|k\|}(t-\tau)^{-M-(\|k\|-j)/(2b)}(D(x))^{j(1-\varepsilon)}E_c^{(0)}(t,x;\tau,\xi)\exp\{g(x)-g(\xi)\},\end{aligned} \tag{2.3.25}$$

$$0\le\tau<t\le T,\quad \{x,\xi\}\subset\mathbb{R}^n,\quad \|k\|\le 2b,$$

where C and c are certain positive constants, g is an arbitrary function satisfying the condition A_{18}.

Proof. According to Levi's method, we seek a FSCP for the equation (2.2.1) in the form

$$Z(t,x;\tau,\xi) = \hat{Z}_0(t,x;\tau,\xi) + \int_\tau^t d\lambda\int_{\mathbb{R}^n}\hat{Z}_0(t,x;\lambda,y)Q(\lambda,y;\tau,\xi)dy, \tag{2.3.26}$$

$$0\le\tau<t\le T,\quad \{x,\xi\}\subset\mathbb{R}^n.$$

Assuming that the unknown function Q is continuous and satisfies the estimates (2.3.13) and (2.3.14), and using Property 2.12, we obtain the integral equation (1.1.47) for Q, in which

$$K(t,x;\tau,\xi) := \sum_{0<\|k\|\leq 2b} a_k(t,x) \sum_{l<k} C_k^l \partial_x^l \partial_y^{k-l} G_0(t,x;\tau,\xi;y)|_{y=x}.$$

Due to the condition for the coefficients a_k, $\|k\| \leq 2b$, and the estimates (2.3.5), we have

$$\begin{aligned}|K(t,x;\tau,\xi)| &\leq C \sum_{0<\|k\|\leq 2b} (D(x))^{2b-\|k\|} \sum_{l<k} (t-\tau)^{-M-(\|l\|+\|k-l\|(1-\varepsilon)/(2b))} \\ &\quad \times E_c^{(0)}(t,x;\tau,\xi) \exp\{-c(t-\tau)(D(x))^{2b}\} \\ &\leq C \sum_{\|k\|\leq 2b} \sum_{l<k} (t-\tau)^{-M-1+(\|k\|-\|l\|-\|k-l\|(1-\varepsilon))/(2b)} E_c^{(0)}(t,x;\tau,\xi) \\ &\quad \times \exp\{-c_1(t-\tau)(D(x))^{2b}\} \leq C(t-\tau)^{-M+\chi-1} E_c^{(0)}(t,x;\tau,\xi), \\ &\quad 0 \leq \tau < t \leq T, \quad \{x,\xi\} \subset \mathbb{R}^n,\end{aligned}$$

where $c_1 \in (0,c)$, $\chi := \min\limits_{j\in\{1,\dots,n\}} (m_j \varepsilon)$.

The last estimate makes it possible, just as in the proof of Theorem 2.2 for the equation (2.1.1) with bounded coefficients, to prove the solvability of the integral equation (1.1.47) and the estimate (2.3.13) for Q. The proof of the estimate (2.3.14) for Q and the estimates (2.3.24) for Z is carried out like the proof of the corresponding estimates in Theorem 2.2. The arguments must be modified to take into account the new conditions upon the coefficients. For the detailed proofs see [179]

Note that the estimates (2.3.24) do not contain the dissipation characteristic D. Therefore they should be refined. For this purpose we introduce in the equation (2.1.1) a new unknown function v setting

$$u(t,x) = e^{g(x)} v(t,x), \quad (t,x) \in \Pi_{(0,T]}, \tag{2.3.27}$$

where the function g satisfies the condition A_{18}. Then we obtain an equation for v:

$$\left(I\partial_t - \sum_{\|k\|\leq 2b} \tilde{a}_k(t,x)\partial_x^k \right) v(t,x) = \tilde{f}(t,x), \quad (t,x) \in \Pi_{(0,T]}, \tag{2.3.28}$$

where

$$\tilde{a}_k(t,x) := \sum_{\substack{l\geq k,\\ \|l\|\leq 2b}} C_l^k a_l(t,x) e^{-g(x)} \partial_x^{l-k} e^{g(x)}, \quad \tilde{f}(t,x) := e^{-g(x)} f(t,x).$$

Let us prove that the coefficients $\tilde{a}_k$, $\|k\| \leq 2b$, satisfy the conditions A_{15}–A_{17}.

Let us verify A_{15}. The functions

$$\tilde{b}_k(t,x) := \tilde{a}_k(t,x)(D(x))^{\|k\|-2b} = \sum_{\substack{l\geq k,\\ \|l\|\leq 2b}} C_l^k b_l(t,x) e^{-g(x)} \partial_x^{l-k} e^{g(x)} (D(x))^{\|k\|-\|l\|},$$

$$(t,x)\in \Pi_{[0,T]}, \quad \|k\|\leq 2b,$$

are bounded, since

$$|\tilde{b}_k(t,x)| \leq C \sum_{\substack{l\geq k,\\ \|l\|\leq 2b}} |b_l(t,x)|(D(x))^{-\|l-k\|\varepsilon} \leq C, \quad \|k\|\leq 2b,$$

due to the boundedness of the functions b_l, $\|l\|\leq 2b$, and the condition A_{18}. For the equation (2.3.28) the characteristic equation from Definition 1.5 has the form

$$\det\Big(Ip - \sum_{\|k\|+k_{n+1}=2b} \tilde{b}_k(t,x)(i\sigma)^k \mu^{k_{n+1}}\Big) = 0$$

or

$$\det\Big(Ip - \sum_{\|k\|+k_{n+1}=2b} b_k(t,x)(i\sigma)^k \mu^{k_{n+1}}$$
$$- \sum_{\|k\|+k_{n+1}=2b} \sum_{\substack{l>k,\\ \|l\|\leq 2b}} C_l^k b_l(t,x)(D(x))^{\|k\|-\|l\|} e^{-g(x)} \partial_x^{l-k} e^{g(x)} (i\sigma)^k \mu^{k_{n+1}}\Big) = 0. \tag{2.3.29}$$

By the condition A_{15} for the equation (2.1.1), the equation (2.3.29) without the third summand satisfies the $2B$-parabolicity condition. Since the terms of the sum from the third summand contain at least one differentiation of the function e^g, by the condition A_{18} they are estimated via the parameter η, so that, for a sufficiently small η, p-roots of the equation (2.3.29) maintain their proper location. Hence, the equation (2.3.28) is a dissipative $\overrightarrow{2b}$-parabolic equation on $\Pi_{[0,T]}$ with the dissipation characteristic D.

Let us verify the condition A_{16}. Due to the condition A_{16} for a_k, $\|k\|\leq 2b$, and the condition A_{18}, the coefficients $\tilde{a}_k$, $\|k\|\leq 2b$, have the continuous derivatives $\partial_x^l \tilde{a}_k$, $\|l\|\leq 2b$, and

$$|\partial_x^l \tilde{a}_k(t,x)| = \Big| \sum_{\substack{m\geq k,\\ \|m\|\leq 2b}} \sum_{p\leq r} \sum_{r\leq l-p} C_m^k C_l^p C_{l-p}^r \partial_x^{l-p-r} a_m(t,x) \partial_x^r e^{-g(x)} \partial_x^{m+p-k} e^{g(x)} \Big|$$
$$\leq C \sum_{\substack{m\geq k,\\ \|m\|\leq 2b}} \sum_{p\leq l} \sum_{r\leq l-p} (D(x))^{2b-\|m\|+\|l-p-r\|(1-\varepsilon)} (D(x))^{\|r\|(1-\varepsilon)} (D(x))^{\|m+p-k\|(1-\varepsilon)}$$

$$= C \sum_{\substack{m \geq k, \\ \|m\| \leq 2b}} (D(x))^{2b - \|k\| + \|l\|(1-\varepsilon)} (D(x))^{-\|m-k\|\varepsilon} \leq C(D(x))^{2b-\|k\|+\|l\|(1-\varepsilon)}.$$

The functions $\tilde{b}_k$, $\|k\| \leq 2b$, are continuous in t uniformly with respect to $x \in \mathbb{R}^n$, since for any $\varepsilon_1 > 0$

$$|\Delta_{t'}^{t''} \tilde{b}_k(t', x)| = \left| \sum_{\substack{l \geq k, \\ \|l\| \leq 2b}} C_l^k \Delta_{t'}^{t''} b_l(t', x) e^{-g(x)} \partial_x^{l-k} e^{g(x)} (D(x))^{\|k\| - \|l\|} \right|$$

$$\leq C\varepsilon' \sum_{\substack{l \geq k, \\ \|l\| \leq 2b}} (D(x))^{\|l-k\|(1-\varepsilon)} (D(x))^{\|k\|-\|l\|} \leq \varepsilon_1, \quad x \in \mathbb{R}^n,$$

as soon as $|t' - t''| < \delta$, $\{t', t''\} \subset [0, T]$, where ε' and δ are from the condition of the uniform, with respect to x, continuity in t of the functions b_l, $\|l\| \leq 2b$.

The condition A_{17} holds for the equation (2.3.28), due to the condition A_{17} for the equation (2.1.1), and the condition A_{18}.

Since the equation (2.3.28) satisfies the conditions A_{15}–A_{17}, this equation possesses a FSCP $\tilde{Z}$ with the estimates (2.3.24). The equality (2.3.27) implies the connection between the FSCPs Z and $\tilde{Z}$:

$$Z(t, x; \tau, \xi) = \exp\{g(x) - g(\xi)\} \tilde{Z}(t, x; \tau, \xi),$$
$$0 \leq \tau < t \leq T, \quad \{x, \xi\} \subset \mathbb{R}^n. \tag{2.3.30}$$

Using this equality, the estimates (2.3.24) for $\partial_x^k \tilde{Z}$, $\|k\| \leq 2b$, and the condition A_{18}, we obtain the estimates (2.3.25). □

Consider a dissipative equation (2.1.1) with Hölder continuous coefficients.

Theorem 2.13. *Let the conditions B_{13}–B_{15} be satisfied. Then there exists a FSCP Z for the equation* (2.1.1), *such that*

$$|\partial_x^k Z(t, x; \tau, \xi)| \leq C_l \Big((t-\tau)^{-M - \|k\|/(2b)} \exp\{-c(t-\tau)(D(\xi))^{2b}\} + (D(\xi))^{-l} \Big)$$
$$\times E_c^{(0)}(t, x; \tau, \xi), \quad \|k\| < 2b, \tag{2.3.31}$$

$$|\partial_x^k Z(t, x; \tau, \xi)| \leq C_l \Big((t-\tau)^{-M - \|k\|/(2b)} \exp\{-c(t-\tau)(D(\xi))^{2b}\} + (D(\xi))^{-l} \Big)$$
$$\times E_c^{(0)}(t, x; \tau, \xi) \Big((D(x))^{2b} + \exp\{2b\varepsilon M (t-\tau)(D(x))^{2b}\} \Big),$$
$$\|k\| = 2b, \tag{2.3.32}$$
$$0 \leq \tau < t \leq T, \quad \{x, \xi\} \subset \mathbb{R}^n,$$

where l is an arbitrary fixed positive number, $C_l > 0$, $c > 0$.

In addition,

$$|\partial_x^k Z(t,x;\tau,\xi)| \le C\sum_{j=0}^{\|k\|}(t-\tau)^{-M-(\|k\|-j)/(2b)}(D(x))^j E_c^{(0)}(t,x;\tau,\xi)\exp\{g(x)-g(\xi)\},$$
$$0\le\tau<t\le T,\quad \{x,\xi\}\subset\mathbb{R}^n,\quad \|k\|<2b, \tag{2.3.33}$$

where $C>0$, $c>0$, and g is an arbitrary function satisfying the condition B_{16}.

Proof. In contrast to the case considered in Theorem 2.12, instead of the formula (2.3.26) for a FSCP Z we shall use the formula (2.1.84), with Z_0 defined in (2.3.8). Assuming the conditions (2.3.17) and (2.3.18), we obtain an integral equation (1.1.47) for Q, with the kernel

$$K(t,x;\tau,\xi) := \sum_{\|k\|\le 2b} \Delta_x^\xi a_k(t,x)\partial_x^k Z_0(t,x;\tau,\xi),\quad 0\le\tau<t\le T,\quad \{x,\xi\}\subset\mathbb{R}^n,$$

where $c_0\in(0,c)$.

If $q(x;\xi)\le 1$, then this inequality, due to the condition B_{15}, implies

$$|K(t,x;\tau,\xi)|\le C_0(t-\tau)^{-M-1+\gamma/(2b)}E_{c_0}^{(0)}(t,x;\tau,\xi)\exp\{-c_0(t-\tau)(D(\xi))^{2b}\}.$$

Let $q(x;\xi)>1$. Then we use the inequality (2.3.21), in which x and ξ are substituted for y and x respectively, while $\varepsilon>0$ is chosen in such a way that $c_1 := c_0-\max\{2b\varepsilon/q'', 2bM\varepsilon\}>0$. Now

$$|K(t,x;\tau,\xi)|\le C_0'(t-\tau)^{-M-1+\gamma/(2b)}E_{c_1}^{(0)}(t,x;\tau,\xi)\exp\{-c_1(t-\tau)(D(\xi))^{2b}\}.$$

Hence, we have the estimate

$$|K(t,x;\tau,\xi)|\le C_1(t-\tau)^{-M-1+\gamma/(2b)}E_{c_1}^{(0)}(t,x;\tau,\xi)\exp\{-c_1(t-\tau)(D(\xi))^{2b}\},$$
$$0\le\tau<t\le T,\quad \{x,\xi\}\subset\mathbb{R}^n. \tag{2.3.34}$$

This estimate is the base for solving the equation (1.1.47) for Q by iterations. We have to give estimates for the iterated kernels K_m, $m\ge 2$. If in the estimate (2.3.34) we drop the factor $\exp\{-c_1(t-\tau)(D(\xi))^{2b}\}$, then we obtain bounds for the kernels K_m just as in Lemma 1.8, and carry out the whole procedure of the Levi method as in Section 2.1.5. The estimates for Z obtained this way are as follows:

$$|\partial_x^k Z(t,x;\tau,\xi)|\le C(t-\tau)^{-M-\|k\|/(2b)}E_c^{(0)}(t,x;\tau,\xi),$$
$$0\le\tau<t\le T,\quad \{x,\xi\}\subset\mathbb{R}^n,\quad \|k\|<2b. \tag{2.3.35}$$

It would be good to maintain the dissipation characteristic in the estimates of the kernels K_m and all the subsequent bounds. That is possible indeed, and we explain the technique by the example of the kernel K_2. Let us write it in the form

$$K_2(t,x;\tau,\xi) = \int_\tau^t d\lambda \int_{B(\xi,1)} K(t,x;\lambda,y)K(\lambda,y;\tau,\xi)dy$$
$$+ \int_\tau^t d\lambda \int_{\mathbb{R}^n\setminus B(\xi,1)} K(t,x;\lambda,y)K(\lambda,y;\tau,\xi)dy =: I_1 + I_2. \quad (2.3.36)$$

Using the estimate (2.3.34), the condition B_{15}, and the inequality (2.1.80), we find that

$$|I_1| \le C_1^2 \int_\tau^t ((t-\lambda)(\lambda-\tau))^{-1+\gamma/(2b)} d\lambda \int_{\mathbb{R}^n} E_{c_1}^{(0)}(t,x;\lambda,y)$$
$$\times E_{c_1}^{(0)}(\lambda,y;\tau,\xi)((t-\lambda)(\lambda-\tau))^{-M} dy \exp\{-c_2(t-\tau)(D(\xi))^{2b}\}$$
$$\le C_{2\delta} B(\gamma/(2b),\gamma/(2b))(t-\tau)^{-M-1+2\gamma/(2b)} E_{c_1(1-\delta)}^{(0)}(t,x;\tau,\xi)$$
$$\times \exp\{-c_2(t-\tau)(D(\xi))^{2b}\},$$
$$0 \le \tau < t \le T, \quad \{x,\xi\} \subset \mathbb{R}^n, \quad (2.3.37)$$

where $c_2 \in (0,c_1)$, $C_{2\delta} > 0$, δ is an arbitrary fixed number from the interval $(0,1)$.

In order to obtain a bound for I_2, we use the estimate

$$|K(\lambda,y;\tau,\xi)| \le C_{1l} E_{c_1(1-\delta)}^{(0)}(\lambda,y;\tau,\xi)(D(\xi))^{-l},$$
$$0 \le \tau < \lambda \le T, \quad y \in \mathbb{R}^n \setminus B(\xi,1), \quad \xi \in \mathbb{R}^n, \quad (2.3.38)$$

where l is an arbitrary fixed positive number. If $\lambda - \tau \ge 1$, then the estimate (2.3.38) is a direct consequence of (2.3.34):

$$|K(\lambda,y;\tau,\xi)| \le C_1 E_{c_1}^{(0)}(\lambda,y;\tau,\xi)\exp\{-c_1(D(\xi))^{2b}\}$$
$$\le C_{1l} E_{c_1}^{(0)}(\lambda,y;\tau,\xi)(D(\xi))^{-l}.$$

For the case where $\lambda - \tau < 1$, we get using the estimate (2.3.34) and the fact that $q(y;\xi) > 1$, that

$$|K(\lambda,y;\tau,\xi)|$$
$$\le C_1(\lambda-\tau)^{-M-1+\gamma/(2b)} E_{c_1(1-\delta)}^{(0)}(\lambda,y;\tau,\xi)$$
$$\times \exp\{-c_1\delta(\lambda-\tau)^{1-q''}\}\exp\{-c_1(\lambda-\tau)(D(\xi))^{2b}\}$$
$$\le C_{1l}(\lambda-\tau)^{-M-1+\gamma/(2b)} E_{c_1(1-\delta)}^{(0)}(\lambda,y;\tau,\xi)((\lambda-\tau)^{1/(2b)}D(\xi))^{-l}$$
$$\times \exp\{-c_1\delta(\lambda-\tau)^{1-q''}\}$$

$$= C_{1l}E^{(0)}_{c_1(1-\delta)}(\lambda, y; \tau, \xi)(D(\xi))^{-l}((\lambda-\tau)^{-M-1+(\gamma-l)/(2b)}\exp\{-c_1\delta(\lambda-\tau)^{1-q''}\})$$
$$\le C_{1l}E^{(0)}_{c_1(1-\delta)}(\lambda, y; \tau, \xi)(D(\xi))^{-l}.$$

Now we use the inequality

$$E^{(0)}_{c_0}(t,x;\lambda,y)E^{(0)}_{c_0}(\lambda,y;\tau,\xi) \le E^{(0)}_{c_0}(t,x;\tau,\xi),$$
$$\tau<\lambda<t, \quad \{x,y,\xi\}\subset\mathbb{R}^n, \quad c_0>0, \tag{2.3.39}$$

a consequence of the inequalities (1.3.13) with $n=1$, $\beta=1/(2b_j)$, $j\in\{1,\dots,n\}$.

From the estimates (2.3.34) and (2.3.38), using (2.1.46) and (2.3.39) we find that

$$|I_2| \le C_1C_{1l}(D(\xi))^{-l}\int_\tau^t (t-\lambda)^{-1+\gamma/(2b)}d\lambda \int_{\mathbb{R}^n} E^{(0)}_{c_1}(t,x;\lambda,y)$$
$$\times E^{(0)}_{c_1(1-\delta)}(\lambda,y;\tau,\xi)(t-\lambda)^{-M}\exp\{-c_1(t-\lambda)(D(y))^{2b}\}dy$$
$$\le C_1C_{1l}(D(\xi))^{-l}E^{(0)}_{c_1(1-\delta)}(t,x;\tau,\xi)$$
$$\times \int_\tau^t (t-\lambda)^{-1+\gamma/(2b)}d\lambda\int_{\mathbb{R}^n} E^{(0)}_{c_1\delta}(t,x;\lambda,y)(t-\lambda)^{-M}dy$$
$$= C_{2l}B(\gamma/(2b),1)(t-\tau)^{\gamma/(2b)}E^{(0)}_{c_1(1-\delta)}(t,x;\tau,\xi)(D(\xi))^{-l},$$
$$0\le\tau<t\le T, \quad \{x,\xi\}\subset\mathbb{R}^n. \tag{2.3.40}$$

From (2.3.36), (2.3.37), and (2.3.40) we get the required estimate

$$|K_2(t,x;\tau,\xi)| \le (C_{2\delta}B(\gamma/(2b),\gamma/(2b))(t-\tau)^{-M-1+2\gamma/(2b)}$$
$$\times\exp\{-c_2(t-\tau)(D(\xi))^{2b}\}$$
$$+C_{2l}B(\gamma/(2b),1)(t-\tau)^{\gamma/(2b)}(D(\xi))^{-l})E^{(0)}_{c_1(1-\delta)}(t,x;\tau,\xi),$$
$$0\le\tau<t\le T, \quad \{x,\xi\}\subset\mathbb{R}^n.$$

The estimates of the kernels K_m, $m\ge3$, which imply the estimate (2.3.17) for Q, as well as the estimates (2.3.18) and (2.3.31), are given in detail in [116] for a more general case of degenerations on the initial hyperplane.

In order to obtain the estimates (2.3.33), we introduce a new unknown function v by the formula (2.3.27), in which the function g satisfies the condition B_{16}. As in the proof of Theorem 2.12, we obtain the equation (2.3.28) and verify, for a sufficiently small η, the conditions B_{13}–B_{15}, provided these conditions are fulfilled for the equation (2.1.1). This gives the equality (2.3.30) for Z, in which $\tilde{Z}$ is a FSCP for the equation (2.3.28). This equality, the estimates (2.3.35) for $\tilde{Z}$, and the condition B_{16} imply the estimates (2.3.33).

Let us prove the estimates (2.3.32). It is sufficient to consider the integral (2.1.85). For $\|k\| = 2b$, due to Properties 2.11 and 2.13, we have

$$\begin{aligned}\partial_x^k W(t,x;\tau,\xi) &= \int_\tau^{t_0} d\lambda \int_{\mathbb{R}^n} \partial_x^k Z_0(t,x;\lambda,y) Q(\lambda,y;\tau,\xi)dy \\ &+ \int_{t_0}^{t} d\lambda \int_{\mathbb{R}^n} \partial_x^k Z_0(t,x;\lambda,y)\Delta_y^x Q(\lambda,y;\tau,\xi)dy \\ &+ \int_{t_0}^{t} \left(\int_{\mathbb{R}^n} \Delta_z^x \partial_x^k G_0(t,x;\lambda,y;z)|_{z=y} dy \right) Q(\lambda,x;\tau,\xi)d\lambda \\ &=: \sum_{j=1}^{3} J_j. \end{aligned} \tag{2.3.41}$$

Using (2.1.46), (2.1.80), (2.3.10), (2.3.17), (2.3.39), and the method of the proof of the estimate (2.3.38), we find that

$$\begin{aligned}|J_1| &\le C_l(t-\tau)^{-1}\Bigg(\int_\tau^{t_0} (\lambda-\tau)^{-1+\gamma/(2b)} d\lambda \int_{B(\xi,1)} E_c^{(0)}(t,x;\lambda,y) \\ &\times E_c^{(0)}(\lambda,y;\tau,\xi)((t-\lambda)(\lambda-\tau))^{-M} \\ &\times \exp\{-c((t-\lambda)(D(y))^{2b} + (\lambda-\tau)(D(\xi))^{2b})\}dy \\ &+ \int_\tau^{t_0} (\lambda-\tau)^{-M-1+\gamma/(2b)} \exp\{-c(\lambda-\tau)(D(\xi))^{2b}\}d\lambda \\ &\times \int_{\mathbb{R}^n\setminus B(\xi,1)} (t-\lambda)^{-M} E_c^{(0)}(t,x;\lambda,y)E_c^{(0)}(\lambda,y;\tau,\xi)dy + \int_\tau^{t_0} (\lambda-\tau)^{\gamma/(2b)} d\lambda \\ &\times \int_{\mathbb{R}^n} (t-\lambda)^{-M} E_c^{(0)}(t,x;\lambda,y)E_c^{(0)}(\lambda,y;\tau,\xi)dy(D(\xi))^{-l} \Bigg) \\ &\le C_l(t-\tau)^{-1}\Bigg((t-\tau)^{-M} E_{c_1}^{(0)}(t,x;\tau,\xi)\exp\{-c_1(t-\tau)(D(\xi))^{2b}\} \\ &\times \int_\tau^{t_0} (\lambda-\tau)^{-1+\gamma/(2b)} d\lambda + E_{c_1}^{(0)}(t,x;\tau,\xi) \int_\tau^{t_0} d\lambda \int_{\mathbb{R}^n} (t-\lambda)^{-M} \\ &\times E_{c-c_0}^{(0)}(t,x;\lambda,y)dy(D(\xi))^{-l} \Bigg) \end{aligned}$$

$$\leq C_l((t-\tau)^{-M-1+\gamma/(2b)}$$
$$\times \exp\{-c_1(t-\tau)(D(\xi))^{2b}\} + (D(\xi))^{-l})E_{c_1}^{(0)}(t,x;\tau,\xi). \tag{2.3.42}$$

It follows from (2.3.20) that

$$|J_2| \leq C_l((t-\tau)^{-M-1+\gamma/(2b)}$$
$$\times \exp\{-c_1(t-\tau)(D(\xi))^{2b}\} + (D(\xi))^{-l})E_{c_1}^{(0)}(t,x;\tau,\xi), \tag{2.3.43}$$

while (2.3.23) implies the estimate

$$|J_3| \leq C_l((t-\tau)^{-M-1+2\gamma/(2b)} \exp\{-c_1(t-\tau)(D(\xi))^{2b}\} + (D(\xi))^{-l})$$
$$\times E_c^{(0)}(t,x;\tau,\xi)((D(x))^{2b} + \exp\{2bM\varepsilon(D(x))^{2b}\}). \tag{2.3.44}$$

Using the equalities (2.1.84) and (2.3.41), the estimates (2.3.10) and (2.3.42)–(2.3.44), we obtain the estimate (2.3.32). □

2.3.3. Fundamental solution of the Cauchy problem for some equations reducible to dissipative ones. Let us consider the equation (2.1.1) with unbounded coefficients under some assumptions, different from those of Section 2.3.2.

Let $b' := \max_{j\in\{1,\dots,n\}} b_j$, $\delta_0 \in (0,1)$, and the function $\eta_0 \in C^{4b'}(\mathbb{R})$ be such that

$$\eta_0(r) := |r| \text{ for } |r| \geq 1+\delta_0, \qquad \eta_0(r) := 1 \text{ for } |r| \leq 1-\delta_0,$$
$$\text{and } \ |\eta_0^{(k)}(r)| \leq C, \qquad r \in \mathbb{R}, \quad k \in \{1,\dots,4b'\}.$$

Consider the functions

$$D_0(x) := \left(\sum_{j=1}^{n} (\eta_0(x_j))^{q_j}\right)^{(1-\varepsilon_0)/(2b)}, \quad x \in \mathbb{R}^n, \tag{2.3.45}$$

and

$$g_0(t,x) := a(1-(Aa)^{2b-1}t)^{-1/(2b-1)} \left(\sum_{j=1}^{n} (\eta_0(x_j))^{q_j}\right)^{1-\varepsilon_0},$$
$$(t,x) \in \Pi_{[0,T]}, \tag{2.3.46}$$

where the number $\varepsilon_0 \in (0,1)$ is chosen in a special way for each specific situation, A is a sufficiently large positive number, and $a > 0$ is such a number that $T \leq (Aa)^{1-2b}/2$.

We shall use the following set of conditions containing the function $D = D_0$ defined in (2.3.45):

(C_{11}) The functions $b_k(t,x) := a_k(t,x)(D_0(x))^{\|k\|-2b}$, $(t,x) \in \Pi_{[0,T]}$, $\|k\| \le 2b$, are bounded, and the equation (2.1.1) is uniformly $\overrightarrow{2b}$-parabolic on $\Pi_{[0,T]}$;

(C_{12}) The conditions A_{16} and A_{17} are satisfied;

(C_{13}) The equation (2.1.1) possesses the adjoint equation (2.1.3) whose coefficients satisfy C_{12}.

Let us consider the FSCPs for the equation (2.1.1) and the adjoint equation (2.1.3).

Theorem 2.14. *Suppose that the conditions C_{11} and C_{12} are fulfilled for the equation (2.1.1). Then there exists a FSCP Z for this equation, and*

$$|\partial_x^k Z(t,x;\tau,\xi)| \le C \sum_{j=0}^{\|k\|} (t-\tau)^{-M-(\|k\|-j)/(2b)} (D_0(x))^{j(1-\varepsilon)} E_c^{(0)}(t,x;\tau,\xi) \times \exp\{g_0(t,x) - g_0(\tau,\xi)\},$$
$$0 \le \tau < t \le T, \quad \{x,\xi\} \subset \mathbb{R}^n, \quad \|k\| \le 2b, \tag{2.3.47}$$

where $C > 0$, $c > 0$, D_0 and g_0 are the functions from (2.3.45) *and* (2.3.46) *respectively.*

If, in addition to the conditions C_{11} and C_{12}, the condition C_{13} is fulfilled, then there exists a FSCP $Z^*(\tau,\xi;t,x)$, $0 \le \tau < t \le T$, $\{\xi,x\} \subset \mathbb{R}^n$, for the adjoint equation (2.1.3), *and Z^*, Z satisfy the equalities* (2.1.111) *and* (2.1.112).

Proof. Let us make in the equation (2.1.1) a change of variables by the formula

$$u(t,x) = e^{g_0(t,x)} v(t,x), \quad (t,x) \in \Pi_{(0,T]},$$

in which g_0 is the function from (2.3.46). As a result we obtain the equation

$$\Bigg(I\partial_t - \sum_{\|k\| \le 2b} a_k(t,x)\partial_x^k + \sum_{\|k\|<2b} \sum_{\substack{l>k,\\ \|l\|\le 2b}} C_l^k a_l(t,x) e^{-g_0(t,x)} \partial_x^{l-k} e^{g_0(t,x)} \partial_x^k - \partial_t g_0(t,x) I \Bigg) v(t,x)$$
$$= e^{-g_0(t,x)} f(t,x), \quad (t,x) \in \Pi_{(0,T]}. \tag{2.3.48}$$

The equation (2.3.48) is a dissipative $\overrightarrow{2b}$-parabolic equation satisfying the conditions A_{15}–A_{17} with the function D_0 defined in (2.3.45). The properties of coefficients of this equation involved in the conditions A_{15}–A_{17} are consequences of the conditions C_{11} and C_{12}, and the inequalities

$$|\partial_x^k g_0(t,x)| \le C_1 (D_0(x))^{\|k\|(1-\varepsilon)}, \quad 0 < \|k\| \le 4b, \tag{2.3.49}$$
$$C_2 A^{2b-1} a^{2b} \le (D_0(x))^{-2b} \partial_t g_0(t,x) \le C_3 A^{2b-1} a^{2b}, \tag{2.3.50}$$
$$|\partial_t \partial_x^l g_0(t,x)| \le C_4 A^{2b-1} a^{2b} (D_0(x))^{\|l\|(1-\varepsilon)}, \quad 0 < \|l\| \le 2b, \tag{2.3.51}$$
$$(t,x) \in \Pi_{[0,T]}.$$

These inequalities are valid under the appropriate choice of the number ε_0 from the equality (2.3.45).

In order to show that the equation (2.3.48) is dissipative, we consider its characteristic equation

$$\det\Bigg(I(p+(D_0(x))^{-2b}\partial_t g_0(t,x)\mu^{2b}) - \sum_{\|k\|+k_{n+1}=2b} b_k(t,x)(i\sigma)^k\mu^{k_{n+1}}$$

$$- \sum_{\|k\|+k_{n+1}=2b} \sum_{\substack{l>k,\\ \|l\|\le 2b}} C_l^k b_l(t,x)e^{-g_0(t,x)}\partial_x^{l-k}e^{g_0(t,x)}(D(x))^{\|k\|-\|l\|}(i\sigma)^k\mu^{k_{n+1}} \Bigg) = 0.$$

By the inequalities (2.3.49)–(2.3.51) and the condition C_{11}, the coefficients of this equation are bounded, and for $\mu = 0$ its p-roots are located properly. Therefore, in order to prove the inequality

$$\operatorname{Re} p(t,x,\sigma,\mu) \le -\delta\left(\sum_{j=1}^{n}\sigma_j^{2b_j} + \mu^{2b}\right),$$

it suffices to verify that $\operatorname{Re} p(t,x,\sigma,\mu) \ne 0$ for all $(t,x) \in \Pi_{[0,T]}$, $\sigma \in \mathbb{R}^n$, and $\mu \in \mathbb{R}$. That can be done, using the inequality (2.3.50) and choosing A large enough.

Hence, by Theorem 2.12 there exists a FSCP $\tilde{Z}$ for the equation (2.3.48), it satisfies the estimates (2.3.24), and it is connected with the FSCP Z for the equation (2.1.1) by the formula (2.3.30), in which $g_0(t,x)$ and $g_0(\tau,\xi)$ are substituted for $g(x)$ and $g(\xi)$ respectively. Using the estimates (2.3.24) and (2.3.49), we obtain the estimates (2.3.47).

The existence of the FSCP Z^* and the equalities (2.1.111), (2.1.112) for Z and Z^* are proved just as in Remark 5 from Section 2.1.5. □

2.3.4. Theorems about the solvability of the Cauchy problem. The results regarding the FSCP expounded in Sections 2.3.2 and 2.3.3 make it possible to study, just as in Section 2.2, properties of the corresponding Poisson integrals and volume potentials, and to investigate on this basis the solvability of the Cauchy problem. Since the above FSCP estimates for equations with growing coefficients are not as exact as those for the case of bounded coefficients (Section 2.1), the subsequent results regarding the Cauchy problem are more unpretentious than the ones from Section 2.2. Below we give only two theorems about the solvability of the Cauchy problem. We do not present here the properties of the Poisson integrals and volume potentials needed there, since that would require essentially a repetition of the proofs from Sections 2.2.2 and 2.2.4.

In order to formulate the theorems, we need norms and spaces similar to those used in Section 2.2. There we employed the weight functions Ψ_ν defined in

(2.2.1). Here we use the following weight functions:

$$\Phi_\nu(t,x) := \Psi_\nu(t,x)e^{\nu g(x)}, \quad \Phi_\nu^0(t,x) := \Psi_\nu(t,x)e^{\nu g_0(t,x)},$$
$$(t,x) \in \Pi_{[0,T]}, \quad x \in \mathbb{R}^n, \quad \nu \in \{-1,1\},$$

where the function Ψ_ν is the same as before, while g and g_0 are the functions from the estimates (2.3.25) and (2.3.47) respectively.

The definitions of the spaces used below,

$$L_p^{\vec{a},g(\cdot)}, \; M^{\vec{a},g(\cdot)}, \; L_1^{-\vec{k}(T,\vec{a}),\,-g(\cdot)} \quad \text{and} \quad C_0^{-\vec{k}(T,\vec{a}),-g(\cdot)},$$

with the norms

$$\|\varphi\|_p^{\vec{a},g(\cdot)}, \; \|\mu\|^{\vec{a},g(\cdot)}, \; \|\varphi\|_1^{-\vec{k}(T,\vec{a}),-g(\cdot)}, \quad \text{and} \quad \|\varphi\|_0^{-\vec{k}(T,\vec{a}),-g(\cdot)}$$

respectively, and also the norms $\|u(t,\cdot)\|_p^{\vec{k}(t,\vec{a}),g(\cdot)}$ are the same as the definition of the corresponding spaces without the index $g(\cdot)$ in Section 2.1.2, only with the weight functions Φ_ν substituted for Ψ_ν. If instead of the functions Ψ_ν we take the functions Φ_ν^0, then we obtain another series of spaces

$$L_p^{\vec{a},g_0(0,\cdot)}, \; M^{\vec{a},g_0(0,\cdot)}, \; L_1^{-\vec{k}(T,\vec{a}),-g_0(T,\cdot)}, \; C_0^{-\vec{k}(T,\vec{a}),-g_0(T,\cdot)},$$

and norms

$$\|\varphi\|_p^{\vec{a},g_0(0,\cdot)}, \qquad \|\mu\|^{\vec{a},g_0(0,\cdot)}, \qquad \|\varphi\|_1^{-\vec{k}(T,\vec{a}),-g_0(T,\cdot)},$$
$$\|\varphi\|_0^{-\vec{k}(T,\vec{a}),-g_0(T,\cdot)}, \qquad \|u(t,\cdot)\|_p^{\vec{k}(t,\vec{a}),g_0(t,\cdot)}.$$

For the inhomogeneous part f of the equation (2.1.1) we use the following conditions:

(D_{15p}) The function $f : \Pi_{(0,T]} \to \mathbb{C}_{N1}$ is continuous, locally Hölder continuous in x uniformly with respect to t, and for any $t \in (0,T]$ the expressions

$$\|f(t,\cdot)\|_p^{\vec{k}(t,\vec{a}),g(\cdot)} \quad \text{and} \quad F_p(t) := \int_0^t \|f(\tau,\cdot)\|_p^{\vec{k}(\tau,\vec{a}),g(\cdot)} d\tau$$

are finite.

(D_{16p}) The function $f : \Pi_{(0,T]} \to \mathbb{C}_{N1}$ is continuous, locally Hölder continuous in x uniformly with respect to t, and for any $t \in (0,T]$ the expressions

$$\|f(t,\cdot)\|_p^{\vec{k}(t,\vec{a}),g_0(t,\cdot)} \quad \text{and} \quad F_p^0(t) := \int_0^t \|f(\tau,\cdot)\|_p^{\vec{k}(\tau,\vec{a}),g_0(\tau,\cdot)} d\tau,$$

where $p \in [1,\infty]$, are finite.

The following theorem holds for dissipative equations with smooth coefficients.

Theorem 2.15. *Suppose that the conditions A_{15}–A_{17} are satisfied. Then:*

1) *If $\varphi \in L_p^{\vec{a},g(\cdot)}$, and the function f satisfies the condition D_{15p}, $p \in [1,\infty]$, then the function* (2.2.92) *is a solution of the equation* (2.1.1), *such that for all $t \in (0,T]$*

$$\|u(t,\cdot)\|_p^{\vec{k}(t,\vec{a}),g(\cdot)} \leq C(\|\varphi\|_p^{\vec{a},g(\cdot)} + F_p(t)),$$

with a certain constant $C > 0$, where the function F_p is the one from the condition D_{15p}, and the relation

$$\lim_{t\to 0} \|u(t,\cdot) - \varphi(\cdot)\|_p^{\vec{k}(t,\vec{a}),g(\cdot)} = 0,$$

holds if $p \in [1,\infty)$, whereas for $p = \infty$ the relation (2.2.22) *holds for any $\psi \in L_1^{-\vec{k}(T,\vec{a}),-g(\cdot)}$.*

2) *If $\mu \in M^{\vec{a},g(\cdot)}$, and the condition D_{151} is satisfied for f, then the function* (2.2.103) *is a solution of the equation* (2.1.1), *and there exists such a constant $C > 0$ that for all $t \in (0,T]$,*

$$\|u(t,\cdot)\|_1^{\vec{k}(t,\vec{a}),g(\cdot)} \leq C(\|\mu\|^{\vec{a},g(\cdot)} + F_1(t)),$$

and the relation (2.2.47) *holds for an arbitrary function $\psi \in C_0^{-\vec{k}(T,\vec{a}),-g(\cdot)}$.*

A similar theorem is valid for the equations from Section 2.3.3, reducible to dissipative ones.

Theorem 2.16. *Let the conditions C_{11} and C_{12} be satisfied. Then:*

1) *If $\varphi \in L_p^{\vec{a},g_0(0,\cdot)}$, and the function f satisfies the condition D_{16p}, $p \in [1,\infty]$, then the formula* (2.2.92) *defines a solution of the equation* (2.1.1), *for which*

$$\|u(t,\cdot)\|_p^{\vec{k}(t,\vec{a}),g_0(t,\cdot)} \leq C(\|\varphi\|_p^{\vec{a},g_0(0,\cdot)} + F_p^0(t)), \quad t \in (0,T];$$

if $p \in [1,\infty)$, the relation

$$\lim_{t\to 0} \|u(t,\cdot) - \varphi(\cdot)\|_p^{\vec{k}(t,\vec{a}),g_0(t,\cdot)} = 0 \quad \textit{holds,}$$

while for $p = \infty$ the relation (2.2.22) *holds for each $\psi \in L_1^{-\vec{k}(T,\vec{a}),-g_0(T,\cdot)}$;*

2) *If $\mu \in M^{\vec{a},g_0(0,\cdot)}$, and the function f satisfies the condition D_{161}, then the formula* (2.2.103) *defines a solution of the equation* (2.1.1), *for which*

$$\|u(t,\cdot)\|_1^{\vec{k}(t,\vec{a}),g_0(t,\cdot)} \leq C(\|\mu\|^{\vec{a},g_0(0,\cdot)} + F_1^0(t)), \quad t \in (0,T],$$

and the relation (2.2.47) *holds for any $\psi \in C_0^{-\vec{k}(T,\vec{a}),-g_0(T,\cdot)}$.*

2.4 Equations with degenerations on the initial hyperplane

2.4.1. Fundamental solution of the Cauchy problem. In this section we give a brief exposition of the results regarding equations from the classes $\mathbf{E}_{13}$ and $\mathbf{E}_{14}$, that is the equations of the form (1.1.11), for which the coefficients a_k with $\|k\| = 2b$ are bounded, while those with $\|k\| < 2b$ are either bounded or may grow as $|x| \to \infty$. The degenerations at $t = 0$ in these equations are caused by the presence of the functions α and β with the properties indicated in Section 1.1.1.

Below we use the following additional notation:

$$\hat{A}(t,\tau) := \int_{\tau}^{t} \frac{d\lambda}{\alpha(\lambda)},$$

$$\hat{B}(t,\tau) := \int_{\tau}^{t} \frac{\beta(\lambda)}{\alpha(\lambda)} d\lambda,$$

$$\hat{E}^d(t,\tau) := \exp\{d\hat{A}(t,\tau)\},$$

$$\hat{E}_c(t,x;\tau,\xi) := \exp\left\{-c\sum_{j=1}^{n} (\hat{B}(t,\tau))^{1-q_j} |x_j - \xi_j|^{q_j}\right\},$$

$$\hat{E}_c^d(t,x;\tau,\xi) := \hat{E}_c(t,x;\tau,\xi)\hat{E}^d(t,\tau),$$

$$\hat{\rho}(t,x;\tau,\xi) := \left((\hat{A}(t,\tau))^{1/b} + (p(x;\xi))^2\right)^{1/2},$$

$$0 \le \tau < t \le T, \quad \{x,\xi\} \subset \mathbb{R}^n,$$

where $c > 0$ and $d \in \mathbb{R}$.

First we consider the equation (1.1.11) for the case where the corresponding equation (2.1.1) without degenerations satisfies the condition A_{11}, and the coefficients a_k, $\|k\| \le 2b$, satisfy A_{12} and possibly also the condition A_{14}.

The degenerations will be classified by $\hat{A}(T,0)$ and $\hat{B}(T,0)$. In the case $\hat{A}(T,0) < \infty$ we say that the equation (1.1.11) is weakly degenerate. In the case $\hat{A}(T,0) = \infty$ it is called strongly degenerate, and very strongly degenerate, if $\hat{A}(T,0) = \infty$ and $\hat{B}(T,0) = \infty$.

It is not always possible to consider the Cauchy problem for the equation (1.1.11) with the initial data at $t = 0$ in the usual sense. However we can define a FSCP as such a function $Z(t,x;\tau,\xi)$, $0 < \tau < t \le T$, $\{x,\xi\} \subset \mathbb{R}^n$, that the formula

$$u(t,x) = \int_{R^n} Z(t,x;\tau,\xi)\varphi(\xi)d\xi, \quad (t,x) \in \Pi_{(\tau,T]},$$

determines a solution of the homogeneous equation (1.1.11) satisfying the condition $u|_{t=\tau} = \varphi$ for any $\tau \in (0,T)$ and an arbitrary continuous bounded function φ.

A FSCP for the equation (1.1.11) under the above conditions is constructed and investigated by the Levi method just as it was done in Sections 2.1.1–2.1.5 for an equation without degenerations. As a parametrix we take the function (1.1.43) where $G_0(\cdot;\cdot,\cdot;\cdot;y)$ is a FSCP for the equation (1.1.48). Note that this equation contains a term without degenerations, the one with the coefficient a_0 at the unknown function. This term belongs actually to the group of leading terms of the equation. Its presence in (1.1.48) facilitates the estimation procedures of the Levi method and makes it possible to obtain exact results regarding the FSCP for the equation (1.1.11). The parametrix Z_0 satisfies the following estimates:

$$|\partial_x^k Z_0(t,x;\tau,\xi)| \le C_k(\hat{B}(t,\tau))^{-M-\|k\|/(2b)}\,\hat{E}_c^d(t,x;\tau,\xi),$$
$$0<\tau<t\le T, \quad \{x,\xi\}\subset\mathbb{R}^n, k\in\mathbb{Z}_+^n,$$

where $C_k>0$, $c>0$ and $d\in\mathbb{R}$.

The main theorem of this section about the FSCP for the equation (1.1.11) is as follows.

Theorem 2.17. *Suppose that the coefficients a_k, $\|k\|\le 2b$, of the equation (1.1.11) satisfy the conditions A_{11} and A_{12}. Then there exists a FSCP Z for this equation, such that*

$$|\partial_x^k Z(t,x;\tau,\xi)| \le C(\hat{B}(t,\tau))^{-M-\|k\|/(2b)}\,\hat{E}_c^d(t,x;\tau,\xi), \tag{2.4.1}$$
$$|\Delta_x^{x'}\partial_x^k Z(t,x;\tau,\xi)| \le C(p(x;x'))^{\gamma}(\hat{B}(t,\tau))^{-M-(\|k\|+\gamma)/(2b)} \times(\hat{E}_c^d(t,x;\tau,\xi)+\hat{E}_c^d(t,x';\tau,\xi)), \tag{2.4.2}$$
$$0<\tau<t\le T, \quad \{x,x',\xi\}\subset\mathbb{R}^n, \quad \|k\|\le 2b,$$

where $C>0$, $c>0$ and $d\in\mathbb{R}$.

Proof. We proceed as in the proof of Theorem 2.2. Therefore we only indicate some differences from the case considered in that theorem.

The second summand from the formula (2.1.84) is taken as

$$W(t,x;\tau,\xi)=\int_\tau^t\frac{d\lambda}{\alpha(\lambda)}\int_{\mathbb{R}^n}Z_0(t,x;\lambda,y)Q(\lambda,y;\tau,\xi)dy,$$
$$0<\tau<t\le T, \quad \{x,\xi\}\subset\mathbb{R}^n.$$

Instead of the integral equation (1.1.47) we have the equation

$$Q(t,x;\tau,\xi)=K(t,x;\tau,\xi)+\int_\tau^t\frac{d\lambda}{\alpha(\lambda)}\int_{\mathbb{R}^n}K(t,x;\lambda,y)Q(\lambda,y;\tau,\xi)dy,$$
$$0<\tau<t\le T, \quad \{x,\xi\}\subset\mathbb{R}^n,$$

in which

$$K(t,x;\tau,\xi):=\Big(\beta(t)\sum_{\|k\|=2b}\Delta_x^\xi a_k(t,x)\partial_x^k+\beta(t)\sum_{\|k\|<2b}a_k(t,x)\partial_x^k+\Delta_x^\xi a_0(t,x)\Big)\times Z_0(t,x;\tau,\xi).$$

The a priori assumptions regarding the function Q are as follows:

$$\begin{aligned}|Q(t,x;\tau,\xi)| &\le C\beta(t)(\hat{B}(t,\tau))^{-M-1+\gamma/(2b)}\hat{E}^d_c(t,x;\tau,\xi),\\ |\Delta_x^{x'}Q(t,x;\tau,\xi)| &\le C\beta(t)(p(x;x'))^{\gamma_1}(\hat{B}(t,\tau))^{-M-1+\gamma_2/(2b)}\\ &\quad\times(\hat{E}^d_c(t,x;\tau,\xi)+\hat{E}^d_c(t,x';\tau,\xi)),\\ &\quad 0<\tau<t\le T,\quad \{x,x',\xi\}\subset\mathbb{R}^n,\end{aligned}$$

where the numbers γ_1 and γ_2 are the same as in (2.1.87).

For the kernel K we obtain the estimate

$$\begin{aligned}|K(t,x;\tau,\xi)| &\le C_1\beta(t)(\hat{B}(t,\tau))^{-M-1+\gamma/(2b)}\hat{E}^{d_1}_{c_1}(t,x;\tau,\xi),\\ &\quad 0<\tau<t\le T,\quad \{x,\xi\}\subset\mathbb{R}^n.\end{aligned}$$

Instead of the inequalities (2.1.61) and (2.1.80) used in the proof of Theorem 2.2, here we use the inequalities

$$\begin{aligned}&(p(x;y))^{\gamma}\hat{E}_c(t,x;\tau,y)\le C(\hat{B}(t,\tau))^{\gamma/(2b)}\hat{E}_{c_1}(t,x;\tau,y),\\ &\int_{\mathbb{R}^n}\hat{E}_c(t,x;\tau,\xi)\hat{E}_c(\tau,\xi;t_0,y)(\hat{B}(t,\tau)\hat{B}(\tau,t_0))^{-M})d\xi\\ &\qquad\le C_\varepsilon(\hat{B}(t,t_0))^{-M}\hat{E}_{c(1-\varepsilon)}(t,x;t_0,y),\\ &\qquad\quad 0<t_0<\tau<t\le T,\quad \{x,y\}\subset\mathbb{R}^n.\end{aligned}$$

We use the following inequalities for the functions $\hat{A}$, $\hat{B}$, and $\hat{E}^d$:

$$(\hat{B}(t,\tau))^p\hat{E}^d(t,\tau)\le(\beta(t))^p(\hat{A}(t,\tau))^p\hat{E}^d(t,\tau)\le(\beta(T))^p\hat{E}^{d_1}(t,\tau),$$

$$0<\tau<t\le T,\quad p>0,\quad d_1>d. \qquad \square$$

Remark 1. If, in addition to the assumptions of Theorem 2.17, the coefficients a_k, $\|k\|\le 2b$, for every $\{t,t'\}\subset(0,T]$, $t<t'$, and $x\in\mathbb{R}^n$ satisfy the condition

$$|\Delta_t^{t'}a_k(t,x)|\le C(\hat{A}(t',t))^{\gamma/(2b)},$$

then, apart from the estimates (2.4.1) and (2.4.2),

$$\begin{aligned}|\Delta_{t,x}^{t',x'}\partial_x^k Z(t,x;\tau,\xi)| &\le C(\hat{p}(t',x';t,x))^{\gamma}\\ &\quad\times\Big((\hat{B}(t,\tau))^{-M-(\|k\|+\gamma)/(2b)}\hat{E}^d_c(t,x;\tau,\xi)\\ &\quad+(\hat{B}(t',\tau))^{-M-(\|k\|+\gamma)/(2b)}\hat{E}^d_c(t',x';\tau,\xi)\Big),\\ &\quad\|k\|\le 2b,\end{aligned}\tag{2.4.3}$$

$$\left| \int_{\mathbb{R}^n} \partial_x^k Z(t,x;\tau,y)dy \right| \le C(\hat{B}(t,\tau))^{-(\|k\|-\gamma)/(2b)} \hat{E}^d(t,\tau),$$

$$0 < \|k\| \le 2b, \tag{2.4.4}$$

$$\left| \Delta_{t,x}^{t',x'} \int_{\mathbb{R}^n} \partial_x^k Z(t,x;\tau,y)dy \right| \le C(\hat{\rho}(t',x';t,x))^{\gamma} \left(\hat{E}^d(t,\tau) + \hat{E}^d(t',\tau) \right),$$

$$0 < \|k\| \le 2b, \tag{2.4.5}$$

where $0 < \tau < t < t' \le T$, $\{x, x', \xi\} \subset \mathbb{R}^n$.

Remark 2. The constant d in the estimates (2.4.3)–(2.4.5) may have any sign or be equal to 0. For example, if for the equation (1.1.11) the estimates (2.4.3)–(2.4.5) hold with $d = d_0 > 0$, then the corresponding estimates for the equation

$$\left(\alpha(t)I\partial_t - \beta(t) \sum_{0<\|k\|\le 2b} a_k(t,x)\partial_x^k - a_0(t,x) + pI \right) u(t,x) = f(t,x),$$

$$(t,x) \in \Pi_{(0,T]}, \tag{2.4.6}$$

with the parameter $p \in \mathbb{C}$, $\operatorname{Re} p \ge d_0$, are valid with $d = d_0 - \operatorname{Re} p \le 0$.

This follows from the formula

$$Z_p(t,x;\tau,\xi) = \exp\{-p\hat{A}(t,\tau)\} Z(t,x;\tau,\xi),$$
$$0 < \tau < t \le T, \quad \{x,\xi\} \subset \mathbb{R}^n,$$

where Z and Z_p are the FSCP for the equations (1.1.11) and (2.4.6) respectively.

Remark 3. In the case of the weak degeneration of the equation (1.1.11) we may set $\tau = 0$ and $d = 0$ in the estimates (2.4.3)–(2.4.5)

Remark 4. A FSCP for the equation (1.1.11) has some properties similar to the case of an equation without degenerations.

In particular, let t_0 be an arbitrary number from the interval $(0,T)$, $Z(t,x;\tau,\xi)$, $0 < \tau < t \le T$, $\{x,\xi\} \subset \mathbb{R}^n$, be a FSCP for the equation (1.1.11). Then the function $Z(t,x;\tau,\xi)$, $t_0 \le \tau < t \le T$, $\{x,\xi\} \subset \mathbb{R}^n$, is a FSCP with the initial condition at $t = t_0$ for the uniformly $\overrightarrow{2b}$-parabolic equation

$$\left(I\partial_t - \frac{\beta(t)}{\alpha(t)} \sum_{0<\|k\|\le 2b} a_k(t,x)\partial_x^k - \frac{a_0(t,x)}{\alpha(t)} \right) u(t,x) = \frac{f(t,x)}{\alpha(t)}, \quad (t,x) \in \Pi_{[t_0,T]}.$$

If we assume, apart from the conditions A_{11} and A_{12}, also the condition A_{14}, then by Remark 5 from Section 2.1.5, for the adjoint equation

$$-\partial_\tau v(\tau,\xi) - \frac{\beta(\tau)}{\alpha(\tau)} \sum_{0<\|k\|\le 2b} (-\partial_\xi)^k (\bar{a}_k'(\tau,\xi)v(\tau,\xi)) - \frac{\bar{a}_0'(\tau,\xi)}{\alpha(\tau)} v(\tau,\xi) = g(\tau,\xi),$$

$$(\tau,\xi) \in \Pi_{[t_0,T]},$$

there exists a FSCP Z^*, and the equalities (2.1.111) and (2.1.112) hold, in which $t_0 \le \tau < \lambda < t \le T$. Since t_0 is an arbitrary fixed number from $(0, T)$, the equalities (2.1.111) and (2.1.112) hold for any $\tau \in (0, t)$, $t \in (0, T]$. This is the meaning of the *normality property* and the *convolution formula* for the FSCP for the equation (1.1.11).

Results on the FSCP similar to those for equations with growing coefficients (Sections 2.3.2 and 2.3.3) are valid for such equations also in the presence of degenerations at the initial hyperplane. Below we formulate such theorems for the equations

$$\Big(\alpha(t)I\partial_t - \beta(t) \sum_{\|k\|\le 2b} a_k(t,x)\partial_x^k\Big)u(t,x) = f(t,x),$$
$$(t,x) \in \Pi_{(0,T]}, \tag{2.4.7}$$

and

$$\Big(\alpha(t)I\partial_t - \beta(t) \sum_{0<\|k\|\le 2b} a_k(t,x)\partial_x^k - b(t,x)\Big)u(t,x) = f(t,x),$$
$$(t,x) \in \Pi_{(0,T]}. \tag{2.4.8}$$

Note that in the equation (2.4.7) we can pass from the variable $t \in (0, T]$ to a new variable τ using the change of variables $\tau = -B(T, t)$. Here τ will belong to the interval $(-B(T, 0), 0]$ which, in the case of a very strong degeneration, that is $B(T, 0) = \infty$, is unbounded. Hence the equation (2.4.7) is reduced to an equation without degeneration which should in general be considered on an unbounded time interval. Since that would require a specific additional investigation, we do not use the above change of variables.

Theorem 2.18. *Suppose that the coefficients a_k, $\|k\| \le 2b$, of the equation (2.4.7) satisfy the conditions A_{15}–A_{17}. Then there exists a FSCP Z for the equation (2.4.7), such that*

$$|\partial_x^k Z(t,x;\tau,\xi)| \le C(\hat{B}(t,\tau))^{-M-\|k\|/(2b)}\hat{E}_c(t,x;\tau,\xi)\exp\{d\hat{B}(t,\tau)\},$$

and

$$|\partial_x^k Z(t,x;\tau,\xi)| \le C\sum_{j=0}^{\|k\|}(\hat{B}(t,\tau))^{-M-(\|k\|-j)/(2b)}(D(x))^{j(1-\varepsilon)}$$
$$\times \hat{E}_c(t,x;\tau,\xi)\exp\{d\hat{B}(t,\tau) + g(x) - g(\xi)\}, \tag{2.4.9}$$
$$0 < \tau < t \le T, \quad \{x,\xi\} \subset \mathbb{R}^n, \quad \|k\| \le 2b,$$

where $C > 0$, $c > 0$, $d \in \mathbb{R}$, and g is an arbitrary function satisfying the condition A_{18}.

Theorem 2.19. *If the coefficients a_k, $0 < \|k\| \le 2b$, and the dissipation characteristic D satisfy the conditions B_{13}–B_{15}, while the function b is bounded, continuous in t, and Hölder continuous in x, with the exponent γ, on $\Pi_{[0,T]}$, then there exists a FSCP Z for the equation* (2.4.8), *such that*

$$|\partial_x^k Z(t,x;\tau,\xi)| \le C_l((\hat{B}(t,\tau))^{-M-\|k\|/(2b)} \exp\{-c\hat{B}(t,\tau)(D(\xi))^{2b}\} + (D(\xi))^{-l})\hat{E}_c^d(t,x;\tau,\xi), \tag{2.4.10}$$

$$|\partial_x^k Z(t,x;\tau,\xi)| \le C\sum_{j=0}^{\|k\|}(\hat{B}(t,\tau))^{-M-(\|k\|-j)/(2b)}(D(x))^j \hat{E}_c^d(t,x;\tau,\xi) \times \exp\{g(x) - g(\xi)\}, \tag{2.4.11}$$

if $\|k\| < 2b$, *and*

$$|\partial_x^k Z(t,x;\tau,\xi)| \le C_l((\hat{B}(t,\tau))^{-M-1} \exp\{-c\hat{B}(t,\tau)(D(\xi))^{2b}\} + (D(\xi))^{-l})\hat{E}_c^d(t,x;\tau,\xi)((D(x))^{2b} + \exp\{2b\varepsilon M\hat{B}(t,\tau)(D(x))^{2b}\}), \tag{2.4.12}$$

if $\|k\| = 2b$.

In the estimates (2.4.10)–(2.4.12), $0 < \tau < t \le T$, $\{x,\xi\} \subset \mathbb{R}^n$, *$l$ is an arbitrary positive number, g is an arbitrary function satisfying the condition B_{16}, $C_l > 0$, $C > 0$, $c > 0$, $d \in \mathbb{R}$.*

Let us consider also the equation (2.4.7) satisfying the conditions C_{11}–C_{13}, and the condition

(C_{14}) The degeneration is weak, that is $B(T,0) < \infty$.

If the condition C_{14} is fulfilled, we consider the function

$$\hat{g}_0(t,x) := a(1 - (Aa)^{2b-1}\hat{B}(t,0))^{-1/(2b-1)}(D_0(x))^{2b}, \quad (t,x) \in \Pi_{[0,T]}, \tag{2.4.13}$$

where D_0 is a function from (2.3.45), A is a sufficiently big positive number, the number $a > 0$ is such that $\hat{B}(T,0) \le (Aa)^{1-2b}/2$.

Theorem 2.20. *If the equation* (2.4.7) *satisfies the conditions C_{11}, C_{12} and C_{14}, then it has a FSCP Z, such that*

$$|\partial_x^k Z(t,x;\tau,\xi)| \le C\sum_{j=0}^{\|k\|}(\hat{B}(t,\tau))^{-M-(\|k\|-j)/(2b)}(D_0(x))^{j(1-\varepsilon)}\hat{E}_c(t,x;\tau,\xi) \times \exp\{\hat{g}_0(t,x) - \hat{g}_0(\tau,\xi)\}, \tag{2.4.14}$$

$$0 \le \tau < t \le T, \quad \{x,\xi\} \subset \mathbb{R}^n, \quad \|k\| \le 2b,$$

where $C > 0$, $c > 0$, the functions D_0 and $\hat{g}_0$ are the ones from (2.3.45) and (2.4.13) respectively.

If, in addition, the condition C_{13} is fulfilled, then the FSCP Z has the normality property, and a convolution property is valid for it.

The proofs of Theorems 2.16–2.18 are given in [115, 116].

2.4.2. The Cauchy problem and the problem without initial conditions. The results regarding the FSCP for equations from the classes $\mathbf{E}_{13}$ and $\mathbf{E}_{14}$ described in Section 2.4.1 make it possible to investigate the correct solvability of problems with or without initial conditions, depending on the character of degeneration.

Instead of the functions Ψ_ν, which are defined by the equalities (2.2.1) and describe the behaviour at $|x| \to \infty$ of the functions from the spaces used in Sections 2.2 and 2.3, for equations with degenerations we consider the functions

$$\hat{\Psi}_\nu(t,x) := \exp\left\{\nu \sum_{j=1}^{n} \hat{k}_j(t,a_j)|x_j|^{q_j}\right\}, \quad (t,x) \in \Pi_{[0,T]}, \quad \nu \in \{-1,1\},$$

where

$$\hat{k}_j(t,a_j) := c_0 a_j (c_0^{2b_j-1} - (T - \hat{B}(T,t))a_j^{2b_j-1})^{1-q_j}$$

for $t \in (0,T]$ and $\hat{k}_j(0,a_j) := 0$, if $\hat{B}(T,0) = \infty$. Here c_0 is a fixed number from the interval $(0,c)$, c is the constant from the estimates of the FSCP, and the numbers a_j are such that $0 \le a_j < c_0 T^{1-q_j}$.

Note that each function $k_j(\cdot, a_j)$, $j \in \{1,\dots,n\}$, increases and

$$\hat{E}_{c_0}(t,x;\tau,\xi)\hat{\Psi}_1(\tau,\xi) \le \hat{\Psi}_1(t,x), \quad 0 < \tau < t \le T, \quad \{x,\xi\} \subset \mathbb{R}^n. \tag{2.4.15}$$

For equations (1.1.11) with the weak degeneration, it is possible to formulate the Cauchy problem with the usual initial condition at $t = 0$. The case of such equations is not essentially different from the case of non-degenerate equations. For weakly degenerate equations (1.1.11) the results are similar to those from Sections 2.2.2–2.2.7 and 2.3.4.

Instead of the spaces

$$L_p^{\vec{a}},\ M^{\vec{a}},\ L_1^{-\vec{k}(T,\vec{a})},\ C_0^{-\vec{k}(T,\vec{a})},\ L_p^{\vec{a},g(\cdot)},\ M^{\vec{a},g(\cdot)},\ L_1^{-\vec{k}(T,\vec{a}),-g(\cdot)},$$
$$C_0^{-\vec{k}(T,\vec{a}),-g(\cdot)},\ L_p^{\vec{a},g_0(0,\cdot)},\ M^{\vec{a},g_0(0,\cdot)},\ L_1^{-\vec{k}(T,\vec{a}),-g_0(T,\cdot)}, \text{ and } C_0^{-\vec{k}(T,\vec{a}),-g_0(T,\cdot)}$$

we use similar spaces with the weight functions $\hat{\Psi}_\nu$ instead of Ψ_ν. Instead of the spaces of continuous, Hölder continuous, and differentiable functions defined in Section 2.2.1, we use similar spaces of functions with a specified behaviour at $t \to 0$. That behaviour is described by the function

$$(\delta(t))^\mu (\Delta(t,0))^r \hat{E}^d(T,t), \quad t \in (0,T],$$

where μ is a non-negative integer, $\{r,d\} \subset \mathbb{R}$, $\delta : [0,T] \to [0,\infty)$ is a continuous non-decreasing function, such that for any $t \in (0,T]$,

$$0 < \delta(t) \leq \beta(t), \quad \Delta(t,0) := \int_0^t \frac{\delta(\lambda)}{\alpha(\lambda)} d\lambda < \infty.$$

Definitions of some of these spaces can be found in [112] where lemmas similar to Lemmas 2.7–2.10 are proved.

In order to obtain more precise results regarding the solvability of the Cauchy problem for a weakly degenerate equation, one has to take into account the relation between the functions α and β. The condition

$$\int_0^t (\hat{B}(t,\tau))^{-1+\gamma_0/(2b)} \frac{d\tau}{\alpha(\tau)} \leq C, \quad t \in (0,T],$$

with some $\gamma_0 \in (0,1)$, was used for this purpose in [168].

If the equation (1.1.11) is strongly degenerate, then it is generally impossible to fulfil the initial condition (2.2.108). The next simple theorem gives conditions of the correct solvability of a strongly degenerate equation (1.1.11) without initial conditions.

Theorem 2.21. *Suppose that the coefficients a_k, $\|k\| \leq 2b$, of the equation* (1.1.11) *satisfy the conditions A_{11}–A_{13}, and $\hat{A}(T,0) = \infty$. Let a function $f : \Pi_{(0,T]} \to \mathbb{C}_{N1}$ be continuous and satisfy the following conditions with the constant d from the estimate* (2.4.1)*:*

1) *For any $R > 0$ there exist such constants $C > 0$ and $\gamma \in (0,1]$ that for all $t \in (0,T]$ and $\{x,\xi\} \subset B_R$,*

$$|\Delta_x^\xi f(t,x)| \leq C\delta(t)\hat{E}^{-d}(T,t)(p(x;\xi))^\gamma,$$

where $\delta : (0,T] \to [0,\infty)$ is a function satisfying the condition

$$\int_0^T (\delta(t)/\alpha(t))dt < \infty;$$

2) *There exists such a constant $C > 0$, that for all $t \in (0,T]$,*

$$\hat{F}(t) := \int_0^t \hat{E}^d(T,\tau)\|f(\tau,\cdot)\|^{\vec{k}(\tau,\vec{a})} \frac{d\tau}{\alpha(\tau)} \leq C,$$

where

$$\|f(\tau,\cdot)\|^{\vec{k}(\tau,\vec{a})} := \sup_{x\in\mathbb{R}^n} (|f(\tau,x)|\hat{\Psi}_{-1}(\tau,x)).$$

Then the formula

$$u(t,x) = \int_0^t \frac{d\tau}{\alpha(\tau)} \int_{\mathbb{R}^n} Z(t,x;\tau,\xi) f(\tau,\xi)d\xi, \quad (t,x) \in \Pi_{(0,T]}, \tag{2.4.16}$$

determines the unique solution of the equation (1.1.11) *satisfying the estimate*

$$\|u(t,\cdot)\|^{\vec{k}(t,\vec{a})} \le C\hat{E}^{-d}(T,t)\hat{F}(t), \quad t \in (0,T]. \tag{2.4.17}$$

Proof. The fact that the function (2.4.16) is a solution of the equation (1.1.11) and satisfies the estimate (2.4.17) is a consequence of properties of the volume potentials established just as for non-degenerate equations. We have only to prove the uniqueness of this solution. It is sufficient to show that a solution of the homogeneous equation satisfying the inequality

$$\|u(t,\cdot)\|^{\vec{k}(t,\vec{a})} \le C\hat{E}^{-d}(T,t)\varepsilon(t), \quad t \in (0,T], \tag{2.4.18}$$

is identical to 0. Here the function $\varepsilon : (0,T] \to (0,\infty)$ is such that $\varepsilon(t) \underset{t\to 0}{\to} 0$.

Let (t,x) be an arbitrary fixed point from $\Pi_{(0,T]}$, and t_0 be a fixed number from the interval $(0,t/2)$. Evidently, our solution u satisfies the condition

$$u(t,x)|_{t=t_0} = u(t_0,x), \quad x \in \mathbb{R}^n,$$

thus, by virtue of the results known for non-degenerate equations, we have the representation

$$u(t,x) = \int_{\mathbb{R}^n} Z(t,x;t_0,\xi)u(t_0,\xi)d\xi. \tag{2.4.19}$$

Since this is true for any $t_0 \in (0,t/2)$, we can pass to the limit as $t_0 \to 0$. The limit of the right-hand side of (2.4.19) equals zero. Indeed, using the inequalities (2.4.1), (2.4.15) and (2.4.18) we find that

$$\begin{aligned}
&\left| \int_{\mathbb{R}^n} Z(t,x;t_0,\xi)u(t_0,\xi)d\xi \right| \\
&\le C\varepsilon(t_0) \int_{\mathbb{R}^n} \hat{E}^d_{c-c_0}(t,x;t_0,\xi)(\hat{E}_{c_0}(t,x;t_0,\xi)\hat{\Psi}_1(t_0,\xi))\hat{E}^{-d}(T,t_0)(\hat{B}(t,t_0))^{-M}d\xi \\
&\le C\varepsilon(t_0)\hat{\Psi}_1(t,x)\hat{E}^{-d}(T,t) \int_{\mathbb{R}^n} \hat{E}_{c-c_0}(t,x;t_0,\xi)(\hat{B}(t,t_0))^{-M}d\xi \\
&= C\varepsilon(t_0)\hat{\Psi}_1(t,x)\hat{E}^{-d}(T,t) \underset{t_0\to 0}{\to} 0.
\end{aligned}$$

Therefore after passing to the limit in (2.4.19) for $t_0 \to 0$ we obtain that $u(t,x) = 0$. It remains to recall that (t,x) is an arbitrary point from $\Pi_{(0,T]}$. □

Remark 1. The homogeneous equation (1.1.11) may have non-trivial solutions from a class wider than the one determined by the inequality (2.4.18).

For example, consider the equation

$$(\alpha(t)\partial_t - \beta(t)\partial_x^2 + 1)u(t,x) = 0, \quad (t,x) \in (0,T] \times \mathbb{R} =: \Pi_T,$$

assuming that $\hat{A}(T,0) = \infty$ and $\hat{B}(T,0) = \infty$. Its FSCP is

$$Z(t,x;\tau,\xi) := (4\pi\hat{B}(t,\tau))^{-1/2}\hat{E}_{1/4}^{-1}(t,x;\tau,\xi), \quad 0 < \tau < t \le T, \quad \{x,\xi\} \subset \mathbb{R}.$$

The functions $u_1(t,x) := \hat{E}^1(T,t),\ (t,x) \in \Pi_T$, and also

$$u_2(t,x) := \hat{E}^{1/2}(T,t)\exp\{x/\sqrt{2}\},\ (t,x) \in \Pi_T \ (\text{if } \beta(t) = 1,\ t \in [0,T])$$

are its solutions. The functions u_1 and u_2 do not belong to the class determined by the inequality (2.4.18).

Remark 2. If the equation (1.1.11) is not very strongly degenerate, that is $\hat{A}(T,0) = \infty$ while $\hat{B}(T,0) < \infty$, the initial condition can be formulated as follows:

$$(u(t,x)\hat{E}^d(T,t))|_{t=0} = \varphi(x), \quad x \in \mathbb{R}^n, \tag{2.4.20}$$

where d is the number from the estimates (2.4.1). A solution of the problem (1.1.11), (2.4.20) is represented as

$$u(t,x) = \int_{\mathbb{R}^n} Z_0(t,x;\xi)\varphi(\xi)d\xi + \int_0^t \frac{d\tau}{\alpha(\tau)}\int_{\mathbb{R}^n} Z(t,x;\tau,\xi)f(\tau,\xi)d\xi,$$
$$(t,x) \in \Pi_{(0,T]},$$

where

$$Z_0(t,x;\xi) := \lim_{\tau\to 0}(Z(t,x;\tau,\xi)\hat{E}^{-d}(T,\tau)).$$

As an example, consider the equation

$$(\alpha(t)\partial_t - \beta(t)(\partial_{x_1}^2 - \partial_{x_2}^4) - d)u = 0,$$

in which $n = 2$, $d \in \mathbb{R}$, and the functions α and β correspond to the case of a not very strong degeneration. The FSCP for this equation is given by the formula

$$Z(t,x;\tau,\xi) = (2\pi)^{-2}\int_{\mathbb{R}^2} \exp\{i(x-\xi,\sigma) - \hat{B}(t,\tau)(\sigma_1^2 + \sigma_2^4) + d\hat{A}(t,\tau)\}d\sigma,$$

$$0 < \tau < t \le T, \quad \{x,\xi\} \subset \mathbb{R}^2,$$

whence (see Section 6.1.2 in [76])

$$Z(t,x;\tau,\xi) = (2\pi\sqrt{2})^{-1}(\hat{B}(t,\tau))^{-3/4}\exp\{d\hat{A}(t,\tau) - (4\hat{B}(t,\tau))^{-1}(x_1-\xi_1)^2\}$$
$$\times \int_0^\infty \exp\{-r^4\}J_{-1/2}(r(\hat{B}(t,\tau))^{-1/4}|x_2-\xi_2|)r^{1/2}dr,$$
$$0<\tau<t\le T, \quad \{x,\xi\}\subset\mathbb{R}^2,$$

where $J_{-1/2}$ is the Bessel function of the first kind and the order $-1/2$. It is easy to see that

$$Z_0(t,x;\xi) = (2\pi\sqrt{2})^{-1}(\hat{B}(t,0))^{-3/4}\exp\{-d\hat{A}(T,t) - (4\hat{B}(t,0))^{-1}(x_1-\xi_1)^2\}$$
$$\times \int_0^\infty \exp\{-r^4\}J_{-1/2}(r(\hat{B}(t,0))^{-1/4}|x_2-\xi_2|)r^{1/2}dr,$$
$$0<t\le T, \quad \{x,\xi\}\subset\mathbb{R}^2.$$

2.5 Comments

$\overrightarrow{2b}$-parabolic equations (systems of equations) constitute one of the possible generalizations of equations (systems of equations) introduced by Petrovsky [180] (the equations from the class $\mathbf{E}_0$). We begin with a brief survey of the results on the Cauchy problem for equations from the class $\mathbf{E}_0$. Keeping in mind the contents of this book, we describe mainly those works where FSCPs were studied or used. As in the majority of the publications, we shall use the term "an equation" only dealing with the scalar case, while vector equations will be referred to as "systems of equations".

A FSCP for second order parabolic equations with bounded variable coefficients (depending on all the variables) was constructed by Dressel [28] for the case of smooth coefficients, and by Pogorzelski [184] and Aronson [5] for the Hölder continuous coefficients. FSCPs for systems of equations of arbitrary orders were constructed by Bruk [21], Eidelman [47, 48, 49, 52, 45], Pogorzelski [185, 186], Slobodeckij [213], and Aronson [5, 6, 7]. Conditions upon the coefficients were weakened considerably by Eidelman and Matiychuk [67]. All the constructions were based on the classical Levi method [152], a universal tool found applicable even for non-Archimedean parabolic equations [131, 132].

The FSCPs constructed and investigated in the above papers found a variety of applications in the studies of properties of parabolic systems, correct solvability of the Cauchy problem in various classes of functions, integral representations of solutions of the Cauchy problem, as well as solutions defined on the open layer $\Pi_{(0,T]}$, local solvability of the Cauchy problem for quasilinear and nonlinear parabolic (in the sense of Petrovsky) systems, extension of solutions of such systems onto larger time intervals, etc. These results are expounded in detail in the monographs [45, 83, 146] and the survey papers [102, 24, 46].

In 1960 Eidelman [51] defined and began to study a new class of systems, the class of $\overrightarrow{2b}$-parabolic systems. As we have mentioned, such systems generalize the systems parabolic in the sense of Petrovsky, to the case where each spatial variable may have its own weight with respect to the time variable, so that a system from the new class has a vector parabolic weight $\overrightarrow{2b} := (2b_1, \dots, 2b_n)$. For such systems, Eidelman constructed FSCPs and studied solvability of the Cauchy problem, under the assumption that the coefficients are bounded continuous functions satisfying, with respect to the spatial variables, the Hölder condition with respect to the so-called $\overrightarrow{2b}$-parabolic distance.

For the case where, instead of the Hölder condition, the coefficients satisfy the Dini condition, the existence of a FSCP and correct solvability of the Cauchy problem were established by Matiychuk [165], and Eidelman and Matiychuk [66].

An important work for the theory of $\overrightarrow{2b}$-parabolic systems was published in 1968 by Eidelman and Ivasyshen [53]. It contained a sufficiently complete and precise investigation of the FSCP Z and potentials generated by Z, correctness classes of the Cauchy problem for linear systems under various assumptions regarding right-hand sides and initial functions, local solvability of the Cauchy problem for nonlinear systems and extension of their solutions to larger time intervals. Inner estimates of solutions were obtained, and the hypoellipticity of $\overrightarrow{2b}$-parabolic systems was proved. These results generalize the facts known for parabolic systems in the sense of Petrovsky [45], complement them and make them more exact.

Note that in the studies of integral representations of solutions of the Petrovsky- and $\overrightarrow{2b}$-parabolic systems in [45, 24, 53] the sufficient conditions imposed on solutions not always were necessary. Ivasyshen [105, 108] was the first to find necessary and sufficient conditions for solutions of homogeneous $\overrightarrow{2b}$-parabolic systems (in particular the ones parabolic in the sense of Petrovsky) defined on the open layer $\Pi_{(0,T]}$, to be represented by the Poisson integrals of functions or generalized measures from special weighted spaces. The sets of such functions or generalized measures are the sets of initial values of the solutions. It was also clarified in [105, 108], in what sense the solutions satisfy those initial conditions.

It is natural to try to obtain analogs of the above results for parabolic systems with some kinds of degenerations. Let us review the papers devoted to Petrovsky- and $\overrightarrow{2b}$-parabolic equations and systems with degenerations on the initial hyperplane.

Kalashnikov [124, 125] found correctness classes containing increasing functions for second order parabolic equations degenerate with respect to the time variable. His techniques did not use a FSCP. Glushak and Shmulevich [90] constructed FSCPs for parabolic equations of arbitrary orders degenerate on the initial hyperplane, and used them to find a correctness class for a problem without initial conditions. The class contains functions increasing both for $|x| \to \infty$ and $t \to 0$.

Glushko [91] showed the impossibility to consider a classical Cauchy problem with initial data at $t = 0$ for a strongly degenerate equation. Thus it is natural to impose an initial condition with a certain weight. Just such solutions, satisfying a special weighted initial conditions, were found in [90].

Gorodetsky and Zhitaryuk [93, 94] studied properties of solutions of the Cauchy problem for parabolic equations with a weak degeneration on the initial hyperplane, and the initial data from special spaces of distributions.

Several papers by Ivasyshen, Voznyak, and Medynsky [118, 110, 111, 112, 166, 167, 168] were devoted to the construction and study of FSCPs for the Petrovsky parabolic systems degenerate on the initial hyperplane. Using properties of the FSCPs they investigated the correct solvability of systems with a usual initial condition, for the case of a weak degeneration, and without an initial condition, if a degeneration is strong. For a weak degeneration, necessary and sufficient conditions were found for the representation of solutions as sums of the Poisson integrals and volume potentials. It was also investigated, in what sense the solutions satisfy the initial conditions , and a description of the set of initial values of solutions is given. This generalizes the corresponding results from [45, 53, 108] to the case of the Petrovsky parabolic systems with bounded Hölder continuous coefficients and the degeneracy on the initial hyperplane. Similar results for $\overrightarrow{2b}$-parabolic systems with degeneracy on the initial hyperplane were obtained in [13, 14, 15, 16, 112, 113]; the most important of them are given in Section 2.4.

Simultaneously with the study of the case of bounded coefficients considered above, the problem of correct solvability of the Cauchy problem and boundary value problems was considered for parabolic systems with increasing coefficients, that is those which may tend to infinity, as $|x| \to \infty$. This is important both for the theory of partial differential equations, and for applications, since such equations appear in many mathematical models of processes of the real world.

Our survey of results regarding the Cauchy problem for parabolic systems with increasing coefficients begins with the papers by Zhitomirsky [233, 234]. In [233] methods of functional analysis were used to prove the existence and uniqueness of a generalized solution of the Cauchy problem for a second order parabolic equation with increasing coefficients. In [234] he considered systems parabolic in the sense of Petrovsky whose lower term coefficients may have a power-like growth at infinity. Its order depends on the orders of the system and the derivative, at which the coefficient stands. The existence and estimates of a FSCP were proved, which led to the existence and uniqueness theorem for a solution of the Cauchy problem in spaces of rapidly increasing functions. The above power growth order cannot in general be essentially increased without violating the uniqueness of a solution of the Cauchy problem.

Almost simultaneously, similar results were obtained by Eidelman [50, 52] who used a different method. Assuming that the leading coefficients are bounded and smooth while lower coefficients grow in an appropriate way (depending on some positive continuous function D), under an additional assumption of parabolicity of a certain auxiliary system, he constructed a FSCP using the method from [47], proved a theorem on the correct solvability of the Cauchy problem, and showed that his results are close to those by Zhitomirsky. The above condition of parabolicity of an auxiliary system defines a class of dissipative systems (with the dissipation characteristic D) [45].

The study of FSCPs for dissipative parabolic systems in the sense of Petrovsky initiated in [50, 52] was continued in [45, 68, 69] (see also [46, 76, 8, 17, 127]). In these works two sets of conditions upon coefficients of a dissipative system and the characteristic of dissipation are used. One of them includes strong requirements for smoothness of coefficients, without additional assumptions regarding the characteristic of dissipation. The second set of conditions deals with coefficients of minimal smoothness, but imposes some additional assumptions on the characteristic of dissipation. If a parabolic system with bounded leading coefficients is not dissipative, but the (unbounded, as $|x| \to \infty$) lower coefficients satisfy some specific growth conditions, then, as it was shown in the above papers, the system can be reduced to a dissipative one, so that the system possesses a FSCP.

The above results by Eidelman were essentially developed and extended to $\overrightarrow{2b}$-parabolic systems with increasing coefficients in the papers by Ivasyshen and Pasichnyk [114, 115, 116, 117, 178, 179]. Their results are presented in Section 2.4.

The results of Section 2.1 were obtained by Eidelman and Ivasyshen in 1960–1963, prepared for publication in 1964, and published for the first time in 1968 [53].

Note that there are many results on the theory of hypoelliptic equations with different properties with respect to different independent variables. Such a class of equations whose construction is similar to that of $\overrightarrow{2b}$-parabolic systems is called the class of quasi-elliptic, or semi-elliptic equations or systems [98].

Volevich [225] obtained general results about inner properties of solutions of quasi-elliptic systems (including $\overrightarrow{2b}$-parabolic systems as a particular case). A big work by Friberg [82] was devoted to inner properties of solutions of partially hypoelliptic equations. Properties of solutions (infinite differentiability, analyticity) of hypoelliptic equations up to a flat part of the boundary were studied by Hörmander [97]. The paper by Cavalluci [23] dealt with assumptions regarding the coefficients and boundary functions of boundary value problems for quasi-elliptic equations with variable coefficients, under which the solutions belong to the Gevrey classes. Mosolov [173] and Mikhailov [171] studied a Dirichlet type problem for a wide class of equations including quasi-elliptic ones.

Chapter 3

Degenerate Equations of the Kolmogorov Type

3.1 Fundamental solution of the Cauchy problem

3.1.1. Equations from the class $\mathbf{E}_{21}$ with coefficients independent of spatial variables. We shall consider the equation (1.1.14) and use the notation from Sections 1.1.1, 1.1.4, and 1.3.4 for this equation. Let us study the case where the coefficients a_{k_1}, $|k_1| \le 2b$, can depend only on the variable t or a parametric point $(\beta, y) \in \Pi_{[0,T]}$.

For an arbitrary fixed number $\tau \in [0,T)$ consider the Cauchy problem with the initial condition (2.1.6) for the equation

$$(S - A)u(t,x) = 0, \quad (t,x) \in \Pi_{(\tau,T]}. \tag{3.1.1}$$

Here

$$S := \partial_t - \sum_{j=1}^{n_2} x_{1j}\partial_{x_{2j}} - \sum_{j=1}^{n_3} x_{2j}\partial_{x_{3j}},$$

while A is a differential expression coinciding either with

$$A(t, \partial_{x_1}) := \sum_{|k_1|\le 2b} a_{k_1}(t)\partial_{x_1}^{k_1}, \tag{3.1.2}$$

or with

$$A(\beta, y, \partial_{x_1}) := \sum_{|k_1|\le 2b} a_{k_1}(\beta, y)\partial_{x_1}^{k_1}. \tag{3.1.3}$$

In this chapter we shall use the distances

$$d(x;\xi) := \sum_{j=1}^{3} |x_j - \xi_j|^{1/(2b(j-1)+1)},$$

$$d(t,x;\tau,\xi) := |t-\tau|^{1/(2b)} + d(x;\xi) \tag{3.1.4}$$

between points x and ξ, (t,x) and (τ,ξ) respectively, where $\{t,\tau\} \subset \mathbb{R}$, $\{x := (x_1,x_2,x_3), \xi := (\xi_1,\xi_2,\xi_3)\} \subset \mathbb{R}^n$. We also use the points

$$X(t) := (X_1(t), X_2(t), X_3(t)), \quad X_1(t) := x_1, \quad X_2(t) := x_2 + t\hat{x}_1,$$
$$X_3(t) := x_3 + tx_2' + 2^{-1}t^2x_1', \quad t \in \mathbb{R}, \tag{3.1.5}$$

constructed from a point $x \in \mathbb{R}^n$. Similarly we construct the points $Y(t)$, $\Xi(t)$, starting from y, ξ.

We shall use the following conditions upon the coefficients of the expressions (3.1.2) and (3.1.3):

(B_{21}) the coefficients $a_{k_1}(t)$, $t \in [0,T]$, $|k_1| \le 2b$, are continuous, and there exists such a constant $\delta > 0$ that for any $t \in [0,T]$ and $\sigma_1 \in \mathbb{R}^{n_1}$,

$$\operatorname{Re} \sum_{|k_1|=2b} a_{k_1}(t)(i\sigma_1)^{k_1} \le -\delta|\sigma_1|^{2b}; \tag{3.1.6}$$

(B_{22}) the coefficients $a_{k_1}(\beta,y)$, $(\beta,y) \in \Pi_{[0,T]}$, $|k_1| \le 2b$, are bounded, and there exists such a constant $\delta > 0$ that for any $(\beta,y) \in \Pi_{[0,T]}$ and $\sigma_1 \in \mathbb{R}^{n_1}$,

$$\operatorname{Re} \sum_{|k_1|=2b} a_{k_1}(\beta,y)(i\sigma_1)^{k_1} \le -\delta|\sigma_1|^{2b};$$

(B_{23}) the coefficients of the expression (3.1.3) are Hölder continuous in β, y in the following sense – there exist such constants $H > 0$ and $\alpha \in (0,1]$ that for any $\{(\beta,y),(\lambda,z)\} \subset \Pi_{[0,T]}$ and $k_1 \in \mathbb{Z}_+^{n_1}$, $|k_1| \le 2b$,

$$|\Delta_{\beta,y}^{\lambda,z} a_{k_1}(\beta,y)| \le H(d(\beta, Y(\beta-\lambda); \lambda, z))^\alpha; \tag{3.1.7}$$

(B_{24}) there exist bounded derivatives

$$\partial_\beta^{l_0}\partial_y^l a_{k_1}(\beta,y), \quad (\beta,y) \in \Pi_{[0,T]}, \quad |k_1| \le 2b, \quad l_0 \le r_0, \quad |l| \le r,$$

where r_0 and r are certain numbers from $\mathbb{N} \cup \{0\}$;

(B_{25}) the derivatives from the condition B_{24} are Hölder continuous in β, y in the sense of (3.1.7).

First we consider the Cauchy problem (2.1.6), (3.1.1) with $A = A(t,\partial_{x_1})$. As in Section 2.1.2, we seek a solution of the form (2.1.10). Since

$$F_{x\to\sigma}\left[\left(\partial_t - \sum_{j=1}^{n_2} x_{1j}\partial_{x_{2j}} - \sum_{j=1}^{n_3} x_{2j}\partial_{x_{3j}}\right)u\right] = \left(\partial_t + \sum_{j=1}^{n_2} \sigma_{2j}\partial_{\sigma_{1j}} + \sum_{j=1}^{n_3} \sigma_{3j}\partial_{\sigma_{2j}}\right)F[u],$$

for the function v we obtain the Cauchy problem with the initial condition (2.1.12) for the equation (1.1.37).

We solve the problem (1.1.37), (2.1.12) by the method of characteristics. The system of ordinary differential equations

$$\begin{aligned} dt = d\sigma_{11}/\sigma_{21} = \cdots &= d\sigma_{1n_2}/\sigma_{2n_2} = d\sigma_{21}/\sigma_{31} \\ = \cdots &= d\sigma_{2n_3}/\sigma_{3n_3} = dv/(A(t, i\sigma_1)v) \end{aligned}$$

corresponding to (1.1.37) contains $n_2 + n_3 + 1$ equations. Let us find its $n_2 + n_3 + 1$ independent first integrals. From the equations $d\sigma_{1j}/\sigma_{2j} = d\sigma_{2j}/\sigma_{3j}$, $j \in \{1, \ldots, n_3\}$, we get

$$2\sigma_{1j}\sigma_{3j} - \sigma_{2j}^2 = C_j', \quad j \in \{1, \ldots, n_3\}; \tag{3.1.8}$$

the equations $d\sigma_{1j}/\sigma_{2j} = dt$, $j \in \{n_3 + 1, \ldots, n_2\}$, give that

$$\sigma_{1j} - t\sigma_{2j} = C_j', \quad j \in \{n_3 + 1, \ldots, n_2\}, \tag{3.1.9}$$

while the equations $d\sigma_{2j}/\sigma_{3j} = dt$, $j \in \{1, \ldots, n_3\}$, and $dv/(A(t, i\sigma_1)v) = dt$ imply

$$\sigma_{2j} - t\sigma_{3j} = C_j'', \quad j \in \{1, \ldots, n_3\}, \tag{3.1.10}$$

$$v = C''' \exp\left\{\int_\tau^t A(\theta, i\sigma_1)d\theta\right\} \tag{3.1.11}$$

respectively. Now it follows from the equalities (3.1.8) and (3.1.10) that

$$\sigma_{1j} = (C_j' + (C_j'' + t\sigma_{3j})^2)/(2\sigma_{3j}), \quad j \in \{1, \ldots, n_3\}, \tag{3.1.12}$$

while (3.1.9) and (3.1.10) give

$$\sigma_{1j} = C_j' + t\sigma_{2j}, \quad j \in \{n_3 + 1, \ldots, n_2\}, \tag{3.1.13}$$

$$\sigma_{2j} = C_j'' + t\sigma_{3j}, \quad j \in \{1, \ldots, n_3\}. \tag{3.1.14}$$

By the equalities (3.1.11)–(3.1.13) we find that

$$\begin{aligned} v = C''' \exp\Bigg\{\int_\tau^t \sum_{|k_1| \le 2b} a_{k_1}(\theta) i^{|k_1|} \prod_{j=1}^{n_3} \left(\frac{C_j' + (C_j'' + \theta\sigma_{3j})^2}{2\sigma_{3j}}\right)^{k_{1j}} \\ \times \prod_{j=n_3+1}^{n_2} (C_j' + \theta\sigma_{2j})^{k_{1j}} \prod_{j=n_2+1}^{n_1} \sigma_{1j}^{k_{1j}} d\theta\Bigg\}. \end{aligned} \tag{3.1.15}$$

Let $\hat{\sigma}_{1j}$, $\hat{\sigma}_{2j}$, and $\hat{v}$ be the values for $t = \tau$ of σ_{1j}, σ_{2j}, and v respectively. Due to the equalities (3.1.12)–(3.1.15), we have: $\sigma_{1j} = (C_j' + (C_j'' + \tau\sigma_{3j})^2)/(2\sigma_{3j})$, $j \in \{1, \ldots, n_3\}$; $\hat{\sigma}_{1j} = C_j' + \tau\sigma_{2j}$, $j \in \{n_3 + 1, \ldots, n_2\}$; $\hat{\sigma}_{2j} = C_j'' + \tau\sigma_{3j}$, $j \in \{1, \ldots, n_3\}$; $\hat{v} = C'''$, and since $\hat{v} = \psi$, we get $C''' = \psi(\hat{\sigma}_1, \hat{\sigma}_2, \hat{\sigma}_3) = \psi((C_1' +$

$(C_1'' + \tau\sigma_{31})^2)/(2\sigma_{31}), \dots, (C_{n3}' + (C_{n3}'' + \tau\sigma_{3n_3})^2)/(2\sigma_{3n_3}), C_{n_3+1}' + \tau\sigma_{2n_3+1}, \dots, C_{n_2}' + \tau\sigma_{2n_2}, \sigma_{1(n_2+1)}, \dots, \sigma_{1n_1}, C_1'' + \tau\sigma_{31}, \dots, C_{n_3}'' + \tau\sigma_{3n_3}, \sigma_{2(n_3+1)}, \dots, \sigma_{2n_2}, \sigma_3)$. Using the expressions for C_j', C_j'', and C''', which follow from the equalities (3.1.8)–(3.1.10) and (3.1.15), we get

$$v(t,\sigma) = \exp\Big\{\int_\tau^t \sum_{|k_1|\le 2b} a_{k_1}(\theta) i^{|k_1|}(\sigma_1' - (t-\theta)\sigma_2' + 2^{-1}(t-\theta)^2\sigma_3)^{k_1'}$$
$$\times (\sigma_1'' - (t-\theta)\sigma_2'')^{k_1''}(\sigma_1''')^{k_1'''}\,d\theta\Big\}$$
$$\times \psi(\sigma_1' - (t-\tau)\sigma_2' + 2^{-1}(t-\tau)^2\sigma_3, \sigma_1''$$
$$- (t-\tau)\sigma_2'', \sigma_1''', \sigma_2' - (t-\tau)\sigma_3, \sigma_2'', \sigma_3), \qquad (t,\sigma) \in \Pi_{[\tau,T]}.$$

Substituting this expression into the formula (2.1.10) we come to the equality

$$u(t,x) = (2\pi)^{-n} \int_{\mathbb{R}^n} \exp\Big\{ i(x,\sigma) + \int_\tau^t \sum_{|k_1|\le 2b} a_{k_1}(\theta) i^{|k_1|}$$
$$\times (\sigma_1' - (t-\theta)\sigma_2' + 2^{-1}(t-\theta)^2\sigma_3)^{k_1'}$$
$$\times (\sigma_1'' - (t-\theta)\sigma_2'')^{k_1''}(\sigma_1''')^{k_1'''}\,d\theta\Big\}\psi(\sigma_1' - (t-\tau)\sigma_2'$$
$$+ 2^{-1}(t-\tau)^2\sigma_3, \sigma_1'' - (t-\tau)\sigma_2'', \sigma_1''', \sigma_2'$$
$$- (t-\tau)\sigma_3, \sigma_2'', \sigma_3)\,d\sigma, \quad (t,x) \in \Pi_{(\tau,T]}. \tag{3.1.16}$$

In the last integral we perform the change of variables

$$\begin{aligned} \sigma_1' - (t-\tau)\sigma_2' + 2^{-1}(t-\tau)^2\sigma_3 &= \eta_1', \\ \sigma_1'' - (t-\tau)\sigma_2'' &= \eta_1'', \\ \sigma_1''' &= \eta_1''', \\ \sigma_2' - (t-\tau)\sigma_3 &= \eta_2', \\ \sigma_2'' &= \eta_2'', \\ \sigma_3 &= \eta_3, \end{aligned} \tag{3.1.17}$$

as a result we come to the equality

$$u(t,x) = (2\pi)^{-n} \int_{\mathbb{R}^n} \exp\Big\{ i(x_1', \eta_1' + (t-\tau)\eta_2' + 2^{-1}(t-\tau)^2\eta_3)$$
$$+ i(x_1'', \eta_1'' + (t-\tau)\eta_2'') + i(x_1''', \eta_1''') + i(x_2', \eta_2' + (t-\tau)\eta_3)$$

$$+ i(x_2'', \eta_2'') + i(x_3, \eta_3) + \int_\tau^t \sum_{|k_1| \le 2b} a_{k_1}(\theta) i^{|k_1|} (\eta_1' + (\theta - \tau)\eta_2'$$

$$+ 2^{-1}(\theta - \tau)^2 \eta_3)^{k_1'} (\eta_1'' + (\theta - \tau)\eta_2'')^{k_1''} (\eta_1''')^{k_1'''} d\theta \Big\} \psi(\eta) d\eta,$$

$$(t, x) \in \Pi_{(\tau, T]}.$$

Using the expression (2.1.9) (with η and ξ instead of σ and x respectively) and changing the order of integration, we obtain the formula (2.1.14), in which

$$G(t, x; \tau, \xi) := (F^{-1}_{\eta \to z}[V(t, \tau, \eta)])(t, \tau, z)|_{z = X(t-\tau) - \xi},$$
$$0 \le \tau < t \le T, \quad \{x, \xi\} \subset \mathbb{R}^n, \tag{3.1.18}$$

where

$$V(t, \tau, \sigma) := \exp\Big\{ \sum_{|k_1| \le 2b} i^{|k_1|} \int_\tau^t a_{k_1}(\theta)(\sigma_1' + (\theta - \tau)\sigma_2' + 2^{-1}(\theta - \tau)^2 \sigma_3)^{k_1'}$$
$$\times (\sigma_1'' + (\theta - \tau)\sigma_2'')^{k_1''} (\sigma_1''')^{k_1'''} d\theta \Big\},$$
$$0 \le \tau \le t \le T, \quad \sigma \in \mathbb{R}^n, \tag{3.1.19}$$

while $X(t)$ was defined in (3.1.5).

In the integrals from (3.1.18) and (3.1.19) we change the variables by $\eta_j = (t - \tau)^{-j+1-1/(2b)} \sigma_j$, $j \in \{1, 2, 3\}$, $\theta - \tau = (t - \tau)\beta$. Then we get a formula for the function G:

$$G(t, x; \tau, \xi) = (t - \tau)^{-M} (F^{-1}_{\sigma \to z}[\tilde{V}(t, \tau, \sigma)])(t, \tau, z)|_{z = (X(t-\tau) - \xi)_{t-\tau}},$$
$$0 \le \tau < t \le T, \quad \{x, \xi\} \subset \mathbb{R}^n, \tag{3.1.20}$$

in which

$$M := \sum_{j=1}^3 (j - 1 + 1/(2b)) n_j,$$
$$x_t := (t^{-1/(2b)} x_1, t^{-1-1/(2b)} x_2, t^{-2-1/(2b)} x_3), \tag{3.1.21}$$

for $x = (x_1, x_2, x_3) \in \mathbb{R}^n$, and

$$\tilde{V}(t, \tau, \sigma) := V(t, \tau, \sigma_{t-\tau}) = \tilde{V}_0(t, \tau, \sigma) \tilde{V}_1(t, \tau, \sigma),$$
$$0 \le \tau \le t \le T, \quad \sigma \in \mathbb{R}^n, \tag{3.1.22}$$

where

$$\tilde{V}_0(t,\tau,\sigma) := \exp\Big\{ \sum_{|k_1|=2b} i^{|k_1|} \int_0^1 a_{k_1}(\tau + (t-\tau)\beta) \times (\sigma_1^{'} + \beta\sigma_2^{'} + 2^{-1}\beta^2\sigma_3)^{k_1^{'}} (\sigma_1^{''} + \beta\sigma_2^{''})^{k_1^{''}} (\sigma_1^{'''})^{k_1^{'''}} \, d\beta \Big\},$$

$$\tilde{V}_1(t,\tau,\sigma) := \exp\Big\{ \sum_{|k_1|<2b} i^{|k_1|}(t-\tau)^{1-|k_1|/(2b)} \times \int_0^1 a_{k_1}(\tau + (t-\tau)\beta)(\sigma_1^{'} + \beta\sigma_2^{'} + 2^{-1}\beta^2\sigma_3)^{k_1^{'}} \times (\sigma_1^{''} + \beta\sigma_2^{''})^{k_1^{''}} (\sigma_1^{'''})^{k_1^{'''}} \, d\beta \Big\}.$$

By the equality (3.1.20), an investigation of the function G is reduced now to the study of the function $\tilde{V}$.

Let us obtain an estimate for $\tilde{V}_0$. We use the condition (3.1.6) which implies the inequality

$$\operatorname{Re} \sum_{|k_1|=2b} a_{k_1}(\tau + (t-\tau)\beta) i^{|k_1|} (\sigma_1^{'} + \beta\sigma_2^{'} + 2^{-1}\beta^2\sigma_3)^{k_1^{'}} (\sigma_1^{''} + \beta\sigma_2^{''})^{k_1^{''}} (\sigma_1^{'''})^{k_1^{'''}}$$
$$\leq -\delta(|\sigma_1^{'} + \beta\sigma_2^{'} + 2^{-1}\beta^2\sigma_3|^{2b} + |\sigma_1^{''} + \beta\sigma_2^{''}|^{2b} + |\sigma_1^{'''}|^{2b}), \qquad \beta \in [0,1], \quad \sigma \in \mathbb{R}^n.$$

Using this inequality we find that

$$|\tilde{V}_0(t,\tau,\sigma)| \leq \exp\Big\{ -\delta|\sigma'|^{2b} \int_0^1 |(\sigma_1^{'}/|\sigma'|) + \beta(\sigma_2^{'}/|\sigma'|) + 2^{-1}\beta^2(\sigma_3/|\sigma'|)|^{2b} d\beta - \delta|\sigma''|^{2b} \int_0^1 |(\sigma_1^{''}/|\sigma''|) + \beta(\sigma_2^{''}/|\sigma''|)|^{2b} d\beta - \delta|\sigma_1^{'''}|^{2b} \Big\},$$
$$0 \leq \tau \leq t \leq T,$$
$$\sigma' := (\sigma_1^{'}, \sigma_2^{'}, \sigma_3) \in \mathbb{R}^{3n_3} \setminus \{0\},$$
$$\sigma'' := (\sigma_1^{''}, \sigma_2^{''}) \in \mathbb{R}^{2(n_2-n_3)} \setminus \{0\}.$$

Note that there exist such constants $c' > 0$ and $c'' > 0$ that

$$\int_0^1 |(\sigma_1^{'}/|\sigma'|) + \beta(\sigma_2^{'}/|\sigma'|) + 2^{-1}\beta^2(\sigma_3/|\sigma'|)|^{2b} d\beta \geq c', \quad \sigma' \in \mathbb{R}^{3n_3} \setminus \{0\},$$

$$\int_0^1 |(\sigma_1''/|\sigma''|) + \beta(\sigma_2''/|\sigma''|)|^{2b} d\beta \geq c'', \quad \sigma'' \in \mathbb{R}^{2(n_2-n_3)} \setminus \{0\}.$$

Now we get the estimate

$$|\tilde{V}_0(t,\tau,\sigma)| \leq \exp\{-\delta_0|\sigma|^{2b}\}, \quad 0 \leq \tau \leq t \leq T, \quad \sigma \in \mathbb{R}^n, \tag{3.1.23}$$

where $\delta_0 > 0$.

Let us give also an estimate for $|\tilde{V}_0(t,\tau,\sigma+i\gamma)|$, $0 \leq \tau \leq t \leq T$, $\{\sigma,\gamma\} \subset \mathbb{R}^n$, where

$$\tilde{V}_0(t,\tau,\sigma+i\gamma) = \tilde{V}_0(t,\tau,\sigma)\exp\left\{\int_0^1 P_0(t,\tau,\beta,\sigma,\gamma)d\beta\right\}, \tag{3.1.24}$$

$$\begin{aligned} P_0(t,\tau,\beta,\sigma,\gamma) := \sum_{(k_1,p_1)} & a_{k_1}(\tau+(t-\tau)\beta)i^{|k_1|+|p_1|}(\sigma_1'+\beta\sigma_2'+2^{-1}\beta^2\sigma_3)^{k_1'-p_1'} \\ & \times(\gamma_1'+\beta\gamma_2'+2^{-1}\beta^2\gamma_3)^{p_1'}(\sigma_1''+\beta\sigma_2'')^{k_1''-p_1''} \\ & \times(\gamma_1''+\beta\gamma_2'')^{p_1''}(\sigma_1''')^{k_1'''-p_1'''}(\gamma_1''')^{p_1'''}. \end{aligned}$$

Here the summation is performed over all the multi-indices $\{k_1 = (k_1', k_1'', k_1''')$, $p_1 = (p_1', p_1'', p_1''')\} \subset \mathbb{Z}_+^{n_1}$ such that $|k_1| = 2b$ and $0 < p_1 \leq k_1$.

Taking into account the fact that the coefficients a_{k_1}, $|k_1| = 2b$, are bounded, we find that

$$\begin{aligned} |P_0(t,\tau,\beta,\sigma,\gamma)| \leq c \sum_{(k_1,p_1)} & |\sigma_1'+\beta\sigma_2'+(\beta^2/2)\sigma_3|^{|k_1'|-|p_1'|}|\gamma_1'+\beta\gamma_2' \\ & +2^{-1}\beta^2\gamma_3|^{|p_1'|}|\sigma_1''+\beta\sigma_2''|^{|k_1''|-|p_1''|}|\gamma_1''+\beta\gamma_2''|^{|p_1''|}|\sigma_1'''|^{|k_1'''|-|p_1'''|}|\gamma_1'''|^{|p_1'''|} \\ \leq c \sum_{(k_1,p_1)} & (|\sigma_1'|+|\sigma_2'|+|\sigma_3|)^{|k_1'|-|p_1'|}(|\gamma_1'|+|\gamma_2'|+|\gamma_3|)^{|p_1'|}(|\sigma_1''| \\ & +|\sigma_2''|)^{|k_1''|-|p_1''|}(|\gamma_1''|+|\gamma_2''|)^{|p_1''|}|\sigma_1'''|^{|k_1'''|-|p_1'''|}|\gamma_1'''|^{|p_1'''|} \\ \leq c \sum_{(k_1,p_1)} & |\sigma|^{|k_1|-|p_1|}|\gamma|^{|p_1|}, \quad 0 \leq \tau \leq t \leq T, \quad 0 \leq \beta \leq 1, \quad \{\sigma,\gamma\} \subset \mathbb{R}^n. \end{aligned}$$

Let ε be a sufficiently small positive number (it will be specified below). Let us consider the following two cases:

$$1) \quad |\gamma| \leq \varepsilon|\sigma|, \qquad\qquad 2) \quad |\sigma| < \varepsilon^{-1}|\gamma|.$$

In the case 1) we have

$$|P_0(t,\tau,\beta,\sigma,\gamma)| \leq c\varepsilon|\sigma|^{2b}$$

and by (3.1.23) and (3.1.24) we come to the inequality

$$|\tilde{V}_0(t,\tau,\sigma+i\gamma)| \le \exp\{-(\delta_0 - c\varepsilon)|\sigma|^{2b}\}.$$

Assume that ε is such that $\delta_0 - c\varepsilon =: \delta_1 > 0$. Then for the case 1) we have

$$|\tilde{V}_0(t,\tau,\sigma+i\gamma)| \le \exp\{-\delta_1|\sigma|^{2b}\}. \tag{3.1.25}$$

In the case 2) there exists such a constant $c_1 > 0$ that

$$|P_0(t,\tau,\beta,\sigma,\gamma)| \le c_1|\gamma|^{2b}$$

so that for this case

$$|\tilde{V}_0(t,\tau,\sigma+i\gamma)| \le \exp\{-\delta_0|\sigma|^{2b} + c_1|\gamma|^{2b}\}. \tag{3.1.26}$$

The inequalities (3.1.25) and (3.1.26) imply the estimate

$$\begin{gathered}|\tilde{V}_0(t,\tau,\sigma+i\gamma)| \le \exp\{-\delta_1|\sigma|^{2b} + c_1|\gamma|^{2b}\},\\ 0 \le \tau \le t \le T, \quad \{\sigma,\gamma\} \subset \mathbb{R}^n.\end{gathered} \tag{3.1.27}$$

Let us give an estimate of the function $\tilde{V}_1$. Since the functions a_{k_1}, $|k_1| < 2b$, are bounded, we find for $0 \le \tau \le t \le T$ and $\{\sigma,\gamma\} \subset \mathbb{R}^n$ that

$$\begin{aligned}|\tilde{V}_1(t,\tau,\sigma+i\gamma)| &\le \exp\Bigg\{c\int_0^1 \sum_{|k_1|<2b} (|\sigma_1' + \beta\sigma_2' + 2^{-1}\beta^2\sigma_3| + |\gamma_1' + \beta\gamma_2'\\ &\quad + 2^{-1}\beta^2\gamma_3|)^{|k_1'|}(|\sigma_1'' + \beta\sigma_2''| + |\gamma_1'' + \beta\gamma_2''|)^{|k_1''|}(|\sigma_1'''| + |\gamma_1'''|)^{|k_1'''|}d\beta\Bigg\}\\ &\le \exp\Bigg\{c\sum_{|k_1|<2b}(|\sigma_1'| + |\sigma_2'| + |\sigma_3| + |\gamma_1'| + |\gamma_2'| + |\gamma_3|)^{|k_1'|}(|\sigma_1''| + |\sigma_2''|\\ &\quad + |\gamma_1''| + |\gamma_2''|)^{|k_1''|}(|\sigma_1'''| + |\gamma_1'''|)^{|k_1'''|}\Bigg\} \le \exp\Bigg\{c(|\sigma|^{2b-1} + |\gamma|^{2b-1})\Bigg\}\\ &\le C_\varepsilon \exp\Bigg\{\varepsilon(|\sigma|^{2b} + |\gamma|^{2b})\Bigg\},\end{aligned} \tag{3.1.28}$$

where ε is an arbitrary positive number. Let us choose ε in such a way that $\delta_1 - \varepsilon =: \delta_2 > 0$ where δ_1 is the constant from the estimate (3.1.27).

Using (3.1.22), (3.1.27), and (3.1.28), we find that

$$\begin{gathered}|\tilde{V}(t,\tau,\sigma+i\gamma)| \le C\exp\Bigg\{-\delta_2|\sigma|^{2b} + c_2|\gamma|^{2b})\Bigg\},\\ 0 \le \tau \le t \le T, \quad \{\sigma,\gamma\} \subset \mathbb{R}^n,\end{gathered} \tag{3.1.29}$$

where $c_2 := c_1 + \varepsilon$.

For the multi-indices $k = (k_1, k_2, k_3) = ((k_1', k_1'', k_1'''), (k_2', k_2''), k_3) \in \mathbb{Z}_+^n$ and $l = (l_1, l_2, l_3) \in \mathbb{Z}_+^n$, set

$$P_{kl}(\sigma) := (i\sigma_1' + i\sigma_2' + i\sigma_3')^{k_1'}(i\sigma_1'' + i\sigma_2'')^{k_1''}(i\sigma_1''')^{k_1'''}(i\sigma_2' + i\sigma_3)^{k_2'} \times (i\sigma_2'')^{k_2''}(i\sigma_3)^{k_3}(-i\sigma_1)^{l_1}(-i\sigma_2)^{l_2}(-i\sigma_3)^{l_3}, \quad \sigma \in \mathbb{R}^n.$$

It follows immediately from (3.1.29) that for any $\{k, l\} \subset \mathbb{Z}_+^n$,

$$|P_{kl}(\sigma + i\gamma)\tilde{V}(t, \tau, \sigma + i\gamma)| \le C_{kl} \exp\left\{ -\delta_3|\sigma|^{2b} + c_3|\gamma|^{2b}) \right\}, \quad 0 \le \tau \le t \le T, \quad \{\sigma, \gamma\} \subset \mathbb{R}^n, \tag{3.1.30}$$

where $0 < \delta_3 < \delta_2$, $c_3 > c_2$.

Note that the equality (3.1.22) and the estimate (3.1.29) imply the estimate

$$|V(t, \tau, \sigma + i\gamma)| \le C \exp\left\{ \sum_{j=1}^{3} (-\delta_4|\sigma_j|^{2b} + c_4|\gamma_j|^{2b})(t - \tau)^{2b(j-1)+1} \right\}, \quad 0 \le \tau \le t \le T, \quad \{\sigma, \gamma\} \subset \mathbb{R}^n, \tag{3.1.31}$$

in which $0 < \delta_4 < \delta_2$ and $c_4 > c_2$.

The estimates (3.1.29) and (3.1.30) constitute the basis of a complete analytic description of the function G from (3.1.20) and the proof of the fact that G is a FSCP for the equation (3.1.1).

Using the equality (3.1.20), and also its consequences, that is

$$\partial_x^k \partial_\xi^l G(t, x; \tau, \xi) = (t - \tau)^{-M - M_{kl}} (F_{\sigma \to z}^{-1}[P_{kl}(\sigma)\tilde{V}(t, \tau, \sigma)])(t, \tau, z)\Big|_{z=(X(t-\tau)-\xi)_{t-\tau}}, \quad 0 \le \tau < t \le T, \quad \{x, \xi\} \subset \mathbb{R}^n, \quad \{k, l\} \subset \mathbb{Z}_+^n,$$

where $M_{kl} := (|k_1| + |l_1|)/(2b) + (1 + 1/(2b))(|k_2| + |l_2|) + (2 + 1/(2b))(|k_3| + |l_3|)$, the estimates (3.1.29), (3.1.30), and Lemma 1.1 with $p_j = q_j = 2b$, $j \in \{1, \dots, n\}$, we can prove the following theorem in the same way as Theorem 2.2.

Theorem 3.1. *If the condition B_{21} is satisfied, then:*

1) *there exists a unique FSCP G for the equation (3.1.1) with $A = A(t, \partial_{x_1})$;*
2) *the function G and its derivatives can be extended onto $\mathbb{C}^n$, and these extensions are such that*

$$\partial_x^k \partial_\xi^l G(t, x + iy; \tau, \xi + i\eta) = (t - \tau)^{-M - M_{kl}} \Omega_{kl}(t, \tau, z)\Big|_{z=(X(t-\tau)-\xi)_{t-\tau}+i(Y(t-\tau)-\eta)_{t-\tau}}, \quad 0 \le \tau < t \le T, \quad \{x, y, \xi, \eta\} \subset \mathbb{R}^n, \quad \{k, l\} \subset \mathbb{Z}_+^n,$$

where $\Omega_{kl}(t,\tau,z)$, $z\in\mathbb{C}^n$, *for fixed* t *and* τ, *are entire functions of* z *with the growth order* $q:=2b/(2b-1)$ *and the same order of decrease for* $z=x\in\mathbb{R}^n$;

3) G *satisfies the estimates*

$$|\partial_x^k\partial_\xi^l G(t,x+iy;\tau,\xi+i\eta)|\le C_{kl}(t-\tau)^{-M-M_{kl}}E_c^{(1)}(t,x;\tau,\xi)E_{-c'}^{(1)}(t,y;\tau,\eta),$$
$$0\le\tau<t\le T,\quad \{x,y,\xi,\eta\}\subset\mathbb{R}^n,\quad \{k,l\}\subset\mathbb{Z}_+^n, \tag{3.1.32}$$

where

$$E_c^{(1)}(t,x;\tau,\xi):=\exp\left\{-c\sum_{j=1}^{3}(t-\tau)^{1-jq}|X_j(t-\tau)-\xi_j|^q\right\}, \tag{3.1.33}$$

C_{kl}, c *and* c' *are positive constants depending only on the numbers* n_1, n_2, n_3, b, T, $\max\limits_{t\in[0,T],\ |k_1|\le 2b}|a_{k_1}(t)|$, *and the constant* δ *from the condition* (3.1.6);

4) *the formula*

$$\int\limits_{\mathbb{R}^n}G(t,x;\tau,\xi)d\xi=\exp\left\{(t-\tau)\int\limits_0^1 a_0(\tau+(t-\tau)\beta)d\beta\right\},$$
$$0\le\tau<t\le T,\quad x\in\mathbb{R}^n, \tag{3.1.34}$$

holds;

5) *for all* $0\le\tau<t\le T$ *and* $x\in\mathbb{R}^n$,

$$\partial_x^k\int\limits_{\mathbb{R}^n}G(t,x;\tau,\xi)d\xi=0,\quad k\in\mathbb{Z}_+^n\setminus\{0\};$$
$$\partial_{x_2}^{k_2}\partial_{x_3}^{k_3}\int\limits_{\mathbb{R}^{n_2+n_3}}G(t,x;\tau,\xi)d\xi_2d\xi_3=0,\quad (k_2,k_3)\in\mathbb{Z}_+^{n_2+n_3}\setminus\{0\};$$
$$\partial_{x_3}^{k_3}\int\limits_{\mathbb{R}^{n_3}}G(t,x;\tau,\xi)d\xi_3=0,\quad k_3\in\mathbb{Z}_+^{n_3}\setminus\{0\}. \tag{3.1.35}$$

Proof. In view of the remarks given before the formulation of this theorem, we have still to prove only the equalities (3.1.34). Using the formula (3.1.20) we make the changes of the variables ξ, ξ_2, and ξ_3 by the formulas $(X(t-\tau)-\xi)_{t-\tau}=\eta$, $X_j(t-\tau)-\xi_j=(t-\tau)^{j-1+1/(2b)}\eta_j$, $j\in\{2,3\}$. Then we obtain respectively that

$$\int\limits_{\mathbb{R}^n}G(t,x;\tau,\xi)d\xi=\int\limits_{\mathbb{R}^n}\left(F^{-1}_{\sigma\to\eta}[\tilde V(t,\tau,\sigma)]\right)(t,\tau,\eta)d\eta$$
$$=F_{\eta\to 0}[F^{-1}_{\sigma\to\eta}[\tilde V(t,\tau,\sigma)]=\tilde V(t,\tau,0),$$

$$\int\limits_{\mathbb{R}^{n_2+n_3}} G(t,x;\tau,\xi)d\xi_2 d\xi_3 = (t-\tau)^{-n_1/(2b)}$$

$$\times \int\limits_{\mathbb{R}^{n_2+n_3}} \left(F^{-1}_{(\sigma_2,\sigma_3)\to(\eta_2,\eta_3)} F^{-1}_{\sigma_1\to z_1}[\tilde{V}(t,\tau,\sigma)] \right)(t,\tau,z_1,\eta_2,\eta_3)d\eta_2 d\eta_3$$

$$= (t-\tau)^{-n_1/(2b)} F_{(\eta_2,\eta_3)\to(0,0)}[F^{-1}_{(\sigma_2,\sigma_3)\to(\eta_2,\eta_3)} F^{-1}_{\sigma_1\to z_1}[\tilde{V}(t,\tau,\sigma)]]$$

$$= (t-\tau)^{-n_1/(2b)} \left(F^{-1}_{\sigma_1\to z_1}[\tilde{V}(t,\tau,(\sigma_1,0,0))] \right)(t,\tau,z_1),$$

$$\int\limits_{\mathbb{R}^{n_3}} G(t,x;\tau,\xi)d\xi_3 = (t-\tau)^{-M+(2+1/(2b))n_3}$$

$$\times \int\limits_{\mathbb{R}^{n_3}} \left(F^{-1}_{\sigma_3\to\eta_3} F^{-1}_{(\sigma_1,\sigma_2)\to(z_1,z_2)}[\tilde{V}(t,\tau,\sigma)] \right)(t,\tau,z_1,z_2,\eta_3)d\eta_3$$

$$= (t-\tau)^{-M+(2+1/(2b))n_3} F_{\eta_3\to 0}[F^{-1}_{\sigma_3\to\eta_3} F^{-1}_{(\sigma_1,\sigma_2)\to(z_1,z_2)}[\tilde{V}(t,\tau,\sigma)]]$$

$$= (t-\tau)^{-M+(2+1/(2b))n_3} (F^{-1}_{(\sigma_1,\sigma_2)\to(z_1,z_2)}[\tilde{V}(t,\tau,(\sigma_1,\sigma_2,0))](t,\tau,z_1,z_2),$$

where $z_1 = (t-\tau)^{-1/(2b)}(x_1-\xi_1)$ and $z_2 = (t-\tau)^{-1-1/(2b)}(x_2+(t-\tau)\hat{x}_1-\xi_2)$. These equalities imply the desired equalities (3.1.34) and (3.1.35). □

Remark 1. If the change of variables (3.1.17) in the integral (3.1.16) is not performed, then we obtain another formula for the function G, namely

$$G(t,x;\tau,\xi) = (t-\tau)^{-M} \left(F^{-1}_{\sigma\to z}[\hat{V}(t,\tau,\sigma)] \right)(t,\tau,z) \Big|_{z=(x-\Xi(\tau-t))_{t-\tau}},$$

$$0 \le \tau < t \le T, \quad \{x,\xi\} \subset \mathbb{R}^n, \tag{3.1.36}$$

where

$$\hat{V}(t,\tau,\sigma) := \exp\Bigg\{ \sum_{|k_1|\le 2b} i^{|k_1|}(t-\tau)^{1-|k_1|/(2b)}$$

$$\times \int_0^1 a_{k_1}(t-(t-\tau)\beta)(\sigma_1' - \beta\sigma_2' + 2^{-1}\beta^2\sigma_3)^{k_1'}(\sigma_1'' - \beta\sigma_2'')^{k_1''}(\sigma_1''')^{k_1'''} d\beta \Bigg\}.$$

Set

$$\hat{E}_c^{(1)}(t,x;\tau,\xi) := \exp\left\{ -c\sum_{j=1}^{3}(t-\tau)^{1-jq}|x_j - \Xi_j(\tau-t)|^q \right\}. \tag{3.1.37}$$

Using the formula (3.1.36) and the above method, we obtain estimates for derivatives of G differing from (3.1.32) only by the substitution of $\hat{E}_c^{(1)}$ for $E_c^{(1)}$. By the way, these estimates can be obtained directly from the estimates (3.1.32), if we use the inequalities

$$E_{c_1}^{(1)}(t,x;\tau,\xi) \le \hat{E}_{c_1}^{(1)}(t,x;\tau,\xi) \le E_{c_2}^{(1)}(t,x;\tau,\xi),$$
$$0 \le \tau < t, \quad \{x,\xi\} \subset \mathbb{R}^n, \tag{3.1.38}$$

in which c_1 and c_2 are such constants that $c_2 < c < c_1$.

The inequalities (3.1.38) can be proved easily with the use of the equalities (3.1.33), (3.1.37) and the inequalities (1.3.9), (1.3.10).

Example 3.1. The FSCP for the equation from Example 1.10 is given by the formula

$$\begin{aligned} G(t,x;\tau,\xi) &= (4\pi a)^{-n/2} 12^{n_2/2} 720^{n_3/2} (t-\tau)^{-(n_1+3n_2+5n_3)/2} \\ &\times \exp\Big\{ -a^{-1}((4(t-\tau))^{-1}|x_1-\xi_1|^2 \\ &+ 3(t-\tau)^{-3}|x_2+2^{-1}(t-\tau)(\hat{x}_1+\hat{\xi}_1)-\xi_2|^2 \\ &+ 180(t-\tau)^{-5}|x_3+2^{-1}(t-\tau)(x_2'+\xi_2') \\ &+ 12^{-1}(t-\tau)^2(x_1'-\xi_1')-\xi_3|^2) \Big\}, \\ & t>\tau, \quad \{x,\xi\} \subset \mathbb{R}^n. \end{aligned} \tag{3.1.39}$$

This formula is a direct consequence of (3.1.20) and (3.1.22).

Similarly we get an expression for the FSCP in the case $n_3 = 0$, that is for the equation

$$\left(\partial_t - \sum_{j=1}^{n_2} x_{1j}\partial_{x_{2j}} - a\sum_{j=1}^{n_1} \partial_{x_{1j}}^2 \right) u(t,x) = 0, \quad t>0, \quad x=(x_1,x_2) \in \mathbb{R}^{n_1+n_2}.$$

This expression has the form

$$\begin{aligned} &G(t,x;\tau,\xi) = (4\pi a)^{-(n_1+n_2)/2} 12^{n_2/2} (t-\tau)^{-(n_1+3n_2)/2} \\ &\times \exp\{-a^{-1}((4(t-\tau))^{-1}|x_1-\xi_1|^2 + 3(t-\tau)^{-3}|x_2+2^{-1}(t-\tau)(\hat{x}_1+\hat{\xi}_1)-\xi_2|^2)\}, \\ &\qquad t>\tau, \quad \{x=(x_1,x_2), \xi=(\xi_1,\xi_2)\} \subset \mathbb{R}^{n_1+n_2}. \end{aligned} \tag{3.1.40}$$

Analyzing the formulas (3.1.39) and (3.1.40) we see that the FSCP for the equation from Example 1.10 is given by the formula

$$G(t,x;\tau,\xi) = G_1(t,x';\tau,\xi')G_2(t,x'';\tau,\xi'')G_3(t,x_1''';\tau,\xi_1'''),$$
$$t>\tau, \quad \{x,\xi\} \subset \mathbb{R}^n,$$

where $x' := (x_1', x_2', x_3)$, $\xi' := (\xi_1', \xi_2', \xi_3)$, $x'' := (x_1'', x_2'')$, $\xi'' := (\xi_1'', \xi_2'')$, while G_1, G_2, and G_3 are the FSCPs for the equations

$$\left(\partial_t - \sum_{j=1}^{n_3}(x_{1j}\partial_{x_{2j}} + x_{2j}\partial_{x_{3j}} + a\partial^2_{x_{1j}})\right)u = 0,$$

$$\left(\partial_t - \sum_{j=n_3+1}^{n_2}(x_{1j}\partial_{x_{2j}} + a\partial^2_{x_{1j}})\right)u = 0$$

and

$$\left(\partial_t - a\sum_{j=n_2+1}^{n_1}\partial^2_{x_{1j}}\right)u = 0$$

respectively.

Remark 2. The FSCP for the equation (3.1.1) of the second order ($b = 1$) can be estimated in terms of the FSCP for the model equation from Example 1.10, that is in terms of the expression (3.1.39).

Set

$$\begin{aligned} E_c^{(2)}(t, x; \tau, \xi) := \exp\{&-c((4(t-\tau))^{-1}|x_1 - \xi_1|^2 \\ &+ 3(t-\tau)^{-3}|x_2 + 2^{-1}(t-\tau)(\hat{x}_1 + \hat{\xi}_1) - \xi_2|^2 \\ &+ 180(t-\tau)^{-5}|x_3 + 2^{-1}(t-\tau)(x_2' + \xi_2') \\ &+ 12^{-1}(t-\tau)^2(x_1' - \xi_1') - \xi_3|^2)\}, \\ &t > \tau, \quad \{x, \xi\} \subset \mathbb{R}^n, \quad c \in \mathbb{R}. \end{aligned} \tag{3.1.41}$$

Using the equalities (3.1.33), (3.1.41), and the inequalities (3.1.9), (3.1.10), we find that

$$\begin{aligned} E_{c_1}^{(1)}(t, x; \tau, \xi) \le E_c^{(2)}(t, x; \tau, \xi) \le E_{c_2}^{(1)}(t, x; \tau, \xi), \\ t > \tau, \quad \{x, \xi\} \subset \mathbb{R}^n, \end{aligned} \tag{3.1.42}$$

where $c_2 < c < c_1$.

Due to these inequalities, for the derivatives of the FSCP G in the case $b = 1$ we obtain the estimates differing from (3.1.32) only by the substitution of $E_c^{(2)}$ for $E_c^{(1)}$.

Note that the above FSCPs G have the normality property and satisfy the convolution formula, that is the following property similar to Property 2.8.

Property 3.1. *Let G be the FSCP for the equation*

$$(Lu)(t, x) := (S - A(t, \partial_{x_1}))u(t, x) = 0, \quad (t, x) \in \Pi_{(0,T]}. \tag{3.1.43}$$

Then the function

$$G^*(\tau, \xi; t, x) := \bar{G}(t, x; \tau, \xi), \quad 0 \le \tau < t \le T, \quad \{x, \xi\} \subset \mathbb{R}^n, \tag{3.1.44}$$

is a FSCP for the adjoint equation

$$(L^*v)(\tau,\xi) := \left(S^* - \sum_{|k_1|\le 2b} \bar{a}_{k_1}(\tau)(-\partial_{\xi_1})^{k_1}\right) v(\tau,\xi) = 0,$$
$$\tau,\xi) \in \Pi_{[0,T)}, \tag{3.1.45}$$

where

$$S^* := -\partial_\tau + \sum_{j=1}^{n_2} \xi_{1j}\partial_{\xi_{2j}} + \sum_{j=1}^{n_3} \xi_{2j}\partial_{\xi_{3j}},$$

and

$$G(t,x;\tau,\xi) = \int_{\mathbb{R}^n} G(t,x;\lambda,y)G(\lambda,y;\tau,\xi)dy,$$
$$0 \le \tau < \lambda < t \le T, \quad \{x,\xi\} \subset \mathbb{R}^n. \tag{3.1.46}$$

Proof. Let $t \in (0,T]$. Consider the problem

$$(L^*v)(\tau,\xi) = 0, \quad (\tau,\xi) \in \Pi_{[0,t)}; \quad v(\tau,\xi)|_{\tau=t} = \varphi(\xi), \quad \xi \in \mathbb{R}^n.$$

Solving it by the same method as the Cauchy problem (2.1.6), (3.1.43), we obtain a formula for the solution

$$v(\tau,\xi) = \int_{\mathbb{R}^n} G^*(\tau,\xi;t,x)\varphi(x)dx, \quad (\tau,\xi) \in \Pi_{[0,t)},$$

where

$$G^*(\tau,\xi;t,x) := (t-\tau)^{-M}(F^{-1}_{\sigma\to z}[\bar{\tilde{V}}(t,\tau,\sigma)])(t,\tau,z)|_{z=(\xi-X(t-\tau))_{t-\tau}}. \tag{3.1.47}$$

Here the function $\tilde{V}$ is the same as in (3.1.22), and the bar, as usual, means the complex conjugation. Comparing the formulas (3.1.20) and (3.1.47) we get the equality (3.1.44).

In order to prove the equality (3.1.46), we use the following Green-Ostrogradsky formula, similar to (2.1.110):

$$\int_{t_1}^{t_2} d\theta \int_{B_R} (\bar{v}Lu - (L^*v)u)(\theta,y)dy$$
$$= \int_{B_R} (\bar{v}u)(\theta,y)\Big|_{\theta=t_1}^{t_2} dy - \int_{t_1}^{t_2} d\theta \int_{\Gamma_R} \left(\sum_{j=1}^{n_2} y_{1j}\mu_{2j} + \sum_{j=1}^{n_3} y_{2j}\mu_{3j}\right)(\bar{v}u)(\theta,y)dS_y$$
$$+ \int_{t_1}^{t_2} d\theta \int_{\Gamma_R} \sum_{j=1}^{n_1} B^j[v,u](\theta,y)\mu_{1j}dS_y, \tag{3.1.48}$$

where $t_1 < t_2$, B_R is the ball $\{y \in \mathbb{R}^n \mid |y| \leq R\}$, Γ_R is its boundary, $(\mu_{11}, \ldots, \mu_{1n_1}, \mu_{21}, \ldots, \mu_{2n_2}, \mu_{31}, \ldots, \mu_{3n_3})$ is the unit vector of the outer normal to Γ_R, L and L^* are the differential expressions from (3.1.43) and (3.1.45), $B^j[v,u]$, $j \in \{1, \ldots, n_1\}$, are bilinear forms containing derivatives in y_1 of v and u of the orders not exceeding $2b - 1$. Setting in (3.1.48) $t_1 = \lambda$, $t_2 = t - \varepsilon$, where $\varepsilon > 0$ is such that $\lambda < t - \varepsilon$, $u(\theta, y) = Z(\theta, y; \tau, \xi)$, $v(\theta, y) = Z^*(\lambda, y; t, x)$, and passing to the limit first as $R \to \infty$, and then as $\varepsilon \to 0$, we come to (3.1.46). Here we have to use the equality (3.1.44). □

Let us consider now the equation (3.1.1) with $A = A(\beta, y, \partial_{x_1})$. If the condition B_{22} is satisfied, then, just as in the preceding case, we construct the FSCP $G(\cdot,\cdot;\cdot,\cdot;\beta,y)$ and give its complete analytic description implying, in particular, the estimates

$$|\partial_x^k \partial_\xi^l G(t,x;\tau,\xi;\beta,y)| \leq C_{kl}(t-\tau)^{-M-M_{kl}} E_c^{(1)}(t,x;\tau,\xi),$$
$$0 \leq \tau < t \leq T, \quad \{x,\xi\} \subset \mathbb{R}^n, \quad (\beta,y) \in \Pi_{[0,T]}, \quad \{k,l\} \subset \mathbb{Z}_+^n, \tag{3.1.49}$$

where the constants C_{kl} and c depend only on $n_1, n_2, n_3, b, T, \delta$ (from the condition B_{22}), and $\max\limits_{|k_1| \leq 2b} \sup\limits_{(\beta,y) \in \Pi_{[0,T]}} |a_{k_1}(\beta, y)|$.

Properties of the function G are similar to Properties 2.1–2.4 of the FSCP for a $\overrightarrow{2b}$-parabolic equation with parameters.

Property 3.2. *Let the conditions B_{22} and B_{23} be satisfied. Then*

$$|\Delta_{\beta,y}^{\lambda,z} \partial_x^k \partial_\xi^l G(t,x;\tau,\xi;\beta,y)|$$
$$\leq C_{kl} H(d(\beta, Y(\beta-\lambda); \lambda, z))^\alpha (t-\tau)^{-M-M_{kl}} E_c^{(1)}(t,x;\tau,\xi),$$
$$0 \leq \tau < t \leq T, \quad \{x,\xi\} \subset \mathbb{R}^n, \quad \{(\beta,y),(\lambda,z)\} \subset \Pi_{[0,T]}, \quad \{k,l\} \subset \mathbb{Z}_+^n.$$

Proof. It follows from the above results that it suffices to obtain, for the equation (3.1.1) without parameters, an estimate of the difference

$$\Delta_{\beta,y}^{\lambda,z} V(t,\tau,s;\beta,y), \quad 0 \leq \tau < t \leq T,$$
$$s = \sigma + i\gamma \in \mathbb{C}^n, \quad \{(\beta,y),(\lambda,z)\} \subset \Pi_{[0,T]},$$

of the function V, defined as in (3.1.19):

$$V(t,\tau,s;\beta,y) := \exp\left\{ \sum_{|k_1| \leq 2b} a_{k_1}(\beta,y) \int_\tau^t Q_{k_1}(\theta-\tau,s)\,d\theta \right\}, \tag{3.1.50}$$

where

$$Q_{k_1}(t,s) := i^{|k_1|}(s_1' + ts_2' + 2^{-1}t^2 s_3)^{k_1'} (s_1'' + ts_2'')^{k_1''} (s_1''')^{k_1'''}.$$

The function (3.1.50) is, obviously, a solution of the problem

$$\left(\frac{d}{dt} - \sum_{|k_1|\le 2b} a_{k_1}(\beta,y)Q_{k_1}(t-\tau,s)\right)V = 0, \quad V|_{t=\tau} = 1.$$

Therefore

$$\left(\frac{d}{dt} - \sum_{|k_1|\le 2b} a_{k_1}(\beta,y)Q_{k_1}(t-\tau,s)\right)\Delta^{\lambda,z}_{\beta,y}V(t,\tau,s;\beta,y)$$
$$= \sum_{|k_1|\le 2b} \Delta^{\lambda,z}_{\beta,y}a_{k_1}(\beta,y)Q_{k_1}(t-\tau,s)V(t,\tau,s;\lambda,z),$$
$$\Delta^{\lambda,z}_{\beta,y}V(t,\tau,s;\beta,y)|_{t=\tau} = 0$$

so that

$$\Delta^{\lambda,z}_{\beta,y}V(t,\tau,s;\beta,y)$$
$$= \int_\tau^t V(t,\theta,s;\beta,y) \sum_{|k_1|\le 2b} \Delta^{\lambda,z}_{\beta,y}a_{k_1}(\beta,y)Q_{k_1}(\theta-\tau,s)V(\theta,\lambda,s;\lambda,z)d\theta.$$

Using the inequality (3.1.7) and the estimate (3.1.31) for $V(\cdot,\cdot,\cdot;\beta,y)$ we find that

$$|\Delta^{\lambda,z}_{\beta,y}V(t,\tau,s;\beta,y)| \le CH(d(\beta,Y(\beta-\lambda);\lambda,z))^\alpha \int_\tau^t \exp\left\{\sum_{j=1}^{3}(-\delta_4|\sigma_j|^{2b} + c_4|\gamma_j|^{2b})\right.$$
$$\left.\times\left((t-\theta)^{2b(j-1)+1} + (\theta-\tau)^{2b(j-1)+1}\right)\right\}d\theta$$
$$\times \sum_{|k_1|\le 2b} (|s_1^{'}| + |s_2^{'}| + |s_3|)^{k_1^{'}}(|s_1^{''}| + |s_2^{''}|)^{k_1^{''}}|s_1^{'''}|^{k_1^{'''}}$$
$$\le CH(d(\beta,Y(\beta-\lambda);\lambda,z))^\alpha \exp\left\{\sum_{j=1}^{3}(-\delta_5|\sigma_j|^{2b} + c_5|\gamma_j|^{2b})(t-\tau)^{2b(j-1)+1}\right\}. \quad \square$$

Property 3.3. *If the conditions B_{22} and B_{24} are satisfied, then*

$$|\partial_x^k\partial_\xi^m\partial_\beta^{l_0}\partial_y^l G(t,x;\tau,\xi;\beta,y)| \le C_{kml_0l}(t-\tau)^{-M-M_{kl}}E_c^{(1)}(t,x;\tau,\xi),$$
$$0\le\tau<t\le T, \quad \{x,\xi\}\subset\mathbb{R}^n, \quad (\beta,y)\in\Pi_{[0,T]},$$
$$\{k,m\}\subset\mathbb{Z}_+^n, \quad l_0\le r_0, \quad |l|\le r.$$

If, in addition, the condition B_{25} *is satisfied, then*

$$
\begin{gathered}
|\Delta^{\lambda,z}_{\beta,y}\partial^k_x\partial^m_\xi\partial^{l_0}_\beta\partial^l_y G(t,x;\tau,\xi;\beta,y)| \\
\leq C_{kml_0l}(d(\beta,Y(\beta-\lambda);\lambda,z))^\alpha(t-\tau)^{-M-M_{kl}}E^{(1)}_c(t,x;\tau,\xi), \\
0\leq\tau<t\leq T,\quad \{x,\xi\}\subset\mathbb{R}^n,\quad \{(\beta,y),(\lambda,z)\}\subset\Pi_{[0,T]}, \\
\{k,m\}\subset\mathbb{Z}^n_+,\quad l_0\leq r_0,\quad |l|\leq r.
\end{gathered}
$$

This property is proved just as Property 2.2, using the method of proof of Property 3.2.

Property 3.4. *The equalities* (3.1.34) *and* (3.1.35) *hold, with* $G(\cdot,\cdot;\cdot,\cdot;\beta,y)$ *and* $a_0(\beta,y)$ *substituted for* $G(\cdot,\cdot;\cdot,\cdot)$ *and* $a_0(\cdot)$ *respectively.*

This is proved just as the assertions 4) and 5) of Theorem 3.1.

Next we consider properties of the function Z_1 from (1.1.44), which will be used as a parametrix for an equation from the class $\mathbf{E}_{21}$.

Property 3.5. *Let the conditions* B_{22} *and* B_{23} *be satisfied. Then*

$$
|\partial^k_x Z_1(t,x;\tau,\xi)|\leq C_k(t-\tau)^{-M-M_{k0}}E^{(1)}_c(t,x;\tau,\xi), \tag{3.1.51}
$$

$$
\begin{gathered}
|\Delta^z_x\partial^k_x Z_1(t,x;\tau,\xi)|\leq C_k(d(x;z))^{\alpha_0}(t-\tau)^{-M-M_{k0}-\alpha_0/(2b)}E^{(1)}_c(t,x;\tau,\xi), \\
(d(x;z))^{2b}\leq t-\tau,\quad k\in\mathbb{Z}^n_+;
\end{gathered} \tag{3.1.52}
$$

$$
\left|\int\limits_{\mathbb{R}^n}\partial^k_x Z_1(t,x;\tau,y)dy\right|\leq C_k(t-\tau)^{-M_{k0}+\alpha/(2b)},\quad k\in\mathbb{Z}^n_+\setminus\{0\}; \tag{3.1.53}
$$

$$
|SZ_1(t,x;\tau,\xi)|\leq C(t-\tau)^{-M-1}E^{(1)}_c(t,x;\tau,\xi), \tag{3.1.54}
$$

$$
\left|\int\limits_{\mathbb{R}^n}SZ_1(t,x;\tau,y)dy\right|\leq C(t-\tau)^{-1+\alpha/(2b)}, \tag{3.1.55}
$$

where $0\leq\tau<t\leq T$, $\{x,z,\xi\}\subset\mathbb{R}^n$, α_0 *is an arbitrary fixed number from the interval* $(0,1]$, α *is the number from the condition* B_{23}.

The *proof* is similar to that of Property 2.5. The estimates (3.1.51) follow from the estimates (3.1.49), the estimates (3.1.52) are deduced from (3.1.51) with the use of the mean value theorem and the inequality

$$
E^{(1)}_c(t,y;\tau,\xi)\leq C_1E^{(1)}_{c_1}(t,x;\tau,\xi),\quad c_1\in(0,c),
$$

where y is a point from the line connecting x and z.

In order to prove (3.1.53), we use the representation

$$
\int\limits_{\mathbb{R}^n}\partial^k_x Z_1(t,x;\tau,y)dy=-\int\limits_{\mathbb{R}^n}\Delta^{\tau,y}_{t,z}\partial^k_x G_0(t,x;\tau,y;t,z)|_{z=x}dy,
$$

valid due to the formula (1.1.44) and Property 3.4. Using Property 3.3, the inequality

$$(d(t, X(t-\tau); \tau, y))^\alpha E_c^{(1)}(t, x; \tau, y) \le C(t-\tau)^{\alpha/(2b)} E_{c_1}^{(1)}(t, x; \tau, y),$$
$$\tau < t, \quad \{x, y\} \subset \mathbb{R}^n, \quad c_1 \in (0, c), \tag{3.1.56}$$

and the fact that

$$\int_{\mathbb{R}^n} (t-\tau)^{-M} E_{c_2}^{(1)}(t, x; \tau, y) dy = C, \quad \tau < t, \quad x \in \mathbb{R}^n, \tag{3.1.57}$$

we obtain the estimates (3.1.53).

The estimates (3.1.54) and (3.1.55) follow from the estimates (3.1.51), (3.1.53), and the equality

$$SZ_1(t, x; \tau, \xi) = \sum_{|k_1|=2b} a_{k_1}(\tau, \xi) \partial_{x_1}^{k_1} Z_1(t, x; \tau, \xi),$$
$$0 \le \tau < t \le T, \quad \{x, \xi\} \subset \mathbb{R}^n. \tag{3.1.58}$$

□

The next property deals with the integral

$$W_1(t, x; \tau, \xi) := \int_\tau^t d\beta \int_{\mathbb{R}^n} Z_1(t, x; \beta, y) Q(\beta, y; \tau, \xi) dy,$$
$$0 \le \tau < t \le T, \quad \{x, \xi\} \subset \mathbb{R}^n. \tag{3.1.59}$$

Having in mind the use of this property to substantiate the Levi method (on the basis of Lemma 1.9 and Lemma 1.10), we assume certain special properties of the function Q, in particular the estimates similar to the estimates (1.3.49) and (1.3.58) for the function R from Lemma 1.9 and Lemma 1.10.

Let $E_c^{(1)}$ and $E_c^{(2)}$ be the functions defined in (3.1.33) and (3.1.41), and

$$E_c^{(0)}(t, x_1; \tau, \xi_1) := \exp\{-c(t-\tau)^{1-q}|x_1 - \xi_1|^q\},$$
$$t > \tau, \quad \{x_1, \xi_1\} \subset \mathbb{R}^{n_1}. \tag{3.1.60}$$

Let us introduce the function F_c, which will be employed in estimates of the function Q. Set

$$F_c(t, x; \tau, \xi)$$
$$:= E_c^{(0)}(t, x_1; \tau, \xi_1) \sum_{j=0}^{\infty} (\hat{C}\Gamma(\alpha/(2b))(t-\tau)^{\alpha/(2b)})^j (\Gamma(j\alpha/(2b)))^{-1} E_{c\delta_0^j}^{(1)}(t, x; \tau, \xi),$$
$$t > \tau, \quad \{x, \xi\} \subset \mathbb{R}^n, \tag{3.1.61}$$

where $\hat{C}$ and δ_0 are certain positive constants, $\delta_0 < 1$, Γ is Euler's Gamma function, α is the number from the condition B_{23}. For the case of an equation (3.1.1) of the second order ($b = 1$) the estimates of the function Q contain the function $E_c^{(2)}$ instead of F_c.

Note that the integral

$$I^{(1)} := \int_{\mathbb{R}^n} E_c^{(1)}(t,x;\beta,y)F_c(\beta,y;\tau,\xi)((t-\beta)(\beta-\tau))^{-M}dy$$

satisfies the inequality

$$I^{(1)} \leq C(t-\tau)^{-M}F_{c_1}(t,x;\tau,\xi), \quad \tau<\beta<t, \quad \{x,\xi\}\subset\mathbb{R}^n, \tag{3.1.62}$$

where $c_1 < c$, while for the integral

$$I^{(2)} := \int_{\mathbb{R}^n} E_c^{(2)}(t,x;\beta,y)E_c^{(2)}(\beta,y;\tau,\xi)((t-\beta)(\beta-\tau))^{-M}dy$$

we have the inequality

$$I^{(2)} = C(t-\tau)^{-M}E_c^{(2)}(t,x;\tau,\xi), \quad \tau<\beta<t, \quad \{x,\xi\}\subset\mathbb{R}^n. \tag{3.1.63}$$

Indeed, by the inequalities (1.3.50), (1.3.53), and (1.3.54) we get

$$I^{(1)} \leq E_{c/2}^{(0)}(t,x_1;\tau,\xi_1)\sum_{j=0}^{\infty}(\hat{C}\Gamma(\alpha/(2b))(t-\tau)^{\alpha/(2b)})^j(\Gamma(j\alpha/(2b)))^{-1}$$
$$\times\int_{\mathbb{R}^n} E_{c/2}^{(1)}(t,x;\beta,y)E_{c\delta_0^j}^{(1)}(\beta,y;\tau,\xi)((t-\beta)(\beta-\tau))^{-M}dy \leq C(t-\tau)^{-M}F_{c_1}(t,x;\tau,\xi).$$

The equality (3.1.63) is a consequence of (1.3.59).

Property 3.6. *Let the conditions B_{22}, B_{23} be satisfied, and a continuous function Q satisfy the conditions*

$$|Q(t,x;\tau,\xi)| \leq C(t-\tau)^{-M-1+\alpha/(2b)}F_c(t,x;\tau,\xi), \tag{3.1.64}$$
$$|\Delta_x^y Q(t,x;\tau,\xi)| \leq C(d(x;y))^{\alpha_1}(t-\tau)^{-M-1+\alpha_2/(2b)} \times (F_c(t,x;\tau,\xi)+F_c(t,y;\tau,\xi)), \tag{3.1.65}$$

where $0\leq\tau<t\leq T$, $\{x,y,z\}\subset\mathbb{R}^n$, $\alpha_1\in(0,\alpha)$ and $\alpha_2=\alpha-\alpha_1$. Then the function (3.1.59) has the continuous derivatives $\partial_{x_1}^{k_1}W_1$, $|k_1|\leq 2b$, and SW_1, such that

$$\partial_{x_1}^{k_1}W_1(t,x;\tau,\xi) = \int_\tau^t d\beta\int_{\mathbb{R}^n}\partial_{x_1}^{k_1}Z_1(t,x;\beta,y)Q(\beta,y;\tau,\xi)dy,$$
$$|k_1| < 2b, \tag{3.1.66}$$

$$\partial_{x_1}^{k_1} W_1(t,x;\tau,\xi) = \int\limits_{\tau}^{t_1} d\beta \int\limits_{\mathbb{R}^n} \partial_{x_1}^{k_1} Z_1(t,x;\beta,y) Q(\beta,y;\tau,\xi)dy$$

$$+ \int\limits_{t_1}^{t} d\beta \int\limits_{\mathbb{R}^n} \partial_{x_1}^{k_1} Z_1(t,x;\beta,y) \Delta_y^{X(t-\beta)} Q(\beta,y;\tau,\xi)dy$$

$$+ \int\limits_{t_1}^{t} \Big(\int\limits_{\mathbb{R}^n} \partial_x^{k_1} Z_1(t,x;\beta,y)dy \Big) Q(\beta, X(t-\beta);\tau,\xi)d\beta,$$

$$|k_1| = 2b, \tag{3.1.67}$$

$$SW_1(t,x;\tau,\xi) = Q(t,x;\tau,\xi) + \int\limits_{\tau}^{t_1} d\beta \int\limits_{\mathbb{R}^n} SZ_1(t,x;\beta,y)Q(\beta,y;\tau,\xi)dy$$

$$+ \int\limits_{t_1}^{t} d\beta \int\limits_{\mathbb{R}^n} SZ_1(t,x;\beta,y)\Delta_y^{X(t-\beta)} Q(\beta,y;\tau,\xi)dy$$

$$+ \int\limits_{t_1}^{t} \Big(\int\limits_{\mathbb{R}^n} SZ_1(t,x;\beta,y)dy \Big) Q(\beta, X(t-\beta);\tau,\xi)d\beta, \tag{3.1.68}$$

$$0 \le \tau < t \le T, \quad \{x,\xi\} \subset \mathbb{R}^n, \quad t_1 := t - (t-\tau)/2.$$

Proof. The fact that the integrals from (3.1.66)–(3.1.68) are convergent is established easily, with the use of the estimates (3.1.51), (3.1.53)–(3.1.55), (3.1.62), (3.1.64), (3.1.65), the inequality (3.1.56), and the inequality

$$F_c(\beta, X(t-\beta);\tau,\xi) \le F_{c_1}(t,x;\tau,\xi),$$
$$\tau < \beta < t, \quad \{x,\xi\} \subset \mathbb{R}^n, \quad c_1 \in (0,c). \tag{3.1.69}$$

The proof of the formulas (3.1.66) and (3.1.67) is similar to that of the corresponding formulas from Properties 2.6 and 2.7. In order to prove (3.1.68), we consider the set of functions

$$W_{1h}(t,x;\tau,\xi) := \int\limits_{\tau}^{t-h} d\beta \int\limits_{\mathbb{R}^n} Z_1(t,x;\beta,y)Q(\beta,y;\tau,\xi)dy, \quad (t,x) \in \Pi_{(\tau,T]},$$

where h is a parameter. Applying to W_{1h} the differential expression S we come to the formula

$$SW_{1h}(t,x;\tau,\xi) = \int\limits_{\mathbb{R}^n} Z_1(t,x;t-h,y)Q(t-h,y;\tau,\xi)dy$$

$$+ \int_{\tau}^{t-h} d\beta \int_{\mathbb{R}^n} SZ_1(t,x;\beta,y)\Delta_y^{X(t-\beta)} Q(\beta,y;\tau,\xi)dy$$

$$+ \int_{\tau}^{t-h} \left(\int_{\mathbb{R}^n} SZ_1(t,x;\beta,y)dy \right) Q(\beta, X(t-\beta);\tau,\xi)d\beta.$$

The existence of the uniform limits, as $h \to 0$, for summands from this formula is proved just as a similar result in the proof of the formula for the derivative in t of the volume potential in Property 2.6. □

3.1.2. Equations from the classes $\mathbf{E}_{21}$ and $\mathbf{E}_{22}$ with coefficients depending on all the variables. Here we present theorems about the FSCPs for the equation (1.1.14) and, in particular, for the equation (1.1.15) for the general case.

Thus, we consider the equation

$$\left(S - \sum_{|k_1| \le 2b} a_{k_1}(t,x)\partial_{x_1}^{k_1} \right) u(t,x) = f(t,x), \quad (t,x) \in \Pi_{(0,T]}, \tag{3.1.70}$$

and the equation

$$\left(S - \sum_{j,l=1}^{n_1} a_{jl}(t,x)\partial_{x_{1j}}\partial_{x_{1l}} - \sum_{j=1}^{n_1} a_j(t,x)\partial_{x_{1j}} - a_0(t,x) \right) u(t,x) = f(t,x),$$
$$(t,x) \in \Pi_{(0,T]}. \tag{3.1.71}$$

For their coefficients we use the following conditions:

(A_{21}) there exists such a constant $\delta > 0$ that for any $(t,x) \in \Pi_{[0,T]}$ and $\sigma_1 \in \mathbb{R}^{n_1}$ the inequality (1.1.17) is valid;

(A_{22}) the coefficients a_{k_1}, $|k_1| \le 2b$, are bounded and Hölder continuous in t, x in the sense of (3.1.7) on $\Pi_{[0,T]}$;

(A_{23}) there exist bounded and Hölder continuous in t, x in the sense of (3.1.7) derivatives $\partial_{x_1}^{k_1} a_{k_1}$, $|k_1| \le 2b$, on $\Pi_{[0,T]}$;

(A_{24}) there exists such a constant $\delta > 0$ that for any $(t,x) \in \Pi_{[0,T]}$ and $\sigma_1 \in \mathbb{R}^{n_1}$ the inequality (1.1.18) is valid;

(A_{25}) the coefficients a_{jl}, a_j, $\{j,l\} \subset \{1,\ldots,n_1\}$, and a_0 are bounded and Hölder continuous in t, x on $\Pi_{[0,T]}$;

(A_{26}) there exist bounded and Hölder continuous in t, x in the sense of (3.1.7) derivatives $\partial_{x_{1j}}\partial_{x_{1l}} a_{jl}$ and $\partial_{x_{1j}} a_j$, $\{j,l\} \subset \{1,\ldots,n_1\}$ on $\Pi_{[0,T]}$.

If the conditions A_{23} and A_{26} are satisfied, then there exist adjoint equations for the equations (3.1.70) and (3.1.71) , of the form

$$S^* v(\tau,\xi) - \sum_{|k_1| \le 2b} (-\partial_{\xi_1})^{k_1} (\bar{a}_{k_1}(\tau,\xi) v(\tau,\xi)) = g(\tau,\xi),$$
$$(\tau,\xi) \in \Pi_{[0,T)}, \tag{3.1.72}$$

and

$$S^*v(\tau,\xi) - \sum_{j,l=1}^{n_1} \partial_{\xi_{1j}}\partial_{\xi_{1l}}(\bar{a}_{jl}(\tau,\xi)v(\tau,\xi))$$
$$+\sum_{j=1}^{n_1} \partial_{\xi_{1j}}(\bar{a}_j(\tau,\xi)v(\tau,\xi)) - a_0(\tau,\xi)v(\tau,\xi) = g(\tau,\xi),$$
$$(\tau,\xi) \in \Pi_{[0,T)} \tag{3.1.73}$$

respectively. Their coefficients satisfy respectively the conditions A_{22} and A_{25}.

Theorem 3.2. *Let the conditions A_{21} and A_{22} be satisfied. Then there exists a FSCP Z for the equation* (3.1.70), *and*

$$|\partial_{x_1}^{k_1} Z(t,x;\tau,\xi)| \le C(t-\tau)^{-M-|k_1|/(2b)} F_c(t,x;\tau,\xi), \quad |k_1| \le 2b, \tag{3.1.74}$$
$$|SZ(t,x;\tau,\xi)| \le C(t-\tau)^{-M-1} F_c(t,x;\tau,\xi), \tag{3.1.75}$$
$$0 \le \tau < t \le T, \quad \{x,\xi\} \subset \mathbb{R}^n.$$

If, in addition, the condition A_{23} is satisfied, then there exists a FSCP Z^* for the adjoint equation (3.1.72), *which is connected with the FSCP Z by the relation*

$$Z^*(\tau,\xi;t,x) = \bar{Z}(t,x;\tau,\xi), \quad 0 \le \tau < t \le T, \quad \{x,\xi\} \subset \mathbb{R}^n, \tag{3.1.76}$$

and the convolution formula (2.1.112) *holds for Z.*

Proof. In accordance with Section 1.1.4, the FSCP Z for the equation (3.1.70) is sought in the form

$$Z(t,x;\tau,\xi) = Z_1(t,x;\tau,\xi) + W_1(t,x;\tau,\xi),$$
$$0 \le \tau < t \le T, \quad \{x,\xi\} \subset \mathbb{R}^n, \tag{3.1.77}$$

where Z_1 is a parametrix with Properties 3.5 and 3.6, while the function W_1 is defined by the formula (3.1.59), in which Q is an unknown function. We assume that Q is continuous and satisfies the estimates (3.1.64) and (3.1.65).

If we apply the differential expression

$$S - \sum_{|k_1|\le 2b} a_{k_1}(t,x)\partial_{x_1}^{k_1}$$

to the function (3.1.77) and use Property 3.6, then we obtain for Q an integral equation (1.1.47), in which

$$K(t,x;\tau,\xi) := \left(\sum_{|k_1|=2b} \Delta_{t,x}^{\tau,\xi} a_{k_1}(t,x)\partial_{x_1}^{k_1} + \sum_{|k_1|<2b} a_{k_1}(t,x)\partial_{x_1}^{k_1} \right) Z_1(t,x;\tau,\xi).$$

By the estimates (3.1.51), the condition A_{22}, and the inequality (3.1.56) we have

$$
\begin{aligned}
&|K(t,x;\tau,\xi)| \\
&\le C\left((d(t,X(t-\tau);\tau,\xi))^\alpha (t-\tau)^{-M-1} + \sum_{|k_1|<2b} (t-\tau)^{-M-|k_1|/(2b)} \right) E_c^{(1)}(t,x;\tau,\xi) \\
&\le C(t-\tau)^{-M-1+\alpha/(2b)} E_{c_1}^{(1)}(t,x;\tau,\xi), \\
&\qquad\qquad 0 \le \tau < t \le T, \quad \{x,\xi\} \subset \mathbb{R}^n. \qquad (3.1.78)
\end{aligned}
$$

This means that the kernel K satisfies the conditions of Lemma 1.9 with $\beta = 1/(2b)$ and $\chi = \alpha/(2b)$. Due to that lemma, the solution Q of the equation (1.1.47) has the estimate (3.1.64).

The function Q satisfies also the estimate (3.1.65). Indeed, for $(d(x;z))^{2b} > (t-\tau)/2$ this estimate follows directly from (3.1.64). For $(d(x;z))^{2b} \le (t-\tau)/2$ its proof is similar to that of the estimate (2.1.87). First, using the condition A_{12}, the inequalities (3.1.51), (3.1.52), and (3.1.56) we get

$$
\begin{aligned}
&|\Delta_x^z K(t,x;\tau,\xi)| \\
&\le \sum_{|k_1|=2b} (|\Delta_x^z a_{k_1}(t,x)||\partial_{x_1}^{k_1} Z_1(t,z;\tau,\xi)| + |\Delta_{t,x}^{\tau,\xi} a_{k_1}(t,z)||\Delta_x^z \partial_{x_1}^{k_1} Z_1(t,x;\tau,\xi)|) \\
&\quad + \sum_{|k_1|<2b} (|\Delta_x^z a_{k_1}(t,x)||\partial_{x_1}^{k_1} Z_1(t,x;\tau,\xi)| + |a_{k_1}(t,z)||\Delta_x^z \partial_{x_1}^{k_1} Z_1(t,x;\tau,\xi)|) \\
&\le C(t-\tau)^{-M-1}(d(x;z))^\alpha \Bigg(1 + (d(t,X(t-\tau);\tau,\xi))^\alpha (t-\tau)^{-\alpha/(2b)} \\
&\quad + \sum_{|k_1|<2b} (t-\tau)^{1-|k_1|/(2b)} (1 + (t-\tau)^{-\alpha/(2b)}) \Bigg) E_c^{(1)}(t,x;\tau,\xi) \\
&\le C(d(x;z))^\alpha (t-\tau)^{-M-1} E_{c_1}^{(1)}(t,x;\tau,\xi) \\
&\le C(d(x;z))^{\alpha_1} (t-\tau)^{-M-1+\alpha_2/(2b)} E_{c_1}^{(1)}(t,x;\tau,\xi), \\
&0 \le \tau < t \le T, \quad \{x,z,\xi\} \subset \mathbb{R}^n, \quad (d(x;z))^{2b} < t-\tau, \quad c_1 \in (0,c). \qquad (3.1.79)
\end{aligned}
$$

Next, on the basis of the equality (1.1.47) we write the representation

$$
\begin{aligned}
\Delta_x^z Q(t,x;\tau,\xi) &= \Delta_x^z K(t,x;\tau,\xi) + \int_\tau^\eta d\lambda \int_{\mathbb{R}^n} \Delta_x^z K(t,x;\lambda,y) Q(\lambda,y;\tau,\xi) dy \\
&+ \int_\eta^t d\lambda \int_{\mathbb{R}^n} K(t,x;\lambda,y) Q(\lambda,y;\tau,\xi) dy - \int_\eta^t d\lambda \int_{\mathbb{R}^n} K(t,z;\lambda,y) Q(\lambda,y;\tau,\xi) dy =: \sum_{j=1}^4 I_j,
\end{aligned}
$$

in which $\eta := t - (d(x;z))^{2b}$, and find bounds for the summands I_j, $j \in \{1,2,3,4\}$.

I_1 satisfies the estimate (3.1.79). Using (3.1.62), (3.1.64), (3.1.78), and (3.1.79) we have

$$
\begin{aligned}
|I_2| \le{} & C(d(x;z))^{\alpha_1} \int_\tau^t (t-\lambda)^{-1+\alpha_2/(2b)}(\lambda-\tau)^{-1+\alpha/(2b)} d\lambda \\
& \times \int_{\mathbb{R}^n} E_{c_1}^{(1)}(t,x;\lambda,y) F_c(\lambda,y;\tau,\xi)((t-\lambda)(\lambda-\tau))^{-M} dy \\
\le{} & C(d(x;z))^{\alpha_1}(t-\tau)^{-M-1+(\alpha+\alpha_2)/(2b)} F_{c_2}(t,x;\tau,\xi),
\end{aligned}
$$

$$
\begin{aligned}
|I_3| \le{} & C \int_\tau^t ((t-\lambda)(\lambda-\tau))^{-1+\alpha/(2b)} d\lambda \\
& \times \int_{\mathbb{R}^n} E_{c_1}^{(1)}(t,x;\lambda,y) F_c(\lambda,y;\tau,\xi)((t-\lambda)(\lambda-\tau))^{-M} dy \\
\le{} & C(d(x;z))^{\alpha}(t-\tau)^{-M-1+\alpha/(2b)} F_{c_2}(t,x;\tau,\xi),
\end{aligned}
$$

where $c_2 \in (0,c_1)$. The estimate for I_4 is similar to the one for I_3.

Hence, we have proved the estimate (3.1.65), so that we have shown the existence of the FSCP Z. Let us prove its estimates (3.1.74). The estimate (3.1.75) is their consequence, since

$$
SZ(t,x;\tau,\xi) = \sum_{|k_1|\le 2b} a_{k_1}(t,x)\partial_{x_1}^{k_1} Z(t,x;\tau,\xi),
$$
$$
0 \le \tau < t \le T, \quad \{x,\xi\} \subset \mathbb{R}^n.
$$

By the formula (3.1.77) and the estimates (3.1.51), in order to obtain the estimates (3.1.74), it suffices to find estimates of the function W_1 and its derivatives. Here we use the inequalities (3.1.51), (3.1.53), (3.1.56), (3.1.62), (3.1.64), (3.1.65), (3.1.69), and the equalities (3.1.57), (3.1.66), (3.1.67).

For $|k_1| < 2b$, we have

$$
\begin{aligned}
|\partial_{x_1}^{k_1} W_1(t,x;\tau,\xi)| \le{} & C \int_\tau^t (t-\beta)^{-|k_1|/(2b)}(\beta-\tau)^{-1+\alpha/(2b)} d\beta \\
& \times \int_{\mathbb{R}^n} E_c^{(1)}(t,x;\beta,y) F_c(\beta,y;\tau,\xi)((t-\beta)(\beta-\tau))^{-M} dy \\
\le{} & C(t-\tau)^{-M-(|k_1|-\alpha)/(2b)} F_{c_1}(t,x;\tau,\xi), \qquad (3.1.80)
\end{aligned}
$$

while for $|k_1| = 2b$,

$$
\begin{aligned}
&|\partial_{x_1}^{k_1} W_1(t,x;\tau,\xi)| \\
&\le C\Bigg(\int_\tau^{t_1} d\beta \int_{\mathbb{R}^n} (t-\beta)^{-M-1} E_c^{(1)}(t,x;\beta,y)(\beta-\tau)^{-M-1+\alpha/(2b)} F_c(\beta,y;\tau,\xi)dy \\
&\quad + \int_{t_1}^{t} d\beta \int_{\mathbb{R}^n} (t-\beta)^{-M-1} E_c^{(1)}(t,x;\beta,y)(d(X(t-\beta);y))^{\alpha_1} \\
&\quad \times (\beta-\tau)^{-M-1+\alpha_2/(2b)} (F_c(\beta,y;\tau,\xi) + F_c(\beta,X(t-\beta);\tau,\xi))dy \\
&\quad + \int_{t_1}^{t} (t-\beta)^{-1+\alpha/(2b)} (\beta-\tau)^{-M-1+\alpha/(2b)} F_c(\beta,X(t-\beta);\tau,\xi)d\beta \Bigg) \\
&\le C\Bigg((t-\tau)^{-1} \int_\tau^{t_1} (\beta-\tau)^{-1+\alpha/(2b)} d\beta \\
&\quad \times \int_{\mathbb{R}^n} E_c^{(1)}(t,x;\beta,y)F_c(\beta,y;\tau,\xi)((t-\beta)(\beta-\tau))^{-M} dy \\
&\quad + (t-\tau)^{-1+\alpha_2/(2b)} \int_{t_1}^{t} (t-\beta)^{-1+\alpha_1/(2b)} d\beta \\
&\quad \times \int_{\mathbb{R}^n} E_{c_1}^{(1)}(t,x;\beta,y)F_c(\beta,y;\tau,\xi)((t-\beta)(\beta-\tau))^{-M} dy \\
&\quad + (t-\tau)^{-M-1+\alpha_2/(2b)} \int_{t_1}^{t} (t-\beta)^{-1+\alpha_1/(2b)} d\beta \\
&\quad \times \int_{\mathbb{R}^n} (t-\beta)^{-M} E_{c_1}^{(1)}(t,x;\beta,y)dy F_{c_2}(t,x;\tau,\xi) \\
&\quad + (t-\tau)^{-M-1+\alpha/(2b)} \int_{t_1}^{t} (t-\beta)^{-1+\alpha/(2b)} d\beta F_{c_2}(t,x;\tau,\xi) \Bigg) \\
&\le C(t-\tau)^{-M-1+\alpha/(2b)} F_{c_2}(t,x;\tau,\xi). \qquad (3.1.81)
\end{aligned}
$$

The estimates (3.1.51), (3.1.80), and (3.1.81) imply the required estimates (3.1.74).

The proof of the existence of the FSCP Z^* for the adjoint equation (3.1.72) is similar to that for the FSCP Z, since this equation becomes an equation of the same type as (3.1.70), if $-\tau$ is taken as the new time variable.

The equality (3.1.76) and the convolution formula for the FSCP Z are proved just as Properties 2.8 and 3.1, with the use of the Green-Ostrogradsky formula (3.1.48), in which L and L^* are the differential expressions from the equations (3.1.70) and (3.1.72). □

Let us formulate separately a theorem about the FSCP for the second order equation (3.1.71). The assertions of Theorem 3.2 are valid for this particular case of the equation (3.1.70). Since for the equation (3.1.71) instead of the function F_c we can take the function $E_c^{(2)}$ defined by (3.1.41), and the latter function satisfies the equality (3.1.63) and Lemma 1.10, the results of Theorem 3.2 can be improved significantly for this equation. Namely, we have the following theorem.

Theorem 3.3. *If the conditions A_{24} and A_{25} are satisfied, then there exists a FSCP Z for the equation* (3.1.71), *with the estimates*

$$|\partial_{x_1}^{k_1} Z(t,x;\tau,\xi)| \le C(t-\tau)^{-(n_1+3n_2+5n_3+|k_1|)/2} E_c^{(2)}(t,x;\tau,\xi),$$
$$|k_1| \le 2, \tag{3.1.82}$$
$$|SZ(t,x;\tau,\xi)| \le C(t-\tau)^{-(n_1+3n_2+5n_3+2)/2} E_c^{(2)}(t,x;\tau,\xi), \tag{3.1.83}$$
$$0 \le \tau < t \le T, \quad \{x,\xi\} \subset \mathbb{R}^n.$$

Under the additional assumption A_{26}, there exists the FSCP Z^* for the adjoint equation (3.1.73) *connected with Z by the relation* (3.1.76), *and the convolution formula* (2.1.112) *holds for Z.*

Remark. Due to the inequalities (3.1.42), in the estimates (3.1.82) and (3.1.83) the function $E_c^{(2)}$ can be substituted by the function $E_c^{(1)}$ with $q = 2$, that is by the function

$$\exp\left\{-c\sum_{j=1}^{3}(t-\tau)^{1-2j}|X_j(t-\tau)-\xi_j|^2\right\}, \quad t > \tau, \quad \{x,\xi\} \subset \mathbb{R}^n.$$

3.1.3. Equations from the class $\mathbf{E}_{23}$. For equations of the form (1.1.16) the results are similar to those of Sections 3.1.1 and 3.1.2. Here we only formulate them for a general equation (1.1.16), and, in order to make special features of such equations understood, give a more detailed exposition for the case where the coefficients do not depend on spatial variables.

For convenience, we recall some notation already used above, and introduce new ones taking into account special features of equations from the class $\mathbf{E}_{23}$.

Let b be the least common multiple of the numbers $b_1, \dots, b_{n_1}$; $m_j = b/b_j$, $q_j = 2b_j/(2b_j-1)$, $j \in \{1,\dots,n_1\}$; $\|k_1\| = \sum_{j=1}^{n_1} m_j k_{1j}$, if $k_1 = (k_{11},\dots,k_{1n_1}) \in \mathbb{Z}_+^{n_1}$. Instead of the expressions (3.1.4), (3.1.5), (3.1.21), and (3.1.61) for M, M_{kl}, x_t, $d(x;\xi)$, $d(t,x;\tau,\xi)$, $X(t)$, and F_c, used in Sections 3.1.1 and 3.1.2, here we use the following expressions:

$$M = \sum_{s=1}^{3} \sum_{j=1}^{n_s} (2b(s-1)+m_j)/(2b),$$

$$M_{kl} = \sum_{s=1}^{3} \sum_{j=1}^{n_s} (2b(s-1)+m_j)(k_{sj}+l_{sj})/(2b),$$

$$x_t = (t^{-1/(2b_1)}x_1, \dots, t^{-1/(2b_{n_1})}x_{1n_1}, t^{-1-1/(2b_1)}x_{21}, \dots, t^{-1-1/(2b_{n_2})}x_{2n_2}, t^{-2-1/(2b_1)}x_{31}, \dots, t^{-2-1/(2bn_3)}x_{3n_3}), d(x;\xi) = \sum_{s=1}^{3} \sum_{j=1}^{n_s} |x_{sj}-\xi_{sj}|^{1/(2b(s-1)+m_j)};$$

$$d(t,x;\tau,\xi) = |t-\tau|^{1/(2b)} + d(x;\xi),$$

$$X(t) = (X_{11}(t), \dots, X_{1n_1}(t), X_{21}(t), \dots, X_{2n_2}(t), X_{31}(t), \dots, X_{3n_3}(t)),$$

$$X_{sj}(t) = \sum_{r=0}^{s-1} \frac{1}{r!} t^r x_{(s-r)j}, \quad j \in \{1,\dots,n_s\}, \quad s \in \{1,2,3\},$$

$$F_c(t,x;\tau,\xi) = \exp\left\{-c\sum_{j=1}^{n_1}(t-\tau)^{1-q_j}|x_{1j}-\xi_{1j}|^{q_j}\right\} \times \sum_{j=0}^{\infty}(\hat{C}\Gamma(\alpha/(2b))(t-\tau)^{\alpha/(2b)})^j(\Gamma(j\alpha/(2b)))^{-1}E^{(3)}_{c\delta_0^j}(t,x;\tau,\xi), \tag{3.1.84}$$

where

$$E_c^{(3)}(t,x;\tau,\xi) := \exp\{-c\rho(t-\tau,x,\xi)\},$$

$$\rho(t,x,\xi) := \sum_{s=1}^{3} \sum_{j=1}^{n_s} t^{1-sq_j}|X_{sj}(t)-\xi_{sj}|^{q_j}. \tag{3.1.85}$$

Consider an equation

$$(Lu)(t,x) := \left(S - \sum_{\|k_1\|\le 2b} a_{k_1}(t,x)\partial_{x_1}^{k_1}\right)u(t,x) = f(t,x), \quad (t,x) \in \Pi_{(0,T]}, \tag{3.1.86}$$

assuming that:

(A_{27}) there exists such a constant $\delta > 0$ that for any $(t,x) \in \Pi_{[0,T]}$ and $\sigma_1 \in \mathbb{R}^{n_1}$ the inequality (1.1.19) holds;

(A_{28}) the coefficients a_{k_1}, $\|k_1\| \le 2b$, are bounded on $\Pi_{[0,T]}$ and Hölder continuous in t, x, in the sense similar to (3.1.7), that is there exist such constants

$H > 0$ and $\alpha \in (0,1]$ that for any $\{(t,x),(\tau,\xi)\} \subset \Pi_{[0,T]}$ and $k_1 \in \mathbb{Z}_+^{n_1}$, $\|k_1\| \le 2b$,

$$|\Delta_{t,x}^{\tau,\xi} a_{k_1}(t,x)| \le H(d(t,X(t-\tau);\tau,\xi))^\alpha; \tag{3.1.87}$$

(A_{29}) there exist bounded and Hölder continuous in t,x, in the sense of (3.1.87), derivatives $\partial_{x_1}^{k_1} a_{k_1}$, $\|k_1\| \le 2b$, on $\Pi_{[0,T]}$.

Here the counterpart of Theorem 3.2 is the following result.

Theorem 3.4. *If the conditions A_{27} and A_{28} are satisfied, then there exists a FSCP Z for the equation* (3.1.86), *such that*

$$\begin{gathered}|\partial_{x_1}^{k_1} Z(t,x;\tau,\xi)| \le C(t-\tau)^{-M-\|k_1\|/(2b)} F_c(t,x;\tau,\xi), \quad \|k_1\| \le 2b,\\ |SZ(t,x;\tau,\xi)| \le C(t-\tau)^{-M-1} F_c(t,x;\tau,\xi),\\ 0 \le \tau < t \le T, \quad \{x,\xi\} \subset \mathbb{R}^n.\end{gathered} \tag{3.1.88}$$

Under an additional assumption A_{29}, there exists a FSCP Z^* for the equation adjoint to (3.1.86), *and Z, Z^* satisfy the equalities* (3.1.76) *and* (3.1.112).

The proof of this theorem, similar to the one of Theorem 3.2, is based on a detailed investigation of properties of the FSCP for an equation with only the leading terms, with coefficients depending only on the parameters β, y. Below we study the FSCP G for the equation (3.1.86) whose coefficients a_{k_1}, $\|k_1\| \le 2b$, may depend only on the variable t.

Consider the equation

$$\Big(S - \sum_{\|k_1\| \le 2b} a_{k_1}(t)\partial_{x_1}^{k_1}\Big) u(t,x) = 0, \quad (t,x) \in \Pi_{(\tau,T]}, \tag{3.1.89}$$

where τ is an arbitrary fixed number from $[0,T)$, assuming that the coefficients a_{k_1}, $\|k_1\| \le 2b$, are continuous on $[0,T]$, and for each $t \in [0,T]$, $\sigma_1 \in \mathbb{R}^{n_1}$ the inequality

$$\operatorname{Re} \sum_{\|k_1\|=2b} a_{k_1}(t)(i\sigma_1)^{k_1} \le -\delta \sum_{j=1}^{n_1} \sigma_{1j}^{2b_j} \tag{3.1.90}$$

holds with some constant $\delta > 0$.

Just as in Section 3.1.1, we solve the problem (2.1.6), (3.1.89). For its solution we obtain the formula (2.1.14), in which G is given by the formula (3.1.20) where

$$\begin{aligned}\tilde V(t,\tau,\sigma) := \exp\Big\{ \sum_{\|k_1\| \le 2b} i^{|k_1|} \int_0^1 a_{k_1}(\tau + (t-\tau)\beta)(\sigma_1' + \beta\sigma_2' + 2^{-1}\beta^2\sigma_3)^{k_1'}\\ \times (\sigma_1'' + \beta\sigma_2'')^{k_1''}(\sigma_1''')^{k_1'''}\, d\beta (t-\tau)^{1-\|k_1\|/(2b)}\Big\},\end{aligned} \tag{3.1.91}$$

while M, $X(t-\tau)$, and $x_{t-\tau}$ were defined in (3.1.84).

Investigating the function $\tilde{V}$ just as in Section 3.1.1, we obtain the following estimates similar to (3.1.29) and (3.1.30):

$$|\tilde{V}(t,\tau,\sigma+i\gamma)| \le C\exp\{-\delta_2 r(\sigma)+c_2 r(\gamma)\},$$
$$|P_{kl}(\sigma+i\gamma)V(t,\tau,\sigma+i\gamma)| \le C_{kl}\exp\{-\delta_3 r(\sigma)+c_3 r(\gamma)\},$$
$$0\le\tau\le t\le T,\quad \{\sigma,\gamma\}\subset\mathbb{R}^n,$$

where $r(\sigma) := \sum_{s=1}^{3}\sum_{j=1}^{n_s}\sigma_{sj}^{2b_j}$.

Using these estimates and Lemma 1.1 just as in the proofs of Theorems 2.1 and 3.1, we prove their analogue for the equation (3.1.89).

Theorem 3.5. *If the coefficients a_{k_1}, $\|k_1\|\le 2b$, of the equation* (3.1.89) *are continuous on* $[0,T]$, *and the condition* (3.1.90) *is satisfied, then there exists a FSCP G for this equation, such that:*

1) *the function G and its derivatives can be extended to $\mathbb{C}^n$, and for these extensions we have*

$$\partial_x^k\partial_\xi^l G(t,x+iy;\tau,\xi+i\eta)$$
$$= (t-\tau)^{-M-M_{kl}}\Omega_{kl}(t,\tau,z)|_{z=(X(t-\tau)-\xi)_{t-\tau}+i(Y(t-\tau)-\eta)_{t-\tau}},$$
$$0\le\tau<t\le T,\quad \{x,y,\xi,\eta\}\subset\mathbb{R}^n,\quad \{k,l\}\subset\mathbb{Z}_+^n,$$

where $\Omega_{kl}(t,\tau,z)$, $z=(z_1,\dots,z_n)\in\mathbb{C}^n$, for fixed t and τ, are entire functions of $z_1,\dots,z_n$ of the respective growth orders $p_1,\dots,p_n$ and the same decrease orders for $z=x\in\mathbb{R}^n$, and $p_j=q_j$ for $j\in\{1,\dots,n_1\}$, $p_j=q_r$, if $j=n_1+r$, $r\in\{1,\dots,n_2\}$, $p_j=q_s$, if $j=n_1+n_2+s$, $s\in\{1,\dots,n_3\}$;

2) *the estimates*

$$|\partial_x^k\partial_\xi^l G(t,x+iy;\tau,\xi+i\eta)| \le C_{kl}(t-\tau)^{-M-M_{kl}}E_c^{(3)}(t,x;\tau,\xi)E_{-c'}^{(3)}(t,y;\tau,\eta),$$
$$0\le\tau<t\le T,\quad \{x,y,\xi,\eta\}\subset\mathbb{R}^n,\quad \{k,l\}\subset\mathbb{Z}_+^n, \tag{3.1.92}$$

hold, with C_{kl}, c, and c' being positive constants, while $E_c^{(2)}$ is defined in (3.1.85);

3) *the equalities* (3.1.34) *and* (3.1.35) *hold;*

4) *the function* (3.1.44) *is a FSCP for the adjoint equation*

$$(L^*v)(\tau,\xi) := \Big(S^* - \sum_{\|k_1\|\le 2b}\bar{a}_{k_1}(\tau)(-\partial_{\xi_1})^{k_1}\Big)v(\tau,\xi) = 0,$$
$$(\tau,\xi)\in\Pi_{[0,T)}, \tag{3.1.93}$$

where S^* is the same as in (3.1.45);

5) *the convolution formula* (3.1.46) *is valid.*

Note that the assertions 4) and 5) of the above theorems are proved just as the similar statements from Property 3.1. The formula (3.1.48) is used, in which the differential expressions L and L^* are defined in (3.1.86) and (3.1.93), the bilinear forms $B^j[v,u]$ contain only the derivatives of the form $\partial_{y_1}^{k_1}v$ and $\partial_{y_1}^{k_1}u$ with $\|k_1\| < 2b$.

Example 3.2. A FSCP for the equation

$$\left(\partial_t - \sum_{j=1}^{2} x_{1j}\partial_{x_{2j}} - x_{21}\partial_{x_3} - \sum_{j=1}^{2} a_j\partial_{x_{1j}}^2 - a_3\partial_{x_{13}}^4\right)u = 0,$$

from the class $\mathbf{E}_{23}$, where $n_1 = 3$, $n_2 = 2$, $n_3 = 1$, $b_1 = b_2 = 1$, $b_3 = 2$, and a_1, a_2, and a_3 are positive constants, is given by the formula

$$G(t,x;\tau,\xi) = G_1(t,x';\tau,\xi')G_2(t,x'';\tau,\xi'')G_3(t,x_{13};\tau,\xi_{13}),$$
$$t > \tau, \quad \{x,\xi\} \subset \mathbb{R}^6, \tag{3.1.94}$$

with $x' := (x_{11}, x_{21}, x_3)$, $\xi' := (\xi_{11}, \xi_{21}, \xi_3)$, $x'' := (x_{12}, x_{22})$, $\xi'' := (\xi_{12}, \xi_{22})$.
Here

$$\begin{aligned}
G_1(t,x';\tau,\xi') :=\ & 3\sqrt{15}(\pi a_1)^{-3/2}t^{-9/2}\exp\{-a_1^{-1}((4(t-\tau))^{-1}(x_{11}-\xi_{11})^2 \\
& + 3(t-\tau)^{-3}(x_{21} + 2^{-1}(t-\tau)(x_{11}+\xi_{11}) - \xi_{21})^2 \\
& + 180(t-\tau)^{-5}(x_3 + 2^{-1}(t-\tau)(x_{21}+\xi_{21}) \\
& + 12^{-1}(t-\tau)^2(x_{11}-\xi_{11}) - \xi_3)^2)\}, \\
G_2(t,x'';\tau,\xi'') :=\ & \sqrt{3}(2a_2(t-\tau))^{-1}\exp\{-a_2^{-1}((4(t-\tau))^{-1}(x_{12}-\xi_{12})^2 \\
& + 3(t-\tau)^{-3}(x_{22} + 2^{-1}(t-\tau)(x_{12}+\xi_{12}) - \xi_{22})^2)\}, \\
G_3(t,x_{13};\tau,\xi_{13}) :=\ & (4\pi^2 a_3(t-\tau))^{-1/4} \\
& \times \int_0^\infty \exp\{-\alpha^4\}\alpha^{1/2}\cos((a_3(t-\tau))^{-1/4}(x_{13}-\xi_{13})\alpha)d\alpha.
\end{aligned} \tag{3.1.95}$$

The formulas (3.1.94), (3.1.95) are obtained by a straightforward calculation from the formulas (3.1.20) and (3.1.91) (see also Examples 2.1 and 3.1).

In conclusion of this subsection we give some results regarding the FSCP for equations of the form (3.1.89) with degenerations on the initial hyperplane, of the same kind as in the equation (1.1.11).

Let us consider an equation of the form

$$\left(\alpha(t)\partial_t - \beta(t)\left(\sum_{s=2}^{3}\sum_{j=1}^{n_s} x_{(s-1)j}\partial_{x_{sj}} + \sum_{0<\|k_1\|\le 2b} a_{k_1}(t)\partial_{x_1}^{k_1}\right) - a_0(t)\right)u(t,x) = 0,$$
$$(t,x) \in \Pi_{(0,T]}, \tag{3.1.96}$$

assuming that the functions α and β are such as in the equation (1.1.11), the coefficients a_{k_1}, $\|k_1\| \leq 2b$, are continuous on $[0,T]$, and the condition (3.1.90) is satisfied.

Let the functions $\hat{A}$ and $\hat{B}$ be the same as in Section 2.4.1,

$$\begin{aligned}\hat{X}(t,\tau) &:= (\hat{X}_{11}(t,\tau),\dots,\hat{X}_{1n_1}(t,\tau),\hat{X}_{21}(t,\tau),\\ &\qquad \dots,\hat{X}_{2n_2}(t,\tau),\hat{X}_{31}(t,\tau),\dots,\hat{X}_{3n_3}(t,\tau)),\\ \hat{X}_{sj}(t,\tau) &:= \sum_{r=0}^{s-1}\frac{1}{r!}(\hat{B}(t,\tau))^r x_{(s-r)j}, j\in\{1,\dots,n_s\}, \quad s\in\{1,2,3\},\\ \hat{E}_c^{(3)}(t,x;\tau,\xi) &:= \exp\left\{-c\sum_{s=1}^{3}\sum_{j=1}^{n_s}(\hat{B}(t,\tau))^{1-sq_j}|\hat{X}_{sj}(t,\tau)-\xi_{sj}|^{q_j}\right\}.\end{aligned}$$

The following theorem is an analogue of Theorem 3.5 for the equation (3.1.96).

Theorem 3.6. *The following assertions are valid:*

1) *there exists a FSCP* $G(t,x;\tau,\xi)$, $0<\tau<t\leq T$, $\{x,\xi\}\subset\mathbb{R}^n$, *for the equation* (3.1.96);
2) $$\begin{aligned}&\partial_x^k\partial_\xi^l G(t,x+iy;\tau,\xi+i\eta)\\ &\quad = (\hat{B}(t,\tau))^{-M-M_{kl}}\Omega_{kl}(t,\tau,z)|_{z=(\hat{X}(t,\tau)-\xi)_{\hat{B}(t,\tau)}+i(\hat{Y}(t,\tau)-\eta)_{\hat{B}(t,\tau)}},\\ &\qquad 0<\tau<t\leq T, \quad \{x,y,\xi,\eta\}\subset\mathbb{R}^n, \quad \{k,l\}\subset\mathbb{Z}_+^n,\end{aligned}$$
   where $\Omega_{kl}(t,\tau,z)$, $z=(z_{11},\dots,z_{1n_1},z_{21},\dots,z_{2n_2},z_{31},\dots,z_{3n_3})\in\mathbb{C}^n$, *are, for fixed* t *and* τ, *entire functions of* z_{sj} *with the growth orders* $p_{sj}=q_j$ *and the same decrease orders for real values of the independent variables;*
3) *the estimates*
   $$\begin{aligned}&|\partial_x^k\partial_\xi^l G(t,x+iy;\tau,\xi+i\eta)|\\ &\leq C_{kl}(\hat{B}(t,\tau))^{-M-M_{kl}}\exp\{d\hat{A}(t,\tau)\}\hat{E}_c^{(3)}(t,x;\tau,\xi)\hat{E}_{-c'}^{(3)}(t,y;\tau,\eta),\\ &\qquad 0<\tau<t\leq T, \quad \{x,y,\xi,\eta\}\subset\mathbb{R}^n, \quad \{k,l\}\subset\mathbb{Z}_+^n, \qquad (3.1.97)\end{aligned}$$
   are valid, with $C_{kl}>0$, $c>0$, $c'>0$, *and* $d\in\mathbb{R}$;
4) *the FSCP G has the normality property, and the convolution formula (3.1.46) with* $\tau>0$ *holds;*
5) *in the case of weak degeneration, that is* $\hat{A}(0,T)<\infty$, *the estimates* (3.1.97) *are valid with* $\tau=0$ *and* $d=0$.

The proof (sketched in [120]) is similar to that of Theorems 3.1 and 3.5.

3.2 Cauchy problem

3.2.1. Correct solvability of the Cauchy problem for equations from the class $\mathbf{E}_{23}^0$. In this section and the next one we study the Cauchy problem for equations from the class $\mathbf{E}_{23}^0$. This class consists of equations from $\mathbf{E}_{23}$ whose coefficients do not depend on spatial variables. The results obtained in Sections 3.2.1 and 3.2.2 are similar to Theorems 2.6, 2.7, and 2.11 for homogeneous $\overrightarrow{2b}$-parabolic equations from the class $\mathbf{E}_{11}$.

Below we define the families of Banach spaces $L_p^{\vec{k}(t,\vec{a})}$, $t \in [0,T]$, $p \in [1,\infty]$, of functions with a rapid growth in spatial variables at infinity. The fact that a solution belongs to such a space is interpreted as a characterization of its evolution in t. A technique of obtaining special estimates is developed in order to take into account a complicated anisotropy of the problem. Section 3.2.1 is devoted to the solvability of the Cauchy problem. In Section 3.2.2 we prove the uniqueness of a solution of the Cauchy problem in the spaces $L_p^{\vec{k}(t,\vec{a})}$, $t \in [0,T]$, and a representation theorem for solutions defined on the layer $\Pi_{(0,T]}$, via their special limit values on the initial hyperplane. Hence, for equations from the class $\mathbf{E}_{23}^0$ we determine the spaces Φ and U, with which the meta-theorems A and B from Section 1.1.3 are valid for this situation.

In Sections 3.2.1 and 3.2.2 we use the notation from Section 3.1.3, the notation q' and q'' for the largest and smallest of the numbers q_j, $j \in \{1,\dots,n_1\}$, and also the following analogue of the distance q from (2.1.2):

$$q(x;\xi) := \left(\sum_{s=1}^{3}\sum_{j=1}^{n_s} |x_{sj}-\xi_{sj}|^{q_j}\right)^{1/q'}, \quad \{x,\xi\} \subset \mathbb{R}^n.$$

In addition, for $\xi \in \mathbb{R}^n$ and $R > 0$ we set $B(\xi,R) := \{x \in \mathbb{R}^n \mid q(x;\xi) \le R\}$, $B_R := B(0,R)$.

Consider an equation

$$(Lu)(t,x) := \left(S - \sum_{\|k_1\|\le 2b} a_{k_1}(t)\partial_{x_1}^{k_1}\right)u(t,x) = 0, \quad (t,x) \in \Pi_{(0,T]}, \tag{3.2.1}$$

assuming that the coefficients $a_{k_1} : [0,T] \to \mathbb{C}$, $\|k_1\| \le 2b$, are continuous, and the condition (3.1.90) is satisfied.

By Theorem 3.5 there exists a FSCP G for the equation (3.2.1) with the properties described in that theorem. In particular,

$$\begin{gathered}|\partial_x^k \partial_\xi^l G(t,x;\tau,\xi)| \le C_{kl}(t-\tau)^{-M-M_{kl}} E_c^{(3)}(t,x;\tau,\xi),\\ 0 \le \tau < t \le T, \quad \{x,\xi\} \subset \mathbb{R}^n, \quad \{k,l\} \subset \mathbb{Z}_+^n.\end{gathered} \tag{3.2.2}$$

Let us define necessary norms and spaces.

We introduce the following sets of functions:

$$\begin{aligned}
\vec{k}(t,\vec{a}) &:= (k_{11}(t,a_{11}),\dots,k_{1n_1}(t,a_{1n_1}),k_{21}(t,a_{21}),\\
&\qquad \dots,k_{2n_2}(t,a_{2n_2}),k_{31}(t,a_{31}),\dots,k_{3n_3}(t,a_{3n_3})),\\
\vec{s}(t) &:= (s_{11}(t),\dots,s_{1n_1}(t),s_{21}(t),\dots,s_{2n_2}(t),s_{31}(t),\dots,s_{3n_3}(t));\\
k_{lj}(t,a_{lj}) &:= c_0 a_{lj}(c_0^{2b_j-1} - a_{lj}^{2b_j-1}t^{2b_j(l-1)+1})^{1-q_j}, \quad j\in\{1,\dots,n_l\},\ l\in\{1,2,3\};\\
s_{1j}(t) &:= k_{1j}(t,a_{1j}) + 2^{q_j-1}\theta(n_2-j)t^{q_j}k_{2j}(t,a_{2j})\\
&\qquad + 2^{q_j-2}\theta(n_3-j)t^{2q_j}k_{3j}(t,a_{3j}), \quad j\in\{1,\dots,n_1\};\\
s_{2j}(t) &:= 2^{q_j-1}k_{2j}(t,a_{2j}) + 4^{q_j-1}\theta(n_3-j)t^{q_j}k_{3j}(t,a_{3j}), \quad j\in\{1,\dots,n_2\};\\
s_{3j}(t) &:= 4^{q_j-1}k_{3j}(t,a_{3j}), \quad j\in\{1,\dots,n_3\}, \quad t\le T,
\end{aligned}$$

where $c_0\in(0,c)$, c is the constant from the estimates (3.2.2), $\vec{a} := (a_{11},\dots,a_{1n_1}, a_{21},\dots,a_{2n_2},a_{31},\dots,a_{3n_3})$ is a collection of such non-negative numbers that

$$T < \min_{\substack{j\in\{1,\dots,n_l\}\\ l\in\{1,2,3\}}} \left(\frac{c_0}{a_{lj}}\right)^{(2b_j-1)/(2b_j(l-1)+1)},$$

$\theta(\tau)=1$ for $\tau\ge 0$, and $\theta(\tau)=0$ for $\tau<0$.

Note that $\vec{k}(0,\vec{a})=\vec{a}$ and $k_{lj}(t,a_{lj})\ge a_{lj}$, $t\in[0,T]$, $j\in\{1,\dots,n_l\}$, $l\in\{1,2,3\}$; in addition,

$$-c_0\rho(t,x,\xi) + [\vec{a},\xi] \le [\vec{k}(t,\vec{a}),X(t)], \quad t\in(0,T], \quad \{x,\xi\}\subset\mathbb{R}^n, \tag{3.2.3}$$

where $\rho(t,x,\xi)$ was defined in (3.1.85). We used the notation

$$[\vec{a},\xi] := \sum_{l=1}^{3}\sum_{j=1}^{n_l} a_{lj}|\xi_{lj}|^{q_j}. \tag{3.2.4}$$

Note also the relations

$$\begin{gathered}
k_{1j}(t-\tau,k_{1j}(\tau,a_{1j})) = k_{1j}(t,a_{1j}), \quad k_{lj}(t-\tau,k_{lj}(\tau,a_{lj})) \le k_{lj}(t,a_{lj}),\\
0\le\tau\le t\le T, \quad j\in\{1,\dots,n_l\}, \quad l\in\{2,3\}.
\end{gathered}$$

Let $p\in[1,\infty]$, and $u(t,x)$, $(t,x)\in\Pi_{[0,T]}$, be a given complex-valued function, measurable in x for any $t\in[0,T]$. For every $t\in[0,T]$ we define the norms

$$\begin{aligned}
\|u(t,\cdot)\|_p^{\vec{k}(t,\vec{a})} &:= \|u(t,x)\exp\{-[\vec{k}(t,\vec{a}),X(t)]\}\|_{L_p(\mathbb{R}^n)},\\
\|u(t,\cdot)\|_p^{\vec{s}(t)} &:= \|u(t,x)\exp\{-[\vec{s}(t),x]\}\|_{L_p(\mathbb{R}^n)}.
\end{aligned}$$

We shall use the following spaces:

- $L_p^{\vec{k}(t,\vec{a})}$, $t \in [0,T]$, $p \in [1,\infty]$, the spaces of measurable functions $\varphi : \mathbb{R}^n \to \mathbb{C}$ with finite norms $\|\varphi\|_p^{\vec{k}(t,\vec{a})}$; $L_p^{\vec{a}} := L_p^{\vec{k}(0,\vec{a})}$;
- $M^{\vec{a}}$, the space of complex-valued generalized Borel measures μ on $\mathbb{R}^n$ satisfying the condition
$$\|\mu\|^{\vec{a}} := \int_{\mathbb{R}^n} \exp\{-[\vec{a}, x]\} d|\mu|(x) < \infty;$$
- $L_1^{-\vec{s}(T)}$, the space of measurable functions $\psi : \mathbb{R}^n \to \mathbb{C}$ with a finite norm
$$\|\psi\|_1^{-\vec{s}(T)} := \|\psi(x) \exp\{[\vec{s}(T), x]\}\|_{L_1(\mathbb{R}^n)};$$
- $C_0^{-\vec{s}(T)}$, the space of continuous functions $\psi : \mathbb{R}^n \to \mathbb{C}$, such that
$$|\psi(x)| \exp\{[\vec{s}(T), x]\} \to 0 \quad \text{as} \quad |x| \to \infty.$$

For $\psi \in C_0^{-\vec{s}(T)}$, we set
$$\|\psi\|_\infty^{-\vec{s}(T)} := \sup_{x \in \mathbb{R}^n} (|\psi(x)| \exp\{[\vec{s}(T), x]\}).$$

Using the inequalities (1.3.9) and (1.3.10) we have
$$\begin{aligned} &|X_{2j}(t)|^{q_j} \leq 2^{q_j - 1}(|x_{2j}|^{q_j} + t^{q_j}|x_{1j}|^{q_j}), \\ &|X_{3j}(t)|^{q_j} \leq 4^{q_j - 1}(|x_{3j}|^{q_j} + t^{q_j}|x_{2j}|^{q_j} + 2^{-q_j} t^{2q_j}|x_{1j}|^{q_j}), \end{aligned} \tag{3.2.5}$$
so that
$$\exp\{-[\vec{s}(t), x]\} \leq \exp\{-[\vec{k}(t, \vec{a}), X(t)]\},$$
hence
$$\|u(t,\cdot)\|_p^{\vec{s}(t)} \leq \|u(t,\cdot)\|_p^{\vec{k}(t,\vec{a})}, \quad t \in [0,T], \quad p \in [1,\infty]. \tag{3.2.6}$$

Note that $s_{1j}(t) \geq a_{1j}$, $s_{lj}(t) > a_{lj}$, $l \in \{2,3\}$, $t \in [0,T]$, therefore for $\varphi \in L_p^{\vec{a}}$ we find that
$$\|\varphi\|_p^{\vec{s}(t)} \leq \|\varphi\|_p^{\vec{a}}, \quad t \in [0,T], \quad p \in [1,\infty]. \tag{3.2.7}$$

Now we shall formulate and prove the main theorem on the correct solvability of the Cauchy problem for equations from the class $\mathbf{E}_{23}^0$.

Theorem 3.7. *For any function $\varphi \in L_p^{\vec{a}}$, $p \in [1,\infty]$, and a generalized measure $\mu \in M^{\vec{a}}$ the formulas*
$$u(t,x) = (P\varphi)(t,x) := \int_{\mathbb{R}^n} G(t,x;0,\xi)\varphi(\xi)d\xi, \quad (t,x) \in \Pi_{(0,T]}, \tag{3.2.8}$$
$$u_0(t,x) = (P\mu)(t,x) := \int_{\mathbb{R}^n} G(t,x;0,\xi)d\mu(\xi), \quad (t,x) \in \Pi_{(0,T]}, \tag{3.2.9}$$

determine the unique solutions of the equation (3.2.1) *on the layer* $\Pi_{(0,T]}$, *possessing the following properties: there exists a constant* $C > 0$, *independent of* φ *and* μ, *and such that for any* $t \in (0, T]$,

$$\|u(t,\cdot)\|_p^{\vec{k}(t,\vec{a})} \leq C\|\varphi\|_p^{\vec{a}}, \tag{3.2.10}$$

$$\|u_0(t,\cdot)\|_1^{\vec{k}(t,\vec{a})} \leq C\|\mu\|^{\vec{a}}; \tag{3.2.11}$$

if $p \in [1, \infty)$, *then*

$$\lim_{t\to 0} \|u(t,\cdot) - \varphi(\cdot)\|_p^{\vec{s}(t)} = 0, \tag{3.2.12}$$

while for $p = \infty$ *and the function* (3.2.9) $u(t,\cdot) \underset{t\to 0}{\overset{w}{\rightarrow}} \varphi$, $u_0(t,\cdot) \underset{t\to 0}{\overset{w}{\rightarrow}} \mu$, *that is for all* ψ *from the spaces* $L_1^{-\vec{s}(T)}$ *and* $C_0^{-\vec{s}(T)}$ *respectively*

$$\lim_{t\to 0} \int_{\mathbb{R}^n} \psi(x)u(t,x)dx = \int_{\mathbb{R}^n} \psi(x)\varphi(x)dx \tag{3.2.13}$$

and

$$\lim_{t\to 0} \int_{\mathbb{R}^n} \psi(x)u_0(t,x)dx = \int_{\mathbb{R}^n} \psi(x)d\mu(x). \tag{3.2.14}$$

Remark. It follows from Theorem 3.7 that the solutions (3.2.8) and (3.2.9) belong, for each fixed $t \in (0, T]$, to the spaces $L_p^{\vec{k}(t,\vec{a})}$ and $L_1^{\vec{k}(t,\vec{a})}$ respectively.

Let us study the Poisson integrals (3.2.8) of the function φ and (3.2.9) of the generalized measure μ and show that they represent solutions of the equation (3.2.1) and have the properties listed in Theorem 3.7. Here we use the technique resembling that of Section 2.2.2.

Lemma 3.1. *If* $\varphi \in L_p^{\vec{a}}$, $p \in [1, \infty]$, *then the function* (3.2.8) *satisfies the estimate*

$$\|u(t,\cdot)\|_p^{\vec{k}(t,\vec{a})} \leq C\|\varphi\|_p^{\vec{a}}, \quad t \in (0, T]. \tag{3.2.15}$$

Proof. Consider first the case $p = \infty$. By the estimates (3.2.2), for $(t, x) \in \Pi_{(0,T]}$ we have

$$\begin{aligned}|u(t,x)| &\leq C \int_{\mathbb{R}^n} t^{-M} \exp\{-c\rho(t,x,\xi)\}(|\varphi(\xi)| \exp\{-[\vec{a},\xi]\}) \exp\{[\vec{a},\xi]\}d\xi \\ &\leq C\|\varphi\|_\infty^{\vec{a}} \int_{\mathbb{R}^n} t^{-M} \exp\{-(c-c_0)\rho(t,x,\xi)\} \exp\{-c_0\rho(t,x,\xi) + [\vec{a},\xi]\}d\xi.\end{aligned}$$

Using the inequality (3.2.3) and the fact that for any $t > 0$, $x \in \mathbb{R}^n$, and $\delta > 0$,

$$\int\limits_{\mathbb{R}^n} t^{-M} \exp\{-\delta\rho(t,x,\xi)\} d\xi = C, \tag{3.2.16}$$

we find that

$$|u(t,x)| \le C\|\varphi\|_\infty^{\vec{a}} \exp\{[\vec{k}(t,\vec{a}), X(t)]\}, \quad (t,x) \in \Pi_{(0,T]},$$

which implies the estimate (3.2.15) for $p = \infty$.

Let us consider the case $p \in (1,\infty)$. Using the estimates (3.2.2) and (3.2.3), and the Hölder inequality, we have

$$\begin{aligned}
|u(t,x)| &\le C \int\limits_{\mathbb{R}^n} (|\varphi(\xi)| \exp\{-[\vec{a},\xi]\}) \exp\{-((c-c_0)/2)\rho(t,x,\xi)\} \\
&\quad \times \exp\{-((c+c_0)/2)\rho(t,x,\xi) + [\vec{a},\xi]\} t^{-M} d\xi \\
&\le C \Bigg(\int\limits_{\mathbb{R}^n} |\varphi(\xi)|^p \exp\{-p[\vec{a},\xi]\} \exp\{-\rho((c-c_0)/2)\rho(t,x,\xi)\} d\xi \Bigg)^{1/p} \\
&\quad \times \Bigg(\int\limits_{\mathbb{R}^n} \exp\{-p'((c+c_0)/2)\rho(t,x,\xi) + p'[\vec{a},\xi]\} d\xi \Bigg)^{1/p'} t^{-M} \\
&\le C \Bigg(\int\limits_{\mathbb{R}^n} |\varphi(\xi)|^p \exp\{-p[\vec{a},\xi]\} \exp\{-p((c-c_0)/2)\rho(t,x,\xi)\} t^{-M} d\xi \Bigg)^{1/p} \\
&\quad \times \exp\{[\vec{k}(t,\vec{a}), X(t)]\} \Bigg(\int\limits_{\mathbb{R}^n} \exp\{-p'((c-c_0)/2)\rho(t,x,\xi)\} t^{-M} d\xi \Bigg)^{1/p'}, \\
&\qquad (t,x) \in \Pi_{(0,T]},
\end{aligned}$$

where the number p' is such that $1/p + 1/p' = 1$. It follows, by the equality (3.2.16), that

$$\begin{aligned}
&\|u(t,\cdot)\|_p^{\vec{k}(t,\vec{a})} \\
&\le C \Bigg(\int\limits_{\mathbb{R}^n} \Bigg(\int\limits_{\mathbb{R}^n} |\varphi(\xi)|^p \exp\{-p[\vec{a},\xi]\} \exp\{-p((c-c_0)/2)\rho(t,x,\xi)\} t^{-M} d\xi \Bigg) dx \Bigg)^{1/p} \\
&= C \Bigg(\int\limits_{\mathbb{R}^n} |\varphi(\xi)|^p \exp\{-p[\vec{a},\xi]\} \Bigg(\int\limits_{\mathbb{R}^n} \exp\{-p((c-c_0)/2)\rho(t,x,\xi)\} t^{-M} dx \Bigg) d\xi \Bigg)^{1/p} \\
&= C\|\varphi\|_p^{\vec{a}}, \qquad t \in (0,T].
\end{aligned}$$

If $p=1$, then using the inequalities (3.2.2) and (3.2.3), we get

$$|u(t,x)| \le C \int_{\mathbb{R}^n} |\varphi(\xi)| \exp\{-[\vec{a},\xi]\} \exp\{-(c-c_0)\rho(t,x,\xi)\}$$
$$\times \exp\{-c_0\rho(t,x,\xi)+[\vec{a},\xi]\}t^{-M}d\xi$$
$$\le C\exp\{[\vec{k}(t,\vec{a}),X(t)]\} \int_{\mathbb{R}^n} |\varphi(\xi)| \exp\{-[\vec{a},\xi]\}$$
$$\times \exp\{-(c-c_0)\rho(t,x,\xi)\}t^{M}d\xi, \quad (t,x)\in \Pi_{(0,T]},$$

whence, by (3.1.16) we obtain for $t\in(0,T]$ that

$$\|u(t,\cdot)\|_1^{\vec{k}(t,\vec{a})} \le C\int_{\mathbb{R}^n} |\varphi(\xi)|\exp\{-[\vec{a},\xi]\}\left(\int_{\mathbb{R}^n} \exp\{-(c-c_0)\rho(t,x,\xi)\}t^{-M}dx\right)d\xi$$
$$= C\|\varphi\|_1^{\vec{a}}. \qquad \square$$

Let us find out in which sense the function (3.2.8) satisfies the initial condition.

Lemma 3.2. *Suppose that* $\varphi \in L_p^{\vec{a}}$, $p\in[1,\infty]$. *Then the function* (3.2.8) *satisfies* (3.2.12) *and* (3.2.13).

Proof. Let $p\in[1,\infty)$. We have to prove that for any $\varepsilon>0$ there exists such $\delta\in(0,T)$ that for all $t\in(0,\delta)$,

$$\|(P\varphi)(t,\cdot)-\varphi(\cdot)\|_p^{\vec{s}(t)} < \varepsilon. \tag{3.2.17}$$

Let $\varphi^{(R)}$ be the function determined by the equalities (2.2.24). We have

$$\|(P\varphi)(t,\cdot)-\varphi(\cdot)\|_p^{\vec{s}(t)}$$
$$\le \|(P(\varphi-\varphi^{(R)}))(t,\cdot)\|_p^{\vec{s}(t)} + \|(P\varphi^{(R)})(t,\cdot)-\varphi^{(R)}(\cdot)\|_p^{\vec{s}(t)} + \|\varphi-\varphi^{(R)}\|_p^{\vec{s}(t)},$$
$$t\in(0,T]. \tag{3.2.18}$$

By Lemma 3.1,

$$\|(P(\varphi-\varphi^{(R)}))(t,\cdot)\|_p^{\vec{k}(t,\vec{a})} \le C\|\varphi-\varphi^{(R)}\|_p^{\vec{a}}, \quad t\in(0,T]. \tag{3.2.19}$$

Using the inequalities (3.2.6), (3.2.7), (3.2.18), and (3.2.19), we find that

$$\|(P\varphi)(t,\cdot)-\varphi(\cdot)\|_p^{\vec{s}(t)} \le (C+1)\|\varphi-\varphi^{(R)}\|_p^{\vec{a}} + \|(P\varphi^{(R)})(t,\cdot)-\varphi^{(R)}(\cdot)\|_p^{\vec{s}(t)},$$
$$t\in(0,T].$$

Given $\varepsilon>0$, we choose $R>0$ in such a way that

$$\|\varphi-\varphi^{(R)}\|_p^{\vec{a}} = \left(\int_{\mathbb{R}^n\setminus B_R} |\varphi(x)|^p\exp\{-p[a,x]\}dx\right)^{1/p} < \varepsilon/(2(C+1)).$$

Since

$$\|(P\varphi^{(R)})(t,\cdot)-\varphi^{(R)}(\cdot)\|_p^{\vec{s}(t)} \le \|(P\varphi^{(R)})(t,\cdot)-\varphi^{(R)}(\cdot)\|_{L_p(\mathbb{R}^n)} =: I^{1/p},$$

in order to prove (3.2.17), it suffices to prove the existence of such $\delta \in (0,T)$ that for any $t \in (0,\delta)$,

$$I^{1/p} < \varepsilon/2. \tag{3.2.20}$$

Let us write I as $I = I_1 + I_2$ where

$$I_1 := \int\limits_{\mathbb{R}^n \setminus B_{2R}} \left| \int\limits_{B_R} G(t,x;0,\xi)\varphi^{(R)}(\xi)d\xi \right|^p dx,$$

$$I_2 := \int\limits_{B_{2R}} \left| \int\limits_{B_R} G(t,x;0,\xi)\varphi^{(R)}(\xi)d\xi - \varphi^{(R)}(x) \right|^p dx.$$

If $p = 1$, we get

$$\begin{aligned} I_1 &\leq C \int\limits_{\mathbb{R}^n \setminus B_{2R}} \left(\int\limits_{B_R} |\varphi^{(R)}(\xi)| \exp\{-c\rho(t,x,\xi)\} t^{-M} d\xi \right) dx \\ &= C \int\limits_{B_R} |\varphi^{(R)}(\xi)| \left(\int\limits_{\mathbb{R}^n \setminus B_{2R}} \exp\{-c\rho(t,x,\xi)\} t^{-M} dx \right) d\xi. \end{aligned} \tag{3.2.21}$$

Note the existence of such $\delta_0 \in (0,1)$ that for every $t \in (0,\delta_0)$, $x \in \mathbb{R}^n \setminus B_{2R}$, and $\xi \in B_R$ we have the inequality

$$\rho(t,x,\xi) \geq t^{1-q''}(R/2)^{q'}. \tag{3.2.22}$$

Indeed, for $t \in (0,1)$,

$$\rho(t,x,\xi) \geq t^{1-q''}(q(X(t);\xi))^{q'} \geq t^{1-q''}|q(X(t);0) - q(\xi;0)|^{q'}. \tag{3.2.23}$$

Next, using the notation

$$Y(t) := (\underbrace{0,\dots,0}_{n_1}, -tx_{11},\dots,-tx_{1n_2}, -tx_{21} - 2^{-1}t^2x_{11},\dots,-tx_{2n_3} - 2^{-1}t^2x_{1n_3}),$$

we obtain for $t \in (0,1)$ that

$$q(X(t);0) = q(x;Y(t)) \geq |q(x;0) - q(Y(t);0)|, \tag{3.2.24}$$

$$\begin{aligned} q(Y(t);0) &\leq \left(\sum_{j=1}^{n_2} |tx_{1j}|^{q_j} + \sum_{j=1}^{n_3} 2^{q_j-1}|tx_{2j}|^{q_j} + \sum_{j=1}^{n_3} 2^{q_j-1}|2^{-1}t^2x_{1j}|^{q_j} \right)^{1/q'} \\ &\leq \left(t^{q''} \sum_{j=1}^{n_2} |x_{1j}|^{q_j} + 2^{q'-1}t^{q''} \sum_{j=1}^{n_3} |x_{2j}|^{q_j} + 2^{-1}t^{2q'} \sum_{j=1}^{n_3} |x_{1j}|^{q_j} \right)^{1/q'} \\ &\leq c_1 t^{q''/q'} q(x;0). \end{aligned} \tag{3.2.25}$$

Let the number $\delta_0 > 0$ be such that $c_1\delta_0^{q''/q'} \leq 1/4$. Then (3.2.24) and (3.2.25) for any $t \in (0, \delta_0)$ and $x \in \mathbb{R}^n \setminus B_{2R}$ imply the inequality $q(X(t); 0) \geq (3/4)q(x; 0) \geq 3R/2$, and then (3.2.22) follows from (3.2.23).

From the inequality (3.2.21), using the estimate (3.2.22), we obtain that

$$\begin{aligned} I_1 &\leq C\exp\{-(c-c_0)t^{1-q''}(R/2)^{q'}\}\int_{\mathbb{R}^n}|\varphi^{(R)}(\xi)|\left(\int_{\mathbb{R}^n}\exp\{-c_0\rho(t,x,\xi)\}t^{-M}dx\right)d\xi \\ &= C\|\varphi^{(R)}\|_{L_1(\mathbb{R}^n)}\exp\{-(c-c_0)t^{1-q''}(R/2)^{q'}\}, \quad t \in (0,\delta_0). \end{aligned} \tag{3.2.26}$$

Here we used the fact that

$$\int_{\mathbb{R}^n}\exp\{-c_0\rho(t,x,\xi)\}t^{-M}dx = C, \quad t > 0, \xi \in \mathbb{R}^n. \tag{3.2.27}$$

If $p > 1$, using (3.2.2), (3.2.16), (3.2.22), and the Hölder inequality we find, for $t \in (0, \delta_0)$ and $x \in \mathbb{R}^n \setminus B_{2R}$, that

$$\begin{aligned} &\left|\int_{B_R} G(t,x;0,\xi)\varphi^{(R)}(\xi)d\xi\right| \\ &\leq C\int_{B_R}\exp\{-c\rho(t,x,\xi)\}|\varphi^{(R)}(\xi)|t^{-M}d\xi \\ &\leq C\left(\int_{B_R}|\varphi^{(R)}(\xi)|^p\exp\{-p((c-c_0)/2)\rho(t,x,\xi)\}t^{-M}d\xi\right)^{1/p} \\ &\quad\times\left(\int_{B_R}\exp\{-p'((c+c_0)/2)\rho(t,x,\xi)\}t^{-M}d\xi\right)^{1/p'} \\ &\leq C\exp\{-((c-c_0)/2)t^{1-q''}(R/2)^{q'}\} \\ &\quad\times\left(\int_{B_R}|\varphi^{(R)}(\xi)|^p\exp\{-p((c-c_0)/2)\rho(t,x,\xi)\}t^{-M}d\xi\right)^{1/p} \\ &\quad\times\left(\int_{\mathbb{R}^n}\exp\{-p'c_0\rho(t,x,\xi)\}t^{-M}d\xi\right)^{1/p'} \\ &= C\exp\{-((c-c_0)/2)t^{1-q''}(R/2)^{q'}\} \\ &\quad\times\left(\int_{B_R}|\varphi^{(R)}(\xi)|^p\exp\{-p((c-c_0)/2)\rho(t,x,\xi)\}t^{-M}d\xi\right)^{1/p}, \end{aligned}$$

whence

$$
\begin{aligned}
I_1 \leq & C \exp\{-p((c-c_0)/2)t^{1-q''}(R/2)^{q'}\} \\
& \times \int\limits_{\mathbb{R}^n \setminus B_{2R}} \left(\int\limits_{B_R} |\varphi^{(R)}(\xi)|^p \exp\{-p((c-c_0)/2)\rho(t,x,\xi)\} t^{-M} d\xi \right) dx \\
\leq & C \|\varphi^{(R)}\|_{L_p(\mathbb{R}^n)} \exp\{-(p(c-c_0)/2)t^{1-q''}(R/2)^{q'}\}, \quad t \in (0, \delta_0). \qquad (3.2.28)
\end{aligned}
$$

Let us give an estimate for I_2. Let $\varphi_h^{(R)}$ be the averaging for $\varphi^{(R)}$. It satisfies the relation (2.2.31) and, just as in the proof of Lemma 2.5, for a fixed $h > 0$, uniformly with respect to $x \in B_{2R}$,

$$
\left| \int\limits_{\mathbb{R}^n} G(t,x;0,\xi)\varphi_h^{(R)}(\xi)d\xi - \varphi_h^{(R)}(x) \right| \underset{t\to 0}{\to} 0. \qquad (3.2.29)
$$

Now

$$
\begin{aligned}
I_2^{1/p} \leq & \left(\int\limits_{B_{2R}} \left| \int\limits_{\mathbb{R}^n} G(t,x;0,\xi)(\varphi^{(R)}(\xi) - \varphi_h^{(R)}(\xi))d\xi \right|^p dx \right)^{1/p} \\
& + \left(\int\limits_{B_{2R}} \left| \int\limits_{\mathbb{R}^n} G(t,x;0,\xi)\varphi_h^{(R)}(\xi)d\xi - \varphi_h^{(R)}(x) \right|^p dx \right)^{1/p} \\
& + \left(\int\limits_{B_{2R}} |\varphi_h^{(R)}(x) - \varphi^{(R)}(x)|^p dx \right)^{1/p}.
\end{aligned}
$$

Repeating the estimates for the first summand, similar to the ones carried out in the proof of Lemma 3.1, and using the relations (2.2.31) and (3.2.29), we find that there exists such a constant $\delta_2 > 0$ that for any $t \in (0, \delta_2)$,

$$
I_2 < (1/2)(\varepsilon/2)^p. \qquad (3.2.30)
$$

The inequalities (3.2.26) and (3.2.28) imply the existence of such $\delta_1 > 0$ that for all $t \in (0, \delta_1)$,

$$
I_1 < (1/2)(\varepsilon/2)^p.
$$

Together with (3.2.30) this implies the inequality (3.2.20) for all $t \in (0, \delta)$ where δ is the smaller of the numbers δ_1 and δ_2.

Let us prove the relation (3.2.13). First we note that the integrals from (3.2.13) make sense for all $\varphi \in L_\infty^{\vec{a}}$, $\psi \in L_1^{-\vec{s}(T)}$, and $t \in (0, T]$ since, by the inequality (3.2.6) and Lemma 3.1 $\|u(t,\cdot)\|_\infty^{\vec{s}(t)} < \infty$ for any $t \in (0, T]$, if $\varphi \in L_\infty^{\vec{a}}$.

Indeed, for any $t \in (0, T]$ $a_{lj} \le s_{lj}(t) \le s_{lj}(T)$, $j \in \{1, \dots, n_l\}$, $l \in \{1, 2, 3\}$, so that

$$\begin{aligned}\left|\int\limits_{\mathbb{R}^n} \psi(x) u(t,x) dx\right| &\le \int\limits_{\mathbb{R}^n} (|\psi(x)| \exp\{[\vec{s}(T), x]\})(|u(t,x)| \exp\{-[\vec{s}(t), x]\}) dx \\ &\le \|\psi\|_1^{-\vec{s}(T)} \|u(t,\cdot)\|_\infty^{\vec{s}(t)} < \infty, \\ \left|\int\limits_{\mathbb{R}^n} \psi(x) \varphi(x) dx\right| &\le \int\limits_{\mathbb{R}^n} (|\psi(x)| \exp\{[\vec{s}(T), x]\})(|\varphi(x)| \exp\{-[\vec{a}, x]\}) dx \\ &\le \|\psi\|_1^{-\vec{s}(T)} \|\varphi\|_\infty^{\vec{a}} < \infty.\end{aligned}$$

Due to (3.2.8), in order to prove (3.2.13), it suffices to show that

$$\int\limits_{\mathbb{R}^n} (v(t,\xi) - \psi(\xi)) \varphi(\xi) d\xi \underset{t \to 0}{\to} 0 \qquad \text{where} \qquad v(t,\xi) := \int\limits_{\mathbb{R}^n} G(t,x;0,\xi) \psi(x) dx.$$

Since $\varphi \in L_\infty^{\vec{a}}$, we have

$$\left|\int\limits_{\mathbb{R}^n} (v(t,\xi) - \psi(\xi)) \varphi(\xi) d\xi\right| \le \|\varphi\|_\infty^{\vec{a}} \int\limits_{\mathbb{R}^n} |v(t,\xi) - \psi(\xi)| \exp\{[\vec{a}, \xi]\} d\xi$$

and (3.2.13) will be proved, if we show that

$$\int\limits_{\mathbb{R}^n} |v(t,\xi) - \psi(\xi)| \exp\{[\vec{a}, \xi]\} d\xi \underset{t \to 0}{\to} 0. \tag{3.2.31}$$

Since $a_{lj} < k_{lj}(T, a_{lj})$, $j \in \{1, \dots, n_l\}$, $l \in \{1, 2, 3\}$, there exists such $\gamma > 0$ that

$$\begin{aligned}s_{lj}(T) &\ge k_{lj}(T, a_{lj}) \ge g_{lj}(t) \\ &:= c_0 k_{lj}(T, a_{lj}) (c_0^{2b_j - 1} + (k_{lj}(T, a_{lj}))^{2b_j - 1} t^{2b_j(l-1)+1})^{1-q_j} \ge a_{lj},\end{aligned}$$

$j \in \{1, \dots, n_l\}$, $l \in \{1, 2, 3\}$, for all $t \in [0, \gamma]$. Therefore

$$\exp\{[\vec{g}(t), \xi]\} \ge \exp\{[\vec{a}, \xi]\}, \quad t \in [0, \gamma], \quad \xi \in \mathbb{R}^n,$$

and for proving (3.2.31) it is sufficient to prove, for any $\varepsilon > 0$, the existence of such $\delta \in (0, \gamma)$ that for all $t \in (0, \delta)$,

$$\|v(t,\cdot) - \psi(\cdot)\|_1^{-\vec{g}(t)} := \|(v(t,\cdot) - \psi(\cdot)) \exp\{[\vec{g}(t), \cdot]\}\|_{L_1(\mathbb{R}^n)} < \varepsilon \tag{3.2.32}$$

where

$$\vec{g}(t) := (g_{11}(t), \dots, g_{1n_1}(t), g_{21}(t), \dots, g_{2n_2}(t), g_{31}(t), \dots, g_{3n_3}(t)).$$

The proof of (3.2.32) is similar to that of (3.2.17). As it was done there, we introduce for $R > 0$ a function $\psi^{(R)}(x)$, $x \in \mathbb{R}^n$, by the equalities (2.2.24), in which ψ is substituted for φ. For $t \in (0,\gamma)$ we have

$$\begin{aligned}\|v(t,\cdot)-\psi(\cdot)\|_1^{-\vec{g}(t)} &\le \left\| \int_{\mathbb{R}^n} G(t,x;0,\xi)(\psi-\psi^{(R)})(x)dx \right\|_1^{-\vec{g}(t)} \\ &+ \left\| \int_{\mathbb{R}^n} G(t,x;0,\xi)\psi^{(R)}(x)dx - \psi^{(R)}(\xi) \right\|_1^{-\vec{g}(t)} \\ &+ \|\psi-\psi^{(R)}\|_1^{-\vec{g}(t)} =: \sum_{j=1}^{3} J_j. \end{aligned} \tag{3.2.33}$$

Let us give an estimate for J_1. Using the estimate (3.2.2) we obtain

$$\begin{aligned} &\left| \int_{\mathbb{R}^n} G(t,x;0,\xi)(\psi-\psi^{(R)})(x)dx \right| \\ &\le C \int_{\mathbb{R}^n} \exp\{-c\rho(t,x,\xi) - [\vec{k}(T,\vec{a}),X(t)]\} |(\psi-\psi^{(R)})(x)| \\ &\qquad \times \exp\{[\vec{k}(T,\vec{a}),X(t)]\} t^{-M} dx \\ &\le C \int_{\mathbb{R}^n} \exp\{-(c-c_0)\rho(t,x,\xi)\} \exp\{-c_0\rho(t,x,\xi) - [\vec{k}(T,\vec{a}),X(t)]\} \\ &\qquad \times (|(\psi-\psi^{(R)})(x)|) \exp\{[\vec{s}(T),x]\} t^{-M} dx \\ &\le C \exp\{-[\vec{g}(t),\xi]\} \int_{\mathbb{R}^n} \exp\{-(c-c_0)\rho(t,x,\xi)\} (|(\psi-\psi^{(R)})(x)| \\ &\qquad \times \exp\{[\vec{s}(T),x]\}) t^{-M} dx. \end{aligned} \tag{3.2.34}$$

Here we used the inequality

$$[\vec{k}(T,\vec{a}),X(t)] \le [\vec{s}(T),x], \quad t \ge 0, \quad x \in \mathbb{R}^n, \tag{3.2.35}$$

valid by virtue of (3.2.5) and the definition of $\vec{s}(T)$, and the inequality

$$-c_0\rho(t,x,\xi) - [\vec{k}(T,\vec{a}),X(t)] \le -[\vec{g}(t),\xi], \quad t \ge 0, \quad \{x,\xi\} \subset \mathbb{R}^n, \tag{3.2.36}$$

which is proved just as (3.2.3).

It follows from (3.2.34) and the equality (3.2.16) that

$$J_1 \le C\|\psi-\psi^{(R)}\|_1^{-\vec{s}(T)}, \quad t \in (0,\gamma). \tag{3.2.37}$$

Since $g_{lj}(t) \le g_{lj}(T)$, $t \in (0,\gamma)$, $j \in \{1,\dots,n_l\}$, $l \in \{1,2,3\}$, we have

$$J_3 \le \|\psi-\psi^{(R)}\|_1^{-\vec{s}(T)}, \quad t \in (0,\gamma),$$

thus

$$J_1 + J_3 \le (C+1)\|\psi - \psi^{(R)}\|_1^{-\vec{s}(T)}, \quad t \in (0,\gamma).$$

Since

$$\|\psi - \psi^{(R)}\|_1^{-\vec{s}(T)} = \int\limits_{\mathbb{R}^n \setminus B_R} |\psi(x)| \exp\{[\vec{s}(T), x]\} dx \underset{R\to\infty}{\longrightarrow} 0,$$

we get

$$J_1 + J_3 \underset{R\to\infty}{\longrightarrow} 0, \quad t \in (0,\gamma). \tag{3.2.38}$$

Consider the expression J_2. We write it as

$$\begin{aligned} J_2 = &\int\limits_{\mathbb{R}^n \setminus B_{2R}} \left| \int\limits_{\mathbb{R}^n} G(t,x;0,\xi)\psi^{(R)}(x)dx \right| \exp\{[\vec{g}(t),\xi]\} d\xi \\ &+ \int\limits_{B_{2R}} \left| \int\limits_{\mathbb{R}^n} G(t,x;0,\xi)\psi^{(R)}(x)dx - \psi^{(R)}(\xi) \right| \exp\{[\vec{g}(t),\xi]\} d\xi \\ =:& J_2' + J_2''. \end{aligned}$$

Just as for (3.2.37), we prove the inequality

$$\left\| \int\limits_{\mathbb{R}^n} G(t,x;0,\xi)\psi^{(R)}(x)dx \right\|_1^{\vec{g}(t)} \le C\|\psi^{(R)}\|_1^{-\vec{s}(T)}, \quad t \in (0,\gamma),$$

whence

$$J_2' \underset{R\to\infty}{\longrightarrow} 0, \quad t \in (0,\gamma). \tag{3.2.39}$$

For J_2'' we have

$$J_2'' \le \int\limits_{B_{2R}} \left| \int\limits_{\mathbb{R}^n} G(t,x;0,\xi)\psi^{(R)}(x)dx - \psi^{(R)}(\xi) \right| d\xi.$$

The last integral is considered like I_2 above, so that $J_2'' \underset{t\to 0}{\longrightarrow} 0$, $R > 0$. Together with the relations (3.2.33), (3.2.38), (3.2.39) this yields (3.2.32). □

Lemma 3.3. *Let $\mu \in M^{\vec{a}}$. Then the function* (3.2.9) *satisfies the estimate* (3.2.11) *and the relation* (3.2.14).

Proof. Using the inequalities (3.2.2) and (3.2.3) we find that

$$|u_0(t,x)| \le C \int_{\mathbb{R}^n} \exp\{-(c-c_0)\rho(t,x,\xi)\} \exp\{-c_0\rho(t,x,\xi) + [\vec{a},\xi]\}$$
$$\times \exp\{-[\vec{a},\xi]\} t^{-M} d|\mu|(\xi)$$
$$\le C \exp\{[\vec{k}(t,\vec{a}), X(t)]\} \int_{\mathbb{R}^n} \exp\{-(c-c_0)\rho(t,x,\xi)\} \exp\{-[\vec{a},\xi]\} d|\mu|(\xi),$$
$$(t,x) \in \Pi_{(0,T]},$$

which implies the estimate (3.2.11) if we take into account the equality (3.2.27) with $c-c_0$ substituted for c_0.

Let us prove the relation (3.2.14). By the definition of the functions s_{lj}, $j \in \{1,\dots,n_l\}$, $l \in \{1,2,3\}$, and the estimate (3.2.11), the integrals from (3.2.14) make sense for any $\psi \in C_0^{-\vec{s}(T)}$, $\mu \in M^{\vec{a}}$, and $t \in (0,T]$. Using the formula (3.2.9) we obtain that

$$\left| \int_{\mathbb{R}^n} \psi(x) u_0(t,x) dx - \int_{\mathbb{R}^n} \psi(x) d\mu(x) \right| \le \|v(t,\cdot) - \psi(\cdot)\|_\infty^{-\vec{a}} \|\mu\|^{\vec{a}}$$

where

$$v(t,\xi) := \int_{\mathbb{R}^n} G(t,x;0,\xi)\psi(x)dx.$$

Therefore it is sufficient to prove that

$$\|v(t,\cdot) - \psi(\cdot)\|_\infty^{-\vec{g}(t)} \underset{t\to 0}{\longrightarrow} 0 \tag{3.2.40}$$

where the function $\vec{g}$ is the same as in Lemma 3.2.

Let $R > 0$, and let the function θ_R be the same as in the proof of Lemma 2.6. Set $\varphi^{(R)} := \theta_R\varphi$. For $t \in (0,\gamma)$ we have

$$\|v(t,\cdot) - \varphi(\cdot)\|_\infty^{-\vec{g}(t)} \le \left\| \int_{\mathbb{R}^n} G(t,x;0,\xi)(\varphi - \varphi^{(R)})(x)dx \right\|_\infty^{-\vec{g}(t)}$$
$$+ \left\| \int_{\mathbb{R}^n} G(t,x;0,\xi)\varphi^{(R)}(x)dx - \varphi^{(R)}(\xi) \right\|_\infty^{-\vec{g}(t)}$$
$$+ \|\varphi^{(R)} - \varphi\|_\infty^{-\vec{g}(t)} =: \sum_{j=1}^3 K_j. \tag{3.2.41}$$

As in the proof of the inequalities (3.2.34), using the equality (3.2.27) with $c - c_0$ instead of c_0 we get

$$\left\| \int_{\mathbb{R}^n} G(t,x;0,\xi)(\varphi - \varphi^{(R)})(x)dx \right\|$$
$$\leq C\|\varphi - \varphi^{(R)}\|_\infty^{-\vec{s}(T)} \int_{\mathbb{R}^n} \exp\{-(c-c_0)\rho(t,x,\xi)\}t^{-M}dx \exp\{-[\vec{g}(t),\xi]\},$$

whence

$$K_1 \leq C\|\varphi - \varphi^{(R)}\|_\infty^{-\vec{s}(T)}, \quad t \in (0,\gamma).$$

By the inequalities $g_{lj}(t) \leq s_{lj}(T)$, $t \in [0,\gamma)$, $j \in \{1,\dots,n_l\}$, $l \in \{1,2,3\}$, we have

$$K_3 \leq \|\varphi - \varphi^{(R)}\|_\infty^{-\vec{s}(T)}, \quad t \in [0,\gamma),$$

so that

$$K_1 + K_3 \leq (C+1)\|\varphi - \varphi^{(R)}\|_\infty^{-\vec{s}(T)}, \quad t \in [0,\gamma).$$

Since

$$\|\varphi - \varphi^{(R)}\|_\infty^{-\vec{s}(T)} \leq \sup_{x \in \mathbb{R}^n \setminus B_{R/2}} (|\varphi(x)| \exp\{[\vec{s}(T),x]\}) \underset{R\to\infty}{\longrightarrow} 0,$$

we get

$$K_1 + K_3 \underset{R\to\infty}{\longrightarrow} 0, \quad t \in (0,\gamma). \tag{3.2.42}$$

Next,

$$\begin{aligned} K_2 \leq & \sup_{\xi \in \mathbb{R}^n \setminus B_{2R}} \left(\left| \int_{\mathbb{R}^n} G(t,x;0,\xi)\varphi^{(R)}(x)dx \right| \exp\{[\vec{g}(t),\xi]\} \right) \\ & + \exp\{\max_{j,l} g_{lj}(T)(2R)^{q'}\} \sup_{\xi \in B_{2R}} \left| \int_{\mathbb{R}^n} G(t,x;0,\xi)\varphi^{(R)}(x)dx - \varphi^{(R)}(\xi) \right| \\ =: & K_2' + K_2''. \end{aligned} \tag{3.2.43}$$

In order to find an estimate for K_2', we use the inequality (3.2.22), for any $t \in (0,\delta_0)$, $x \in B_R$, and $\xi \in \mathbb{R}^n \setminus B_{2R}$. This inequality, for such points, is proved like in the case considered in the proof of Lemma 3.2. Instead of the inequalities (3.2.23) and (3.2.24), we use the inequalities

$$\rho(t,x,\xi) \geq t^{1-q''}|q(\xi;0) - q(X(t);0)|^{q'},$$

and

$$q(X(t);0) \leq q(x;0) + q(Y(t);0),$$

respectively; then we use (3.2.25) choosing δ_0 in such a way that $c_1\delta_0^{q''/q'} \leq 1/2$. As a result, we get

$$q(X(t);0) \leq (3/2)q(x;0) \leq 3R/2,$$
$$\rho(t,x,\xi) \geq t^{1-q''}(2R-3R/2)^{q'} = t^{1-q''}(R/2)^{q'}.$$

Using the estimate (3.2.2), the inequalities (3.2.22), (3.2.35), and (3.2.36), the equality (3.2.27), we have

$$\begin{aligned} K_2' \leq & \sup_{\xi\in\mathbb{R}^n\setminus B_{2R}} \Bigg(C\int_{B_R} \exp\{-(c-c_0)\rho(t,x,\xi)\}\exp\{-c_0\rho(t,x,\xi) - [\vec{k}(t,\vec{a}),X(t)]\} \\ & \times (|\varphi^{(R)}(x)|\exp\{[\vec{s}(T),x]\})t^{-M}dx\exp\{[\vec{g}(t),\xi]\}\Bigg) \\ \leq & \, C\exp\{-((c-c_0)/2)t^{1-q''}(R/2)^{q'}\}\|\varphi^{(R)}\|_\infty^{-\vec{s}(T)} \\ & \times \int_{\mathbb{R}^n} \exp\{-((c-c_0)/2)\rho(t,x,\xi)\}t^{-M}dx \\ = & \, C\|\varphi\|_\infty^{-\vec{s}(T)}\exp\{-((c-c_0)/2)t^{1-q''}(R/2)^{q'}\} \underset{t\to 0}{\longrightarrow} 0, \quad R>0. \end{aligned} \tag{3.2.44}$$

Since $\varphi^{(R)}$ is a continuous and bounded function, by a property of the FSCP for the equation adjoint to (3.2.1), we obtain that

$$K_2'' \underset{t\to 0}{\longrightarrow} 0, \quad R>0. \tag{3.2.45}$$

The relations (3.2.41)–(3.2.45) imply the required relation (3.2.40). □

It follows from Lemmas 3.1–3.3 that the functions (3.2.8) and (3.2.9) satisfy the estimates (3.2.10), (3.2.11) and the relations (3.2.12)–(3.2.14). Since the FSCP G is a solution of the equation (3.2.1), the estimates (3.2.2) show that the functions (3.2.8) and (3.2.9) are solutions of this equation. Thus the formulas (3.2.8) and (3.2.9) indeed determine solutions of the equation (3.2.1) on the layer $\Pi_{(0,T]}$, and these solutions have the properties listed in Theorem 3.7. The uniqueness of these solutions is a consequence of Theorem 3.8 on the representation of solutions of the Cauchy problem by Poisson integrals, which will be proved in the next subsection.

3.2.2. Integral representations of solutions for equations from the class $\mathbf{E}_{23}^0$. Here we give some theorems on the Poisson integral representation of solutions of the Cauchy problem and solutions defined on an open layer, for equations from the class $\mathbf{E}_{23}^0$. The proofs are similar to those for the corresponding assertions from Lemmas 2.17, 2.18 and Theorem 2.11 for equations from the class $\mathbf{E}_{11}$.

Theorem 3.8. *Let $\varphi \in L_p^{\vec{a}}$, $p \in [1,\infty]$, and $\mu \in M^{\vec{a}}$. Then:*

1) *a solution u of the equation (3.2.1) satisfying the condition*

$$\|u(t,\cdot)\|_p^{\vec{k}(t,\vec{a})} \le C, \quad t \in (0,T], \tag{3.2.46$_p$}$$

with some constant $C > 0$, and the conditions (3.2.12) and (3.2.13) for $p \in [1,\infty)$ and $p = \infty$ respectively, can be represented in the form (3.2.8);

2) *a solution u of the equation (3.2.1) satisfying the condition (3.2.46$_1$) and the relation (3.2.14), can be represented in the form (3.2.9).*

Proof. 1) Let u be a solution satisfying the above conditions. As in the proof of Lemma 2.16, using the Green-Ostrogradsky formula (3.1.48) we obtain the following formula for the solution u:

$$\begin{aligned} u(t,x) &= \int_{\mathbb{R}^n} G(t,x;h,\xi)\zeta_R(\xi)u(h,\xi)d\xi \\ &\quad - \int_h^t d\tau \int_{B_{3R/4}\setminus B_{R/2}} \overline{L^*(G^*(\tau,\xi;t,x)\zeta_R(\xi))}u(\tau,\xi)d\xi \\ &=: I_1^{(R)} + I_2^{(R)} \end{aligned} \tag{3.2.47}$$

where the function ζ_R is the same as in the proof of Lemma 2.16, (t,x) is a fixed point from $\Pi_{(0,T]}$, R is a sufficiently big positive number, $0 < h < t/2$.

Let us pass in (3.2.47) to the limit, as $R \to \infty$. We shall prove that $I_1^{(R)}$ tends to the integral

$$I_1 := \int_{\mathbb{R}^n} G(t,x;h,\xi)u(h,\xi)d\xi.$$

Using the estimate (3.2.2) we get

$$\begin{aligned} |I_1 - I_1^{(R)}| &= \left| \int_{\mathbb{R}^n} G(t,x;h,\xi)(1-\zeta_R(\xi))u(h,\xi)d\xi \right| \\ &\le C(t-h)^{-M} \int_{\mathbb{R}^n\setminus B_{R/2}} \exp\{-(c-c_0)\rho(t-h,x,\xi)\} \exp\{-c_0\rho(t-h,x,\xi) \\ &\quad + [\vec{k}(h,\vec{a}),\Xi(h)]\}(|u(h,\xi)| \exp\{-[\vec{k}(h,\vec{a}),\Xi(h)]\})d\xi \end{aligned} \tag{3.2.48}$$

where

$$[\vec{k}(h,\vec{a}),\Xi(h)] \equiv \sum_{l=1}^{3}\sum_{j=1}^{n_l} k_{lj}(h,a_{lj})|\Xi_{lj}(h)|^{q_j},$$

$$\Xi_{lj}(h) := \sum_{r=0}^{l-1} \frac{1}{r!} h^r \xi_{(l-r)j}, \quad j \in \{1,\dots,n_l\}, \quad l \in \{1,2,3\}.$$

Now we use the inequalities

$$-c_0\rho(t-h,x,\xi)+[\vec{k}(h,\vec{a}),\Xi(h)]\leq -c_0\rho(t-h,x,\xi)+[\vec{s}(h),\xi]$$
$$\leq[\vec{k}(t-h,\vec{s}(h)),X(t-h)]\leq c_1; \qquad (3.2.49)$$

here h is a sufficiently small number, such that

$$0<t-h\leq T<H(h):=\min_{l,j}\left(\frac{c_0}{s_{lj}(h)}\right)^{(2b_j-1)/(2b_j(l-1)+1)},$$
$$x\in B_{R_0/4},\quad \xi\in\mathbb{R};$$
$$\rho(t-h,x,\xi)\geq(t-h)^r(R/4)^{q'},$$
$$0<t-h\leq T,\quad x\in B_{R_0/4},\quad \xi\in\mathbb{R}^n\setminus B_{R/2}, \qquad (3.2.50)$$

where $c_1>0$, $r=1-q''$ for $0<t-h\leq 1$, and $r=1-3q'$ for $t-h>1$, R is a sufficiently large number, R_0 is a fixed number, such that $0<R_0<R$.

The second inequality from (3.2.49) is proved in the same way as the inequality (3.2.3), while the first and third ones are obvious due to the inequalities (3.2.5) and the definition of the functions s_{lj}, $j\in\{1,\dots,n_l\}$, $l\in\{1,2,3\}$. The proof of the inequality (3.2.50) is similar to that of (3.2.22). Just as we saw there, in the present case we obtain that

$$\rho(t-h,x,\xi)\geq(t-h)^r|q(\xi;0)-q(X(t-h);0)|^{q'},$$
$$q(X(t-h);0)\leq q(x;0)+q(Y(t-h);0)$$
$$\leq q(x;0)+\left(\sum_{j=1}^{n_2}T^{q_j}|x_{1j}|^{q_j}+\sum_{j=1}^{n_3}(2^{q_j-1}T^{q_j}|x_{2j}|^{q_j}+2^{-1}T^{2q_j}|x_{1j}|^{q_j})\right)^{1/q'}$$
$$\leq(1+c_2)q(x;0)\leq(1+c_2)R_0/4,$$
$$c_2:=\max\{(2T)^{q_j},\ T^{2q_j},\ j\in\{1,\dots,n_2\}\},$$

so that

$$\rho(t-h,x,\xi)\geq(t-h)^r|R/2-(1+c_2)R_0/4|^{q'}\geq(t-h)^r(R/4)^{q'}$$

for all $R\geq(1+c_2)R_0$.

It follows from the inequalities (3.2.48)–(3.2.50) that

$$|I_1-I_1^{(R)}|\leq C\exp\{-((c-c_0)/2)(t-h)^r(R/4)^{q'}\}(t-h)^{-M}$$
$$\times\int_{\mathbb{R}^n}\exp\{-((c-c_0)/2)\rho(t-h,x,\xi)\}(|u(h,\xi)|$$
$$\times\exp\{-[\vec{k}(h,\vec{a}),\Xi(h)]\})d\xi. \qquad (3.2.51)$$

If $p = 1$, it follows immediately that for a fixed h,

$$|I_1 - I_1^{(R)}| \le C(t-h)^{-M} \exp\{-((c-c_0)/2)(t-h)^r (R/4)^{q'}\} \|u(h,\cdot)\|_1^{\vec{k}(h,\vec{a})} \underset{R\to\infty}{\longrightarrow} 0.$$

If $p \subset (1,\infty)$, then by the Hölder inequality

$$|I_1 - I_1^{(R)}| \le C(t-h)^{-M/p} \exp\{-((c-c_0)/2)(t-h)^r (R/4)^{q'}\} \|u(h,\cdot)\|_p^{\vec{k}(h,\vec{a})}$$
$$\times \left(\int_{\mathbb{R}^n} \exp\{-p'((c-c_0)/2)\rho(t-h,x,\xi)\}(t-h)^{-M} d\xi \right)^{1/p'}$$
$$= C(t-h)^{-M/p} \|u(h,\cdot)\|_p^{\vec{k}(h,\vec{a})} \exp\{-((c-c_0)/2)(t-h)^r (R/4)^{q'}\} \underset{R\to\infty}{\longrightarrow} 0$$

for a fixed h. If $p = \infty$, for a fixed h we have

$$|I_1 - I_1^{(R)}| \le C\|u(h,\cdot)\|_\infty^{\vec{k}(h,\vec{a})} \exp\{-((c-c_0)/2)(t-h)^r (R/4)^{q'}\} \underset{R\to\infty}{\longrightarrow} 0.$$

Now we shall prove that $I_2^{(R)} \underset{R\to\infty}{\longrightarrow} 0$. Note that

$$L^*(G^*(\tau,\xi;t,x)\zeta_R(\xi)) = \zeta_R(\xi) L^* G^*(\tau,\xi;t,x) + \sum_{l=2}^{3}\sum_{j=1}^{n_l} \xi_{(l-1)j} \partial_{\xi_{lj}} \zeta_R(\xi) G^*(\tau,\xi;t,x)$$
$$- \sum_{\|k_1\|\le 2b} \bar{a}_{k_1}(\tau) \sum_{0<l_1\le k_1} C_{k_1}^{l_1} (-\partial_{\xi_1})^{l_1} \zeta_R(\xi) (-\partial_{\xi_1})^{k_1-l_1} G^*(\tau,\xi;t,x). \tag{3.2.52}$$

Since $L^*G^*(\tau,\xi;t,x) = 0$ for $\tau < t$, the first summand from (3.2.52) vanishes, and by the properties of the function ζ_R the whole expression (3.2.52) equals zero on $\mathbb{R}^n \setminus (B_{3R/4} \setminus B_{R/2})$. Using the fact that for $\xi \in B_{3R/4} \setminus B_{R/2}$ we have

$$\left| \sum_{l=2}^{3}\sum_{j=1}^{n_l} \xi_{(l-1)j} \partial_{\xi_{lj}} \zeta_R(\xi) \right| \le C R^{(q'/q'')-1},$$

$$|\partial_{\xi_1}^{l_1} \zeta_R(\xi)| \le C R^{-|l_1|}, \quad |l_1| := \sum_{j=1}^{n_1} l_{1j},$$

the normality of the FSCP, and the estimates (3.2.2), for $R \ge 1$ we find that

$$|L^*(G^*(\tau,\xi;t,x)\zeta_R(\xi))| \le C(t-\tau)^{-M-1+\alpha} \exp\{-c\rho(t-\tau,x,\xi)\}, \quad \alpha > 0.$$

Using this estimate we prove, just as for $I_1 - I_1^{(R)}$ above, that

$$\left| \int_{B_{3R/4}\setminus B_{R/2}} L^*(G^*(\tau,\xi;t,x)\zeta_R(\xi)) u(\tau,\xi) d\xi \right|$$
$$\le C\|u(\tau,\cdot)\|_p^{\vec{k}(\tau,\vec{a})} \exp\{-((c-c_0)/2)(t-\tau)^r (R/4)^{q'}\} (t-\tau)^{-\beta}$$

where $\beta = M+1-\alpha$ for $p=1$, $\beta = (M/p)+1-\alpha$ for $1<p<\infty$, and $\beta = 1-\alpha$ for $p=\infty$. Now, taking into account the inequality

$$\exp\{-((c-c_0)/2)(t-\tau)^r(R/4)^{q'}\}(t-\tau)^{-\beta} \le C\exp\{-\varepsilon(t-\tau)^r R^{q'}\},$$
$$0\le \tau < t, \quad R\ge 1, \quad \varepsilon > 0,$$

and the condition $(3.2.46_p)$, we get

$$|I_2^{(R)}| \le C\exp\{-\varepsilon t^r R^{q'}\} \underset{R\to\infty}{\to} 0.$$

Thus, after passing to the limit in (3.2.47), as $R\to\infty$, we find that

$$u(t,x) = \int\limits_{\mathbb{R}^n} G(t,x;h,\xi)u(h,\xi)d\xi. \tag{3.2.53}$$

Let us pass to the limit in (3.2.53), as $h\to 0$, and prove that

$$\lim_{h\to 0}\int\limits_{\mathbb{R}^n} G(t,x;h,\xi)u(h,\xi)d\xi = \int\limits_{\mathbb{R}^n} G(t,x;0,\xi)\varphi(\xi)d\xi. \tag{3.2.54}$$

This will imply the representation (3.2.8).

Let us write the difference

$$\Delta_h := \int\limits_{\mathbb{R}^n} G(t,x;h,\xi)u(h,\xi)d\xi - \int\limits_{\mathbb{R}^n} G(t,x;0,\xi)\varphi(\xi)d\xi$$

as

$$\Delta_h = J_1^{(h)} + J_2^{(h)}, \tag{3.2.55}$$

where

$$J_1^{(h)} := \int\limits_{\mathbb{R}^n} \Delta_h^0 G(t,x;h,\xi)u(h,\xi)d\xi, \tag{3.2.56}$$

while

$$J_2^{(h)} := \int\limits_{\mathbb{R}^n} G(t,x;0,\xi)(u(h,\xi)-\varphi(\xi))d\xi,$$

if $p\in[1,\infty)$, and

$$J_2^{(h)} := \int\limits_{\mathbb{R}^n} G(t,x;0,\xi)u(h,\xi)d\xi - \int\limits_{\mathbb{R}^n} G(t,x;0,\xi)\varphi(\xi)d\xi$$

if $p=1$.

We obtain directly, if $p\in\{1,\infty\}$, or using the Hölder inequality, if $p\in(1,\infty)$, that

$$|J_1^{(h)}| \le J_{1p}^{(h)}\|u\|_p^{\vec{k}(h,\vec{a})}, \tag{3.2.57}$$
$$J_{1p}^{(h)} := \|\Delta_h^0 G(t,x;h,\xi)\exp\{[\vec{k}(h,\vec{a}),\Xi(h)]\}\|_{L_{p'}(\mathbb{R}^n)},$$

where $p' = \infty$ for $p = 1$, $p' = p/(p-1)$ for $p \in (1, \infty)$, and $p' = 1$ for $p = \infty$. Similarly, for $p \in [1, \infty)$ we get

$$|J_2^{(h)}| \leq J_{2p}^{(h)} \|u(h, \cdot) - \varphi(\cdot)\|_p^{\vec{s}(h)}, \tag{3.2.58}$$
$$J_{2p}^{(h)} := \|G(t, x; 0, \cdot) \exp\{[\vec{s}(h), \cdot]\}\|_{L_{p'}(\mathbb{R}^n)}.$$

Let us prove that

$$J_l^{(h)} \underset{h \to 0}{\longrightarrow} 0, \quad l \in \{1, 2\}. \tag{3.2.59}$$

Due to the condition (3.2.46$_p$), the relation (3.2.59) for $l = 1$ will be proved if we show that

$$J_{1p}^{(h)} \underset{h \to 0}{\longrightarrow} 0. \tag{3.2.60}$$

The proof of this relation will be based on the following fact: for arbitrary fixed $t \in (0, T]$ and $x \in \mathbb{R}^n$, there exist such constants $C_1 > 0$, $c_1 \in (c_0, c)$ (c is the constant from the estimate (3.2.2)), and $h_0 > 0$, that for any $h \in (0, h_0)$ and $\xi \in \mathbb{R}^n$,

$$|G(t, x; h, \xi)| \leq C_1 t^{-M} \exp\{-c_1 \rho(t, x, \xi)\}, \tag{3.2.61}$$

Let us prove that. By the estimate (3.2.2), we have

$$|G(t, x; h, \xi)| \leq C t^{-M} \exp\left\{-c \sum_{l=1}^{3} \sum_{j=1}^{n_l} t^{1 - l q_j} |X_{lj}(t-h) - \xi_{lj}|^{q_j}\right\}.$$

Since for $c > 0$ there exist such constants $C_1 > 0$ and $c_1 \in (c_0, c)$ that for each $\{u, v\} \subset \mathbb{R}$, $|v| \leq 1$,

$$\exp\{-c|u - v|^q\} \leq C_1 \exp\{-c_1 |u|^q\},$$

where $q \in (1, 2]$, we get

$$\begin{aligned}
&\exp\{-c t^{1 - 2q_j} |X_{2j}(t-h) - \xi_{2j}|^{q_j}\} \\
&\quad = \exp\{-c|t^{-2+1/q_j}(X_{2j}(t) - \xi_{2j}) - t^{-2+1/q_j} h x_{1j}|^{q_j}\} \\
&\quad \leq C_1 \exp\{-c_1 t^{1-2q_j} |X_{2j}(t) - \xi_{2j}|^{q_j}\},
\end{aligned}$$

if h is taken such that $h t^{-2+1/q_j} |x_{1j}| \leq 1$, and

$$\begin{aligned}
&\exp\{-c t^{1-3q_j} |X_{3j}(t-h) - \xi_{3j}|^{q_j}\} \\
&\quad = \exp\{-c|t^{-3+1/q_j}(X_{3j}(t) - \xi_{3j}) - t^{-3+1/q_j}(h x_{2j} + h t x_{1j} - h^2 x_{1j}/2)|^{q_j}\} \\
&\quad \leq C_1 \exp\{-c_1 t^{1-3q_j} |X_{3j}(t) - \xi_{3j}|^{q_j}\},
\end{aligned}$$

if h is taken such that $(h(t|x_{1j}| + |x_{2j}|) + h^2 |x_{1j}|/2) t^{-3+1/q_j} \leq 1$.

By the estimates (3.2.2), (3.2.61) and the estimates (3.2.49) with t substituted for $t-h$, we have

$$\begin{aligned}
&|\Delta_h^0 G(t,x;h,\xi)\exp\{[\vec{k}(h,\vec{a}),\Xi(h)]\}| \\
&\quad \le Ct^{-M}\exp\{-c_1\rho(t,x,\xi)+[\vec{k}(h,\vec{a}),\Xi(h)]\} \\
&\quad \le Ct^{-M}\exp\{-(c_1-c_0)\rho(t,x,\xi)\}\exp\{-c_0\rho(t,x,\xi)+[\vec{s}(h),\xi]\} \\
&\quad \le Ct^{-M}\exp\{-(c_1-c_0)\rho(t,x,\xi)\}\exp\{[\vec{k}(t,\vec{s}(h)),X(t)]\} \\
&\quad \le C\exp\{-(c_1-c_0)\rho(t,x,\xi)\}
\end{aligned}$$

where h is an arbitrary number from $(0,h_0)$, such that $T<H(h)$ ($H(h)$ is taken from (3.2.49)). Using this inequality and the fact that, by the continuity of the function G, $\Delta_h^0 G(t,x;h,\xi)\exp\{[\vec{k}(h,\vec{a}),\Xi(h)]\} \underset{h\to 0}{\to} 0$, from the dominated convergence theorem we obtain the relation (3.2.60) for $p\in[1,\infty]$.

The relation (3.2.59) for $l=2$ and $p\in[1,\infty)$ follows from the equality (3.2.12) and the fact that the inequalities (3.2.2), (3.2.49) and the equality (3.2.16) imply the following inequalities valid for any sufficiently small h:

$$\begin{aligned}
J_{2p}^{(h)} &\le C\Bigg(\int_{\mathbb{R}^n}\exp\{-p'(c-c_0)\rho(t,x,\xi)+p'(-c_0\rho(t,x,\xi)+[\vec{s}(h),\xi])\}t^{-Mp'}d\xi\Bigg)^{1/p'} \\
&\le C\exp\{[\vec{k}(t,\vec{s}(h)),X(t)]\}t^{-M/p} \\
&\le Ct^{-M/p}, \qquad p\in(1,\infty); \\
J_{2p}^{(h)} &\le C\sup_{\xi\in\mathbb{R}^n}(t^{-M}\exp\{-(c-c_0)\rho(t,x,\xi)+(-c_0\rho(t,x,\xi)+[\vec{s}(h),\xi])\}) \\
&\le Ct^{-M}\exp\{[\vec{k}(t,\vec{s}(h)),X(t)]\} \\
&\le Ct^{-M}, \qquad p=1.
\end{aligned}$$

For $0<t\le T_0<H(T)$, $T_0\le T$, the function $\psi(\xi):=G(t,x;0,\xi)$, $\xi\in\mathbb{R}^n$, due to the estimate (3.2.2) and the inequality

$$-c_0\rho(t,x,\xi)+[\vec{s}(T),\xi]\le[\vec{k}(t,\vec{s}(T)),X(t)],$$

satisfies the inequality

$$|\psi(\xi)|\exp\{[\vec{s}(T),\xi]\}\le Ct^{-M}\exp\{-(c-c_0)\rho(t,x,\xi)+[\vec{k}(t,\vec{s}(T)),X(t)]\}, \qquad \xi\in\mathbb{R}^n, \tag{3.2.62}$$

so that $\psi\in L_1^{-\vec{s}(T)}$. Therefore, by the condition (3.2.13), we obtain for $0<t\le T_0$ and $p=\infty$ the equality

$$\lim_{h\to 0} J_2^{(h)}=0. \tag{3.2.63}$$

The relations (3.2.55), (3.2.59), (3.2.63) imply (3.2.54), so that we come to the formula (3.2.8) for each point (t,x) of the layer $\Pi_{(0,T]}$, for the case $p \in [1,\infty)$, and the layer $\Pi_{(0,T_0]}$, if $p=\infty$. In order to verify this formula for $p=\infty$, for every point $(t,x) \in \Pi_{(0,T]}$, we have to use the formula (3.2.53) for $h \le T_0$, the formula (3.2.8) with $t=h$, and the convolution formula (3.1.46).

2) If a solution of the equation (3.2.1) satisfies the condition (3.2.46$_1$), then, as we showed in the proof of the first part of the theorem, it satisfies the formula (3.2.53). Let us pass in it to the limit, as $h \to 0$. For this purpose we consider the difference

$$\int_{\mathbb{R}^n} G(t,x;h,\xi)u(h,\xi)d\xi - \int_{\mathbb{R}^n} G(t,x;0,\xi)d\mu(\xi) = J_1^{(h)} + J_3^{(h)} \tag{3.2.64}$$

where $J_1^{(h)}$ was defined in (3.2.56), and

$$J_3^{(h)} := \int_{\mathbb{R}^n} G(t,x;0,\xi)u(h,\xi)d\xi - \int_{\mathbb{R}^n} G(t,x;0,\xi)d\mu(\xi).$$

For $J_1^{(h)}$ we have the equality (3.2.59). It follows from the inequality (3.2.62) that for each $t \in (0,T_0]$ the function $\psi(\xi) := G(t,x;0,\xi)$, $\xi \in \mathbb{R}^n$, belongs to the space $C_0^{-\vec{s}(T)}$. Therefore, by the condition (3.2.14), $\lim\limits_{h\to 0} J_3^{(h)} = 0$ which, together with (3.2.53), (3.2.62), and (3.2.64) implies the formula (3.2.9) for $(t,x) \in \Pi_{(0,T_0]}$. This formula is valid for any $(t,x) \in \Pi_{(0,T]}$. This follows from the formulas (3.2.53) with $h \le T_0$, (3.2.9) with $t=h$, and (3.1.46). □

Theorem 3.8 implies the uniqueness of a solution of the Cauchy problem, which completes the proof of Theorem 3.7. The next result is, in a certain sense, converse to this theorem and similar to Theorem 2.11 for the homogeneous equation (2.1.1).

Theorem 3.9. *Let u be a solution of the equation* (3.2.1) *satisfying the condition* (3.2.46$_p$) *with some $C>0$ and $p \in [1,\infty]$. Then, for $p \in (1,\infty]$, there exists the unique function $\varphi \in L_p^{\vec{a}}$, whereas for $p=1$ there exists the unique generalized measure $\mu \in M^{\vec{a}}$, such that the solution u can be represented in the form* (3.2.8) *and* (3.2.9) *respectively.*

Proof. Let $p \in (1,\infty]$. It follows from the condition (3.2.46$_p$) that the sequence of functions

$$\{u(1/\nu,x)\exp\{-[\vec{k}(1/\nu,\vec{a}),X(1/\nu)]\}, x \in \mathbb{R}^n : \nu \ge 1\} \tag{3.2.65}$$

is bounded in the space $L_p(\mathbb{R}^n)$. Therefore, as in the proof of Theorem 2.11, there exists its subsequence

$$\{u(1/\nu(r),x)\exp\{-[\vec{k}(1/\nu(r),\vec{a}),X(1/\nu(r))]\}, x \in \mathbb{R}^n : r \ge 1\} \tag{3.2.66}$$

and a function $\chi \in L_p(\mathbb{R}^n)$, such that for any $\psi \in L_{p'}(\mathbb{R}^n)$,

$$\lim_{r\to\infty} \int_{\mathbb{R}^n} \bar{\psi}(\xi) \exp\{-[\vec{k}(1/\nu(r), \vec{a}), \Xi(1/\nu(r))]\} u(1/\nu(r), \xi) d\xi$$
$$= \int_{\mathbb{R}^n} \bar{\psi}(\xi) \chi(\xi) d\xi. \tag{3.2.67}$$

Set $\varphi(\xi) := \chi(\xi) \exp\{[\vec{a}, \xi]\}$, $\xi \in \mathbb{R}^n$. Then $\varphi \in L_p^{\vec{a}}$, and the relation (3.2.67) is written as

$$\lim_{r\to\infty} \int_{\mathbb{R}^n} \bar{\psi}(\xi) \exp\{-[\vec{k}(1/\nu(r), \vec{a}), \Xi(1/\nu(r))]\} u(1/\nu(r), \xi) d\xi$$
$$= \int_{\mathbb{R}^n} \bar{\psi}(\xi) \exp\{-[\vec{a}, \xi]\} \varphi(\xi) d\xi. \tag{3.2.68}$$

Let (t, x) be a fixed point of the layer $\Pi_{(0,T]}$, and

$$\psi(\xi) := \bar{G}(t, x; 0, \xi) \exp\{[\vec{a}, \xi]\}, \quad \xi \in \mathbb{R}^n. \tag{3.2.69}$$

It follows from the estimate

$$|\psi(\xi)| \le Ct^{-M} \exp\{-(c - c_0)\rho(t, x, \xi) + [\vec{k}(t, \vec{a}), X(t)]\}, \quad \xi \in \mathbb{R}^n, \tag{3.2.70}$$

obtained from the inequalities (3.2.2) and (3.2.3) that $\psi \in L_{p'}(\mathbb{R}^n)$. Therefore, due to the equality (3.2.68), we have

$$\lim_{r\to\infty} \int_{\mathbb{R}^n} G(t, x; 0, \xi) \exp\{-[\vec{k}(1/\nu(r), \vec{a}), \Xi(1/\nu(r))] + [\vec{a}, \xi]\} u(1/\nu(r), \xi) d\xi$$
$$= \int_{\mathbb{R}^n} G(t, x; 0, \xi) \varphi(\xi) d\xi. \tag{3.2.71}$$

Suppose that $1/\nu(r) \le t/2$, $r \ge 1$. By (3.2.53),

$$u(t, x) = \int_{\mathbb{R}^n} G(t, x; 1/\nu(r), \xi) u(1/\nu(r), \xi) d\xi. \tag{3.2.72}$$

This means that

$$u(t, x) - \int_{\mathbb{R}^n} G(t, x; 0, \xi) \varphi(\xi) d\xi$$
$$= \int_{\mathbb{R}^n} \Delta_h^0 G(t, x; h, \xi)|_{h=1/\nu(r)} u(1/\nu(r), \xi) d\xi$$

$$+\int_{\mathbb{R}^n} G(t,x;0,\xi)(1-\exp\{-[\vec{k}(1/\nu(r),\vec{a}),\Xi(1/\nu(r))]+[\vec{a},\xi]\})u(1/\nu(r),\xi)d\xi$$

$$+\Bigg(\int_{\mathbb{R}^n} G(t,x;0,\xi)\exp\{-[\vec{k}(1/\nu(r),\vec{a}),\Xi(1/\nu(r))]+[\vec{a},\xi]\}u(1/\nu(r),\xi)d\xi$$

$$-\int_{\mathbb{R}^n} G(t,x;0,\xi)\varphi(\xi)d\xi\Bigg)$$

$$=:\sum_{j=1}^{3} K_j^{(r)}, \quad r\geq 1. \tag{3.2.73}$$

In order to obtain the representation (3.2.8), it suffices to show that

$$\lim_{r\to\infty} K_j^{(r)}=0 \tag{3.2.74}$$

for $j\in\{1,2,3\}$.

The relation (3.2.71) implies (3.2.74) for $j=3$. Let us prove (3.2.74) for $j=2$. Using the Hölder inequality and the condition (3.2.46$_p$) we get

$$|K_2^{(r)}|\leq \|u(1/\nu(r),\,\cdot\,)\|_p^{\vec{k}(1/\nu(r),\vec{a})}\Bigg(\int_{\mathbb{R}^n} F_r(\xi)d\xi\Bigg)^{1/p'}\leq C\Bigg(\int_{\mathbb{R}^n} F_r(\xi)d\xi\Bigg)^{1/p'} \tag{3.2.75}$$

where

$$F_r(\xi):=|G(t,x;0,\xi)|^{p'}\Big|\exp\{\,[\vec{k}(1/\nu(r),\vec{a}),\Xi(1/\nu(r))]\,\}-\exp\{\,[\vec{a},\xi]\,\}\Big|^{p'},$$

$$\xi\in\mathbb{R}^n,\quad r\geq 1.$$

Let us study properties of the functions F_r, $r\geq 1$. The estimate (3.2.2) and the inequality (3.2.3) imply the inequalities

$$\begin{aligned}(F_r(\Xi))^{1/p'}\leq & Ct^{-M}\exp\{-(c-c_0)\rho(t,x,\xi)\}\\ &\times\Big(\exp\{-c_0\rho(t,x,\xi)+[\vec{s}(1/\nu(r)),\xi]\}+\exp\{-c_0\rho(t,x,\xi)+[\vec{a},\xi]\}\Big)\\ \leq & Ct^{-M}\exp\{-(c-c_0)\rho(t,x,\xi)\}\\ &\times\Big(\exp\{[\vec{k}(t,\vec{s}(1/\nu(r))),X(t)]\}+\exp\{[\vec{k}(t,\vec{a}),X(t)]\}\Big).\end{aligned}$$

Here $r\geq r_0$ where r_0 is taken in such a way that $t<H(1/\nu(r_0))$. Since

$$k_{lj}(t,s_{lj}(1/\nu(r)))\leq k_{lj}(t,s_{lj}(1/\nu(r_0))),$$

for any $r \geq r_0$, $j \in \{1, \dots, n_l\}$ and $l \in \{1, 2, 3\}$, we have

$$(F_r(\Xi))^{1/p'} \leq Ct^{-M} \exp\{-(c - c_0)\rho(t, x, \xi)\}$$
$$\times \left(\exp\{[\vec{k}(t, \vec{s}(1/\nu(r_0))), X(t)]\} + \exp\{[\vec{k}(t, \vec{a}), X(t)]\}\right),$$
$$\xi \in \mathbb{R}^n, \quad r \geq r_0,$$

so that the sequence $\{F_r, r \geq r_0\}$ has an integrable majorant. For each $\xi \in \mathbb{R}^n$ we have $\lim_{r\to\infty} F_r(\xi) = 0$, and by the dominated convergence theorem

$$\lim_{r\to\infty} \int_{\mathbb{R}^n} F_r(\xi) d\xi = 0.$$

Together with (3.2.75) this implies (3.2.74) for $j = 2$.

The equality (3.2.74) for $j = 1$ follows from the relation (3.2.59) for $l = 1$, because $K_1^{(r)} = J_1^{(h)}|_{h=1/\nu(r)}$.

Let us consider the case $p = 1$. Just as in the proof of Theorem 2.11, in this case we prove the existence of such a generalized measure $\mu \in M^{\vec{a}}$ that for any function $\psi \in C_0(\mathbb{R}^n)$

$$\lim_{r\to\infty} \int_{\mathbb{R}^n} \bar{\psi}(\xi) \exp\{-[\vec{k}(1/\nu(r), \vec{a}), \Xi(1/\nu(r))]\} u(1/\nu(r), \xi) d\xi$$
$$= \int_{\mathbb{R}^n} \bar{\psi}(\xi) \exp\{-[\vec{a}, \xi]\} d\mu(\xi). \tag{3.2.76}$$

It follows from the estimate (3.2.70) that the function (3.2.69) belongs to the space $C_0(\mathbb{R}^n)$ for every fixed point $(t, x) \in \Pi_{(0,T]}$. Therefore we obtain by (3.2.76) that

$$\lim_{r\to\infty} \int_{\mathbb{R}^n} G(t, x; 0, \xi) \exp\{-[\vec{k}(1/\nu(r), \vec{a}), \Xi(1/\nu(r))] + [\vec{a}, \Xi]\} u(1/\nu(r), \xi) d\xi$$
$$= \int_{\mathbb{R}^n} G(t, x; 0, \xi) d\mu(\xi). \tag{3.2.77}$$

The subsequent reasoning is the same as in the case $p > 1$. Using the formula (3.2.72) we write the equality

$$u(t, x) - \int_{\mathbb{R}^n} G(t, x; 0, \xi) d\mu(\xi) = K_1^{(r)} + K_2^{(r)} + \tilde{K}_3^{(r)}$$

where $K_1^{(r)}$ and $K_2^{(r)}$ are the same as in (3.2.73), while

$$\tilde{K}_3^{(r)} := \int\limits_{\mathbb{R}^n} G(t,x;0,\xi)\exp\{-[\vec{k}(1/\nu(r),\vec{a}),\Xi(1/\nu(r))] + [\vec{a},\xi]\}u(1/\nu(r),\xi)d\xi$$
$$- \int\limits_{\mathbb{R}^n} G(t,x;0,\xi)d\mu(\xi).$$

Taking into account the equalities (3.2.74) for $j \in \{1,2\}$ and the fact that $\lim\limits_{r\to\infty} \tilde{K}_3^{(r)} = 0$, due to (3.2.77), we deduce the representation (3.2.9) from (3.2.78).

Hence, we have proved the existence of a function $\varphi \in L_p^{\vec{a}}$, for $p \in (1,\infty]$, and a generalized measure $\mu \in M^{\vec{a}}$, for $p = 1$ such that the given solution of the equation (3.2.1) satisfying the condition (3.2.46$_p$) is a Poisson integral of the function φ or the generalized measure μ respectively. The uniqueness of φ and μ follows immediately from Theorem 3.7. □

Corollary. It follows from Theorems 3.7 and 3.9 that the spaces $L_p^{\vec{a}}$, $p \in (1,\infty]$, and $M^{\vec{a}}$ are the sets of initial values of solutions of the equation (3.2.1), if and only if the solutions satisfy the condition (3.2.46$_p$) with $p \in (1,\infty]$ and $p = 1$ respectively. This condition is necessary and sufficient for the solutions to be represented in the forms (3.2.8) and (3.2.9).

Remark. The problem about conditions upon solutions on a domain to guarantee the existence of their limit values at the boundary, is an important classical problem in the theory of analytic or harmonic functions. For solutions defined on the layer $\Pi_{(0,T]}$ and belonging, for each t, to the weight spaces $L_p^{\vec{k}(t,\vec{a})}$, was solved in Chapter 2 for equations from the class $\mathbf{E}_{11}$. For equations from the class $\mathbf{E}_{23}^0$ (in particular, those from the class $\mathbf{E}_{21}$, with coefficients which do not depend on spatial variables), its solution was given above in this section. Note that similar results have been proved also for the equation (3.1.96) with a weak degeneration on the initial hyperplane. They can be found in the paper [120] by Ivasyshen and Voznyak.

3.2.3. Correct solvability of the Cauchy problem and integral representations of solutions for equations from the class $\mathbf{E}_{22}$. Very precise results of the previous sections were based on such properties of a FSCP as its exact estimates, normality, convolution formulas, and also on a good choice of functional spaces. By Theorem 3.3 the FSCP for a general second order equation (3.1.71) has all these properties. Therefore results like Theorems 3.7–3.9 are true for this equation too, and their proofs are identical to the ones for Theorems 3.7–3.9. Below we present such theorems for a more general case, namely for an inhomogeneous equation (3.1.71). Here an additional investigation of the corresponding volume potential is needed.

Let us consider the equation (3.1.71) assuming the conditions A_{24}–A_{26}. By Theorem 3.3 and the remark after it, under these assumptions there exists a FSCP

Z for the equation (3.1.71), such that

$$
\begin{aligned}
|\partial_{x_1}^{k_1} Z(t,x;\tau,\xi)| &\le C(t-\tau)^{-M-|k_1|/2} \exp\{-c\rho(t-\tau,x,\xi)\}, \quad |k_1|\le 2,\\
|SZ(t,x;\tau,\xi)| &\le C(t-\tau)^{-M-1} \exp\{-c\rho(t-\tau,x,\xi)\},\\
& 0\le \tau < t \le T, \{x,\xi\}\subset \mathbb{R}^n, \qquad (3.2.78)
\end{aligned}
$$

where

$$
M := (n_1 + 3n_2 + 5n_3)/2, \quad \rho(t,x,\xi) := \sum_{l=1}^{3} t^{1-2l} |X_l(t) - \xi_l|^2.
$$

The FSCP Z is normal, that is the function (3.1.76) is a FSCP for the adjoint equation (3.1.73), and the convolution formula (2.1.112) holds for Z.

Using the equality (3.1.77), the estimates (3.1.53), (3.1.55) and the estimates of the derivatives $\partial_{x_1}^{k_1} W_1$ similar to (3.1.80), (3.1.81), we obtain also the following estimates for the equation (3.1.71):

$$
\begin{aligned}
\left|\int_{\mathbb{R}^n} \partial_{x_1}^{k_1} Z(t,x;\tau,\xi) d\xi\right| &\le C(t-\tau)^{-(|k_1|-\alpha)/2}, \quad 0<|k_1|\le 2,\\
\left|\int_{\mathbb{R}^n} SZ(t,x;\tau,\xi) d\xi\right| &\le C(t-\tau)^{-1+\alpha/2}, \quad 0\le\tau<t\le T, \quad x\in\mathbb{R}^n. \qquad (3.2.79)
\end{aligned}
$$

The appropriate norms and spaces are special cases of the norms and spaces defined in Section 3.2.1. Here the collections $\vec{k}(t,\vec{a})$ and $\vec{s}(t)$, $t\in[0,T]$, consist respectively of the functions

$$
\begin{aligned}
k_l(t,a_l) &:= c_0 a_l (c_0 - a_l t^{2l-1}), \quad l\in\{1,2,3\};\\
s_1(t) &:= k_1(t,a_1) + 2t^2 k_2(t,a_2) + t^4 k_3(t,a_3),\\
s_2(t) &:= 2k_2(t,a_2) + 4t^2 k_3(t,a_3),\\
s_3(t) &:= 4k_3(t,a_3), \quad t\in[0,T],
\end{aligned}
$$

where $c_0 \in (0,c)$, c is the constant from the estimates (3.2.78), and the form (3.2.4) is

$$
[\vec{a},\xi] := \sum_{l=1}^{3} a_l |\xi_l|^2.
$$

Note that the functions k_l, just as the functions k_{lj} from Section 3.2.1, satisfy the inequalities

$$
k_l(t-\tau, k_l(\tau,a_l)) \le k_l(t,a_l), \quad 0\le\tau\le t\le T, \quad l\in\{1,2,3\}. \qquad (3.2.80)
$$

Let us consider properties of the volume potential

$$u(t,x) := \int_0^t d\tau \int_{\mathbb{R}^n} Z(t,x;\tau,\xi) f(\tau,\xi) d\xi, \quad (t,x) \in \Pi_{(0,T]}, \tag{3.2.81}$$

generated by a FSCP Z for the equation (3.1.71). For a function $f : \Pi_{(0,T]} \to \mathbb{C}$ we use the following conditions:

(D_{21}) a function f is continuous and locally Hölder continuous in x with respect to the distance

$$d(x;\xi) := \sum_{l=1}^{3} |x_l - \xi_l|^{1/(2l-1)}, \quad \{x,\xi\} \subset \mathbb{R}^n, \tag{3.2.82}$$

uniformly with respect to t, and for any $t \in (0,T]$ the expressions

$$\|f(t,\cdot)\|_0^{\vec{k}(t,\vec{a})} := \sup_{x \in \mathbb{R}^n} \left(|u(t,x)| \exp\{-[\vec{k}(t,\vec{a}), x]\}\right)$$

and $$F_0(t) := \int_0^t \|f(\tau,\cdot)\|_0^{\vec{k}(\tau,\vec{a})} d\tau$$

are finite;

(D_{22p}) a function f is continuous and locally Hölder continuous in x with respect to the distance (3.2.82), uniformly with respect to t, and for any $t \in (0,T]$ the expressions

$$\|f(t,\cdot)\|_p^{\vec{k}(t,\vec{a})} \quad \text{and} \quad F_p(t) := \int_0^t \|f(\tau,\cdot)\|_p^{\vec{k}(\tau,\vec{a})} d\tau,$$

where $p \in [1,\infty]$, are finite.

Lemma 3.4. *If a function f satisfies the condition D_{21} or D_{22p}, then the function (3.2.81) has continuous derivatives $\partial_{x_1}^{k_1} u$, $|k_1| \le 2$, and Su, for which*

$$\partial_{x_1}^{k_1} u(t,x) = \int_0^t d\tau \int_{\mathbb{R}^n} \partial_{x_1}^{k_1} Z(t,x;\tau,\xi) f(\tau,\xi) d\xi, \quad |k_1| = 1,$$

$$\partial_{x_1}^{k_1} u(t,x) = \int_0^t d\tau \int_{\mathbb{R}^n} \partial_{x_1}^{k_1} Z(t,x;\tau,\xi) \Delta_\xi^{X(t-\tau)} f(\tau,\xi) d\xi$$

$$+ \int_0^t \left(\int_{\mathbb{R}^n} \partial_{x_1}^{k_1} Z(t,x;\tau,\xi) d\xi \right) f(\tau, X(t-\tau)) d\tau, \quad |k_1| = 2,$$

$$Su(t,x) = f(t,x) + \int_0^t d\tau \int_{\mathbb{R}^n} SZ(t,x;\tau,\xi)\Delta_\xi^{X(t-\tau)} f(\tau,\xi)d\xi$$
$$+ \int_0^t \left(\int_{\mathbb{R}^n} SZ(t,x;\tau,\xi)d\xi \right) f(\tau, X(t-\tau))d\tau, \quad (t,x) \in \Pi_{(0,T]}.$$

The proof is carried out using the estimates (3.2.78) and (3.2.79) by a standard method employed in the proofs of Properties 2.6, 2.7, 3.6, and Lemma 2.11.

Lemma 3.5. *Let a function f satisfy the condition D_{21}. Then the function* (3.2.81) *is a solution of the equation* (3.1.71), *for which*

$$\|u(t,\cdot)\|_0^{\vec{k}(t,\vec{a})} \le CF_0(t), \quad t \in (0,T], \tag{3.2.83}$$

and for any compact set $K \subset \mathbb{R}^n$

$$u(t,x) \underset{t\to 0}{\to} 0 \quad \text{uniformly w.r.t. } x \in K. \tag{3.2.84}$$

Proof. The fact that the function (3.2.81) is a solution of the equation (3.1.71) follows from Lemma 3.4 and the fact that Z is a solution of the corresponding homogeneous equation.

Using the estimates (3.2.3), (3.2.79), (3.2.80), and the equality (3.2.16) we have

$$|u(t,x)| \le C \int_0^t d\tau \int_{\mathbb{R}^n} (t-\tau)^{-M} \exp\{-(c-c_0)\rho(t-\tau,x,\xi)\} \exp\{-c_0\rho(t-\tau,x,\xi)$$
$$+ [\vec{k}(\tau,\vec{a}),\xi]\}(|f(\tau,\xi)| \exp\{-[\vec{k}(\tau,\vec{a}),\xi]\})d\xi$$
$$\le C \exp\{[\vec{k}(t,\vec{a}), X(t)]\}F_0(t), \quad (t,x) \in \Pi_{(0,T]},$$

which implies the estimate (3.2.83).

The relation (3.2.84) is a consequence of the fact that $F_0(t) \underset{t\to 0}{\to} 0$, and the inequality

$$\sup_{x\in K} |u(t,x)| \le C \sup_{(t,x)\in(0,T]\times K} \exp\{[\vec{k}(t,\vec{a}), X(t)]\}F_0(t) \le CC_0F_0(t),$$
$$C_0 := \sup_{x\in K} \exp\{[\vec{s}(T), x]\}.$$
□

Lemma 3.6. *If a function f satisfies the condition D_{22p}, then the function* (3.2.81) *is a solution of the equation* (3.1.71), *with the properties:*

$$\|u(t,\cdot)\|_p^{\vec{k}(t,\vec{a})} \le CF_p(t), \quad t \in (0,T], \tag{3.2.85}$$
$$\lim_{t\to 0} \|u(t,\cdot)\|_p^{\vec{s}(t)} = 0. \tag{3.2.86}$$

Proof. The fact that the function (3.2.81) is a solution of the equation (3.1.71) is verified just as in Lemma 3.5. In order to prove the lemma, we have only to verify the estimate (3.2.85). Due to the inequality (3.2.6) and the relation $F_p(t) \underset{t\to 0}{\to} 0$, it implies (3.2.86).

Note the inequality

$$\|u(t,\cdot)\|_p^{\vec{k}(t,\vec{a})} \leq \int_0^t \|P(t,\tau,\cdot)\|_p^{\vec{k}(t,\vec{a})} d\tau, \quad t \in (0,T], \tag{3.2.87}$$

where

$$P(t,\tau,x) := \int_{\mathbb{R}^n} Z(t,x;\tau,\xi) f(\tau,\xi) d\xi.$$

For $p \in \{1,\infty\}$ the inequality (3.2.87) is obvious, while for $p \in (1,\infty)$ it follows from the Minkowski inequality. Just as for the estimate (3.2.15), using the inequalities (3.2.3), (3.2.78), and (3.2.80) we prove the estimate

$$\|P(t,\tau,\cdot)\|_p^{\vec{k}(t,\vec{a})} \leq C\|f(\tau,\cdot)\|_p^{\vec{k}(\tau,\vec{a})}, \quad 0 \leq \tau < t \leq T.$$

Together with the inequality (3.2.87), it implies the estimate (3.2.85). □

Let us formulate the main theorems for the equation (3.1.71), analogs of Theorems 2.6, 2.7, and 2.11 for an inhomogeneous equation, and Theorems 3.7 and 3.9 for a homogeneous equation.

Theorem 3.10. *Let the conditions A_{24}–A_{26} be satisfied. Then:*

1) *for any function $\varphi \in L_p^{\vec{a}}$ and any function f satisfying the condition D_{22p}, $p \in [1,\infty]$, the formula*

$$u(t,x) = \int_{\mathbb{R}^n} Z(t,x;0,\xi)\varphi(\xi)d\xi + \int_0^t d\tau \int_{\mathbb{R}^n} Z(t,x;\tau,\xi) f(\tau,\xi) d\xi,$$
$$(t,x) \in \Pi_{(0,T]}, \tag{3.2.88}$$

determines the unique solution of the equation (3.1.71) *satisfying the estimate*

$$\|u(t,\cdot)\|_p^{\vec{k}(t,\vec{a})} \leq C(\|\varphi\|_p^{\vec{a}} + F_p(t)), \quad t \in (0,T],$$

and the relations (3.2.12) *and* (3.2.13) *for $p \in [1,\infty)$ and $p = \infty$ respectively;*

2) *for any generalized measure $\mu \in M^{\vec{a}}$ and any function f satisfying the condition D_{22_1}, the formula*

$$u(t,x) = \int_{\mathbb{R}^n} Z(t,x;0,\xi)d\mu(\xi) + \int_0^t d\tau \int_{\mathbb{R}^n} Z(t,x;\tau,\xi) f(\tau,\xi) d\xi,$$
$$(t,x) \in \Pi_{(0,T]}, \tag{3.2.89}$$

determines the unique solution of the equation (3.1.71) *satisfying the estimate*

$$\|u(t,\cdot)\|_1^{\vec{k}(t,\vec{a})} \le C(\|\mu\|^{\vec{a}} + F_1(t)), \quad t \in (0,T],$$

and the relation (3.2.14).

Theorem 3.11. *Suppose that the coefficients of the equation* (3.1.71) *satisfy the conditions* A_{24}–A_{26}, *the right-hand side function* f *satisfies the condition* D_{22p}, *and a solution* u *on the layer* $\Pi_{(0,T]}$ *is such that*

$$\|u(t,\cdot)\|_p^{\vec{k}(t,\vec{a})} \le C, \quad t \in (0,T],$$

with a certain constant $C > 0$ *and* $p \in [1,\infty]$. *Then, if* $p \in (1,\infty]$, *there exists the unique function* $\varphi \in L_p^{\vec{a}}$, *and if* $p = 1$, *there exists the unique generalized measure* $\mu \in M^{\vec{a}}$, *such that the solution* u *can be represented in the form* (3.2.88) *or* (3.2.89) *respectively.*

The proofs of Theorems 3.10 and 3.11 repeat (taking into account Lemma 3.6) the proofs of Theorems 2.6, 2.7, 2.11, 3.7, and 3.9.

Remark. Theorems 3.10 and 3.11 have the same implications as the corollaries of the corresponding theorems in the final parts of Sections 2.2.7 and 3.2.2. Hence, we have proved the important problem, mentioned in the remark at the end of Section 3.2.2, also for the equation (3.1.71) with coefficients depending on all the variables.

The next theorem deals with the correct solvability of the Cauchy problem for the equation (3.1.71), under the assumption that the initial function belongs to the space $C_0^{\vec{a}}$ of continuous functions $\varphi : \mathbb{R}^n \to \mathbb{C}$ with a finite norm

$$\|\varphi\|_0^{\vec{a}} := \sup_{x\in\mathbb{R}^n} (|\varphi(x)| \exp\{-[\vec{a},x]\}).$$

Theorem 3.12. *If coefficients of the equation* (3.1.71) *satisfy* A_{24}–A_{26}, *the function* f *satisfies the condition* D_{21}, *and* $\varphi \in C_0^{\vec{a}}$, *then the formula* (3.2.88) *determines the unique solution of the equation* (3.1.71), *for which*

$$\|u(t,\cdot)\|_0^{\vec{k}(t,\vec{a})} \le C(\|\varphi\|_0^{\vec{a}} + F_0(t)), \quad t \in (0,T],$$

and for every compact set $K \subset \mathbb{R}^n$

$$u(t,x) \underset{t\to 0}{\to} \varphi(x) \quad \text{uniformly} \quad \text{w.r.t.} \quad x \in K. \tag{3.2.90}$$

The proof is similar to that of Theorem 2.3. Instead of Lemma 2.12, here we use Lemma 3.5 for the volume potential (3.2.81), and instead of Lemmas 2.4 and 2.16, similar lemmas for the equation (3.1.71) are used.

3.2.4. On the Cauchy problem for equations from the classes $\mathbf{E}_{21}$ and $\mathbf{E}_{23}$. The estimates of the FSCPs for the equations (3.1.70) and (3.1.86) with $b > 1$, given by Theorems 3.2 and 3.4, are rather complicated, so that we cannot establish such complete results for these cases as the above ones for equations from the classes $\mathbf{E}_{23}^0$ and $\mathbf{E}_{22}$. Nevertheless, using the FSCP estimates (3.1.74), (3.1.75), and (3.1.88), we can prove the correct solvability of the Cauchy problem in the spaces of functions having the same growth orders, for $|x| \to \infty$, as in Sections 3.2.1 and 3.2.2, but with the minimal type with respect to the groups of variables x_2 and x_3. Below we give such results for the equation (3.1.86) and, in particular, for the equation (3.1.70). We deal with spaces of functions which grow rapidly, as $|x_1| \to \infty$, and are bounded in x_2 and x_3.

Let us use the notation from Section 3.1.3. In the notation from Section 3.2.1 we set $a_{lj} = 0$, $j \in \{1, \dots, n_l\}$, $l \in \{2, 3\}$, so that, instead of the collections of functions $\vec{k}(t, \vec{a})$ and $\vec{s}(t)$ we use the collection $\vec{k}_1(t, \vec{a}_1) := (k_{11}(t, a_1), \dots, k_{1n_1}(t, a_{n_1}))$, while the norms $\|u(t,\cdot)\|_p^{\vec{k}(t,\vec{a})}$ are replaced by the norms

$$\|u(t,\cdot)\|_p^{\vec{k}_1(t,\vec{a}_1)} := \|u(t,x)\exp\{-[\vec{k}_1(t,\vec{a}_1), x_1]\}\|_{L_p(\mathbb{R}^n)}$$

where

$$\vec{a}_1 := (a_{11}, \dots, a_{1n_1}), \quad [\vec{k}_1(t,\vec{a}_1), x_1] := \sum_{j=1}^{n_1} k_{1j}(t, a_j)|x_{1j}|^{q_j}.$$

Instead of the spaces $L_p^{\vec{a}}$, $M^{\vec{a}}$, $L_1^{-\vec{s}(T)}$, and $C_0^{-\vec{s}(T)}$, we shall use respectively the spaces $L_p^{\vec{a}_1}$, $M^{\vec{a}_1}$, $L_1^{-\vec{k}_1(T,\vec{a}_1)}$, and $C_0^{-\vec{k}_1(T,\vec{a}_1)}$, defined similarly, with the obvious changes.

We introduce also the norm

$$\|u(t,\cdot)\|_0^{\vec{k}_1(t,\vec{a}_1)} := \sup_{x\in\mathbb{R}^n} (|u(t,x)|\exp\{-[\vec{k}_1(t,\vec{a}_1), x_1]\}),$$

and the space $C_0^{\vec{a}_1}$ of continuous functions φ on $\mathbb{R}^n$ having the finite norms

$$\|\varphi\|_0^{\vec{a}_1} := \sup_{x\in\mathbb{R}^n} (|\varphi(x)|\exp\{-[\vec{a}_1, x_1]\}).$$

For the right-hand side f of the equation (3.1.86) we use the following conditions:

(D_{23}) the function f is continuous, locally Hölder continuous in x with respect to the distance $d(x;\xi)$, $\{x,\xi\} \subset \mathbb{R}^n$, defined in the beginning of Section 3.1.3, uniformly with respect to t, and for any $t \in (0, T]$ the expressions

$$\|f(t,\cdot)\|_0^{\vec{k}_1(t,\vec{a}_1)} \text{ and } F_0^{(1)}(t) := \int_0^t \|f(\tau,\cdot)\|_0^{\vec{k}_1(\tau,\vec{a}_1)} d\tau \text{ are finite;}$$

(D_{24p}) the function f is continuous, Hölder continuous in x, in the same sense as in the condition D_{23}, and for all $t \in (0,T]$ the expressions $\|f(t,\cdot)\|_p^{\vec{k}_1(t,\vec{a}_1)}$ and $F_p^{(1)}(t) := \int_0^t \|f(\tau,\cdot)\|_p^{\vec{k}_1(\tau,\vec{a}_1)} d\tau$, where $p \in [1,\infty]$, are finite.

The following results are similar to Theorems 3.10 and 3.12. They are based on the formulas (3.2.88) and (3.2.89), in which Z is a FSCP for the equation (3.1.86).

Theorem 3.13. *Suppose that the conditions A_{27}–A_{29} are satisfied. Then:*

1) *for each function $\varphi \in L_p^{\vec{a}_1}$ and every function f satisfying the condition D_{24p}, $p \in [1,\infty]$, the formula* (3.2.88) *determines the unique solution of the equation* (3.1.86), *for which*

$$\|u(t,\cdot)\|_p^{\vec{k}_1(t,\vec{a}_1)} \le C(\|\varphi\|_p^{\vec{a}_1} + F_p^{(1)}(t)), \quad t \in (0,T],$$

and

$$\lim_{t\to 0} \|u(t,\cdot) - \varphi(\cdot)\|_p^{\vec{k}_1(t,\vec{a}_1)} = 0,$$

if $p \in [1,\infty)$, or

$$\lim_{t\to 0} \int_{\mathbb{R}^n} \psi(x)u(t,x)dx = \int_{\mathbb{R}^n} \psi(x)\varphi(x)dx,$$

for any function $\psi \in L_1^{-\vec{k}_1(T,\vec{a}_1)}$, if $p = \infty$;

2) *for each generalized measure $\mu \in M^{\vec{a}_1}$ and every function f satisfying the condition D_{24_1}, the formula* (3.2.89) *determines the unique solution of the equation* (3.1.86), *for which*

$$\|u(t,\cdot)\|_1^{\vec{k}_1(t,\vec{a}_1)} \le C(\|\mu\|^{\vec{a}_1} + F_1^{(1)}(t)), \quad t \in (0,T],$$

and

$$\lim_{t\to 0} \int_{\mathbb{R}^n} \psi(x)u(t,x)dx = \int_{\mathbb{R}^n} \psi(x)d\mu(x)$$

for any function $\psi \in C_0^{-\vec{k}_1(T,\vec{a}_1)}$.

Theorem 3.14. *If the conditions A_{27}–A_{29} are satisfied, as well as the condition D_{23} for the function f, and if $\varphi \in C_0^{\vec{a}_1}$, then the formula (3.2.88) determines the unique solution of the equation* (3.1.86), *for which*

$$\|u(t,\cdot)\|_0^{\vec{k}_1(t,\vec{a}_1)} \le C(\|\varphi\|_0^{\vec{a}_1} + F_0^{(1)}(t)), \quad t \in (0,T],$$

and for any compact set $K \subset \mathbb{R}^n$ the relation (3.2.90) *holds.*

The method of proof of Theorems 3.13 and 3.14 is similar to the one used in the proofs of Theorems 2.5–2.7 and 3.7.

3.3 Properties of solutions of the Fokker-Planck-Kolmogorov equations

3.3.1. The Fokker-Planck-Kolmogorov equations. Here we shall consider equations from the class $\mathbf{E}_{22}$ with real-valued coefficients. It has been mentioned that such equations belong to the class of ultra-parabolic equations and often appear in the investigations of various physical phenomena in the so-called diffusion approximation. In many important cases (for example, in the study of the Brownian motion) these mathematical models describe natural phenomena adequately and in a relatively simple way. Usually such equations, used systematically from the beginning of the last century, are called the Fokker-Planck equations. These equations often determine the continuous time evolution of a Markov stochastic process. Studying models of the Brownian motion, Kolmogorov began to investigate Markov processes whose evolution is determined by equations from the class $\mathbf{E}_{22}$ with real-valued coefficients. Thus it is natural to call them the Fokker-Planck-Kolmogorov equations.

The main object of our study, the FSCP $Z(t,x;\tau,\xi)$, $0 \le \tau < t \le T$, $\{x,\xi\} \subset \mathbb{R}^n$, can be interpreted in a natural way as a transition density of a Markov diffusion process with values from the phase space $\mathbb{R}^n$ with three different groups of phase coordinates $x_1 = (x_{11},\dots,x_{1n_1})$, $x_2 = (x_{21},\dots,x_{2n_2})$, and $x_3 = (x_{31},\dots,x_{3n_3})$, $n_1+n_2+n_3 = n$, $1 \le n_3 \le n_2 \le n_1$.

Let us present, in a convenient form, the notation and assumptions used in this section. All functions below are real-valued.

We consider the homogeneous equation (3.1.71), that is the equation

$$\begin{aligned}(Lu)(t,x) := \Bigg(\partial_t - \sum_{j=1}^{n_2} x_{1j}\partial_{x_{2j}} - \sum_{j=1}^{n_3} x_{2j}\partial_{x_{3j}} - \sum_{j,l=1}^{n_1} a_{jl}(t,x)\partial_{x_{1j}}\partial_{x_{1l}} \\ - \sum_{j=1}^{n_1} a_j(t,x)\partial_{x_{1j}} - a_0(t,x)\Bigg) u(t,x) = 0, \\ (t,x) \in \Pi_{(0,T]}, \end{aligned} \tag{3.3.1}$$

and the adjoint equation

$$\begin{aligned}(L^*v)(\tau,\xi) := \Bigg(- \partial_\tau + \sum_{j=1}^{n_2} \xi_{1j}\partial_{\xi_{2j}} + \sum_{j=1}^{n_3} \xi_{2j}\partial_{\xi_{3j}}\Bigg) v(\tau,\xi) \\ - \sum_{j,l=1}^{n_1} \partial_{\xi_{1j}}\partial_{\xi_{1l}}(a_{jl}(\tau,\xi)v(\tau,\xi)) \\ + \sum_{j=1}^{n_1} \partial_{\xi_{1j}}(a_j(\tau,\xi)v(\tau,\xi)) - a_0(\tau,\xi)v(\tau,\xi) = 0, \\ (\tau,\xi) \in \Pi_{[0,T)}. \end{aligned} \tag{3.3.2}$$

For the (real-valued) coefficients of the equation (3.3.1) we use the conditions A_{24}–A_{26} from Section 3.1.2, and also the following conditions:

(C_{21}) the coefficients a_{jl}, a_j, $\{j,l\} \subset \{1,\ldots,n_1\}$, and a_0 are continuous functions on $\Pi_{[0,T]}$, and for all $(t,x) \in \Pi_{[0,T]}$ and $\sigma_1 \in \mathbb{R}^{n_1}$ they satisfy the inequality

$$\sum_{j,l=1}^{n_1} a_{jl}(t,x)\sigma_{1j}\sigma_{1l} \geq \delta_0|\sigma_1|^2, \tag{3.3.3}$$

and the estimates

$$|a_{jl}(t,x)| \leq C_0(|x|^2+1), \quad |a_j(t,x)| \leq C_0(|x|+1), \quad a_0(t,x) \leq C_0$$

with some constants $\delta_0 > 0$ and $C_0 > 0$;

(C_{22}) the coefficients a_{jl}, a_j, $\{j,l\} \subset \{1,\ldots,n_1\}$, and a_0 are continuous, bounded on $\Pi_{[0,T]}$, and the condition (3.3.3) is satisfied.

We shall use the Green-Ostrogradsky formula (3.1.48) which, for the present situation, takes the form

$$\begin{aligned}
&\int_{t_1}^{t_2} d\theta \int_{B_R} (vLu - uL^*v)(\theta,y)dy \\
&= \int_{B_R} (vu)(\theta,y)\Big|_{\theta=t_1}^{t_2} dy - \int_{t_1}^{t_2} d\theta \int_{\Gamma_R} \left(\sum_{j=1}^{n_2} y_{1j}\mu_{2j} + \sum_{j=1}^{n_3} y_{2j}\mu_{3j}\right)(vu)(\theta,y)dS_y \\
&\quad + \int_{t_1}^{t_2} d\theta \int_{\Gamma_R} \sum_{j=1}^{n_1} B^j[v,u](\theta,y)\mu_{1j}dS_y,
\end{aligned} \tag{3.3.4}$$

where t_1, t_2, B_R, Γ_R, and $(\mu_{11},\ldots,\mu_{1n_1},\mu_{21},\ldots,\mu_{2n_2},\mu_{31},\ldots,\mu_{3n_3})$ are such as in (3.1.48), while L and L^* are the differential expressions from (3.3.1) and (3.3.2),

$$B^j[v,u] := -\sum_{l=1}^{n_1}(a_{jl}\partial_{y_{1l}}uv - u\partial_{y_{1l}}(a_{jl}v)) + a_juv, \quad j \in \{1,\ldots,n_1\}.$$

Passing to the limit, as $R \to \infty$, we obtain, for appropriate functions u and v, that

$$\int_{t_1}^{t_2} d\theta \int_{\mathbb{R}^n} (vLu - uL^*v)(\theta,y)dy = \int_{\mathbb{R}^n} (vu)(\theta,y)\Big|_{\theta=t_1}^{t_2} dy. \tag{3.3.5}$$

This formula will be used frequently.

3.3.2. The maximum principle. An exceptional role of the maximum principle and its modifications in the theory of analytic functions, as well as the theory of the second order elliptic and parabolic equations (linear and quasilinear, non-degenerate and degenerate), is well-known. Versions of the maximum principle for equations from the class $\mathbf{E}_{22}$ were discussed in [161, 35]. Here we present some facts from this useful theory. Their formulations and proofs are close to the ones for uniformly parabolic equations [213, 83]. Therefore our exposition will be brief enough.

We begin with the following simple result.

Lemma 3.7. *Let D be a bounded domain in $\mathbb{R}^n$; $u : [0,T] \times \bar{D} \to \mathbb{R}$ be a continuous function having on $(0,T] \times D$ the continuous derivatives involved in the equation* (3.3.1)*; suppose that the coefficients of the equation* (3.3.1) *satisfy the condition C_{21}. If $(Lu)(t,x) \geq 0$, $a_0(t,x) < 0$, $(t,x) \in (0,T] \times D$, $u(t,x) \geq 0$, $(t,x) \in \partial([0,T] \times D) \setminus \{t = T\}$, then $u(t,x) \geq 0$, $(t,x) \in [0,T] \times \bar{D}$.*

Remark. If the function a_0 is bounded from above on $\Pi_{(0,T]}$, then the substitution $u(t,x) = v(t,x)e^{\lambda t}$, $\lambda > \sup\limits_{(t,x)\in\Pi_{(0,T]}} a_0(t,x)$, leads to an equation for v of the form (3.3.1) with a strictly negative on $\Pi_{(0,T]}$ coefficient at v. Therefore the assumption C_{21} makes it possible not to demand that $a_0(t,x) < 0$, $(t,x) \in (0,T] \times D$.

A deeper and more important result is the so-called **strong maximum principle** proved in [83] for uniformly parabolic equations and in [161] for equations from the class $\mathbf{E}_{22}$. Let us describe this maximum principle.

Consider a domain $\Omega \subset \mathbb{R}^{n+1}$. Let $P^0 := (t^0, x^0)$ be an arbitrary point from Ω. Denote by $S(P^0)$ the set of all those points $Q \in \Omega$ which can be connected with P^0 by a simple continuous curve in Ω such that the coordinate t does not decrease along the curve from Q to P^0. Denote by $C(P^0)$ the component of the intersection $\Omega \cap \{t = t^0\}$ containing P^0. Note that $C(P^0) \subset S(P^0)$.

Theorem 3.15. *Let coefficients of the equation* (3.3.1) *be continuous on Ω, the inequalities* (3.3.3) *and $a_0(t,x) \leq 0$ hold for all $(t,x) \in \Omega$, and let the function u be continuous, together with all its derivatives involved in the equation* (3.3.1)*. If $Lu \leq 0$ ($Lu \geq 0$) on Ω, and the function u has a positive maximum (a negative minimum) on Ω attained at a point P^0, then $u(P) = u(P^0)$ for all the points $P \in S(P^0)$.*

Let us present some special results of the maximum principle type for the case of unbounded domains.

Theorem 3.16. *Let the condition C_{21} be satisfied and $u : (0,T] \times \Omega \to \mathbb{R}$ be a function, continuous together with its derivatives involved in the equation* (3.3.1)*. Here $\Omega = \mathbb{R}^n \setminus B_{R_0}$, $R_0 > 0$, or $\Omega = \mathbb{R}^n$. If*

1) *$(Lu)(t,x) \geq 0$, $(t,x) \in (0,T] \times \Omega$;*
2) *$\liminf\limits_{(t,x)\to(t^0,x^0)} u(t,x) \geq 0$ for any point $(t^0,x^0) \in \partial((0,T] \times \Omega) \setminus \{t = T\}$;*
3) *uniformly with respect to $t \in (0,T)$, there exists $\liminf\limits_{|x|\to\infty} u(t,x) \geq 0$, then $u(t,x) \geq 0$, $(t,x) \in (0,T] \times \Omega$.*

Proof. The continuity of the function u and the conditions 2) and 3) guarantee that u is bounded from below, that is there exists such a constant $M_0 > 0$ that $u(t,x) > -M_0$, $(t,x) \in (0,T] \times \Omega$. Set $u(t^0,x^0) := \liminf_{(t,x)\to(t^0,x^0)} u(t,x)$, $(t^0,x^0) \in \partial((0,T]\times\Omega) \setminus \{t = T\}$ and consider on a compact set $[0,T] \times \bar{D}_R$, $D_R := \Omega \cap B_R$, $R > R_0$, or $R > 0$, if $\Omega = \mathbb{R}^n$, an auxiliary function $w(t,x) := M_0 R^{-2}(\beta t + |x|^2)e^{\alpha t} + u(t,x)$, where α and β are positive constants.

By the condition C_{21} we have

$$Lw = Lu + M_0 R^{-2} e^{\alpha t}\Bigg(\alpha(\beta t + |x|^2) + \beta - 2\sum_{j=1}^{n_2} x_{1j}x_{2j} - 2\sum_{j=1}^{n_3} x_{2j}x_{3j} - 2\sum_{j=1}^{n_1} a_{jj}(t,x) - 2\sum_{j=1}^{n_1} a_j(t,x)x_{1j} - a_0(t,x)(\beta t + |x|^2)\Bigg)$$

$$\geq M_0 R^{-2} e^{\alpha t}((\alpha - 2C_0 n_1(1+n_1) - C_0 - 2)|x|^2 + \beta t(\alpha - C_0) + \beta - 2C_0 n_1(1+n_1)).$$

If α and β are chosen so that $\alpha > 2C_0 n_1(1+n_1) + C_0 + 2$, $\beta > 2C_0 n_1(1+n_1)$, then $(Lw)(t,x) > 0$, $(t,x) \in (0,T] \times D_R$. Other conditions of Lemma 3.7 are also satisfied: $w(0,x) = M_0 R^{-2}|x|^2 + u(0,x) \geq 0$; if $|x| = R$, then $w(t,x) \geq M_0 + u(t,x) \geq 0$; if $|x| = R_0$ for the case $\Omega = \mathbb{R}^n \setminus B_{R_0}$ we have $w(t,x) \geq u(t,x) \geq 0$. Therefore $w(t,x) \geq 0$, $(t,x) \in [0,T] \times \bar{D}_R$.

An arbitrary point $(t,x) \in (0,T] \times \Omega$ belongs to $(0,T] \times D_R$ if R is large enough. At this point $w(t,x) \geq 0$. After passing to the limit, as $R \to \infty$, we obtain that $u(t,x) \geq 0$. □

A result similar to Theorem 3.17 is valid for the adjoint equation, which can be written as

$$(L^*u)(t,x) := \Bigg(-\partial_t + \sum_{j=1}^{n_2} x_{1j}\partial_{x_{2j}} + \sum_{j=1}^{n_3} x_{2j}\partial_{x_{3j}} - \sum_{j,l=1}^{n_1} a_{jl}(t,x)\partial_{x_{1j}}\partial_{x_{1l}} - \sum_{j=1}^{n_1} a_j^*(t,x)\partial_{x_{1j}} - a_0^*(t,x)\Bigg)u(t,x) = 0, \quad (t,x) \in \Pi_{[0,T)},$$

if it is assumed that the equation (3.3.1) satisfies the condition C_{22}, and there exist continuous bounded derivatives in x_1 of the second order for a_{jl}, and those of the first order for a_j.

Theorem 3.17. *Let $u : [0,T) \times \Omega \to \mathbb{R}$ be a function, continuous together with the derivatives involved in the equation* (3.3.1). *Here $\Omega = \mathbb{R}^n \setminus B_{R_0}$, $R_0 > 0$ or $\Omega = \mathbb{R}^n$. If*

1) $(L^*u)(t,x) \leq 0$, $(t,x) \in [0,T) \times \Omega$;

2) $\limsup_{(t,x)\to(t^0,x^0)} u(t,x) \le 0$ *for any point* $(t^0,x^0) \in \partial([0,T) \times \Omega) \setminus \{t=0\}$;

3) *uniformly in* $t \in [0,T)$, *there exists* $\limsup_{|x|\to\infty} u(t,x) \le 0$,

then $u(t,x) \le 0$, $(t,x) \in [0,T) \times \Omega$.

Let us give a modification of the maximum principle for functions with a bounded decrease.

Theorem 3.18. *Suppose that the condition* C_{22} *is satisfied, and* $u : \Pi_{(0,T]} \to \mathbb{R}$ *is a function continuous together with the derivatives involved in the equation* (3.3.1), *such that* $(Lu)(t,x) \ge 0$, $(t,x) \in \Pi_{(0,T]}$, *and for some numbers* $B > 0$ *and* $b_l \ge 0$, $l \in \{1,2,3\}$,

$$u(t,x) \ge -B\exp\left\{\sum_{l=1}^{3} b_l|x_l|^2\right\}, \quad (t,x) \in \Pi_{(0,T]}.$$

If $\lim_{t\to 0} u(t,x) \ge 0$, $x \in \mathbb{R}^n$, *then* $u(t,x) \ge 0$, $(t,x) \in \Pi_{(0,T]}$.

Proof. Consider the function

$$H(t,x) := \exp\left\{R_1(1-\mu t)^{-1}\sum_{j=1}^{n_1} x_{1j}^2 + R_2(1-\mu t)^{-3}\sum_{j=1}^{n_2}(x_{2j} - ((1-\mu t)/\mu)x_{1j})^2\right.$$
$$\left. + R_3(1-\mu t)^{-5}\sum_{j=1}^{n_3}(x_{3j} - ((1-\mu t)/\mu)x_{2j} + ((1-\mu t)^2/(2\mu^2))x_{1j})^2 + \alpha t\right\},$$
$$(t,x) \in \Pi_{[0,1/(2\mu)]},$$

where $\mu, \alpha, R_l, l \in \{1,2,3\}$, are positive constants. Denote

$$A_j := x_{2j} - ((1-\mu t)/\mu)x_{3j}, \quad j \in \{1,\dots,n_2\};$$
$$B_j := x_{3j} - ((1-\mu t)/\mu)x_{2j} + ((1-\mu t)^2/(2\mu^2))x_{1j}, \quad j \in \{1,\dots,n_3\};$$
$$P := -\sum_{j,l=1}^{n_1} a_{jl}(t,x)\partial_{x_{1j}}\partial_{x_{1l}}H(t,x).$$

We have

$$P \ge -3C_0H\left(4R_1^2(1-\mu t)^{-2}\sum_{j=1}^{n_1} x_{1j}^2 + 4R_2^2\mu^{-2}(1-\mu t)^{-4}\sum_{j=1}^{n_2} A_j^2\right.$$
$$\left. + R_3^2\mu^{-4}(1-\mu t)^{-6}\sum_{j=1}^{n_3} B_j^2\right) - C_0H(4R_1 + 4R_2\mu^{-2} + R_3\mu^{-4}),$$

$$LH/H = \sum_{j=1}^{n_3}(\mu R_1(1-\mu t)^{-2}x_{1j}^2 + 2R_2(1-\mu t)^{-3}x_{1j}A_j) + 3\mu R_3(1-\mu t)^{-4}A_j^2$$
$$+ 2R_3(1-\mu t)^{-5}x_{2j}B_j - 2R_3(1-\mu t)^{-4}x_{1j}B_j + 5\mu R_3(1-\mu t)^{-6}B_j^2$$

$$
\begin{aligned}
&- (2R_2(1-\mu t)^{-3}x_{1j}A_j + 2R_3\mu^{-1}(1-\mu t)^{-4}x_{1j}B_j) - 2R_3(1-\mu t)^{-5}x_{2j}B_j \\
&- a_j(t,x)(2R_1(1-\mu t)^{-1}x_{1j} - 2R_2\mu^{-1}(1-\mu t)^{-2}A_j + R_3\mu^{-2}(1-\mu t)^{-3}B_j)) \\
&+ \sum_{j=n_3+1}^{n_2} (2R_3(1-\mu t)^{-3}x_{2j}A_j + 3\mu R_2(1-\mu t)^{-4}A_j^2 - 2R_2(1-\mu t)^{-3}x_{1j}A_j \\
&- a_j(t,x)(2R_1(1-\mu t)^{-1}x_{1j} - 2R_2\mu^{-1}(1-\mu t)^{-2}A_j)) \\
&+ \sum_{j=n_2+1}^{n_1} (\mu R_1(1-\mu t)^{-2}x_{1j}^2 - a_j(t,x)2R_1(1-\mu t)^{-1}x_{1j}) + \alpha - a_0(t,x) + P/H \\
&\geq (\mu R_1 - 1)(1-\mu t)^{-2}\sum_{j=1}^{n_1} x_{1j}^2 + (3\mu R_2 - 1)(1-\mu t)^{-4}\sum_{j=1}^{n_2} A_j^2 \\
&+ (5\mu R_3 - 1)(1-\mu t)^{-6}\sum_{j=1}^{n_3} B_j^2 + \alpha - C_0 \\
&- C_0^2(R_1^2 n_1 + \mu^{-2}R_2^2 n_2 + 4^{-1}\mu^{-4}R_3^2 n_3) + P/H \\
&\geq 4(\mu R_1 - 12C_0R_1^2 - 1)\sum_{j=1}^{n_1} x_{1j}^2 + 16(3\mu R_2 - 12\mu^{-2}C_0R_2^2 - 1)\sum_{j=1}^{n_2} A_j^2 \\
&+ 64(5\mu R_3 - 3\mu^{-4}C_0R_3^3 - 1)\sum_{j=1}^{n_3} B_j^2 + \alpha - C_0(4R_1 + 4\mu^{-2}R_2 + \mu^{-4}R_3 + 1) \\
&- C_0^2(R_1^2 n_1 + \mu^{-2}R_2^2 + 4^{-1}\mu^{-4}R_3^2 n_3), \quad (t,x) \in \Pi_{[0,1/(2\mu)]},
\end{aligned}
$$

if coefficients of the equation (3.3.1) are bounded by the constant $C_0 > 0$. Hence, for any R_l, $l \in \{1,2,3\}$, we can choose so large μ and α that $LH/H \geq 0$ on $\Pi_{[0,1/(2\mu)]}$. Choose the constants R_l, $l \in \{1,2,3\}$, μ, and α in such a way that, in addition,

$$
\begin{aligned}
R_1 - R_2\mu^{-1}(1-(4\mu)^{-1}) - R_3\mu^{-2}(2^{-1} + (2\mu)^{-1} - (64\mu^2)^{-1}) &> b_1, \\
R_2(1-\mu^{-1}) - R_3\mu^{-1}(1 + (2\mu^2)^{-1} - (4\mu)^{-1}) &> b_2, \\
R_3 - R_3\mu^{-1}(1 + (2\mu)^{-1}) &> b_3,
\end{aligned}
$$

that is $\ln H - \alpha t > \sum_{l=1}^{3} b_l|x_l|^2$. Now, if we introduce the function $v := u/H$ on the set $\Pi_{(0,1/(2\mu)]}$, then, uniformly with respect to $t \in (0,1/(2\mu))$, there exists $\liminf_{|x|\to\infty} v(t,x) \geq 0$. The function v satisfies the equation

$$
\begin{aligned}
&\left(\partial_t - \sum_{j=1}^{n_2} x_{1j}\partial_{x_{2j}} - \sum_{j=1}^{n_3} x_{2j}\partial_{x_{3j}} - \sum_{j,l=1}^{n_1} a_{jl}(t,x)\partial_{x_{1j}}\partial_{x_{1l}}\right. \\
&\left. - \sum_{j=1}^{n_1} a_j^*(t,x)\partial_{x_{1j}} - a_0^*(t,x)\right) v(t,x) = f(t,x), \quad (t,x) \in \Pi_{(0,1/(2\mu)]},
\end{aligned}
$$

where

$$a_j^* := a_j + (2/H)\sum_{l=1}^{n_1} a_{jl}\partial_{x_{1l}} H, \quad j \in \{1, \dots, n_1\},$$

$$a_0^* := -LH/H, \quad f := Lu/H.$$

Since $f \geq 0$, $a_0^* \leq 0$, $\lim_{t\to 0} v(t,x) \geq 0$, $x \in \mathbb{R}^n$, by Theorem 3.16 we have $v(t,x) \geq 0$, so that $u(t,x) \geq 0$, $(t,x) \in \Pi_{(0,1/(2\mu))}$. Similarly we prove the non-negativity of u on the layers $\Pi_{(1/(2\mu),2/(2\mu)]}$, $\Pi_{(2/(2\mu),3/(2\mu)]}$ etc. □

3.3.3. Properties of the function Z. By Theorem 3.3, if the conditions A_{24} and A_{25} are satisfied, there exists a FSCP Z for the equation (3.3.1). If, in addition, A_{26} is satisfied, then there exists a FSCP Z^* for the adjoint equation (3.3.2). Let us consider some properties of the function Z. In all the properties below except Property 3.12 we assume A_{24}–A_{26}, whereas only the conditions A_{24} and A_{25} are needed for Property 3.12. Note that Properties 3.7 and 3.8 were mentioned in Theorem 3.3.

Property 3.7 (normality). *Let Z be a FSCP for the initial equation* (3.3.1), *and Z^* be a FSCP for the adjoint equation* (3.3.2). *Then*

$$Z^*(\tau,\xi;t,x) = Z(t,x;\tau,\xi), \quad 0 \leq \tau < t \leq T, \quad \{\xi, x\} \subset \mathbb{R}^n. \tag{3.3.6}$$

A FSCP Z satisfying this equality is called a *normal* FSCP (NFSCP).

Proof. By virtue of the estimates (3.2.78) for Z and similar estimates for Z^*, the formula (3.3.5) is valid. In this formula we set $u(\theta,y) = Z(\theta,y;\tau,\xi)$, $v(\theta,y) = Z^*(\theta,y;t,x)$, $t_1 = \tau+\varepsilon$, and $t_2 = t-\varepsilon$ where ε is a sufficiently small positive number. Then we obtain the equality

$$\int_{\mathbb{R}^n} Z^*(\tau+\varepsilon,y;t,x)Z(\tau+\varepsilon,y;\tau,\xi)dy = \int_{\mathbb{R}^n} Z^*(t-\varepsilon,y;t,x)Z(t-\varepsilon,y;\tau,\xi)dy,$$

which, after passing to the limit, as $\varepsilon \to 0$, implies the required equality (3.3.6). □

Property 3.8 (the convolution formula). *The function Z is a solution of the functional equation*

$$Z(t,x;\tau,\xi) = \int_{\mathbb{R}^n} Z(t,x;\lambda,y)Z(\lambda,y;\tau,\xi)dy,$$

$$0 \leq \tau < t \leq T, \quad \{x,\xi\} \subset \mathbb{R}^n. \tag{3.3.7}$$

Proof. Just as in the proof of Property 3.7, we come to the equality

$$\int_{\mathbb{R}^n} Z^*(\lambda,y;t,x)Z(\lambda,y;\tau,\xi)dy = \int_{\mathbb{R}^n} Z^*(t-\varepsilon,y;t,x)Z(t-\varepsilon,y;\tau,\xi)dy. \tag{3.3.8}$$

The equality (3.3.7) is obtained when we pass to the limit in (3.3.8), as $\varepsilon \to 0$, and use the formula (3.3.6). □

The equation (3.3.7) is called the *Chapman-Kolmogorov equation.* It expresses an important fact that a stochastic process is of the Markov type (a process without aftereffect). An investigation of the process can be based on the equation (3.3.7).

Property 3.9 (the uniqueness of the NFSCP). *There exists only one NFSCP for the equation* (3.3.1) *satisfying the estimates* (3.2.78).

Proof. Let Z_1 and Z_2 be two NFSCPs for the equation (3.3.1), both satisfying (3.2.78). We use the formula (3.3.5) setting $u(\theta, y) = Z_1(\theta, y; \tau, \xi)$, $v(\theta, y) = Z_2(t, x; \theta, y)$. Then

$$\int_{\mathbb{R}^n} Z_1(t_2, y; \tau, \xi) Z_2(t, x; t_2, y) dy = \int_{\mathbb{R}^n} Z_1(t_1, y; \tau, \xi) Z_2(t, x; t_1, y) dy.$$

Since t_1 and t_2 from the interval (τ, t) are arbitrary, the last equality means that the function

$$\int_{\mathbb{R}^n} Z_1(\theta, y; \tau, \xi) Z_2(t, x; \theta, y) dy, \quad \theta \in (\tau, t), \quad \{x, \xi\} \subset \mathbb{R}^n,$$

does not depend on θ. Denote this function by $\Phi(t, x; \tau, \xi)$. Thus,

$$\Phi(t, x; \tau, \xi) = \int_{\mathbb{R}^n} Z_1(\theta, y; \tau, \xi) Z_2(t, x; \theta, y) dy. \tag{3.3.9}$$

Letting in the equality (3.3.9) first $\theta \to \tau$, and then $\theta \to t$, we find that

$$\Phi(t, x; \tau, \xi) = Z_2(t, x; \tau, \xi) = Z_1(t, x; \tau, \xi), \quad 0 \le \tau < t \le T, \quad \{x, \xi\} \subset \mathbb{R}^n. \quad \square$$

Our diffusion process is characterized by the diffusion matrix $A(t, x) := (a_{jl}(t, x))_{j=1,l=1}^{n_1,n_1}$, $(t, x) \in \Pi_{(0,T]}$, and the drift vector $a(t, x) := (a_1(t, x), \ldots, a_{n_1}(t, x))$, $(t, x) \in \Pi_{(0,T]}$, whose elements are the appropriate coefficients of the equation (3.3.1). Let us express these important characteristics via the transition density of the stochastic process, that is the FSCP Z.

Property 3.10 (a representation of the diffusion matrix and the drift vector via the function Z). *The following formulas are valid:*

$$a(t, x) = \lim_{\tau \to t} \left((t - \tau)^{-1} \int_{\mathbb{R}^n} (y_1 - x_1) Z(t, x; \tau, y) dy \right),$$

$$A(t, x) = \left(2^{-1} \lim_{\tau \to t} \left((t - \tau)^{-1} \int_{\mathbb{R}^n} (y_{1j} - x_{1j})(y_{1l} - x_{1l}) Z(t, x; \tau, y) dy \right) \right)_{j=1,l=1}^{n_1,n_1},$$

$$(t, x) \in \Pi_{(0,T]}.$$

Proof. It is sufficient to give the proof for the first coordinate a_1 of the vector a and the element a_{11} of the matrix A. Let us prove, for $(t,x) \in \Pi_{(0,T]}$, the formulas

$$a_1(t,x) = \lim_{\tau\to t}\left((t-\tau)^{-1}\int_{\mathbb{R}^n}(y_{11}-x_{11})Z(t,x;\tau,y)dy\right), \tag{3.3.10}$$

$$a_{11}(t,x) = 2^{-1}\lim_{\tau\to t}\left((t-\tau)^{-1}\int_{\mathbb{R}^n}(y_{11}-x_{11})^2 Z(t,x;\tau,y)dy\right). \tag{3.3.11}$$

Their proof is based on the formula (3.3.5), in which we set $u(\theta,y) = y_{11}-x_{11}$ and $v(\theta,y) = Z(t,x;\theta,y)$. Then we obtain the equality

$$-\int_{t_1}^{t_2} d\theta \int_{\mathbb{R}^n} Z(t,x;\theta,y)(a_1(\theta,y) + a_0(\theta,y)(y_{11}-x_{11}))dy$$
$$= \int_{\mathbb{R}^n}(y_{11}-x_{11})Z(t,x;\theta,y)\Big|_{\theta=t_1}^{t_2} dy. \tag{3.3.12}$$

Next in (3.3.12) we put $t_1 = \tau$, $t_2 = t-\varepsilon$, pass to the limit, as $\varepsilon \to 0$, divide the result by $t-\tau$ and obtain the equality

$$(t-\tau)^{-1}\int_{\tau}^{t} d\theta \int_{\mathbb{R}^n} Z(t,x;\theta,y)a_1(\theta,y)dy = (t-\tau)^{-1}\int_{\mathbb{R}^n}(y_{11}-x_{11})Z(t,x;\tau,y)dy$$
$$-(t-\tau)^{-1}\int_{\tau}^{t} d\theta \int_{\mathbb{R}^n} Z(t,x;\theta,y)a_0(\theta,y)(y_{11}-x_{11})dy. \tag{3.3.13}$$

The equality (3.3.10) follows directly from (3.3.13), since the limit of the left-hand side, as $\tau \to t$, coincides with $a_1(t,x)$ due to properties of the FSCP Z and the mean value theorem for integrals, while the second summand of the right-hand side of the equality (3.3.13) tends to zero by our assumptions regarding the function a_0.

In order to prove (3.3.11), in the formula (3.3.5) we set $u(\theta,y) = (y_{11}-x_{11})^2$ and $v(\theta,y) = Z(t,x;\theta,y)$. Then this formula can be re-written as

$$-\int_{t_1}^{t_2} d\theta \int_{\mathbb{R}^n} Z(t,x;\theta,y)(2a_{11}(\theta,y) + 2a_1(\theta,y)(y_{11}-x_{11}) + a_0(\theta,y)(y_{11}-x_{11})^2)dy$$
$$= \int_{\mathbb{R}^n}(y_{11}-x_{11})^2 Z(t,x;\theta,y)\Big|_{\theta=t_1}^{t_2} dy.$$

Then we repeat the reasoning from the previous case and come to the equality

$$(t-\tau)^{-1}\int_{\tau}^{t} d\theta \int_{\mathbb{R}^n} Z(t,x;\theta,y)a_{11}(\theta,y)dy = (2(t-\tau))^{-1}\int_{\mathbb{R}^n}(y_{11}-x_{11})^2 Z(t,x;\tau,y)dy$$

$$-(t-\tau)^{-1}\int_{\tau}^{t} d\theta \int_{\mathbb{R}^n}(a_1(\theta,y)(y_{11}-x_{11})+2^{-1}a_0(\theta,y)(y_{11}-x_{11})^2)Z(t,x;\theta,y)dy,$$

which, after the above arguments, implies the formula (3.3.11). □

Similar reasoning leads to a representation via the function Z for the function $-a_0$, the absorption coefficient of the diffusion process whose evolution is described by the Fokker-Planck-Kolmogorov equation (3.3.1).

Property 3.11 (a representation of the coefficient a_0 via the function Z). *The following formula is valid:*

$$a_0(t,x) = \lim_{\tau\to t}\left((t-\tau)^{-1}\left(\int_{\tau}^{t} d\theta \int_{\mathbb{R}^n} Z(t,x;\theta,y)dy - 1\right)\right), \quad (t,x)\in \Pi_{(0,T]}.$$

Property 3.12 (positivity of the function Z). *The following inequality is valid*

$$Z(t,x;\tau,\xi) > 0, \quad 0 \le \tau < t \le T, \quad \{x,\xi\}\subset \mathbb{R}^n.$$

Proof. Let g be a continuous non-negative function with a compact support. Consider the function

$$v(t,x) := \int_{\mathbb{R}^n} Z(t,x;\tau,\xi)g(\xi)d\xi, \quad (t,x)\in \Pi_{(\tau,T]}. \tag{3.3.14}$$

The function v satisfies the conditions of Theorem 3.16. Indeed, it is a solution of the equation (3.3.1), that is $(Lv)(t,x)=0$, $(t,x)\in\Pi_{(\tau,T]}$, satisfies the conditions $v(\tau,x) = g(x) \ge 0$, $x\in\mathbb{R}^n$, and $\lim_{|x|\to\infty} v(t,x) \ge 0$, $t\in(\tau,T]$. Therefore, by Theorem 3.16 we have $v(t,x)\ge 0$, $(t,x)\in\Pi_{(\tau,T]}$.

Now we take a delta-like sequence of functions g_ν, $\nu\ge 1$, such that $g_\nu(\xi)=0$ for $|\xi-\xi^0| > 1/\nu$, and $\int_{\mathbb{R}^n} g_\nu(\xi)d\xi = 1$, and denote by v_ν the function (3.3.14) corresponding to $g=g_\nu$ and $\tau=\tau^0$. Then for arbitrary fixed points (t^0,x^0) and (τ^0,ξ^0), with $\tau^0 < t^0$, we have $\lim_{\nu\to\infty} v_\nu(t^0,x^0) = Z(t^0,x^0;\tau^0,\xi^0)$. Since $v_\nu(t^0,x^0)\ge 0$ for all $\nu \ge 1$, we find that $Z(t^0,x^0;\tau^0,\xi^0)\ge 0$. Finally, observing that for every fixed point (τ,ξ) the function $Z(t,x;\tau,\xi)$, $(t,x)\in\Pi_{(\tau,T]}$, is not a constant identically, and using the strong maximum principle, we come to the strict positivity of the function Z. □

Property 3.13 (a lower estimate of the function Z). *There exists such a number $\Delta \in (0,T)$ that for any $t_0 \in [0, T-\Delta]$, $(t,x) \in \Pi_{(t_0,t_0+\Delta]}$, and $\delta \in (0, t-t_0)$ there exist such numbers $\omega > 0$ and $\gamma > 0$ that*

$$Z(t,x;\tau,\xi) \geq \omega \exp\{-\gamma|\xi|^2\}, \quad (\tau,\xi) \in \Pi_{[t_0,t-\delta]}. \tag{3.3.15}$$

Proof. Set

$$v(\tau,\xi) := \exp\Bigg\{-\beta\Bigg((4s)^{-1}\sum_{j=1}^{n_1}|x_{1j}-\xi_{1j}|^2 + 3s^{-3}\sum_{j=1}^{n_2}|x_{2j}+2^{-1}s(x_{1j}+\xi_{1j})-\xi_{2j}|^2$$
$$+180s^{-5}\sum_{j=1}^{n_3}|x_{3j}+2^{-1}s(x_{2j}+\xi_{2j})+12^{-1}s^2(x_{1j}-\xi_{1j})-\xi_{3j}|^2\Bigg)e^{\varepsilon s}+\alpha/s\Bigg\}$$

where $s := t - \tau - \delta/2$, $\alpha > 0$, $\beta > 1$, $\varepsilon > 1$. It is easy to verify that

$$(4s)^{-1}\sum_{j=1}^{n_1}|x_{1j}-\xi_{1j}|^2 + 3s^{-3}\sum_{j=1}^{n_2}|x_{2j}+2^{-1}s(x_{1j}+\xi_{1j})-\xi_{2j}|^2$$
$$+ 180s^{-5}\sum_{j=1}^{n_3}|x_{3j}+2^{-1}s(x_{2j}+\xi_{2j})+12^{-1}s^2(x_{1j}-\xi_{1j})-\xi_{3j}|^2$$
$$\geq s^{-1}\Bigg((1/4 - 3s/2 - 63s^2/4 - 45s^3/2)\sum_{j=1}^{n_1}|x_{1j}-\xi_{1j}|^2$$
$$+ (3 - 189s/2 - 45s^2 - 15s^3/2)\sum_{j=1}^{n_2}|x_{2j}-\xi_{2j}|^2$$
$$+ (180 - 270s - 15s^2)\sum_{j=1}^{n_3}|x_{3j}-\xi_{3j}|^2$$
$$- 3s\sum_{j=1}^{n_2}|x_{1j}|^2 - 180(s+12^{-1}s^3)\sum_{j=1}^{n_3}|x_{2j}|^2\Bigg),$$

if $0 < s < 1$. Choose $\Delta \in (0,T)$ in such a way that for any $s \in (0,\Delta)$

$$1/4 - 3s/2 - 63s^2/4 - 45s^3/2 \geq \delta_1 > 0, \quad 3 - 189s/2 - 45s^2 - 15s^3/2 \geq \delta_1,$$
$$180 - 270s - 15s^2 \geq \delta_1.$$

Hence, for arbitrary $t_0 \in [0, T-\Delta]$, $(t,x) \in \Pi_{(t_0,t_0+\Delta]}$ and $\delta \in (0, t-t_0)$ we may consider the function v on the layer $\Pi_{[t_0,t-\delta/2)}$, and

$$v(\tau,\xi) \leq \exp\{-(\delta_1/s)|x-\xi|^2 + 195|x|^2 + \alpha/s\}, \quad (\tau,\xi) \in \Pi_{[t_0,t-\delta/2)}, \tag{3.3.16}$$
$$v(\tau,\xi) \geq \exp\{-\gamma_1(|x|^2+|\xi|^2)\}, \quad (\tau,\xi) \in \Pi_{[t_0,t-\delta)}, \tag{3.3.17}$$

with some γ_1 which depends on δ.

Depending on the coefficients of the equation (3.3.2), we can choose such α and β that

$$(L^*v)(\tau,\xi)\le 0,\quad (\tau,\xi)\in\Pi_{[t_0,t-\delta/2)}. \tag{3.3.18}$$

Indeed,

$$\sum_{j,l=1}^{n_1} a_{jl}(\tau,\xi)\partial_{\xi_{1j}}\partial_{\xi_{1l}}v(\tau,\xi)=\beta e^{\varepsilon s}v(\tau,\xi)\left(\beta e^{\varepsilon s}\sum_{j,l=1}^{n_1}a_{jl}(\tau,\xi)\sigma_j\sigma_l-\sum_{j=1}^{n_1}a_{jj}(\tau,\xi)\tilde{\sigma}_j\right)$$

where

$$\begin{aligned}
&\sigma_j:=(2s)^{-1}(x_{1j}-\xi_{1j})+3s^{-2}A_j+30s^{-3}B_j,\quad \tilde{\sigma}_j:=9/(2s),\quad j\in\{1,\dots,n_3\};\\
&\sigma_j:=(2s)^{-1}(x_{1j}-\xi_{1j})+3s^{-2}A_j,\quad \tilde{\sigma}_j:=2/s,\quad j\in\{n_3+1,\dots,n_2\};\\
&\sigma_j:=(2s)^{-1}(x_{1j}-\xi_{1j}),\quad \tilde{\sigma}_j:=1/(2s),\quad j\in\{n_2+1,\dots,n_1\};\\
&A_j:=x_{2j}+2^{-1}s(x_{1j}+\xi_{1j})-\xi_{2j},\quad j\in\{1,\dots,n_2\};\\
&B_j:=x_{3j}+2^{-1}s(x_{2j}+\xi_{2j})+12^{-1}s^2(x_{1j}-\xi_{1j})-\xi_{3j},\quad j\in\{1,\dots,n_3\}.
\end{aligned}$$

By the condition (3.3.3),

$$\sum_{j,l=1}^{n_1} a_{jl}(\tau,\xi)\partial_{\xi_{1j}}\partial_{\xi_{1l}}v(\tau,\xi)\ge\beta e^{\varepsilon s}v(\tau,\xi)\left(\beta e^{\varepsilon s}\delta_0|\sigma|^2-\sum_{j=1}^{n_1}a_{jj}(\tau,\xi)\tilde{\sigma}_j\right).$$

Choose $\beta>1/\delta_0$ and take into account that $\varepsilon>1$ i $e^{\varepsilon s}>1$. Then

$$\begin{aligned}
&(\beta e^{\varepsilon s}v(\tau,\xi))^{-1}(L^*v)(\tau,\xi)\\
&\le-\Bigg(-(2s)^{-2}\sum_{j=1}^{n_1}|x_{1j}-\xi_{1j}|^2+3s^{-3}\sum_{j=1}^{n_2}(x_{1j}+\xi_{1j})A_j-9s^{-4}\sum_{j=1}^{n_2}A_j^2\\
&\quad+180s^{-5}\sum_{j=1}^{n_3}(x_{2j}+\xi_{2j})B_j+60s^{-4}\sum_{j=1}^{n_3}(x_{1j}-\xi_{1j})B_j\\
&\quad-900s^{-6}\sum_{j=1}^{n_3}B_j^2+(4s)^{-1}\varepsilon\sum_{j=1}^{n_1}|x_{1j}-\xi_{1j}|^2+3\varepsilon s^{-3}\sum_{j=1}^{n_2}A_j^2+180\varepsilon s^{-5}\sum_{j=1}^{n_3}B_j^2\Bigg)\\
&\quad+6s^{-3}\sum_{j=1}^{n_2}\xi_{1j}A_j-180s^{-4}\sum_{j=1}^{n_3}\xi_{1j}B_j+360s^{-5}\sum_{j=1}^{n_3}\xi_{2j}B_j\\
&\quad+\sum_{j=1}^{n_3}\Bigg(\Bigg(-2\sum_{l=1}^{n_1}\partial_{\xi_{1l}}a_{jl}(\tau,\xi)+a_j(\tau,\xi)\Bigg)((2s)^{-1}(x_{1j}-\xi_{1j})-3s^{-2}A_j+30s^{-3}B_j)\\
&\quad-\beta e^{\varepsilon s}\delta_0((2s)^{-1}(x_{1j}-\xi_{1j})-3s^{-2}A_j+30s^{-3}B_j)^2+a_{jj}(\tau,\xi)9(2s)^{-1}\Bigg)
\end{aligned}$$

$$
+\sum_{j=n_3+1}^{n_2}\left(\left(-2\sum_{l=1}^{n_1}\partial_{\xi_{1l}}a_{jl}(\tau,\xi)+a_j(\tau,\xi)\right)((2s)^{-1}(x_{1j}-\xi_{1j})-3s^{-2}A_j)\right.
$$

$$
\left.-\beta e^{\varepsilon s}\delta_0((2s)^{-1}(x_{1j}-\xi_{1j})-3s^{-2}A_j)^2+2s^{-1}a_{jj}(\tau,\xi)\right)
$$

$$
+\sum_{j=n_2+1}^{n_1}\left(\left(-2\sum_{l=1}^{n_1}\partial_{\xi_{1l}}a_{jl}(\tau,\xi)+a_j(\tau,\xi)\right)(2s)^{-1}(x_{1j}-\xi_{1j})\right.
$$

$$
\left.-\beta e^{\varepsilon s}\delta_0(2s)^{-2}(x_{1j}-\xi_{1j})^2+a_{jj}(\tau,\xi)(2s)^{-1}\right)
$$

$$
-(\beta e^{\varepsilon s}s^2)^{-1}\left(\alpha-s^2\left(\sum_{j,l=1}^{n_1}\partial_{\xi_{1j}}\partial_{\xi_{1l}}a_{jl}(\tau,\xi)-\sum_{j=1}^{n_1}\partial_{\xi_{1j}}a_j(\tau,\xi)+a_0(\tau,\xi)\right)\right)
$$

$$
\leq\sum_{j=1}^{n_1}(2s)^{-2}(|x_{1j}-\xi_{1j}|^2-|x_{1j}-\xi_{1j}|^2)
$$

$$
+\sum_{j=1}^{n_2}(9s^{-4}A_j^2-9s^{-4}A_j^2-3s^{-3}(x_{1j}+\xi_{1j})A_j+6s^{-3}\xi_{1j}A_j-3s^{-3}(x_{1j}-\xi_{1j})A_j)
$$

$$
+\sum_{j=1}^{n_3}(900s^{-6}B_j^2-900s^{-6}B_j^2-180s^{-5}(x_{2j}+\xi_{2j})B_j-60s^{-4}(x_{1j}-\xi_{1j})B_j
$$

$$
-180s^{-4}\xi_{1j}B_j+360s^{-5}\xi_{2j}B_j-30s^{-4}(x_{1j}-\xi_{1j})B_j+180s^{-5}A_jB_j)
$$

$$
+\sum_{j=1}^{n_1}\left((1-\varepsilon)(4s)^{-1}(x_{1j}-\xi_{1j})^2+9(4s)^{-1}\left(-2\sum_{l=1}^{n_1}\partial_{\xi_{1l}}a_{jl}(\tau,\xi)+a_j(\tau,\xi)\right)^2\right.
$$

$$
\left.+9(2s)^{-1}a_{jj}(\tau,\xi)\right)+\sum_{j=1}^{n_2}(1-\varepsilon)3s^{-3}A_j^2+\sum_{j=1}^{n_3}(1-\varepsilon)180s^{-5}B_j^2
$$

$$
-(\beta e^{\varepsilon s}s^2)^{-1}\left(\alpha-s^2\left(\sum_{j,l=1}^{n_1}\partial_{\xi_{1j}}\partial_{\xi_{1l}}a_{jl}(\tau,\xi)-\sum_{j=1}^{n_1}\partial_{\xi_{1j}}a_j(\tau,\xi)+a_0(\tau,\xi)\right)\right)
$$

$$
\leq-(\varepsilon-1)\left(\sum_{j=1}^{n_1}(4s)^{-1}(x_{1j}-\xi_{1j})^2+\sum_{j=1}^{n_2}3s^{-3}(x_{2j}+2^{-1}s(x_{1j}+\xi_{1j})-\xi_{2j})^2\right.
$$

$$
\left.+\sum_{j=1}^{n_3}180s^{-5}(x_{3j}+2^{-1}s(x_{2j}+\xi_{2j})+12^{-1}s^2(x_{1j}-\xi_{1j})-\xi_{3j})^2\right)
$$

$$
-(\beta e^{\varepsilon s}s^2)^{-1}\left(\alpha-s^2\left(\sum_{j,l=1}^{n_1}\partial_{\xi_{1j}}\partial_{\xi_{1l}}a_{jl}(\tau,\xi)-\sum_{j=1}^{n_1}\partial_{\xi_{1j}}a_j(\tau,\xi)+a_0(\tau,\xi)\right)\right)
$$

$$+9(4s)^{-1}\sum_{j=1}^{n_1}\left(2a_{jj}(\tau,\xi)+\left(-2\sum_{l=1}^{n_1}\partial_{\xi_{1l}}a_{jl}(\tau,\xi)+a_j(\tau,\xi)\right)^2\right).$$

If we take into account the condition A_{26}, choosing $\alpha > 0$ appropriately we can make the last sum negative. This proves the inequality (3.3.18).

Set

$$\lambda := \min\{Z(t,x;\tau,\xi)|\xi \in B(x,R), \tau \in [t_0, t-\delta/2]\} \exp\{-2\alpha/\delta - 195|x|^2\},$$
$$R > \sqrt{\alpha/\delta_1}.$$

It follows from positivity of Z that $\lambda > 0$. Now we use Theorem 3.17 for the equation (3.3.2), for the function $u(\tau,\xi) := \lambda v(\tau,\xi) - Z(t,x;\tau,\xi)$, $(\tau,\xi) \in \Pi_{[t_0,t-\delta/2)}$ on $[t_0, t-\delta/2) \times (\mathbb{R}^n \setminus B(x,R))$. By the inequality (3.3.18) and the equality (3.3.6), $(L^*u)(\tau,\xi) \le 0$, $(\tau,\xi) \in \Pi_{[t_0,t-\delta/2)}$. The condition $\limsup\limits_{(\tau,\xi)\to(\tau^0,\xi^0)} u(\tau,\xi) \le 0$, for each point $(\tau^0,\xi^0) \in \partial([t_0,t-\delta/2) \times (\mathbb{R}^n \setminus B(x,R)) \setminus \{t = t_0\}$, is satisfied, since, firstly, by the inequality (3.3.16) and the definition of λ,

$$u(\tau,\xi) \le \lambda \exp\{195|x|^2\} - Z(t,x;\tau,\xi) \le 0, \quad |x-\xi| = R, \quad \tau \in [t_0, t-\delta/2);$$

secondly, by (3.3.16),

$$\lim_{(\tau,\xi)\to(t-\delta/2,\xi^0)} v(\tau,\xi) = 0 \qquad \text{and} \qquad \lim_{(\tau,\xi)\to(t-\delta/2,\xi^0)} u(\tau,\xi) \le 0,$$

$\xi^0 \in \mathbb{R}^n \setminus B(x,R)$. Finally, it follows from (3.2.78) and (3.3.16) that

$$\limsup_{|\xi|\to\infty} u(\tau,\xi) = 0, \quad \tau \in [t_0, t-\delta/2).$$

Thus, we obtain that $u(\tau,\xi) \le 0$, $(\tau,\xi) \in [t_0, t-\delta/2) \times (\mathbb{R}^n \setminus B(x,R))$. By the definition of λ and (3.3.16), $u \le 0$ also on $[t_0, t-\delta] \times B(x,R)$. Therefore $u \le 0$ on $\Pi_{[t_0,t-\delta]}$, or $Z(t,x;\tau,\xi) \ge \lambda v(\tau,\xi)$, $(\tau,\xi) \in \Pi_{[t_0,t-\delta]}$, and together with the inequality (3.3.17) this implies the required estimate (3.3.15). □

3.3.4. Uniqueness of solutions of the Cauchy problem. Widder's theorem. Theorems 3.10 and 3.12 described correctness classes, hence also uniqueness classes of solutions of the Cauchy problem for the equation (3.1.71). Here we consider three other types of the uniqueness theorems – a Tikhonov type theorem with pointwise bounds for solutions, a Tikhonov type theorem with weighted L_1-spaces, and a generalized Widder's theorem.

In the paper [229] Widder proved the following theorem: a non-negative solution of the heat equation satisfying the zero initial condition equals zero identically. In fact this theorem can be strengthened as follows: if two non-negative solutions of the heat equation satisfy the same initial conditions, then the solutions coincide. Such a strengthened version of Widder's theorem will be proved for the Fokker-Planck-Kolmogorov equation.

Theorem 3.19. *Suppose that the condition C_{22} is satisfied, B is a positive number, and b_l, $l \in \{1,2,3\}$, are some non-negative numbers. Then a solution u of the equation* (3.3.1) *satisfying the zero initial condition, as well as the condition*

$$|u(t,x)| \leq B \exp\left\{\sum_{l=1}^{3} b_l |x_l|^2\right\}, \quad (t,x) \in \Pi_{(0,T]}, \tag{3.3.19}$$

equals zero identically.

Proof. Applying Theorem 3.18 to solutions u and $-u$ of the equation (3.3.1), which satisfy the zero initial condition and the condition (3.3.19), we find that $u(t,x) = 0$, $(t,x) \in \Pi_{(0,T]}$. □

Theorem 3.20. *Suppose that the conditions A_{24}–A_{26} are satisfied, and b is a non-negative number. Then a solution u of the equation* (3.3.1) *satisfying the zero initial condition, as well as the condition*

$$\int_0^T dt \int_{\mathbb{R}^n} \exp\{-b|x|^2\}|u(t,x)|dx < \infty, \tag{3.3.20}$$

equals zero identically.

Proof. First we shall prove the existence of such a number $\mu \in (0,T]$ depending on the number b from the condition (3.3.20) and on the coefficients of the equation (3.3.1) that $u = 0$ on $\Pi_{(0,\mu]}$.

As in the proof of Lemma 2.16 and Theorem 3.8, using the Green-Ostrogradsky formula (3.3.4), we obtain the following formula similar to (3.2.47):

$$\begin{aligned} u(t,x) &= \int_{\mathbb{R}^n} Z(t,x;h,\xi)\zeta_R(\xi)d\xi - \int_h^t d\tau \int_{K_R} L^*(Z^*(\tau,\xi;t,x)\zeta_R(\xi))u(\tau,\xi)d\xi \\ &=: I_{1h}^{(R)} + I_{2h}^{(R)}, \quad (t,x) \in (0,T] \times B_{R_0/4}. \end{aligned} \tag{3.3.21}$$

Here R_0 is an arbitrary fixed positive number, $R \geq R_0$; $K_R := B_{3R/4} \setminus B_{R/2}$; $\zeta_R(\xi) = \zeta(|\xi|/R)$ where ζ is a non-increasing infinitely differentiable function on $[0,\infty)$, such that $\zeta = 1$ on $[0,1/2]$ and $\zeta = 0$ on $[3/4,\infty)$.

Due to the estimate (3.2.78) and the fact that the integration in $I_{1h}^{(R)}$ is performed actually over $B_{3R/4}$, we have $\lim\limits_{h\to 0} I_{1h}^{(R)} = 0$. Therefore, passing in (3.3.21) to the limit, as $h \to 0$, we find that

$$\begin{aligned} u(t,x) &= -\int_0^t d\tau \int_{K_R} L^*(Z^*(\tau,\xi;t,x)\zeta_R(\xi))u(\tau,\xi)d\xi \\ &=: -I_2^{(R)}, \qquad (t,x) \in (0,T] \times B_{R_0/4}. \end{aligned} \tag{3.3.22}$$

Let us prove that

$$\lim_{R\to\infty} I_2^{(R)} = 0, \quad (t,x) \in (0,\mu] \times B_{R_0/4}, \tag{3.3.23}$$

if $\mu \in (0,T]$ is chosen appropriately.

By the equality (3.3.6), the estimates (3.2.78), and properties of the function ζ_R for $R \geq 1$, we get the estimate

$$|L^*(Z^*(\tau,\xi;t,x)\zeta_R(\xi))| \leq C(t-\tau)^{-M-1/2}\exp\{-c\rho(t-\tau,x,\xi)\}. \tag{3.3.24}$$

For the function ρ we use the inequality

$$\rho(t-\tau,x,\xi) \geq (t-\tau)^r (R/4)^2,$$
$$0 \leq \tau < t \leq T, \quad x \in B_{R_0/4}, \quad \xi \in \mathbb{R}^n \setminus B_{R/2}, \tag{3.3.25}$$

where $r = -1$, if $t-\tau \leq 1$, and $r = -5$, if $t-\tau > 1$. This inequality is proved like (3.2.50).

Using the inequalities (3.3.24) and (3.3.25) we have

$$|I_2^{(R)}| \leq C \int_0^t d\tau \int_{K_R} (t-\tau)^{-M-1/2}\exp\{-c(t-\tau)^r(R/4)^2\}|u(\tau,\xi)|d\xi.$$

Since

$$\begin{aligned}
&(t-\tau)^{-M-1/2}\exp\{-c(t-\tau)^r(R/4)^2\}\\
&\quad = (R/4)^{(2M+1)/r}((t-\tau)^{r/2}(R/4))^{-(2M+1)/r}\exp\{-c(t-\tau)^r(R/4)^2\}\\
&\quad \leq C\exp\{-c_1\mu^r R^2\}, \qquad 0 \leq \tau < t \leq \mu, \quad R \geq 1, \quad c_1 > 0,
\end{aligned}$$

we get

$$\begin{aligned}
|I_2^{(R)}| &\leq C\exp\{-c_1\mu^r R^2\} \int_0^t d\tau \int_{K_R} \exp\{-b|\xi|^2\}|u(\tau,\xi)|\exp\{b|\xi|^2\}d\xi\\
&\leq C\exp\{-(c_1\mu^r - 9b/16)R^2\} \int_0^t d\tau \int_{\mathbb{R}^n} \exp\{-b|\xi|^2\}|u(\tau,\xi)|d\xi.
\end{aligned}$$

By the condition (3.3.20), this implies the relation (3.3.23), if

$$\mu < \min((16c_1/9b)^{-1/r}, T).$$

Passing in (3.3.22) to the limit, as $R \to \infty$, using (3.3.23) and the fact that $R_0 > 0$ is arbitrary, we find that $u = 0$ on $\Pi_{(0,\mu]}$. Similarly we prove that $u = 0$ on $\Pi_{(\mu,2\mu]}$, $\Pi_{(2\mu,3\mu]}$, and so on. □

Theorem 3.21. *Suppose that the conditions A_{24}–A_{26} are satisfied. Then the Cauchy problem for the equation* (3.3.1) *cannot have more than one non-negative solution.*

Proof. Let u be a non-negative solution of the equation (3.3.1). Let us prove that

$$\int_0^{T_0} dt \int_{\mathbb{R}^n} \exp\{-b|x|^2\}|u(t,x)|dx < \infty, \tag{3.3.26}$$

with some $b > 0$ and $T_0 \in (0,T)$.

Set

$$v_R(t,x) := \int_{\mathbb{R}^n} Z(t,x;\tau,\xi)\zeta_R(\xi)u(\tau,\xi)d\xi, \quad (t,x) \in \Pi_{(\tau,T]},$$

where $0 \le \tau < T$, $R > 0$, ζ_R is the function from the proof of Theorem 3.20. Let us verify the conditions for the validity of the maximum principle in the form of Theorem 3.16 for the difference $u - v_R =: U_R$. Obviously, $(LU_R)(t,x) \geq 0$, $(t,x) \in \Pi_{(\tau,T]}$. The condition that $\lim\limits_{(t,x)\to(t^0,x^0)} U_R(t,x) \geq 0$, for every point $(t^0,x^0) \in \partial(\Pi_{(\tau,T]}) \setminus \{t = T\}$, is satisfied because

$$\lim_{t\to\tau} \int_{\mathbb{R}^n} Z(t,x;\tau,\xi)\zeta_R(\xi)u(\tau,\xi)d\xi = \zeta_R(x)u(\tau,x) \le u(\tau,x),$$

$0 \le \tau < T$, $x \in \mathbb{R}^n$. Since the integration in v_R is performed only over $B_{3R/4}$, using the estimate (3.2.78) we get

$$\lim_{|x|\to\infty} \int_{\mathbb{R}^n} Z(t,x;\tau,\xi)\zeta_R(\xi)u(\tau,\xi)d\xi = 0, \quad t \in (\tau,T].$$

This means that the last condition is fulfilled: uniformly with respect to $t \in (0,T)$, there exists $\liminf\limits_{|x|\to\infty} U_R(t,x) \geq 0$.

Hence, by Theorem 3.16, $U_R \geq 0$ on $\Pi_{(\tau,T]}$, that is for each $\tau \in [0,T)$,

$$u(t,x) \geq \int_{\mathbb{R}^n} Z(t,x;\tau,\xi)\zeta_R(\xi)u(\tau,\xi)d\xi, \quad (t,x) \in \Pi_{(\tau,T]}.$$

Since this inequality is valid for any $R > 0$, and the integrand is non-negative (see Property 3.12), for $R \to \infty$ we obtain that

$$u(t,x) \geq \int_{\mathbb{R}^n} Z(t,x;\tau,\xi)u(\tau,\xi)d\xi, \quad (t,x) \in \Pi_{(\tau,T]}. \tag{3.3.27}$$

Let us integrate the inequality (3.3.27) with respect to $\tau \in (t_0, t-\delta)$, $\delta \in (0, t-t_0)$. If we take $t_0 = 0$, $t = \Delta$, and use the estimate (3.3.15), we get

$$(\Delta - \delta)u(\Delta, x) \geq \int_0^{\Delta-\delta} d\tau \int_{\mathbb{R}^n} \omega \exp\{-\gamma|\xi|^2\} u(\tau, \xi) d\xi,$$

that is

$$\int_0^{\Delta-\delta} d\tau \int_{\mathbb{R}^n} \exp\{-\gamma|\xi|^2\} u(\tau, \xi) d\xi < \infty.$$

Now, if we set $t_0 = \Delta - \delta$, $t = 2\Delta - \delta$, similarly we prove the convergence of the integral, in which the integration in τ is performed on the interval $(\Delta - \delta, 2(\Delta - \delta))$. We continue this process until $t = j\Delta - (j-1)\delta < T$, $j \geq 1$. Finally, we set $t_0 = T - \Delta$ and $t = T$. Combining the above results we find that

$$\int_0^{T-\delta} d\tau \int_{\mathbb{R}^n} \exp\{-\gamma|\xi|^2\} u(\tau, \xi) d\xi < \infty,$$

that is (3.3.26) holds with $T_0 = T - \delta$, $b = \gamma$.

Using Theorem 3.20 we obtain the uniqueness on $\Pi_{(0,T_0]}$ of a non-negative solution of the equation (3.3.1). Since δ can be chosen arbitrarily small, this implies the uniqueness of a solution on $\Pi_{(0,T]}$. □

3.4 Comments

The class of equations studied in this chapter is a natural generalization of the classical Kolmogorov equation of diffusion with inertia [135]. This equation appears in the study of models for the Brownian motion, and we begin the comments with a brief exposition of a fragment of this important theory (see also the comment "Brownian Motion" by A. M. Yaglom in [136]).

In the classical theory of the Brownian motion developed by Einstein and Smoluchowski [77, 78] the inertia of a Brownian particle is neglected, that is the mass of a particle is actually assumed equal to 0. Therefore a Brownian particle in the Einstein-Smoluchowski theory cannot have a finite velocity. For example, for the special case of the Brownian motion of a free particle Wiener [177] proved rigorously that a Brownian path is, with probability 1, a continuous nowhere differentiable curve. Note that a model of the Brownian motion of a physical system in the Einstein-Smoluchowski approximation is a continuous Markov process on the coordinate space (the Wiener process, for the case of a free particle).

The fact of nondifferentiability of Brownian paths in the Einstein-Smoluchowski theory is closely connected with an idealization made in this theory (the neglect of inertia) making it invalid on *very small* time intervals Δt. For the simplest case of the Brownian motion of a free particle, a theory taking inertia into account was developed as early as 1930 by Uhlenbeck and Ornstein [224] (see also the papers by Doob [26] and Chandrasekar [25]). In this more precise theory paths are already differentiable (but do not have the second derivative, so that now the acceleration becomes infinite). In fact the same generalization is contained in the paper [135] by Kolmogorov who considered the general case of the Brownian motion for an arbitrary physical system with n degrees of freedom. According to Kolmogorov, inertia is taken into account if a state of the system is described by values of n coordinates $q_1, \dots, q_n$ and their time derivatives (velocities) $\dot{q}_1, \dots, \dot{q}_n$. Here the model of the Brownian motion is a continuous Markov process in the $2n$-dimensional phase space of coordinates and velocities.

Let us give a brief account of this paper [135] by Kolmogorov which motivated our studies. It is based essentially on two of his earlier papers [133, 134] devoted to the classical theory of the continuous time Markov processes. The first version of the modern theory of such processes was developed by Doob [27] and Dynkin [40, 41].

Thus, we consider a system with n degrees of freedom. Let $q_1, \dots, q_n$ be the coordinates. It is assumed that, whenever we know the values of $q = (q_1, \dots, q_n)$ and $\dot{q} = (\dot{q}_1, \dots, \dot{q}_n)$ at an instant of time t, we can find the probability density $G(t, q, \dot{q}; t', q', \dot{q}')$ of possible values q and $\dot{q}$ of the coordinates and their time derivatives at an arbitrary instant $t' > t$. It is assumed that G does not depend on the behaviour of the system before the instant t (there is no aftereffect, the process is of the Markov type). It is proved that the function G is a FSCP for the Fokker-Planck differential equation

$$\partial_{t'} g = -\sum_{j=1}^{n} \dot{q}'_j \partial_{q'_j} g - \sum_{j=1}^{n} \partial_{\dot{q}'_j} \left(f_j(t', q', \dot{q}') g\right) + \sum_{j,l=1}^{n} \partial_{\dot{q}'_j} \partial_{\dot{q}'_l} \left(k_{jl}(t', q', \dot{q}') g\right), \quad (3.4.1)$$

in which the elements k_{jl}, $\{j, l\} \subset \{1, \dots, n\}$, of the diffusion matrix, and the coordinates f_j, $j \in \{1, \dots, n\}$, of the drift vector are determined from the equalities

$$\begin{aligned} \mathbf{M} (\Delta \dot{q}_j)^2 &= k_{jj}(t, q, \dot{q}) \Delta t + o(\Delta t), \\ \mathbf{M} (\Delta \dot{q}_j \Delta \dot{q}_l)^2 &= k_{jl}(t, q, \dot{q}) \Delta t + o(\Delta t), \\ \mathbf{M} (\Delta \dot{q}_j) &= f_j(t, q, \dot{q}) \Delta t + o(\Delta t), \end{aligned}$$

where $\mathbf{M}$ denotes the expectation.

If $n = 1$, then this equation has the form

$$\partial_{t'} g = -\dot{q}' \partial_{q'} g - \partial_{\dot{q}'} \left(f(t', q', \dot{q}') g\right) + \partial^2_{\dot{q}'} \left(k(t', q', \dot{q}') g\right). \quad (3.4.2)$$

If f and k are constants, then, as Kolmogorov showed, a FSCP for the equation (3.4.2) is given by the formula

$$\begin{aligned} G(t,q,\dot{q};t',q',\dot{q}') = 2\sqrt{3}\pi^{-1} k^{-2} (t'-t)^{-2} \exp\{ -(4k(t'-t))^{-1} (\dot{q}' - \dot{q} - f(t'-t))^2 \\ -3k^{-1}(t'-t)^{-3}(q'-q-2^{-1}(\dot{q}'+\dot{q})(t'-t))^2 \}, \quad t<t', \ \{q,\dot{q},q',\dot{q}'\} \subset \mathbb{R}. \quad (3.4.3) \end{aligned}$$

This is a classical formula for a FSCP for Kolmogorov's equation of diffusion with inertia. A large part of the material of Section 3 is based on formulas similar to (3.4.3).

The Uhlenbeck-Ornstein theory of the Brownian motion for a free particle is obtained from the general Kolmogorov's theory, if $n = 1$ (so that the main equation of that theory has the form (3.4.2)), $f = -\alpha q$ (where $\alpha = \beta/m$, m is the mass of a particle, β is the coefficient at the viscous friction force, equal, for a spherical particle of the radius σ, to $6\pi\sigma\eta$, η is the viscosity coefficient), while $k = k_0T/m\beta$.

Il'in and Khasminsky [103] considered a physical problem solved by the techniques close to [135]. It deals with the motion of a particle with a mass m in a force field F. It is assumed that the medium is filled with identical particles of a mass μ. The particle under consideration may collide with them, changing its velocity under the elastic collision law. The velocity distribution of the medium particles is given and does not depend on the motion of the above particle, while the random number of collisions for time t is assumed to be a Poisson process with a parameter a. Under these assumptions, the position $X_\mu(t)$ and velocity $\dot{X}_\mu(t)$ at an instant t together form a Markov process $\{X_\mu(t), \dot{X}_\mu(t)\}$. It is proved that, for a constant medium temperature T and $a \to \infty$, $\mu \to \frac{A}{2} = \text{const}$, this process tends to a Brownian motion process in the phase space whose transition density is a FSCP for the equation

$$\partial_t u = \sum_{j=1}^{n} \dot{x}_j \partial_{x_j} u + \sum_{j=1}^{n} F_j(x) \partial_{\dot{x}_j} u + \frac{A}{m}\left(\frac{T}{9m}\sum_{j=1}^{n} \partial_{\dot{x}_j}^2 u - \frac{1}{3}\sum_{j=1}^{n} \dot{x}_j \partial_{\dot{x}_j} u\right). \quad (3.4.4)$$

The existence of a FSCP $G(t, x, y; \xi, \eta)$ is proved in [103] for the equation

$$\partial_t u = y\partial_x u + a\partial_y^2 u + g(x,y)\partial_y u, \quad \{x, y\} \subset \mathbb{R}, \quad (3.4.5)$$

where a is a positive constant, the function g is bounded and has continuous and bounded derivatives in x and y. Estimates for the function G and its derivatives are obtained. It is proved that G is a transition density of a Markov process $\{X(T), Y(t)\}$ defined as a solution of a system of the Ito stochastic differential equations [104]

$$\begin{cases} dX = Y(t)\, dt, \\ dY = g(X(t), Y(t))\, dt + \sqrt{a}\, d\xi(t) \end{cases}$$

where $\xi(t)$ is a Wiener process.

In addition, in [103] the asymptotic behaviour of solutions of the Cauchy problem for the equation (3.4.5) is investigated under various assumptions regarding the quantities A, T, and m. Terms of asymptotic expansions with respect to small parameters are found, in particular for $A \to \infty$ and $m/A \to 0$.

It is known [27, 41] that non-degenerate diffusion processes with sufficiently regular characteristics possess smooth transition densities. In general, a degeneration of the diffusion matrix leads to the non-existence of a density of the transition probability. However there exist some classes of degenerate processes with smooth transition densities. As it was mentioned above, the processes of the Brownian motion with inertia considered by Kolmogorov, Il'in, and Khasminsky are among such processes.

Sonin [216] studied a natural generalization of the process of diffusion with inertia. For the class of processes considered in [216], the transition densities are constructed as FSCPs for the equations

$$\partial_t u = \left(\frac{1}{2}\sum_{j,l=1}^{n} a_{jl}(t,x,y,z)\partial_{y_j}\partial_{y_l} + \sum_{j=1}^{n} a_j(t,x,y,z)\partial_{y_j} + \sum_{j=1}^{n} b_j(t,x,y,z)\partial_{x_j} + \sum_{j=1}^{n} c_j(t,x,y,z)\partial_{z_j}\right) u, \quad \{x,y,z\} \subset \mathbb{R}^n, \tag{3.4.6}$$

where $b_j(t,x,y,z)$ and $c_j(t,x,y,z)$ behave "approximately" like y_j and x_j respectively.

All the above partial differential equations (3.4.1), (3.4.2), (3.4.4)–(3.4.6) are degenerate parabolic equations. With respect to a part of spatial variables (corresponding to coordinates of a Brownian object in (3.4.1), (3.4.2), (3.4.4), and (3.4.5), or to coordinates and velocities in (3.4.6)) they contain, in the sense of the theory of parabolic equations, only lower order terms. Such equations are called ultraparabolic or elliptic-parabolic.

In this book we consider ultraparabolic equations (3.4.1), (3.4.4), (3.4.6) and their generalizations. It is natural to call them ultraparabolic equations of the Kolmogorov type. Here we give a survey of results regarding the Cauchy problem and, in particular the FSCP, for ultraparabolic equations of the Kolmogorov type and their generalizations.

As it has been mentioned, an explicit formula (3.4.3) for a FSCP for the equation (3.4.2) with constant f and k was found in [135]. Subsequent investigations aimed at finding as weak conditions as possible for the existence of a FSCP, obtaining its precise estimates, considering equations with a more complicated structure. For example, an equation

$$\left(\partial_t - \sum_{|k_1|\le 2b} a_{k_1}(t,x)\partial_{x_1}^{k_1} - \sum_{j=1}^{n_2} b_j(t,x)\partial_{x_{2j}} - \sum_{j=1}^{n_3} c_j(t,x)\partial_{x_{3j}}\right) u(t,x) = f(t,x), \quad (t,x) \in \Pi_{(0,T]}, \tag{3.4.7}$$

was considered.

Here b is a natural number, the differential expression $\partial_t - \sum\limits_{|k_1|\le 2b} a_{k_1}(t,x)\partial_{x_1}^{k_1}$ is uniformly parabolic on $\Pi_{[0,T]}$ in the sense of Petrovsky, the functions b_j and c_j are such that the equation (3.4.7) can be transformed by a non-degenerate change of variables to an equation of the form (1.1.14). The above equation contains three groups of spatial variables x_1, x_2, and x_3. If there are only two groups x_1 and x_2, then the sum with the derivatives in x_3 is absent. We shall assume that such equations can be obtained formally from (3.4.7) and (1.1.14) by setting $n_3 = 0$.

Let us describe the papers which were devoted to, or used, FSCPs for the above classes of equations.

Weber [227] constructed, on an arbitrary open subset of the phase space, a fundamental solution of the equation (3.4.7) with $b = 1$, $n_1 = n_2$, and $n_3 = 0$.

We have reviewed above the paper [103]. In a subsequent paper [101] Il'in stated the existence, uniqueness, and positivity of a FSCP for the equation (3.4.7) with $b = 1$, $n_1 = n_2$, and $n_3 = 0$, provided the coefficients a_{k_1} and b_j are continuous and bounded together with their derivatives of the orders up to 2 and 4 respectively, as well as a theorem on the uniqueness of a solution of the Cauchy problem in the Tikhonov class, and inner estimates of solutions.

Sonin [216] used the Levi method to construct a FSCP for the equation (3.4.6) for the case where the coefficients a_{jl} and a_j are bounded on $[0,T] \times \mathbb{R}^{3n}$ together with their derivatives in all the variables of the orders up to 2 and 1 respectively, while b_j and c_j (the coefficients c_j do not depend on y) have bounded derivatives of the orders up to 3.

The main results of the papers [208, 209] are the smoothness of solutions of an inhomogeneous equation, inner a priori estimates of solutions on bounded domains. Shatyro proved that the second derivatives of a solution of a second order equation (of the form (3.4.7) with $n_3 = 0$ in [208], and of the form (1.1.15) with an arbitrary number of groups of spatial variables and constant coefficients in [209]) with respect to spatial variables from the first group are Hölder continuous, with the same exponent as the right-hand side of the equation. The Hölder exponent for the first derivatives from the l-th group of spatial variables ($l \geq 2$) is smaller by $(2l-3)/(2l-1)$ than that of the right-hand side. It was shown by examples, that the result was exact. The proofs used FSCPs for model equations.

In a series of interesting papers [142, 143, 144, 145], Kuptsov studied equations of the form

$$\left(\partial_t - \sum_{j,l=1}^{n} a_{jl}(t) \partial_{x_j} \partial_{x_l} + \sum_{j,l=1}^{n} b_{jl}(t) x_j \partial_{x_l} \right) u(t,x) = 0, \quad (t,x) \in \Pi_{(0,T]}, \quad (3.4.8)$$

where $(a_{jl}(t))_{j=1,l=1}^{n,n}$ is a non-negative symmetric matrix of a rank n_1 $(1 \leq n_1 \leq n)$, the matrix $(b_{jl}(t))_{j=1,l=1}^{n,n}$ has a special structure and properties. For the equation (3.4.8) with constant coefficients, in [142, 143, 145] a FSCP was constructed, a maximum principle was established, and a mean value theorem was proved. A FSCP for general equations (3.4.8) was constructed in [144].

At the beginning of the seventies Eidelman and Malitskaya [64, 154] initiated the investigation of equations of an arbitrary order with the Kolmogorov structure (equations from the class $\mathbf{E}_{21}$). For such equations, for the case where $n_1 = n_2 = n_3$, and the coefficients may depend only on the time variable, a FSCP was constructed and studied, which resulted in existence and uniqueness theorems for solutions of the Cauchy problem, as well as some results on qualitative properties of solutions. More or less at the same time a similar class of equations (the equations (3.4.7) with $n_3 = 0$) was considered independently by Kato [126] who found hypoellipticity conditions for this class of equations.

Similarly to the case of non-degenerate parabolic equations, for the Kolmogorov type equations with constant coefficients or coefficients depending only on the time variable, it is possible to obtain a complete analytic description of a FSCP, which leads to very precise results on the correct solvability of the Cauchy problem and integral representations for solutions. For equations from the class $\mathbf{E}_{21}$ that was done by Androsova and Ivasyshen [106, 2, 3]. These results are contained in Section 3.1.1 and, for a more general situation, in Sections 3.2.1, 3.2.2.

If coefficients of a Kolmogorov type equation depend on all the variables, it is much more complicated to study its FSCP. In addition to usual difficulties of the Levi method, new ones are caused by the degeneracy of the equations. Various modifications of the Levi parametrix method were proposed, in order to construct and study FSCPs for equations from the classes $\mathbf{E}_{21}$ and $\mathbf{E}_{22}$, in the papers by Eidelman, Ivasyshen, Malitskaya, and Tychinskaya [73, 155, 158, 62, 61]. In [64, 159, 74] equations of the Kolmogorov type with increasing (as $|x| \to \infty$) coefficients were also considered.

Eidelman and Ivasyshen [54, 56, 57, 58, 59] introduced a new class of degenerate equations (the class $\mathbf{E}_{23}$). For the case where the coefficients do not depend on the spatial variables (the class $\mathbf{E}_{23}^0$), they constructed and investigated a FSCP, proved theorems on the correct solvability of the Cauchy problem and integral representations for solutions. These results are expounded in Sections 3.1.3, 3.2.1, and 3.2.2. Section 3.1.3 contains also Theorem 3.6 on the FSCPs for equations from the class $\mathbf{E}_{23}^0$ with degeneracy on the initial hyperplane. This theorem, as well as the results on the Cauchy problem for such equations with a weak degeneracy on the initial hyperplane (mentioned at the end of Section 3.2.2) were obtained by Ivasyshen and Voznyak [120]. Note that similar results for equations from the class $\mathbf{E}_{21}^0$ were published in [119], while in [121] the unique solvability was established, and a localization property was proved for solutions of the Cauchy problem for equations of the form (3.1.96) with generalized initial data.

A relatively recent series of papers [20, 147, 148, 149, 162, 163, 172, 187, 188, 189, 190, 191] is devoted to various aspects of the theory of ultraparabolic equations of the Fokker-Planck-Kolmogorov type. The closest to our Chapter 3 is the paper by Polidoro [187] where the equation

$$Lu := \left(\partial_t - \sum_{j,l=1}^{n_1} a_{jl}(t,x)\partial_{x_j}\partial_{x_l} - \sum_{j,l=1}^{n} b_{jl}x_j\partial_{x_l} \right) u = 0 \qquad (3.4.9)$$

is considered, where $1 \le n_1 < n$, $(a_{jl}(t,x))_{j=1,l=1}^{n_1,n_1}$ is a symmetric matrix, positive definite for each $(t,x) \in \mathbb{R}^{n+1}$, the matrix $B := (b_{jl})_{j=1,l=1}^{n,n}$ has constant real elements. Under certain assumptions regarding the matrix B, the operator L is invariant with respect to a certain dilation group. A special notion of the B-Hölder continuity of functions correspond to the above group invariance. For the case where the coefficients a_{jl} are B-Hölder continuous, a FSCP for the equation (3.4.9) is constructed by the Levi method, and its estimate is given. Note that, under the conditions imposed upon B, the well-known Hörmander's hypoellipticity condition [99] is satisfied by the operator L with the coefficients a_{jl} fixed at any point (t, x). Note also that Hörmander [100] proposed an interesting generalization of the classical Kolmogorov equation and investigated the problem of its hypoellipticity.

In Chapter 3, modifying in an appropriate way Polidoro's notion of the B-Hölder continuity for equations from the classes $\mathbf{E}_{21} - \mathbf{E}_{23}$ we find conditions for the coefficients, under which FSCPs for such equations are constructed, their estimates are obtained and used to prove the correct solvability of the Cauchy problem. These results for equations from the classes $\mathbf{E}_{21} - \mathbf{E}_{23}$ (obtained by Ivasyshen) are published here for the first time.

In the important paper by Scornazzani [207], for the equation (1.1.15) with real-valued coefficients and $n_1 = n_2 = 1$, $n_3 = 0$, the positivity of the FSCP was proved, its lower estimates was obtained, as well as a Widder type theorem (see [229]) on the uniqueness of a non-negative solution of the Cauchy problem, an integral representation of non-negative solutions, and a Fatou type theorem. Some of these results were extended to the general equation (1.1.15) by Dron' and Ivasyshen [38]; for their exposition see Section 3.3. These results are based on the modifications of the maximum principle proposed by Dron' [35] and Malitskaya [161] given in Section 3.3.2. Note also the papers [39, 36, 37] devoted to the Cauchy problem for equations from the class $\mathbf{E}_{22}$.

Boundary value problems for ultraparabolic equations, including those of the Fokker-Planck-Kolmogorov type, have been studied concurrently with the Cauchy problem, starting from the work by Piskunov [181]; see, in particular, [89, 210, 156, 157, 160, 196]. Boundary value problems for equations studied in Chapter 3 constitute a separate research subject and go beyond the scope of this book.

Chapter 4

Pseudo-Differential Parabolic Equations with Quasi-Homogeneous Symbols

4.1 Fundamental solution of the Cauchy problem

4.1.1. Main assumptions. We shall consider the Cauchy problem

$$\partial_t u(t,x) + (Au)(t,x) + \sum_{k=1}^{m} (A_k u)(t,x) = f(t,x), \quad (t,x) \in \Pi_{(0,T]}, \tag{4.1.1}$$

$$u(0,x) = \varphi(x), \tag{4.1.2}$$

where $A, A_1, \dots, A_m$ are pseudo-differential operators with the symbols $a(t,x,\xi)$, $a_1(t,x,\xi)$, ..., $a_m(t,x,\xi)$, that is, e.g.,

$$(Au)(t,x) = (2\pi)^{-n} \int\limits_{\mathbb{R}^n} e^{i(x,\xi)} a(t,x,\xi) \widehat{u}(t,\xi)\, d\xi, \tag{4.1.3}$$

where

$$\widehat{u}(t,\xi) = \int\limits_{\mathbb{R}^n} e^{-i(y,\xi)} u(t,y)\, dy.$$

The principal symbol $a(t,x,\xi)$ is assumed homogeneous of the degree $\gamma \geq 1$ with respect to the variable ξ, and elliptic, the symbols $a_1(t,x,\xi), \dots, a_m(t,x,\xi)$ have the degrees of homogeneity γ_k, $0 < \gamma_k < \gamma$, f and φ are bounded continuous functions. In addition, f is Hölder continuous in x, uniformly with respect to t:

$$|f(t,x) - f(t,y)| \leq C|x-y|^{\lambda}$$

where C, λ do not depend on t. We may assume without loss of generality that

$$\max_{1 \leq k \leq m} \gamma_k < \gamma - \lambda.$$

We assume the following conditions regarding the symbols of the pseudo-differential operators involved in (4.1.1). The principal symbol is such that

(A_{41}) $\operatorname{Re} a(t,x,\sigma) \geq a_0 > 0$, $(t,x) \in \Pi_{(0,T]}$, $|\sigma| = 1$;

(A_{42}) $a(t,x,\sigma)$ has N continuous derivatives in σ for $\sigma \neq 0$, and

$$|\partial_\sigma^\varkappa a(t,x,\sigma)| \leq C_N |\sigma|^{\gamma-|\varkappa|},$$
$$|\partial_\sigma^\varkappa [a(t,x,\sigma) - a(\tau,y,\sigma)]| \leq C_N \left(|x-y|^\lambda + |t-\tau|^{\lambda/\gamma}\right) |\sigma|^{\gamma-|\varkappa|}$$

for all $|\varkappa| \leq N$, $x,y,\sigma \in \mathbb{R}^n$ $(\sigma \neq 0)$, $t,\tau \in [0,T]$. Here $\lambda \in (0,1)$ is a constant, N is such a natural number that $N \geq 2n + 2[\gamma] + 1$. Other symbols satisfy the condition

(A_{43}) If $\xi \neq 0$, $|\varkappa| \leq N$, $x,y \in \mathbb{R}^n$, $t,\tau \in [0,T]$, then

$$\left|\partial_\xi^\varkappa a_k(t,x,\xi)\right| \leq C|\xi|^{\gamma_k - |\varkappa|},$$
$$\left|\partial_\xi^\varkappa [a_k(t,x,\xi) - a_k(\tau,y,\xi)]\right| \leq C\left(|x-y|^\lambda + |t-\tau|^{\lambda/\gamma}\right) |\xi|^{\gamma_k-|\varkappa|},$$

where C does not depend on x,y,ξ,t,τ.

Some additional requirements are imposed on the symbols of an integer order. Namely, a symbol of an even order is assumed to be a polynomial in ξ, and a symbol of an odd order is assumed to be either a polynomial or an even function of ξ. Finally, if γ is an odd integer, we assume that in the expansion of the function $[a(t,x,\xi)]^{-1}$, $|\xi| = 1$, in spherical harmonics (see below) the coefficients at $Y_{2\nu,\mu}$ with $\gamma = n + 2\nu + 2k$, $k = 0,1,2,\ldots$, are equal to zero. This condition is automatically satisfied if $\gamma < n$, or if n is even.

The expression (4.1.3) can be used only for smooth rapidly decreasing functions, and our first task is to reformulate it in terms of hyper-singular integrals.

Our brief exposition of the theory of hyper-singular integrals will be based on properties of spherical harmonics.

4.1.2. Spherical harmonics. A *spherical harmonic* of a degree $\nu \geq 0$ is a restriction to the sphere S^{n-1} of a homogeneous harmonic polynomial on $\mathbb{R}^n$ of the degree ν. The set of all spherical harmonics of a degree ν is a finite-dimensional subspace $H_\nu \subset L_2(S^{n-1})$. Let $\delta_\nu = \dim H_\nu$. Then

$$\delta_\nu \leq C\nu^{n-2}. \tag{4.1.4}$$

For each ν we choose in H_ν an orthonormal basis $\{Y_{\nu\mu}\}$, $\mu = 1,\ldots,\delta_\nu$. The functions $Y_{\nu\mu}$ are even for an even ν, and odd otherwise. The system $\{Y_{\nu\mu}\}$ is an orthonormal basis of $L_2(S^{n-1})$. We shall use the estimates

$$|Y_{\nu\mu}(\sigma)| \leq C\nu^{(n-2)/2}, \quad \sigma \in S^{n-1}; \tag{4.1.5}$$

$$\left|\partial_x^\varkappa Y_{\nu\mu}\left(\frac{x}{|x|}\right)\right| \leq C\nu^{|\varkappa|+(n-2)/2}|x|^{-|\varkappa|}, \quad x \in \mathbb{R}^n. \tag{4.1.6}$$

If $\varphi \in C^{2k}(S^{n-1})$, $k = 0, 1, \dots$, then its Fourier coefficients

$$\varphi_{\nu\mu} = \int_{S^{n-1}} \varphi(\sigma)\overline{Y_{\nu\mu}(\sigma)}\, d\sigma$$

satisfy the estimate

$$|\varphi_{\nu\mu}| \le C_{n,k}\nu^{-2k} \sup_{\sigma \in S^{n-1}, |\varkappa| \le 2k} |D^{\varkappa}\varphi(\sigma)|. \tag{4.1.7}$$

If $\operatorname{Re}\lambda > -n$, then the function $x \mapsto |x|^\lambda Y_{\nu\mu}(\frac{x}{|x|})$ is locally integrable, and we can define an analytic family of tempered distributions

$$\left\langle |x|^\lambda Y_{\nu\mu}\left(\frac{x}{|x|}\right), \Phi(x)\right\rangle = \int_0^\infty r^{\lambda+n-1}\, dr \int_{S^{n-1}} Y_{\nu\mu}(\sigma)\Phi(r\sigma)\, d\sigma$$

where $d\sigma$ means the surface Lebesgue measure. In fact this family admits an analytic continuation to the whole complex plane except the points $\lambda = -\nu - n, -\nu - n - 2, \dots$. The regularization of the function $r^\lambda Y_{\nu\mu}(\frac{x}{|x|})$, $r = |x|$, with $\operatorname{Re}\lambda \le -n$ (and λ being outside the exceptional set) as a distribution from $\mathcal{S}'(\mathbb{R}^n)$ is carried out by subtracting a partial sum of the Taylor series of a test function.

The Fourier transform of the above distribution has the form

$$\widehat{\left(r^\lambda Y_{\nu\mu}\right)} = (-i)^\nu \pi^{n/2} 2^{n+\lambda} \frac{\Gamma\left(\frac{\lambda+\nu+n}{2}\right)}{\Gamma\left(\frac{\nu-\lambda}{2}\right)} r^{n-\lambda} Y_{\nu\mu} \tag{4.1.8}$$

for $\lambda \ne -\nu - n, -\nu - n - 2, \dots$.

The proofs of the above results, as well as additional information about spherical harmonics, can be found in [150, 175, 205].

4.1.3. Hyper-singular integrals. Let us consider the hyper-singular integral (HSI)

$$\left(\mathbf{D}^\alpha_\Omega f\right)(x) = \frac{1}{d_{n,l}(\alpha)} \int_{\mathbb{R}^n} \Omega\left(x, \frac{h}{|h|}\right) \frac{\left(\mathbf{\Delta}^l_h f\right)(x)}{|h|^{n+\alpha}}\, dh, \quad x \in \mathbb{R}^n, \tag{4.1.9}$$

$$\left(\mathbf{\Delta}^l_h f\right)(x) = \sum_{k=0}^{l} (-1)^k \binom{l}{k} f(x - kh),$$

of an order $\alpha > 0$ with a characteristic Ω. We assume that $f(x)$ and $\Omega(x, \theta)$ are bounded continuous complex-valued functions. Below we give a summary of some results by Samko [202, 200, 201] regarding properties of HSIs. Note that in expositions of the HSI theory the characteristic Ω usually does not depend on x; however the appropriate results carry over to our more general situation.

Suppose first that the number α is not an integer. The integral (4.1.9) is absolutely convergent if $l > \alpha$, and the function f has bounded derivatives up to the order $[\alpha]+1$. The absolute convergence over the ball $\{|h|<\varepsilon\}$ is a consequence of the formula

$$\left(\boldsymbol{\Delta}_h^l f\right)(x)=\sum_{|\varkappa|=\mu}\sum_{\nu=0}^{l}\frac{(-\nu h)^{\varkappa}(-1)^{\nu}}{\varkappa!}\binom{l}{\nu}(D^{\varkappa}f)(x-\theta_{\nu}\nu h), \tag{4.1.10}$$

where $0<\theta_\nu<1$, $l\geq\mu$, and the usual notation for operations with multi-indices is employed. The formula (4.1.10) follows from the Taylor formula.

The restricted hyper-singular integral

$$\left(\mathbf{D}^{\alpha}_{\Omega,\varepsilon}f\right)(x)=\frac{1}{d_{n,l}(\alpha)}\int\limits_{|h|>\varepsilon}\Omega\left(x,\frac{h}{|h|}\right)\frac{\left(\boldsymbol{\Delta}_h^l f\right)(x)}{|h|^{n+\alpha}}\,dh,\quad x\in\mathbb{R}^n,$$

is absolutely convergent if, for example, f is bounded. The above convergence conditions can be easily extended to the case when $f(x)$ and its derivatives grow not too rapidly as $|x|\to\infty$. We shall use also conditionally convergent hyper-singular integrals of bounded functions:

$$\left(\mathbf{D}^{\alpha}_{\Omega}f\right)(x)=\lim_{\varepsilon\to 0}\left(\mathbf{D}^{\alpha}_{\Omega,\varepsilon}f\right)(x), \tag{4.1.11}$$

if the limit (4.1.11) exists for all $x\in\mathbb{R}^n$.

If the characteristic $\Omega(x,\sigma)$ is even in σ, that is $\Omega(x,-\sigma)=\Omega(x,\sigma)$, $x\in\mathbb{R}^n$, $\sigma\in S^{n-1}$, then the hyper-singular integral (4.1.9) makes sense also for $l>2[\alpha/2]$, by the formula

$$\begin{aligned}\lim_{\varepsilon\to 0}\left(\mathbf{D}^{\alpha}_{\Omega,\varepsilon}f\right)(x)=&-\frac{1}{d_{n,l}(\alpha)}\sum_{k=1}^{(l-1)/2}\int\limits_{\mathbb{R}^n}\Omega\left(x,\frac{h}{|h|}\right)\frac{\left(\boldsymbol{\Delta}_h^{l+1}f\right)(x+kh)}{|h|^{n+\alpha}}\,dh\\&-\frac{1}{2d_{n,l}(\alpha)}\int\limits_{\mathbb{R}^n}\Omega\left(x,\frac{h}{|h|}\right)\frac{\left(\boldsymbol{\Delta}_h^{l+1}f\right)\left(x+\frac{l+1}{2}h\right)}{|h|^{n+\alpha}}\,dh,\end{aligned} \tag{4.1.12}$$

where this time $l+1>\alpha$.

Next, let α be an integer. If α is even, then $(\mathbf{D}^{\alpha}_{\Omega}f)$ can be defined as before. In this case the hyper-singular integral operator $\mathbf{D}^{\alpha}_{\Omega}$ is actually a differential operator of the order α. If α is odd, then for $l>\alpha$ the integral in (4.1.9) vanishes identically for any function f. In this case $\mathbf{D}^{\alpha}_{\Omega}$ can be defined only for an even characteristic Ω by the formula (4.1.12) with $l=\alpha$.

Let $f\in C^l(\mathbb{R}^n)$, $|f(x)|\leq C(1+|x|)^{-N_1}$, $|(D^{\varkappa}f)(x)|\leq C(1+|x|)^{-N_2}$, $|\varkappa|=l$, where $N_1>\alpha$, $N_2>n$. Then

$$\left|\left(\mathbf{D}^{\alpha}_{\Omega}f\right)(x)\right|\leq C(1+|x|)^{-\min(\alpha+N_1,N_2,\alpha+n)}. \tag{4.1.13}$$

Moreover, if Ω depends on a parameter t being bounded uniformly with respect to t, then the constant in (4.1.13) does not depend on t. The same estimate, with a constant independent of ε, holds for the restricted hyper-singular integral $\mathbf{D}^{\alpha}_{\Omega,\varepsilon}f$.

Let us consider the function $f_\xi(x) = e^{i\xi\cdot x}$. Obviously,

$$\left(\boldsymbol{\Delta}_h^l f_\xi\right)(x) = e^{i(\xi,x)} \sum_{k-0}^{l} (-1)^k \binom{l}{k} e^{-ik(\xi,h)} = e^{i(\xi,x)} \left(1 - e^{-ik(\xi,h)}\right)^l,$$

whence

$$\left(\mathbf{D}_\Omega^\alpha f_\xi\right)(x) = \widetilde{\Omega}(x,\xi) f_\xi(x), \tag{4.1.14}$$

where

$$\widetilde{\Omega}(x,\xi) = \frac{1}{d_{n,l}(\alpha)} \int\limits_{\mathbb{R}^n} \frac{\left(1 - e^{-i(\xi,h)}\right)^l}{|h|^{n+\alpha}} \Omega\left(x, \frac{h}{|h|}\right) dh.$$

The function $\widetilde{\Omega}$ is called *the symbol* of the hyper-singular integral $\mathbf{D}_\Omega^\alpha f$. The normalization constants $d_{n,l}(\alpha)$ are chosen in such a way that the symbol (hence the hyper-singular integral) does not depend on l. Their explicit form can be found in [202, 205]. Here we only note that they do not depend on Ω, and $d_{n,l}(\alpha) > 0$ if $0 < \alpha < 2$. The symbol $\widetilde{\Omega}(x,\xi)$ is a homogeneous function of the degree α; for example, if $\Omega(x,\sigma) \equiv 1$, then $\widetilde{\Omega}(x,\xi) = |\xi|^\alpha$ (the corresponding hyper-singular integral operator is often called a Riesz derivative). The symbol can be represented as

$$\widetilde{\Omega}(x,\xi) = C(n,\alpha) \int\limits_{S^{n-1}} \Omega(x,\sigma)(i(\xi,\sigma))^\alpha \, d\sigma. \tag{4.1.15}$$

Note also that the symbol $\widetilde{\Omega}$ is even if and only if the characteristic Ω is even.

It follows from the formula (4.1.14) that on functions from the Schwartz space $\mathcal{S}(\mathbb{R}^n)$ (and more generally, on smooth functions, which decrease rapidly enough) the operator $\mathbf{D}_\Omega^\alpha$ coincides with the pseudo-differential operator (ΨDO) with the symbol $\widetilde{\Omega}$. Indeed, if $f \in \mathcal{S}(\mathbb{R}^n)$, then

$$f(x) = (2\pi)^{-n} \int\limits_{\mathbb{R}^n} e^{i(x,\xi)} \widehat{f}(\xi) \, d\xi,$$

and by the Fubini theorem

$$\left(\mathbf{D}_\Omega^\alpha f\right)(x) = (2\pi)^{-n} \int\limits_{\mathbb{R}^n} e^{i(x,\xi)} \widetilde{\Omega}(x,\xi) \widehat{f}(\xi) \, d\xi.$$

Conversely, a ΨDO with a homogeneous sufficiently smooth symbol can be represented as a HSI-operator. In order to obtain such a representation, we need an expression for the symbol of the HSI whose characteristic is a spherical harmonic.

Let $\widetilde{Y_{\nu\mu}}(\xi)$ be the symbol of the hyper-singular integral $\mathbf{D}_{Y_{\nu\mu}}^\alpha f$. If α is not an integer, or α is an odd integer and ν is even, then

$$\widetilde{Y_{\nu\mu}}(\xi) = |\xi|^\alpha \lambda(\nu,\alpha) Y_{\nu\mu}\left(\frac{\xi}{|\xi|}\right),$$

where

$$\lambda(\nu,\alpha) = E_{\nu,\alpha} i^{\nu} \frac{\Gamma\left(\frac{n+\alpha}{2}\right)\Gamma\left(1+\frac{\alpha}{2}\right)}{\Gamma\left(\frac{n+\alpha+\nu}{2}\right)\Gamma\left(1+\frac{\alpha-\nu}{2}\right)},$$

$$E_{\nu,\alpha} = (-1)^{\nu} \frac{\cos\frac{\alpha+\nu}{2}\pi}{\cos\frac{\alpha\pi}{2}} \quad \text{(if } \alpha \text{ is not an integer)},$$

$$E_{\nu,\alpha} = (-1)^{\nu/2} \quad \text{(if } \alpha \text{ is an odd integer)}.$$

Let us consider a ΨDO of the form (4.1.3), where the symbol $a(t,x,\xi)$ is continuous and homogeneous in ξ of the degree α, and either α is not an integer, or the symbol is even in ξ and α is an odd integer. Suppose that

$$\left|\partial_{\xi}^{\varkappa} a(t,x,\xi)\right| \le C, \quad |\xi| = 1, x \in \mathbb{R}^n, t \ge 0, |\varkappa| \le N,$$

where $N > 2n + \alpha - 1$. Expanding the symbol by spherical harmonics,

$$a(t,x,\xi) = \sum_{\nu=0}^{\infty}\sum_{\mu=1}^{\delta_\nu} c_{\nu\mu}(t,x) Y_{\nu\mu}(\xi), \quad |\xi| = 1,$$

we set

$$\Omega(t,x,\xi) = \sum_{\nu=0}^{\infty}\sum_{\mu=1}^{\delta_\nu} \frac{c_{\nu\mu}(t,x)}{\lambda(\nu,\alpha)} Y_{\nu\mu}(\xi). \tag{4.1.16}$$

According to (4.1.7), $|c_{\nu\mu}(t,x)| \le C\nu^{-N+1}$. Then, since

$$\left|\cos\left(\frac{\alpha+\nu}{2}\pi\right)\right| = \left|\cos\frac{\alpha\pi}{2}\right|, \quad \text{or} \quad \left|\cos\left(\frac{\alpha+\nu}{2}\pi\right)\right| = \left|\sin\frac{\alpha\pi}{2}\right|,$$

depending on whether ν is even or odd, we find that

$$\left|\frac{1}{\lambda(\nu,\alpha)}\right| \le C\left|\Gamma\left(\frac{n+\alpha+\nu}{2}\right)\Gamma\left(1-\frac{\nu-\alpha}{2}\right)\right|.$$

If α is an odd integer, then $c_{\nu\mu}(x,t) = 0$ for odd ν. Let us use the identity [9]

$$\Gamma\left(1-\frac{\nu-\alpha}{2}\right) = \frac{\pi}{\sin\left(\frac{\nu-\alpha}{2}\pi\right)\Gamma\left(\frac{\nu-\alpha}{2}\right)}.$$

Since $\left|\sin\left(\frac{\nu-\alpha}{2}\pi\right)\right|$ is equal to $\left|\cos\frac{\alpha\pi}{2}\right|$ or to $\left|\sin\frac{\alpha\pi}{2}\right|$ (in particular, it is equal to 1 if α is an odd integer and ν is even), and

$$\left|\frac{\Gamma\left(\frac{n+\alpha+\nu}{2}\right)}{\Gamma\left(\frac{\nu-\alpha}{2}\right)}\right| \le C\nu^{\frac{n}{2}+\alpha}$$

(see [9]), we have

$$\left|\frac{1}{\lambda(\nu,\alpha)}\right| \leq C\nu^{\frac{n}{2}+\alpha},$$

which implies, together with (4.1.5), the estimate

$$\left|\sum_{\mu=1}^{\delta_\nu} \frac{c_{\nu\mu}(t,x)}{\lambda(\nu,\alpha)} Y_{\nu\mu}(\xi)\right| \leq C\nu^{-N+2n+\alpha-2},$$

so that the series (4.1.16) is uniformly convergent.

Let us consider the hyper-singular integral operator $\mathbf{D}^{\alpha}_{\Omega}$. Its symbol is

$$\widetilde{\Omega}(t,x,\xi) = \sum_{\nu=0}^{\infty}\sum_{\mu=1}^{\delta_\nu} \frac{c_{\nu\mu}(t,x)}{\lambda(\nu,\alpha)} \widetilde{Y_{\nu\mu}}(\xi) = |\xi|^{\alpha} \sum_{\nu=0}^{\infty}\sum_{\mu=1}^{\delta_\nu} c_{\nu\mu}(t,x) Y_{\nu\mu}\left(\frac{\xi}{|\xi|}\right) = a(t,x,\xi)$$

(it follows from (4.1.15) that the termwise transition from the characteristics to the symbols was legitimate). Thus a pseudo-differential operator A can be represented on $\mathcal{S}(\mathbb{R}^n)$ as the hyper-singular integral operator with the characteristic Ω.

4.1.4. Parametrix. The parametrix of the problem (4.1.1)–(4.1.2) corresponding to "freezing" the symbol at the point $x = \xi$, $t = \theta$ is defined by

$$Z_0(t-\mu, x-y; \xi, \theta) = (2\pi)^{-n} \int_{\mathbb{R}^n} \exp\{i(x-y)\sigma - a(\theta,\xi,\sigma)(t-\mu)\}\, d\sigma,$$

$$x, y \in \mathbb{R}^n,\ t > \mu,\ \theta \geq 0. \tag{4.1.17}$$

Let us prove several estimates of oscillatory integrals, which will imply estimates of the function Z_0 and its derivatives.

Let ν be a non-negative integer, $P_\nu(\sigma)$ a homogeneous polynomial of the degree ν,

$$\Phi_\nu(z,\theta,\xi) = \int_{\mathbb{R}^n} P_\nu(\sigma) \exp\{i(z,\sigma) - a(\theta,\xi,\sigma)\}\, d\sigma.$$

Lemma 4.1. *If* $N \geq 2n + \nu + [\gamma]$, *then*

$$|\Phi_\nu(z,\theta,\xi)| \leq C(1+|z|)^{-n-\gamma-\nu}, \tag{4.1.18}$$

where C *does not depend on* ξ, θ .

Proof. Let $\eta_0 \in C_0^\infty(\mathbb{R}^n)$, $\eta_0(\sigma) = 1$ as $|\sigma| \leq 1$, $\eta_0(\sigma) = 0$ as $|\sigma| \geq 2$. Denote $\eta_1(\sigma) = 1 - \eta_0(\sigma)$,

$$\Phi_\nu^k(z,\theta,\xi) = \int_{\mathbb{R}^n} \eta_k(\sigma) P_\nu(\sigma) \exp\{i(z,\sigma) - a(\theta,\xi,\sigma)\}\, d\sigma, \quad k = 0, 1.$$

Evidently, $\Phi_\nu = \Phi_\nu^0 + \Phi_\nu^1$.

Let us use the representation $e^{-t} = 1-t+t^2h(t)$, where $h(t) = t^{-2}(e^{-t}-1+t)$. It is easy to see that $h(t)$ has bounded derivatives of all orders on $[0,\infty)$. We have

$$\Phi_\nu^0(z,\theta,\xi) = \int\limits_{\mathbb{R}^n} \eta_0(\sigma)P_\nu(\sigma)e^{i(z,\sigma)}\,d\sigma - \int\limits_{\mathbb{R}^n} \eta_0(\sigma)P_\nu(\sigma)a(\theta,\xi,\sigma)e^{i(z,\sigma)}\,d\sigma$$

$$+ \int\limits_{\mathbb{R}^n} \eta_0(\sigma)P_\nu(\sigma)[a(\theta,\xi,\sigma)]^2 h(a(\theta,\xi,\sigma))e^{i(z,\sigma)}\,d\sigma$$

$$\stackrel{\text{def}}{=} \Phi_\nu^{0,1}(z) + \Phi_\nu^{0,2}(z,,\theta,\xi) + \Phi_\nu^{0,3}(z,\theta,\xi).$$

It is clear that $\Phi_\nu^{0,1} \in \mathcal{S}(\mathbb{R}^n)$. Let $a_\nu(\theta,\xi,\sigma) = P_\nu(\sigma)a(\theta,\xi,\sigma)$. Then a_ν is a homogeneous function of σ of the degree $\gamma+\nu$ of the same smoothness as a.

Let $\varphi \in C_0^\infty(\mathbb{R}^n)$, $\varphi(z) = 0$ as $|z| \le 1$, $\psi(z) \equiv \varphi(-z)$. Then

$$\langle \Phi_\nu^{0,2}(\cdot,\theta,\xi),\varphi\rangle = \int\limits_{\mathbb{R}^n}\int\limits_{\mathbb{R}^n} e^{i(z,\sigma)}\eta_0(\sigma)a_\nu(\theta,\xi,\sigma)\varphi(z)\,d\sigma\,dz$$

$$= \int\limits_{\mathbb{R}^n} \eta_0(\sigma)a_\nu(\theta,\xi,\sigma)\widehat{\psi}(\sigma)\,d\sigma = \int\limits_{\mathbb{R}^n} a_\nu(\theta,\xi,\sigma)\widehat{\psi}(\sigma)\,d\sigma$$

$$- \int\limits_{\mathbb{R}^n} \eta_1(\sigma)a_\nu(\theta,\xi,\sigma)\widehat{\psi}(\sigma)\,d\sigma = \langle \widehat{a_\nu}(\theta,\xi,\cdot),\psi\rangle$$

$$- \int\limits_{\mathbb{R}^n} \eta_1(\sigma)a_\nu(\theta,\xi,\sigma)\,d\sigma \int\limits_{\mathbb{R}^n} \left(L^N e^{i(\sigma,z)}\right)\varphi(z)\,dz,$$

where $\widehat{a_\nu}$ is the Fourier transform of a_ν in the sense of distributions from $\mathcal{S}'(\mathbb{R}^n)$,

$$L = -i|z|^{-2}\sum_{k=1}^n z_k\frac{\partial}{\partial\sigma_k}, \quad Le^{i(\sigma,z)} = e^{i(\sigma,z)}.$$

Let us expand the function $a_\nu(\theta,\xi,\zeta)$, $|\zeta| = 1$, by spherical harmonics:

$$a_\nu(\theta,\xi,\zeta) = \sum_{l=0}^\infty\sum_{\mu=1}^{\delta_l} g_{l\mu}(\theta,\xi)Y_{l\mu}(\zeta).$$

Then for all $\sigma \in \mathbb{R}^n$, $\sigma \ne 0$,

$$a_\nu(\theta,\xi,\sigma) = |\sigma|^{\gamma+\nu}\sum_{l=0}^\infty\sum_{\mu=1}^{\delta_l} g_{l\mu}(\theta,\xi)Y_{l\mu}\left(\frac{\sigma}{|\sigma|}\right).$$

It is known [150] that the Fourier transform of the function $|\sigma|^{\gamma+\nu}Y_{l\mu}\left(\frac{\sigma}{|\sigma|}\right)$ equals

$$(-i)^l 2^{n+\gamma+\nu}\pi^{n/2}\frac{\Gamma\left(\frac{n+l+\gamma+\nu}{2}\right)}{\Gamma\left(\frac{l-\gamma-\nu}{2}\right)}\cdot\frac{Y_{l\mu}\left(\frac{z}{|z|}\right)}{|z|^{n+\gamma+\nu}}, \tag{4.1.19}$$

if $l-\gamma-\nu\neq 0,-2,-4,\dots$ (see [87] regarding the interpretation of the homogeneous function (4.1.19) as an element of $\mathcal{S}'(\mathbb{R}^n)$). Otherwise (that is possible only for a finite number of values of l) the above Fourier transform is a distribution concentrated at the origin. If $|z|\geq 1$, we have

$$\widehat{a_\nu}(\theta,\xi,z)=\sum_{l=l_0}^{\infty}\sum_{\mu=1}^{\delta_l}\left\{g_{l\mu}(\theta,\xi)(-i)^l 2^{n+\gamma+\nu}\pi^{n/2}\frac{\Gamma\left(\frac{n+l+\gamma+\nu}{2}\right)}{\Gamma\left(\frac{l-\gamma-\nu}{2}\right)}\cdot\frac{Y_{l\mu}\left(\frac{z}{|z|}\right)}{|z|^{n+\gamma+\nu}}\right\},$$
$$l_0\geq 0. \tag{4.1.20}$$

The series (4.1.20) is a priori convergent in the distribution sense. However by virtue of (4.1.4), (4.1.5) it converges uniformly for $|z|=1$, so that for $|z|\geq 1$,

$$|\widehat{a_\nu}(\theta,\xi,z)|\leq C|z|^{-n-\gamma-\nu}, \tag{4.1.21}$$

where the constant does not depend on ξ,θ.

Next,

$$\int_{\mathbb{R}^n}\left(L^N e^{i(\sigma,z)}\right)\varphi(z)\,dz=(-i)^N\sum_{|\varkappa|=N}\partial_\sigma^\varkappa\int_{\mathbb{R}^n}|z|^{-2N}z^\varkappa e^{i(\sigma,z)}\varphi(z)\,dz$$

(recall that $\varphi(z)=0$ for $|z|\leq 1$). Taking into account that $\eta_1(\sigma)=0$ for $|\sigma|\leq 1$ we find after integration by parts that

$$\int_{\mathbb{R}^n}\eta_1(\sigma)a_\nu(\theta,\xi,\sigma)\int_{\mathbb{R}^n}\left(L^N e^{i(\sigma,z)}\right)\varphi(z)\,dz$$
$$=i^N\sum_{|\varkappa|=N}\int_{\mathbb{R}^n}\partial_\sigma^\varkappa[\eta_1(\sigma)a_\nu(\theta,\xi,\sigma)]\,d\sigma\int_{\mathbb{R}^n}|z|^{-2N}z^\varkappa e^{i(\sigma,z)}\varphi(z)\,dz.$$

Obviously,

$$\partial_\sigma^\varkappa[\eta_1(\sigma)a_\nu(\theta,\xi,\sigma)]=\eta_1(\sigma)\partial_\sigma^\varkappa a_\nu(\theta,\xi,\sigma)+R(\theta,\xi,\sigma),$$

where $R(\theta,\xi,\sigma)=0$ for $|\sigma|\leq 1$ and $|\sigma|\geq 2$. The function $\partial_\sigma^\varkappa a_\nu(\theta,\xi,\sigma)$ is homogeneous in σ of the degree $\gamma+\nu-N<-n$. By the Fubini theorem

$$\langle\Phi_\nu^{0,2}(\cdot,\theta,\xi),\varphi\rangle=\int_{\mathbb{R}^n}\widehat{a_\nu}(\theta,\xi,-z)\varphi(z)\,dz$$
$$-i^N\sum_{|\varkappa|=N}\int_{\mathbb{R}^n}\left\{\int_{\mathbb{R}^n}[\eta_1(\sigma)\partial_\sigma^\varkappa a_\nu(\theta,\xi,\sigma)+R(\theta,\xi,\sigma)]\,e^{i(\sigma,z)}\,d\sigma\right\}|z|^{-2N}z^\varkappa\varphi(z)\,dz,$$

whence for $|z| \geq 1$

$$\Phi_\nu^{0,2}(z,\theta,\xi) = \widehat{a_\nu}(\theta,\xi,-z) - i^N |z|^{-2N} \times \sum_{|\varkappa|=N} z^\varkappa \int_{\mathbb{R}^n} [\eta_1(\sigma)\partial_\sigma^\varkappa a_\nu(\theta,\xi,\sigma) + R(\theta,\xi,\sigma)]\, e^{i(\sigma,z)}\, d\sigma,$$

$$\left|\Phi_\nu^{0,2}(z,\theta,\xi)\right| \leq |\widehat{a_\nu}(\theta,\xi,-z)| + C|z|^{-N},$$

and due to (4.1.21)

$$\left|\Phi_\nu^{0,2}(z,\xi,\theta)\right| \leq C|z|^{-n-\gamma-\nu}, \quad |z| \geq 1, \tag{4.1.22}$$

where C does not depend on ξ, θ.

Let us find a bound for $\Phi_\nu^{0,3}(z,\theta,\xi)$, $|z| \geq 1$. Let $z = (z_1, \ldots, z_n)$; consider such a number j that $|z_j| \geq |z_k|$, $k = 1, \ldots, n$. Then $|z| \leq \sqrt{n}|z_j|$. Integrating $N_1 = n + \nu + [\gamma] + 1$ times by parts with respect to the variable σ_j we find that

$$\Phi_\nu^{0,3}(z,\theta,\xi) = \frac{(-1)^{N_1}}{(iz_j)^{N_1}} \int_{\mathbb{R}^n} e^{i(\sigma,z)} \partial_{\sigma_j}^{N_1} \left\{\eta_0(\sigma) P_\nu(\sigma) [a(\theta,\xi,\sigma)]^2 h(a(\theta,\xi,\sigma))\right\} d\sigma.$$

Since $\eta_0(\sigma) = 1$ as $|\sigma| \leq 1$, all the terms emerging in the process of differentiating, which contain derivatives of η_0, are continuous and have compact supports. Next, computing the derivative

$$\partial_{\sigma_j}^{N_1} \left\{\eta_0(\sigma) P_\nu(\sigma) [a(\theta,\xi,\sigma)]^2 h(a(\theta,\xi,\sigma))\right\}$$

with the use of the Leibnitz formula and the Faà di Bruno formula for higher derivatives of a superposition of functions [95], it is not hard to calculate that the maximal order of singularity at the point $\sigma = 0$ does not exceed $n + [\gamma] + 1 - 3\gamma < n$. Therefore for $|z| \geq 1$ we have

$$\left|\Phi_\nu^{0,3}(z,\theta,\xi)\right| \leq C|z_j|^{-N_1} \leq C|z|^{-n-\gamma-\nu}, \quad |z| \geq 1,$$

that is $\left|\Phi_\nu^0(z,\theta,\xi)\right| \leq C|z|^{-n-\gamma-\nu}$.

The function $\Phi_\nu^1(z,\theta,\xi)$ is estimated in the same way as $\Phi_\nu^{0,3}(z,\theta,\xi)$, due to the fact that $\eta_1(\sigma) = 0$ as $|\sigma| \leq 1$. We get

$$\left|\Phi_\nu^1(z,\theta,\xi)\right| \leq C|z|^{-n-\gamma-\nu}, \quad |z| \geq 1,$$

so that $|\Phi_\nu(z,\theta,\xi)| \leq C|z|^{-n-\gamma-\nu}$, $|z| \geq 1$. In order to prove (4.1.18), it remains to notice that

$$\left|\Phi_\nu^1(z,\theta,\xi)\right| \leq \int_{\mathbb{R}^n} |P_\nu(\sigma)| e^{-a_0|\sigma|^\gamma}\, d\sigma$$

as $|z| \leq 1$. □

Let $Q_\alpha(r,\sigma)$ be a homogeneous function in σ of the degree $\alpha > 0$ depending on a parameter r of an arbitrary nature, which has, as $\sigma \neq 0$, N continuous derivatives in σ, and

$$|\partial_\sigma^\varkappa Q_\alpha(r,\sigma)| \leq CB_N(r)|\sigma|^{\alpha-|\varkappa|},$$

whenever $|\varkappa| \leq N$, $\sigma \neq 0$, for all values of r. Let us consider the function

$$\Psi_\alpha(z,\theta,\xi,r) = \int_{\mathbb{R}^n} Q_\alpha(r,\sigma)\exp\{(iz,\sigma) - a(\theta,\xi,\sigma)\}\,d\sigma.$$

Lemma 4.2. *If* $N \geq 2n + [\alpha] + [\gamma] + 1$, *then*

$$|\Psi_\alpha(z,\theta,\xi,r)| \leq CB(r)(1+|z|)^{-n-\alpha}, \tag{4.1.23}$$

where C *does not depend on* z,ξ,θ,r.

The *proof* consists essentially of repetition of the arguments from the proof of Lemma 4.1. The only new point that appears is the following. In the decomposition of the integral Ψ_α the term

$$\Psi_\alpha^{0,1}(z,r) = \int_{\mathbb{R}^n} \eta_0(\sigma)Q_\alpha(r,\sigma)\exp\{i(z,\sigma)\}\,d\sigma$$

(in contrast to the function $\Phi_\nu^{0,1}$ from the proof of Lemma 4.1) need not belong to $\mathcal{S}(\mathbb{R}^n)$. The technique of Lemma 4.1 gives the estimate

$$\left|\Psi_\alpha^{0,1}(z,r)\right| \leq CB(r)|z|^{-n-\alpha}, \quad |z| \geq 1,$$

which leads to the inequality (4.1.23). □

Let us consider the function

$$\Phi_0(z,\theta,\xi) = \int_{\mathbb{R}^n} \exp\{i(z,\sigma) - a(\theta,\xi,\sigma)\}\,d\sigma.$$

Lemma 4.3. *If* $N \geq 2n + [\gamma]$, *then for any* $z,\xi_1,\xi_2 \in \mathbb{R}^n$, $\theta > 0$,

$$|\Phi_0(z,\theta,\xi_1) - \Phi_0(z,\theta,\xi_2)| \leq C|\xi_1 - \xi_2|^\lambda(1+|z|)^{-n-\gamma}.$$

The *proof* is similar to that of Lemma 4.1. Some technical differences are caused by the necessity to obtain estimates of expressions of the form

$$\partial_{\sigma_j}^l[h(a(\theta,\xi_1,\sigma)) - h(a(\theta,\xi_2,\sigma))]$$

in two situations: for $h(s) = e^{-s}$, $|\sigma| \geq 1$, and for a function h about which we know only that it has bounded derivatives of any order, for $|\sigma| \leq 2$.

By the Faà di Bruno formula

$$\partial^l_{\sigma_j}[h(a(\theta,\xi_1,\sigma)) - h(a(\theta,\xi_2,\sigma))] = \sum C_{m_1\dots m_\mu}\Big\{ h^{(k)}(a(\theta,\xi_1,\sigma))$$
$$\times \left[\partial_{\sigma_j} a(\theta,\xi_1,\sigma)\right]^{m_1}\left[\partial^2_{\sigma_j} a(\theta,\xi_1,\sigma)\right]^{m_2}\cdots\left[\partial^\mu_{\sigma_j} a(\theta,\xi_1,\sigma)\right]^{m_\mu}$$
$$- h^{(k)}(a(\theta,\xi_2,\sigma))\left[\partial_{\sigma_j} a(\theta,\xi_2,\sigma)\right]^{m_1}\left[\partial^2_{\sigma_j} a(\theta,\xi_2,\sigma)\right]^{m_2}\cdots\left[\partial^\mu_{\sigma_j} a(\theta,\xi_2,\sigma)\right]^{m_\mu}\Big\}, \tag{4.1.24}$$

where the summation is over all tuples $(m_1,\dots,m_\mu)$ of natural numbers, such that $m_1+2m_2+\dots+\mu m_\mu = l$ while μ varies from 1 to l; here $k = m_1+\dots+m_\mu$. Let us use the elementary formulas

$$b_1\cdots b_\nu - b'_1\cdots b'_\nu = (b_1-b'_1)b_2\cdots b_\nu + b'_1(b_2-b'_2)b_3\cdots b_\nu + \dots + b'_1b'_2\cdots b'_{\nu-1}(b_\nu - b'_\nu);$$
$$b_\nu - (b')^\nu = (b-b')(b^{\nu-1} + b^{\nu-2}b' + \dots + (b')^{\nu-1}.$$

Each summand in (4.1.24) is a sum of expressions containing one of the differences

$$h^{(k)}(a(\theta,\xi_1,\sigma)) - h^{(k)}(a(\theta,\xi_2,\sigma))$$

or

$$\left[\partial^\nu_{\sigma_j} a(\theta,\xi_1,\sigma)\right]^{m_\nu} - \left[\partial^\nu_{\sigma_j} a(\theta,\xi_2,\sigma)\right]^{m_\nu}$$

multiplied by a bounded function and by a homogeneous function of the degree $m_1(\gamma-1)+m_2(\gamma-2)+\dots+m_\nu(\gamma-\nu) = \gamma k - l$ (in the first case), or $\gamma k - l - m_\nu(\gamma-\nu)$ (in the second case). Further,

$$\left|h^{(k)}(a(\theta,\xi_1,\sigma)) - h^{(k)}(a(\theta,\xi_2,\sigma))\right| \le C|a(\theta,\xi_1,\sigma) - a(\theta,\xi_2,\sigma)| \le C|\xi_1-\xi_2|^\lambda|\sigma|^\gamma;$$

$$\left|\left[\partial^\nu_{\sigma_j} a(\theta,\xi_1,\sigma)\right]^{m_\nu} - \left[\partial^\nu_{\sigma_j} a(\theta,\xi_2,\sigma)\right]^{m_\nu}\right| \le C|\xi_1-\xi_2|^\lambda|\sigma|^{m_\nu(\gamma-\nu)}.$$

Thus for $|\sigma|\le 2$

$$\left|\partial^l_{\sigma_j}[h(a(\theta,\xi_1,\sigma)) - h(a(\theta,\xi_2,\sigma))]\right| \le C|\xi_1-\xi_2|^\lambda|\sigma|^{\gamma-l};$$

for $h(s) = e^{-s}$, $|\sigma|\ge 1$, we get

$$\left|\partial^l_{\sigma_j}[e^{-a(\theta,\xi_1,\sigma)} - e^{-a(\theta,\xi_2,\sigma)}]\right| \le C|\xi_1-\xi_2|^\lambda|\sigma|^{\gamma+(\gamma-l)l}e^{-a_0|\sigma|^\gamma}.$$

With these estimates taken into account the proof of Lemma 4.3 goes just like that of Lemma 4.1. □

The next lemma can be proved in a similar way.

Lemma 4.4. *Assume the conditions of Lemma 4.2. Then*

$$|\Psi_\alpha(z,\theta,\xi_1,r) - \Psi_\alpha(z,\theta,\xi_2,r)| \le CB(r)|\xi_1-\xi_2|^\lambda(1+|z|)^{-n-\gamma-\alpha},$$

where C does not depend on ξ_1,ξ_2,θ,r.

Under the above conditions (A_{41}), (A_{42}) we obtain from Lemmas 4.1 – 4.4 the following estimates of the parameterized fundamental solution Z (uniformly with respect to the parameters):

$$|\partial_x^\varkappa Z_0(t-\mu,x-y,\theta,\xi)| \le C_\varkappa(t-\mu)\left[(t-\mu)^{1/\gamma}+|x-y|\right]^{-n-\gamma-|\varkappa|}, \qquad (4.1.25)$$
$$|\varkappa| \le N-2n-[\gamma],$$

$$|\partial_t Z_0(t-\mu,x-y,\theta,\xi)| \le C\left[(t-\mu)^{1/\gamma}+|x-y|\right]^{-n-\gamma}; \qquad (4.1.26)$$

$$|Z_0(t-\mu,x-y,\theta,\xi_1) - Z_0(t-\mu,x-y,\theta,\xi_2)|$$
$$\le C(t-\mu)|\xi_1-\xi_2|^\lambda\left[(t-\mu)^{1/\gamma}+|x-y|\right]^{-n-\gamma}; \qquad (4.1.27)$$

$$|\partial_t Z_0(t-\mu,x-y,\theta,\xi_1) - \partial_t Z_0(t-\mu,x-y,\theta,\xi_2)|$$
$$\le C(t-\mu)|\xi_1-\xi_2|^\lambda\left[(t-\mu)^{1/\gamma}+|x-y|\right]^{-n-2\gamma}. \qquad (4.1.28)$$

We have the equation

$$\int\limits_{\mathbb{R}^n} Z_0(t-\mu,x-y,\theta,\xi)\,dy = 1, \quad z,\xi\in\mathbb{R}^n,\ t>\mu,\ \theta\ge 0, \qquad (4.1.29)$$

which follows from (4.1.17) and the Fourier inversion. Using (4.1.26) we get from (4.1.29) that

$$\int\limits_{\mathbb{R}^n} \partial_t Z(t-\mu,x-y,\theta,\xi)\,dy = 0. \qquad (4.1.30)$$

The inequality

$$\left|\int\limits_{\mathbb{R}^n} \partial_t Z(t-\mu,x-y,\mu,y)\,dy\right| \le C(t-\mu)^{-1+\lambda/\gamma} \qquad (4.1.31)$$

follows from (4.1.30) and (4.1.28).

4.1.5. An example. As we see from the above estimates, the fundamental solution has at least a power-like decay at infinity. The next example shows that the estimates are exact, so that, in contrast to the parabolic differential equations considered in Chapter 2, here the FSCP does not possess an exponential decay.

Let $n = 1$, $a(t, x, \xi) \equiv |\xi|^\gamma$, $1 < \gamma < 2$, $a_1 = \cdots = a_m = 0$. We shall consider the asymptotics of the function

$$Z_0(t, x) = \frac{1}{\pi} \operatorname{Re} \int_0^\infty e^{-ix\xi - |\xi|^\gamma t} \, d\xi, \quad t > 0, \tag{4.1.32}$$

as $x \to \infty$ and t is fixed. Since the function Z_0 is even in x, it is sufficient to consider $x \to +\infty$. We also may set $t = 1$; the general case is obtained by a change of variables.

Let us calculate the Mellin transform

$$(Mg)(s) = \int_0^\infty x^{s-1} g(x) \, dx, \quad 0 < \operatorname{Re} s < 1,$$

of the function $g(x) = \frac{1}{\pi} \int_0^\infty e^{-ix\xi - |\xi|^\gamma} \, d\xi$. The path of integration in (4.1.32) may be replaced by a ray $\arg \xi = -\delta$, $0 < \delta < \pi/2\gamma$, so that

$$(Mg)(s) = \int_0^\infty dx \int_0^\infty G(x, r, s) \, dr \tag{4.1.33}$$

with

$$G(x, r, s) = \frac{1}{\pi} x^{s-1} e^{-i\delta} \exp(-irxe^{-i\delta}) \exp(-r^\gamma e^{-i\gamma\delta}).$$

For $0 < \delta < \pi/2\gamma$ we obtain that

$$\int_0^\infty dr \int_0^\infty |G(x, r, s)| \, dx = \frac{1}{\pi\gamma} a^{-s'} \Gamma(s') b^{(s'-1)/\gamma} \Gamma\left(\frac{1 - s'}{\gamma}\right),$$

with $s' = \operatorname{Re} s$, $a = \sin \delta > 0$, $b = \cos(\gamma\delta) > 0$. Hence Fubini's theorem may be applied to the integral (4.1.33). Setting $z = ix \exp\{-i\delta\}$ we rewrite the integral in x as an integral along the ray $\arg z = \frac{\pi}{2} - \delta$. The latter integral can be replaced by an integral along the positive half-axis and evaluated in a closed form. The remaining integration in r is performed similarly – by the change of variables $z = r^\gamma e^{-i\gamma\delta}$ we obtain an integral along the ray $\arg z = -\gamma\delta$, and then replace it by the integral along the positive half-axis. As a result, we find that

$$(Mg)(s) = \frac{1}{\pi\gamma} (-i)^s \Gamma(s) \Gamma\left(\frac{1 - s}{\gamma}\right).$$

If $f(x) = Z_0(1, x)$, then for a real s we have $(Mf(s) = \operatorname{Re}(Mg)(s)$. Using the identity $\Gamma(z)\Gamma(1 - z) = \dfrac{\pi}{\sin \pi z}$ we find that

$$(Mf)(s) = \frac{1}{\gamma} \cdot \frac{\Gamma(s)\Gamma\left(\frac{1-s}{\gamma}\right)}{\Gamma\left(\frac{1-s}{2}\right)\Gamma\left(\frac{1+s}{2}\right)}. \tag{4.1.34}$$

By analytic continuation, the identity (4.1.34) is valid on the strip $0 < \operatorname{Re} s < 1$. The inverse Mellin transform is given by

$$Z_0(1,x) = \frac{1}{2\pi i} \int\limits_{c-i\infty}^{c+i\infty} x^{-s}(Mf)(s)\, ds$$

where the path of integration is the straight line from $c - i\infty$ to $c + i\infty$ with $0 < c < 1$. Replacing s by $-s$ and inserting (4.1.34) we get

$$Z_0(1,x) = \frac{1}{2\pi i\gamma} \int\limits_{-c-i\infty}^{-c+i\infty} \frac{\Gamma(-s))\Gamma\left(\frac{1+s}{\gamma}\right)}{\Gamma\left(\frac{1-s}{2}\right)\Gamma\left(\frac{1+s}{2}\right)} x^s\, ds. \tag{4.1.35}$$

The path of integration in (4.1.35) may be deformed into one running clockwise around $\mathbb{R}_+ - c$. Now the formula (4.1.35) is actually the Fox function representation (see Appendix)

$$Z_0(1,x) = \frac{1}{\gamma} H_{22}^{11}\left[x \left| \begin{matrix} \left(1 - \frac{1}{\gamma}, \frac{1}{\gamma}\right), & \left(\frac{1}{2}, \frac{1}{2}\right) \\ (0,1), & \left(\frac{1}{2}, \frac{1}{2}\right) \end{matrix} \right. \right]. \tag{4.1.36}$$

The representation (4.1.36), together with the asymptotic formula (A.17) from the Appendix, shows that

$$Z_0(1,x) \sim \frac{1}{\pi}\Gamma(1+\gamma)\sin\frac{\pi\gamma}{2} \cdot x^{-1-\gamma}, \quad x \to \infty.$$

This asymptotic relation means that the estimate (4.1.25) is exact.

4.1.6. Heat potential. Let us consider the heat potential

$$u(t,x,\tau) = \int\limits_{\tau}^{t} d\mu \int\limits_{\mathbb{R}^n} Z_0(t-\mu, x-y; \mu, y) f(\mu, y)\, dy. \tag{4.1.37}$$

We assume the above conditions (A_{41}), (A_{42}) with $N \geq 2n + 2[\gamma] + 1$. Suppose also that

$$|f(\mu,y)| \leq C(\mu-\tau)^{-\rho}, \quad |f(\mu,x) - f(\mu,y)| \leq C|x-y|^{\lambda}(\mu-\tau)^{-\rho},$$

where $0 \leq \rho < 1$, $0 < \lambda \leq 1$. The integral in (4.1.37) is absolutely convergent due to (4.1.25). It also follows from (4.1.25) that the function $u(x,t,\tau)$ has continuous derivatives in x of any order $< \gamma$, and they can be obtained by differentiating under the integral sign.

Just as in the case of parabolic partial differential equations (see, e.g., Property 2.6), we obtain the formula

$$\partial_t u(t,x,\tau) = f(t,x) + \int_\tau^t d\theta \int_{\mathbb{R}^n} \partial_t Z_0(t-\theta, x-\xi; \theta, \xi)[f(\theta,\xi) - f(\theta,x)]\, d\xi$$
$$+ \int_\tau^t f(\theta,x)\, d\theta \int_{\mathbb{R}^n} \partial_t Z_0(t-\theta, x-\xi; \theta,\xi)\, d\xi. \tag{4.1.38}$$

The proof is based on the estimates (4.1.26), (4.1.28), and (4.1.31).

Now we shall consider the action of HSI operators $\mathbf{D}_\Omega^\alpha$ of the form (4.1.9) on the potential (4.1.37). For brevity we do not indicate the dependence on the parameter t in the characteristics and symbols of HSIs.

Let us begin with the simplest case when $\alpha < \gamma$. Take α not an integer, and let $l \geq [\alpha]+1$. It follows from (4.1.10) and (4.1.25) that for $|h| \leq (t-\mu)^{1/\gamma}$,

$$\left|\left(\mathbf{\Delta}_h^l Z_0\right)(t-\mu, x-y; \mu, y)\right|$$
$$\leq C|h|^{[\alpha]+1}(t-\mu) \sum_{\nu=0}^{l} \left[(t-\mu)^{1/\gamma} + |x - \theta_\nu \nu h - y|\right]^{-n-\gamma-[\alpha]-1},$$
$$0 < \theta_\nu < 1. \tag{4.1.39}$$

By (4.1.25),

$$\left|\left(\mathbf{\Delta}_h^l Z_0\right)(t-\mu, x-y; \mu, y)\right| \leq C(t-\mu) \sum_{\nu=0}^{l} \left[(t-\mu)^{1/\gamma} + |x-y-\nu h|\right]^{-n-\gamma}, \tag{4.1.40}$$

for $|h| \geq (t-\mu)^{1/\gamma}$. For $\alpha < \gamma$ from these estimates and the Fubini theorem we get that the HSI $\mathbf{D}_\Omega^\alpha$ converges absolutely, and

$$\left(\mathbf{D}_\Omega^\alpha u\right)(t,x,\tau) = \int_\tau^t d\mu \int_{\mathbb{R}^n} Z_\Omega(t-\mu, x, x-y, \mu, y) f(\mu, y)\, dy \tag{4.1.41}$$

where $Z_\Omega = \mathbf{D}_\Omega^\alpha Z_0$, that is

$$Z_\Omega(s,x,z,p,\beta) = (2\pi)^{-n} \int_{\mathbb{R}^n} \widetilde{\Omega}(x,\sigma) \exp\{i(z,\sigma) - a(p,\beta,\sigma)s\}\, d\sigma,$$

$\widetilde{\Omega}$ being the symbol of the HSI under consideration. By using (4.1.12), it is easy to see that (4.1.41) remains in force also in the case of an odd integer $\alpha < \gamma$ and an even characteristic Ω.

The case $\alpha = \gamma$ is much more complicated. Assume that the symbol $\widetilde{\Omega}(x,\sigma)$ has $N \geq 2n + 2[\gamma] + 1$ continuous derivatives in $\sigma \neq 0$, and that

$$\left|\partial_\sigma^\varkappa \widetilde{\Omega}(x,\sigma)\right| \leq C_N |\sigma|^{\gamma-|\varkappa|},$$
$$\left|\partial_\sigma^\varkappa \left[\widetilde{\Omega}(x,\sigma) - \widetilde{\Omega}(y,\sigma)\right]\right| \leq C_N |x-y|^\lambda |\sigma|^{\gamma-|\varkappa|}, \qquad |\varkappa| \leq N.$$

Let also $\widetilde{\Omega}(x,\sigma) \neq 0$ if $\sigma \neq 0$. If γ is an integer, and the symbol $\widetilde{\Omega}(x,\sigma)$ is not a polynomial in σ (which is possible only for an odd γ and an even characteristic Ω), we assume in addition that in the expansion

$$\left[\widetilde{\Omega}(x,\sigma)\right]^{-1} = \sum_{\nu=0}^{\infty} \sum_{\mu=1}^{\delta_{2\nu}} c_{2\nu,\mu} Y_{2\nu,\mu}(\sigma), \quad |\sigma| = 1,$$

$c_{2\nu,\mu}(x) = 0$ if $\gamma = n + 2\nu + 2k$, $k = 0, 1, 2, \ldots$. We shall not consider the case when $\widetilde{\Omega}$ is a polynomial in σ because it is covered by the classical theory of parabolic differential equations.

Lemma 4.5. *Under the above conditions the hyper-singular integral $\mathbf{D}_\Omega^\gamma u$ exists in the sense of the conditional convergence* (4.1.11), *and*

$$\left(\mathbf{D}_\Omega^\gamma u\right)(t,x,\tau) = \int_\tau^t d\mu \int_{\mathbb{R}^n} Z_\Omega(t-\mu, x, x-y, \mu, y)[f(\mu,y) - f(\mu,x)]\, dy$$
$$+ \int_\tau^t f(\mu,x)\, d\mu \int_{\mathbb{R}^n} [Z_\Omega(t-\mu, x, x-y, \mu, y) - Z_\Omega(t-\mu, x, x-y, \mu, x)]\, dy, \tag{4.1.42}$$

where $Z_\Omega = \mathbf{D}_\Omega^\gamma Z_0$, that is

$$Z_\Omega(t,x,z,\theta,\xi) = (2\pi)^{-n} \int_{\mathbb{R}^n} \widetilde{\Omega}(x,\sigma) \exp\{i(z,\sigma) - a(\theta,\xi,\sigma)t\}\, d\sigma,$$

and $\widetilde{\Omega}(x,\sigma)$ is the symbol of the hyper-singular integral operator $\mathbf{D}_\Omega^\gamma$.

Proof. Denote

$$u_s(t,x,\tau) = \int_\tau^{t-s} d\mu \int_{\mathbb{R}^n} Z_0(t-\mu, x-y; \mu, y) f(\mu,y)\, dy, \quad 0 < s < t-\tau.$$

Using (4.1.39) and (4.1.40) we prove the absolute convergence of the hyper-singular integral $\mathbf{D}_\Omega^\gamma u_s$, and the formula

$$\left(\mathbf{D}_\Omega^\gamma u_s\right)(t,x,\tau) = \int_\tau^{t-s} d\mu \int_{\mathbb{R}^n} Z_\Omega(t-\mu, x, x-y, \mu, y) f(\mu,y)\, dy.$$

We shall need the following properties of the function Z_Ω:

$$\int_{\mathbb{R}^n} Z_\Omega(t,x,z,\theta,\xi)\,dz = 0; \tag{4.1.43}$$

$$|Z_\Omega(t-\mu,x,x-y,\mu,y)| \le C\left[(t-\mu)^{1/\gamma}+|x-y|\right]^{-n-\gamma}; \tag{4.1.44}$$

$$\begin{aligned}&|Z_\Omega(t-\mu,x,x-y,\mu,y)-Z_\Omega(t-\mu,x,x-y,\mu,x)|\\ &\qquad \le C|x-y|^\lambda(t-\mu)\left[(t-\mu)^{1/\gamma}+|x-y|\right]^{-n-2\gamma}.\end{aligned} \tag{4.1.45}$$

The formula (4.1.43) is a consequence of (4.1.29); we use (4.1.39), (4.1.40), and the Fubini theorem. The inequality (4.1.44) follows from Lemma 4.2, while (4.1.45) is a consequence of Lemma 4.4.

According to (4.1.43),

$$\begin{aligned}\left(\mathbf{D}^\gamma_\Omega u_s\right)(t,x,\tau) &= \int_\tau^{t-s} d\mu \int_{\mathbb{R}^n} Z_\Omega(t-\mu,x,x-y,\mu,y)[f(\mu,y)-f(\mu,x)]\,dy\\ &+ \int_\tau^{t-s} f(\mu,x)\,d\mu \int_{\mathbb{R}^n}[Z_\Omega(t-\mu,x,x-y,\mu,y)-Z_\Omega(t-\mu,x,x-y,\mu,x)]\,dy.\end{aligned} \tag{4.1.46}$$

Let $\Phi(t,x,\tau)$ be the function in the right-hand side of (4.1.42) (the integrals converge due to (4.1.44), (4.1.45)). Taking into account (4.1.44), (4.1.45) we see from (4.1.46) that uniformly with respect to $x\in\mathbb{R}^n$,

$$\lim_{s\to 0}\left(\mathbf{D}^\gamma_\Omega u_s\right)(t,x,\tau) = \Phi(t,x,\tau). \tag{4.1.47}$$

Let us consider the restricted hyper-singular integral

$$\left(\mathbf{D}^\gamma_{\Omega,\varepsilon} u_s\right)(t,x,\tau) = \int_\tau^{t-s} d\mu \int_{\mathbb{R}^n} Z_{\Omega,\varepsilon}(t-\mu,x,x-y,\mu,y)f(\mu,y)\,dy, \tag{4.1.48}$$

where $\varepsilon>0$, $Z_{\Omega,\varepsilon}=\mathbf{D}^\gamma_{\Omega,\varepsilon}Z$. Our immediate goal is to obtain an integral representation expressing $\mathbf{D}^\gamma_{\Omega,\varepsilon}u_s$ in terms of $\mathbf{D}^\gamma_\Omega u_s$. We introduce the constant characteristic $\Omega_\xi(\sigma)=\Omega(\xi,\sigma)$ depending on $\xi\in\mathbb{R}^n$ as a parameter. Let $\widetilde{\Omega_\xi}$ be the corresponding symbol. Fixing the rest of the variables we denote $F(x)=Z(t-\mu,x-y;\mu,y)$ and express F via $\mathbf{D}^\gamma_{\Omega_\xi}F$. Since the Fourier transform of the function $G_\xi=\mathbf{D}^\gamma_{\Omega_\xi}F$ equals $\widetilde{\Omega_\xi}(\zeta)\widehat{F}(\zeta)$, we find that

$$F(x)=(2\pi)^{-n}\int_{\mathbb{R}^n}\frac{e^{i\zeta\cdot x}}{\widetilde{\Omega_\xi}(\zeta)}\widehat{G_\xi}(\zeta)\,d\zeta=(2\pi)^{-n}\int_{\mathbb{R}^n}\frac{\widehat{G_{x,\xi}}(\zeta)}{\widetilde{\Omega_\xi}(\zeta)}\,d\zeta,$$

where $G_{x,\xi}(z)=\left(\mathbf{D}^\gamma_{\Omega_\xi}F\right)(x+z)$.

Let us expand the homogeneous function $1/\widetilde{\Omega}_\xi$ in a series of spherical harmonics:

$$\frac{1}{\widetilde{\Omega}_\xi(\zeta)} = |\zeta|^{-\gamma} \sum_{\nu=0}^{\infty} \sum_{\mu=1}^{\delta_\nu} g_{\nu\mu}(\xi) Y_{\nu\mu}\left(\frac{\zeta}{|\zeta|}\right).$$

For each fixed $\xi \in \mathbb{R}^n$ the series converges uniformly with respect to ζ, so that

$$F(x) = (2\pi)^{-n} \sum_{\nu=0}^{\infty} \sum_{\mu=1}^{\delta_\nu} g_{\nu\mu}(\xi) \int_{\mathbb{R}^n} \widehat{G_{x,\xi}}(\zeta) Y_{\nu\mu}\left(\frac{\zeta}{|\zeta|}\right) |\zeta|^{-\gamma}\, d\zeta. \qquad (4.1.49)$$

The Fourier transform (in the sense of the space $\mathcal{S}'(\mathbb{R}^n)$) of the distribution $|\zeta|^{-\gamma} Y_{\nu\mu}\left(\frac{\zeta}{|\zeta|}\right)$ equals

$$i^\nu \pi^{n/2} 2^{n-\gamma} \frac{\Gamma((n+\nu-\gamma)/2)}{\Gamma((\nu+\gamma)/2)} |z|^{\gamma-n} \check{Y}_{\nu\mu}\left(\frac{z}{|z|}\right)$$

(see (4.1.8)), where $\check{Y}_{\nu\mu}(\sigma) = \check{Y}_{\nu\mu}(-\sigma)$. Strictly speaking, we cannot use this to transform the integral in (4.1.49), since $G_{x,\xi} \notin \mathcal{S}(\mathbb{R}^n)$. However if $n/2 < \operatorname{Re}\lambda < n$, then the formula

$$\int_{\mathbb{R}^n} \widehat{G_{x,\xi}}(\zeta) Y_{\nu\mu}\left(\frac{\zeta}{|\zeta|}\right) |\zeta|^{-\lambda}\, d\zeta$$

$$= i^\nu \pi^{n/2} 2^{n-\lambda} \frac{\Gamma((n+\nu-\lambda)/2)}{\Gamma((\nu+\lambda)/2)} \int_{\mathbb{R}^n} G_{x,\xi}(z) |z|^{\lambda-n} \check{Y}_{\nu\mu}\left(\frac{z}{|z|}\right) dz \qquad (4.1.50)$$

is nevertheless valid, since in this case the function $|\zeta|^{-\lambda} Y_{\nu\mu}\left(\frac{\zeta}{|\zeta|}\right)$ can be represented as a sum of functions from $L_1(\mathbb{R}^n)$ and $L_2(\mathbb{R}^n)$. An explicit form of the function $\widehat{G_{x,\xi}}$ and the estimate for the function $G_{x,\xi}$ (see [150]) enable us to perform in the standard way [87, 150] the analytic continuation in λ and to prove (4.1.50) for $\lambda = \gamma \neq n + \nu + 2k$, $k = 0, 1, 2, \ldots$.

Denote

$$H(\xi, z) = \pi^{n/2} 2^{n-\gamma} |z|^{\gamma-n} \frac{\Gamma((n+\nu-\gamma)/2)}{\Gamma((\nu+\gamma)/2)}$$

$$\times \sum_{\nu=0}^{\infty} \sum_{\mu=1}^{\delta_\nu} i^\nu \frac{\Gamma((n+\nu-\gamma)/2)}{\Gamma((\nu+\gamma)/2)} g_{\nu\mu}(\xi) Y_{\nu\mu}\left(\frac{z}{|z|}\right).$$

The series converges uniformly with respect to z. Moreover, it follows from (4.1.5), (4.1.6), and (4.1.7) that the function $H(\xi, z)$ is smooth in z for $z \neq 0$: $H(\xi, \cdot) \in C^{[\gamma]+1}(S^{n-1})$. Now

$$F(x) = (2\pi)^{-n} \int_{\mathbb{R}^n} H(\xi, z) \left(\mathbf{D}^\gamma_{\Omega_\xi} F\right)(x+z)\, dz,$$

whence

$$\left(\mathbf{D}^{\gamma}_{\Omega_\xi,\varepsilon}F\right)(x) = (2\pi)^{-n}\left[d_{n,l}(\gamma)\right]^{-1}\int\limits_{\mathbb{R}^n}\left(\mathbf{D}^{\gamma}_{\Omega_\xi}F\right)(\zeta)\,d\zeta$$

$$\times\int\limits_{|h|>\varepsilon}\frac{\mathbf{\Delta}^l_h H_1(\xi,x-\zeta)}{|h|^{n+\gamma}}\Omega_\xi\left(\frac{h}{|h|}\right)dh,\quad H_1(\xi,\sigma)=H(\xi,-\sigma).$$

Writing out the expression for the difference $\mathbf{\Delta}^l_h$ in detail and making the substitution $h=|\zeta|y$, $\zeta=\varepsilon\eta$, after elementary transformations we get

$$\left(\mathbf{D}^{\gamma}_{\Omega_\xi,\varepsilon}F\right)(x) = \int\limits_{\mathbb{R}^n}K(\xi,\eta)\left(\mathbf{D}^{\gamma}_{\Omega_\xi}F\right)(x-\varepsilon\eta)\,d\eta, \tag{4.1.51}$$

where

$$K(\xi,\eta) = (2\pi)^{-n}\left[d_{n,l}(\gamma)\right]^{-1}|\eta|^{-n}\int\limits_{|y|>|\eta|^{-1}}\frac{\left(\mathbf{\Delta}^l_y H_1\right)\left(\xi,\frac{\eta}{|\eta|}\right)}{|y|^{n+\gamma}}\Omega_\xi\left(\frac{y}{|y|}\right)dy. \tag{4.1.52}$$

We need the following auxiliary result.

Lemma 4.6. *The function $K(\xi,\eta)$ is integrable with respect to η, and*

$$\int\limits_{\mathbb{R}^n}K(\xi,\eta)\,d\eta = 1. \tag{4.1.53}$$

Proof. An estimate of K for $|\eta|\to 0$ immediately follows from (4.1.52):

$$|K(\xi,\eta)| \le \begin{cases} C|\eta|^{-n+\gamma}, & \text{if } \gamma<n,\\ C, & \text{if } \gamma\ge n,\end{cases}\qquad |\eta|\to 0. \tag{4.1.54}$$

In order to obtain an estimate for $|\eta|\to\infty$, we first prove that

$$\left(\mathbf{D}^{\gamma}_{\Omega_\xi}H_1\right)(\eta) = \left[d_{n,l}(\gamma)\right]^{-1}\int\limits_{\mathbb{R}^n}\frac{\left(\mathbf{\Delta}^l_y H_1\right)(\xi,\eta)}{|y|^{n+\gamma}}\Omega_\xi\left(\frac{y}{|y|}\right)dy = 0 \tag{4.1.55}$$

for any $\eta\neq 0$.

Let us consider the functional

$$\langle R,\varphi\rangle = \lim_{\varepsilon\to 0}\int\limits_{\mathbb{R}^n}\varphi(x)\left(\mathbf{D}_{\Omega_\xi,\varepsilon}H_1\right)(x)\,dx$$

defined on real-valued $\varphi \in \mathcal{S}(\mathbb{R}^n)$. Writing the difference operator explicitly, after interchanging integrals we find that

$$\langle R, \varphi \rangle = \lim_{\varepsilon \to 0} \int_{\mathbb{R}^n} H_1(\xi, \eta) \left(\mathbf{D}^{\gamma}_{\Omega^*_\xi, \varepsilon} \varphi \right) (\eta)\, d\eta, \quad \Omega^*_\xi(\theta) = \Omega_\xi(-\theta).$$

By (4.1.13),

$$\left| \left(\mathbf{D}^{\gamma}_{\Omega^*_\xi, \varepsilon} \varphi \right) (\eta) \right| \leq C(1 + |\eta|)^{-n-\gamma} \tag{4.1.56}$$

where the constant does not depend on $\varepsilon \geq 0$. Therefore

$$\langle R, \varphi \rangle = \int_{\mathbb{R}^n} H_1(\xi, \eta) \left(\mathbf{D}^{\gamma}_{\Omega^*_\xi} \varphi \right) (\eta)\, d\eta. \tag{4.1.57}$$

The functional R is continuous, that is $R \in \mathcal{S}'(\mathbb{R}^n)$. Indeed, if $\varphi_m \to 0$ in $\mathcal{S}(\mathbb{R}^n)$, we have

$$|\langle R, \varphi_m \rangle| \leq \int_{|\eta| \leq 1} \left| H_1(\xi, \eta) \left(\mathbf{D}^{\gamma}_{\Omega^*_\xi} \varphi_m \right) (\eta) \right| d\eta + \int_{|\eta| > 1} \left| H_1(\xi, \eta) \left(\mathbf{D}^{\gamma}_{\Omega^*_\xi} \varphi_m \right) (\eta) \right| d\eta.$$

The second summand tends to 0, as $m \to \infty$, by the estimate (4.1.56) with $\varepsilon = 0$ and the constant independent of m. In the first summand the HSI $\mathbf{D}^{\gamma}_{\Omega^*_\xi} \varphi_m$ is estimated directly with the use of (4.1.10).

Let us calculate the Fourier transform $\widehat{R}$,

$$\langle \widehat{R}, \varphi \rangle = (2\pi)^n \langle R, \widetilde{\varphi} \rangle,$$

where $\widetilde{\varphi}$ is the inverse Fourier transform of φ. By (4.1.57),

$$\langle \widehat{R}, \varphi \rangle = \int_{\mathbb{R}^n} H_1(\xi, \eta) \left(\mathbf{D}^{\gamma}_{\Omega^*_\xi} \widetilde{\varphi} \right) (\eta)\, d\eta.$$

Since the symbol of the HSI $\mathbf{D}^{\gamma}_{\Omega^*_\xi}$ equals $\widetilde{\Omega}_\xi(-\zeta)$, we find that

$$\langle \widehat{R}, \varphi \rangle = (2\pi)^n \int_{\mathbb{R}^n} H_1(\xi, \eta) F^{-1}_{\zeta \to \eta} \left[\widetilde{\Omega}_\xi(-\zeta) \varphi(-\zeta) \right] d\eta.$$

As before (see (4.1.50)) we can throw the Fourier transform over H_1, and since H_1 emerged from the expansion of $1/\widetilde{\Omega}$ in spherical harmonics, we come to the equality

$$\langle \widehat{R}, \varphi \rangle = \int_{\mathbb{R}^n} \varphi(\zeta)\, d\zeta,$$

that is $R = \delta$.

Suppose that the support of the function φ is contained in a layer $\nu < |x| < N$. Direct estimates show that $\left|\left(\mathbf{D}^{\gamma}_{\Omega_\xi,\varepsilon} H_1\right)(x)\right| \le C$ for $\nu < |x| < N$ where C depends on ν and N, but does not depend on ε. Then we may pass to the limit, as $\varepsilon \to 0$, in the integral defining the functional R, and since $\langle R, \varphi\rangle = \varphi(0) = 0$, we find that

$$\int_{\mathbb{R}^n} \varphi(x) \left(\mathbf{D}^{\gamma}_{\Omega_\xi} H_1\right)(x) = 0.$$

Since φ is arbitrary, this implies (4.1.55).

Using (4.1.55) we can rewrite (4.1.52) for $\eta \ne 0$ as

$$K(\xi,\eta) = -(2\pi)^{-n}\left[d_{n,l}(\gamma)\right]^{-1} |\eta|^{-n} \int_{|y|\le|\eta|^{-1}} \frac{\left(\mathbf{\Delta}^l_y H_1\right)\left(\xi, \frac{\eta}{|\eta|}\right)}{|y|^{n+\gamma}} \Omega_\xi\left(\frac{y}{|y|}\right) dy.$$

Suppose first that γ is not an integer. Since $H_1 \in C^{[\gamma]+1}(S^{n-1})$, we find that

$$\left|\left(\mathbf{\Delta}^l_y H_1\right)(\xi,\eta)\right| \le C|y|^{[\gamma]+1},$$

and an elementary estimate based on (4.1.10) shows that

$$|K(\xi,\eta)| \le C|\eta|^{-n-1+\{\gamma\}}, \quad |\eta| \to \infty, \tag{4.1.58}$$

where $\{\gamma\}$ is the fractional part of γ.

Thus, for the case, where γ is not an integer, the inequalities (4.1.54) and (4.1.58) imply the integrability of K.

Let γ be an odd integer, and the characteristic Ω be even. Then our HSI can be defined by the formula (4.1.12) with $l = \gamma$. For this case

$$K(\xi,\eta) = (2\pi)^{-n}\left[d_{n,l}(\gamma)\right]^{-1} |\eta|^{-n} \times \sum_{\mu=1}^{(l+1)/2} c_\mu \int_{|y|\le|\eta|^{-1}} \frac{\left(\mathbf{\Delta}^{l+1}_y H_1\right)\left(\xi, \frac{\eta}{|\eta|} - \mu y\right)}{|y|^{n+\gamma}} \Omega_\xi\left(\frac{y}{|y|}\right) dy \tag{4.1.59}$$

where $c_\mu = 1$ for $\mu = 1, 2, \ldots, (l-1)/2$, and $c_{(l+1)/2} = 1/2$. It follows from (4.1.59) and (4.1.10) that in this case

$$|K(\xi,\eta)| \le C|\eta|^{-n-1}, \quad |\eta| \to \infty. \tag{4.1.60}$$

Again, (4.1.54) and (4.1.60) imply the integrability of K.

We have actually proved the relation (4.1.51), with an integrable function K, for a general class of functions F, for example, for any non-zero $F \in \mathcal{S}(\mathbb{R}^n)$ whose Fourier transform vanishes on a neighbourhood of the origin. For such a function F, $\mathbf{D}^{\gamma}_{\Omega_\xi} F \in \mathcal{S}(\mathbb{R}^n)$ and does not vanish identically. In this situation we may pass to the limit in (4.1.51), as $\varepsilon \to 0$, which implies (4.1.53). □

Continuation of the proof of Lemma 4.5. Now we return to

$$F(x) = Z_0(t-\mu, x-y; \mu, y).$$

Setting $\xi = x$ in (4.1.51) and returning to the previous notation we find that

$$Z_{\Omega,\varepsilon}(t-\mu, x, x-y, \mu, y) = \int_{\mathbb{R}^n} K(x,\eta) Z_\Omega(t-\mu, x, x-\varepsilon\eta - y, \mu, y)\, d\eta.$$

Substituting in (4.1.48) we get that

$$\left(\mathbf{D}^\gamma_{\Omega,\varepsilon} u_s\right)(t,x,\tau) = \int_{\mathbb{R}^n} K(x,\eta)\left(\mathbf{D}^\gamma_\Omega u_s\right)(t, x-\varepsilon\eta, \tau)\, d\eta. \tag{4.1.61}$$

As $s \to 0$, we have that $u_s(t,x,\tau) \to u(t,x,\tau)$, uniformly with respect to $x \in \mathbb{R}^n$. Hence for $s \to 0$,

$$\left(\mathbf{D}^\gamma_{\Omega,\varepsilon} u_s\right)(t,x,\tau) \longrightarrow \left(\mathbf{D}^\gamma_{\Omega,\varepsilon} u\right)(t,x,\tau).$$

Using (4.1.47) we pass to the limit in (4.1.61) as $s \to 0$. Then

$$\left(\mathbf{D}^\gamma_{\Omega,\varepsilon} u\right)(t,x,\tau) = \int_{\mathbb{R}^n} K(x,\eta)\Phi(t, x-\varepsilon\eta, \tau)\, d\eta. \tag{4.1.62}$$

It follows from the uniform convergence (4.1.47) that the function $\Phi(t,x,\tau)$ is continuous in x. Now from (4.1.62), (4.1.53), and the dominated convergence theorem we get the existence of the limit

$$\left(\mathbf{D}^\gamma_\Omega u\right)(t,x,\tau) = \lim_{\varepsilon\to 0}\left(\mathbf{D}^\gamma_{\Omega,\varepsilon} u\right)(t,x,\tau) = \Phi(t,x,\tau). \qquad \square$$

4.2 Cauchy problem

4.2.1. Existence of a solution. By a solution of the problem (4.1.1), (4.1.2) we mean a bounded function $u(t,x)$, jointly continuous on $\Pi_{[0,T]}$, which satisfies the initial condition (4.1.2), and satisfies the equation (4.1.1) with A and A_k replaced by the appropriate conditionally convergent hyper-singular integral operators $\mathbf{D}^\gamma_\Omega$ and $\mathbf{D}^{\gamma_k}_{\Omega_k}$, or by differential operators understood in the classical sense, if corresponding symbols are polynomials in ξ.

Theorem 4.1. *The Cauchy problem* (4.1.1), (4.1.2) *possesses a solution*

$$u(t,x) = \int_0^t d\tau \int_{\mathbb{R}^n} \Gamma(t,x;\tau,\xi) f(\tau,\xi)\, d\xi + \int_{\mathbb{R}^n} \Gamma(t,x;0,\xi)\varphi(\xi)\, d\xi, \tag{4.2.1}$$

where a fundamental solution $\Gamma(t,x;\tau,\xi)$, $x,\xi\in\mathbb{R}^n$, $0\le\tau<t\le T$, *is of the form*

$$\Gamma(t,x;\tau,\xi)=Z_0(t-\tau,x-\xi;\tau,\xi)+W(t,x;\tau,\xi), \tag{4.2.2}$$

$$|W(t,x;\tau,\xi)|\le C\Big\{(t-\tau)^{(\gamma+\lambda)/\gamma}\Big[(t-\tau)^{1/\gamma}+|x-\xi|\Big]^{-n-\gamma} + (t-\tau)\sum_{k=1}^{n+1}\Big[(t-\tau)^{1/\gamma}+|x-\xi|\Big]^{-n-\gamma_k}\Big\},$$

$$\gamma_{m+1}=\gamma-\lambda, \tag{4.2.3}$$

while Z_0 *satisfies the estimates* (4.1.25)–(4.1.28).

Proof. According to the usual scheme of the Levi method, the FSCP is sought in the form (4.2.2), where

$$W(t,x;\tau,\xi)=\int_\tau^t d\mu\int_{\mathbb{R}^n} Z_0(t-\mu,x-\eta;\mu,\eta)\Phi(\mu,\eta;\tau,\xi)\,d\eta, \tag{4.2.4}$$

and Φ is determined from the integral equation

$$\Phi(t,x;\tau,\xi)=P(t,x;\tau,\xi)+\int_\tau^t d\mu\int_{\mathbb{R}^n} P(t,x;\mu,\eta)\Phi(\mu,\eta;\tau,\xi)\,d\eta, \tag{4.2.5}$$

in which

$$P(t,x;\tau,\xi)=(2\pi)^{-n}\int_{\mathbb{R}^n}\Big[a(\tau,\xi,\sigma)-a(t,x,\sigma)-\sum_{k=1}^m a_k(t,x,\sigma)\Big]\exp\{i(x-\xi,\sigma) - a(\tau,\xi,\sigma)(t-\tau)\}\,d\sigma.$$

By Lemma 4.2,

$$|P(t,x;\tau,\xi)|\le C\Big\{(t-\tau)^{-\frac{n}{\gamma}-1}\Big[|x-\xi|^\lambda+(t-\tau)^{\lambda/\gamma}\Big]\Big[1+(t-\tau)^{-1/\gamma}|x-\xi|\Big]^{-n-\gamma} + \sum_{k=1}^m (t-\tau)^{-(n+\gamma_k)/\gamma}\Big[1+(t-\tau)^{-1/\gamma}|x-\xi|\Big]^{-n-\gamma_k}\Big\}, \tag{4.2.6}$$

whence

$$|P(t,x;\tau,\xi)|\le C\Big\{\Big[(t-\tau)^{1/\gamma}+|x-\xi|\Big]^{-n-\gamma+\lambda}+\sum_{k=1}^m\Big[(t-\tau)^{1/\gamma}+|x-\xi|\Big]^{-n-\gamma_k}\Big\}. \tag{4.2.7}$$

The integral equation (4.2.5) can be solved by the method of successive approximations:

$$\Phi(t,x;\tau,\xi) = \sum_{\nu=1}^{\infty} P_\nu(t,x;\tau,\xi), \tag{4.2.8}$$

where $P_1 = P$,

$$P_{\nu+1}(t,x;\tau,\xi) \int_{\tau}^{t} d\mu \int_{\mathbb{R}^n} P(t,x;\mu,\eta)P(\mu,\eta;\tau,\xi)\,d\eta, \quad \nu = 1,2,\ldots.$$

We get estimates of the iterated kernels by induction from (4.2.7) and Lemma 1.11:

$$|P_\nu(t,x;\tau,\xi)| \le CM^\nu (t-\tau)^{(\nu-1)\lambda/\gamma} \frac{[\Gamma(\lambda/\gamma)]^\nu}{\Gamma(\nu\lambda/\gamma)} \left\{ \left[(t-\tau)^{1/\gamma} + |x-\xi|\right]^{-n-\gamma+\lambda} + \sum_{k=1}^{m} \left[(t-\tau)^{1/\gamma} + |x-\xi|\right]^{-n-\gamma_k} \right\}.$$

From this and the Stirling formula we deduce the convergence of the series (4.2.8), along with an estimate of the solution Φ:

$$|\Phi(t,x;\tau,\xi)| \le C\left\{ \left[(t-\tau)^{1/\gamma} + |x-\xi|\right]^{-n-\gamma+\lambda} + \sum_{k=1}^{m} \left[(t-\tau)^{1/\gamma} + |x-\xi|\right]^{-n-\gamma_k} \right\}. \tag{4.2.9}$$

The inequality (4.2.3) follows from (4.2.9) and Lemma 1.11.

Let $P = P_1' - P_2'$ where

$$P_1'(t,x;\tau,\xi) = (2\pi)^{-n} \int_{\mathbb{R}^n} [a(\tau,\xi,\sigma) - a(t,x,\sigma)] \exp\{i(x-\xi,\sigma) - a(\tau,\xi,\sigma)(t-\tau)\}\,d\sigma,$$

$$P_2'(t,x;\tau,\xi) = (2\pi)^{-n} \sum_{k=1}^{m} \int_{\mathbb{R}^n} a_k(t,x,\sigma) \exp\{i(x-\xi,\sigma) - a(\tau,\xi,\sigma)(t-\tau)\}\,d\sigma.$$

We consider the function

$$\begin{aligned} R(x,y,t,\xi,\tau) &= P(t,x;\tau,\xi) - P(t,y;\tau,\xi) \\ &= R_1(x,y,t,\xi,\tau) - R_2(x,y,t,\xi,\tau), \quad |x-y| \le 1, \end{aligned}$$

where

$$R_j(x,y,t,\xi,\tau) = P_j'(t,x;\tau,\xi) - P_j'(t,x;\tau,\xi), \quad j = 1,2.$$

Let us estimate $|R_1|$. Assume first that $|x-y| \geq (t-\tau)^{1/\gamma}$. Let λ', $0<\lambda'<\lambda$, be fixed. Carrying out an estimate as in (4.2.6), we get that

$$
\begin{aligned}
|R_1(x,y,t,\xi,\tau)| &\leq C(t-\tau)^{-(n+\gamma-\lambda)/\gamma} \Big\{ \Big[1+|x-\xi|(t-\tau)^{-1/\gamma}\Big]^{-n-\gamma+\lambda} \\
&\qquad + \Big[1+|y-\xi|(t-\tau)^{-1/\gamma}\Big]^{-n-\gamma+\lambda} \Big\} \\
&\leq C|x-y|^{\lambda'}(t-\tau)^{-(n+\gamma-\lambda+\lambda')/\gamma} \Big[1+|\widetilde{\xi}-\xi|(t-\tau)^{-1/\gamma}\Big]^{-n-\gamma+\lambda}
\end{aligned}
$$

where $\widetilde{\xi}$ is the one of the points x or y that is closest to ξ.

Now assume that $|x-y| \leq (t-\tau)^{1/\gamma}$. We write R_1 in the form

$$
\begin{aligned}
&R_1(x,y,t,\xi,\tau) \\
&= (2\pi)^{-n/2} \int_{\mathbb{R}^n} \{\exp[i(x-\xi,\sigma)] - \exp[i(y-\xi,\sigma)]\} \\
&\qquad\qquad \times [a(\tau,\xi,\sigma) - a(t,x,\sigma)] \exp[-a(\tau,\xi,\sigma)(t-\tau)]\, d\sigma \\
&\quad + (2\pi)^{-n/2} \int_{\mathbb{R}^n} [a(\tau,y,\sigma) - a(t,x,\sigma)] \exp[i(y-\xi,\sigma) - a(\tau,\xi,\sigma)(t-\tau)]\, d\sigma \\
&\equiv R_{11} + R_{12}.
\end{aligned}
$$

Using Lemma 4.2 and the fact that either $|y-\xi| \geq |x-\xi|$, or

$$
|x-\xi|^\lambda \leq 2^\lambda \left(|x-y|^\lambda + |y-\xi|^\lambda\right) \leq 2^\lambda \left[(t-\tau)^{\lambda/\gamma} + |y-\xi|^\lambda\right],
$$

we arrive at the estimate

$$
|R_{11}(x,y,t,\xi,\tau)| \leq C(t-\tau)^{-(n+\gamma-\lambda)/\gamma} \Big[1+|\widetilde{\xi}-\xi|(t-\tau)^{-1/\gamma}\Big]^{-n-\gamma+\lambda}. \tag{4.2.10}
$$

On the other hand,

$$
\begin{aligned}
&\exp\{i(x-\xi,\sigma)\} - \exp\{i(y-\xi,\sigma)\} \\
&\qquad = i[(x-y,\sigma)] \int_0^1 \exp\{i(y-\xi,\sigma(1-\rho)) + i(x-\xi,\sigma\rho)\}\, d\rho,
\end{aligned}
$$

whence

$$
\begin{aligned}
R_{11}(x,y,t,\xi,\tau) &= i \int_{\mathbb{R}^n} (x-y,\sigma) \left\{ \int_0^1 \exp\{i(y-\xi,\sigma(1-\rho)) + i(x-\xi,\sigma\rho)\}\, d\rho \right\} \\
&\qquad \times [a(\tau,\xi,\sigma) - a(t,x,\sigma)] \exp[-a(\tau,\xi,\sigma)(t-\tau)]\, d\sigma, \\
|R_{11}(x,y,t,\xi,\tau)| &\leq C|x-y| \int_{\mathbb{R}^n} |\sigma|^{1+\gamma} \exp\left[-a_0|\sigma|^\gamma(t-\tau)\right]\, d\sigma,
\end{aligned}
$$

so that

$$|R_{11}(x,y,t,\xi,\tau)| \le C|x-y|(t-\tau)^{-(n+\gamma+1)/\gamma}. \tag{4.2.11}$$

We raise both sides of (4.2.11) to the power $\varepsilon \in (0,1)$ and both sides of (4.2.10) to the power $1-\varepsilon$, and multiply the resulting inequalities getting

$$|R_{11}(x,y,t,\xi,\tau)| \le C|x-y|^{\varepsilon}(t-\tau)^{-(n+\gamma-\lambda+\varepsilon(\lambda+1))/\gamma}\left[1+|\tilde{\xi}-\xi|(t-\tau)^{-1/\gamma}\right]^{-(n+\gamma-\lambda)(1-\varepsilon)}.$$

An estimate of R_{12} follows at once from Lemma 4.2:

$$|R_{12}(x,y,t,\xi,\tau)| \le C|x-y|^{\lambda}(t-\tau)^{-(n+\gamma))/\gamma}\left[1+|y-\xi|(t-\tau)^{-1/\gamma}\right]^{-n-\gamma}$$
$$\le C|x-y|^{\lambda'}(t-\tau)^{-(n+\gamma-\lambda+\lambda'))/\gamma}\left[1+|\tilde{\xi}-\xi|(t-\tau)^{-1/\gamma}\right]^{-n-\gamma+\lambda}.$$

Suppose that

$$\varepsilon' < \min\left(\frac{\lambda}{\lambda+1}, \frac{\gamma-\lambda}{n+\gamma-\lambda}\right), \quad \bar{\lambda} \le \min(\lambda', \varepsilon'), \quad \hat{\lambda} = \max(\lambda', \varepsilon'(\lambda+1)),$$

and $\tilde{\lambda} = \lambda + (n+\gamma-\lambda)\varepsilon'$. Combining the above estimates, we get

$$|R_1(x,y,t,\xi,\tau)| \le C|x-y|^{\bar{\lambda}}(t-\tau)^{-(n+\gamma-\lambda+\hat{\lambda})/\gamma}\left[1+|\tilde{\xi}-\xi|(t-\tau)^{-1/\gamma}\right]^{-(n+\gamma-\tilde{\lambda})} \tag{4.2.12}$$

where $0 < \bar{\lambda} < \lambda$, $0 < \hat{\lambda} < \lambda$, and $\tilde{\lambda} < \gamma$.

An estimate of R_2 is obtained similarly:

$$|R_2(x,y,t,\xi,\tau)| \le C|x-y|^{\bar{\lambda}}\sum_{k=1}^{m}\Big\{(t-\tau)^{-(n+\gamma_k+\varepsilon)/\gamma}\left[1+|\tilde{\xi}-\xi|(t-\tau)^{-1/\gamma}\right]^{-n-\gamma_k+\varepsilon(n+\gamma_k)}$$
$$+(t-\tau)^{-(n+\gamma_k)/\gamma}\left[1+|\tilde{\xi}-\xi|(t-\tau)^{-1/\gamma}\right]^{-n-\gamma_k}\Big\}, \tag{4.2.13}$$

where $\varepsilon \in (0,1)$ can be chosen arbitrarily (and the constants C and $\bar{\lambda}$ depend, of course, on ε).

We consider a function $u(t,x)$ of the form (4.2.1) and see that it is a solution of the problem (4.1.1), (4.1.2). Let

$$u_1(t,x) = \int_0^t d\tau \int_{\mathbb{R}^n} \Gamma(t,x;\tau,\xi)f(\tau,\xi)\,d\xi,$$

$$u_2(t,x) = \int_{\mathbb{R}^n} \Gamma(t,x;0,\xi)\varphi(\xi)\,d\xi.$$

Then, by (4.2.2) and (4.2.4),

$$u_1(t,x)=\int_0^t d\tau\int_{\mathbb{R}^n} Z(t-\tau,x-\xi;\tau,\xi)f(\tau,\xi)d\xi+\int_0^t d\mu\int_{\mathbb{R}^n} Z(t-\mu,x-\eta;\mu,\eta)F(\mu,\eta)d\eta, \tag{4.2.14}$$

$$u_2(t,x)=\int_{\mathbb{R}^n} Z(t,x-\xi;0,\xi)\varphi(\xi)d\xi+\int_0^t d\mu\int_{\mathbb{R}^n} Z(t-\mu,x-\eta;\mu,\eta)G(\mu,\eta)d\eta, \tag{4.2.15}$$

where

$$F(\mu,\eta)=\int_0^\mu d\tau\int_{\mathbb{R}^n}\Phi(\mu,\eta;\tau,\xi)f(\tau,\xi)\,d\xi,\quad G(\mu,\eta)=\int_{\mathbb{R}^n}\Phi(\mu,\eta;0,\xi)\varphi(\xi)\,d\xi.$$

The interchange of integrals is justified with the help of (4.1.25), (4.2.9), Lemma 1.11, and Fubini's theorem.

Let us study properties of F and G. According to (4.2.5),

$$\Phi(t,x;\tau,\xi)-\Phi(t,y;\tau,\xi)=R(x,y,t,\xi,\tau)+\int_\tau^t d\mu\int_{\mathbb{R}^n} R(x,y,t,\eta,\mu)\Phi(\mu,\eta;\tau,\xi)\,d\eta.$$

Using (4.2.9), (4.2.12), (4.2.13), and Lemma 1.11, after elementary transformations we get

$$\begin{aligned}&|F(\mu,\eta)|\le C, && |F(\mu,\eta_1)-F(\mu,\eta_2)|\le C|\eta_1-\eta_2|^\Lambda,\\ &|G(\mu,\eta)|\le C\mu^{-\rho}, && |G(\mu,\eta_1)-G(\mu,\eta_2)|\le C|\eta_1-\eta_2|^\Lambda\mu^{-\rho},\end{aligned}$$

where $\rho<1$ and $0<\Lambda<1$.

Thus, (4.2.14) and (4.2.15) give a representation of $u(t,x)$ as a sum of volume potentials to each of which Lemma 4.5 is applicable. Computing the derivative with respect to t according to (4.1.38), and the HSIs $\mathbf{D}^{\gamma_k}_{\Omega_k}u$ and $\mathbf{D}^{\gamma}_{\Omega}u$ according to (4.1.41) and (4.1.42), we see by a direct calculation that $u(t,x)$ satisfies (4.1.1).

Next, it is obvious that $u_1(0,x)=0$. Moreover, it follows from (4.1.25) and (4.2.3) that $u_1(t,x)\to 0$ as $t\to 0$, uniformly with respect to $x\in\mathbb{R}^n$. It also follows from (4.2.3) that

$$\int_{\mathbb{R}^n} W(t,x;0,\xi)\varphi(\xi)\,d\xi\longrightarrow 0,\quad \text{as } t\to 0,$$

uniformly with respect to $x\in\mathbb{R}^n$. Consider the function

$$v(t,x)=\int_{\mathbb{R}^n} Z(t,x-\xi;0,\xi)\varphi(\xi)\,d\xi=:v_1(t,x)+v_2(t,x)$$

where

$$v_1(t,x) = \int_{\mathbb{R}^n} Z(t, x-\xi; 0, x)\varphi(\xi)\, d\xi,$$

$$v_2(t,x) = \int_{\mathbb{R}^n} [Z(t, x-\xi; 0, \xi) - Z(t, x-\xi; 0, x)]\varphi(\xi)\, d\xi.$$

Using (4.1.27) we see that $v_2(t,x) \to 0$ as $t \to 0$, uniformly with respect to $x \in \mathbb{R}^n$. Finally, by (4.1.29),

$$v_1(t,x) = \varphi(x) + \int_{\mathbb{R}^n} Z(t, x-\xi; 0, x)[\varphi(\xi) - \varphi(x)]\, d\xi. \tag{4.2.16}$$

It follows from the definition (4.1.17) of the function Z that

$$Z(t, x-\xi; 0, x) = t^{-n/\gamma} Z(1, t^{-1/\gamma}(x-\xi); 0, x).$$

After the substitution in (4.2.16) and the change of variables $y = t^{-1/\gamma}(x-\xi)$ we find that

$$v_1(t,x) = \varphi(x) + \int_{\mathbb{R}^n} Z(1, y; 0, x)[\varphi(x - t^{1/\gamma}y) - \varphi(x)]\, dy.$$

Using (4.1.25) and the dominated convergence theorem and taking into account that φ is continuous and bounded, we get that $v_1(t,x) \to \varphi(x)$ as $t \to 0$, for all $x \in \mathbb{R}^n$. Moreover, if $x_k \underset{k\to\infty}{\longrightarrow} x_0$, $t \to 0$, then $v_1(t, x_k) \to \varphi(x_0)$. Combining this with what was proved above, we come to the conclusion that $u(0,x) = \varphi(x)$, and $u(t,x)$ is jointly continuous on $[0,T] \times \mathbb{R}^n$. □

There is also a version of (4.2.1) for the case when the equation (4.1.1) is considered for $t > \tau$, with the initial condition $u(\tau, x) = \varphi(x)$. Then a solution of the Cauchy problem is given by the formula

$$u(t,x) = \int_\tau^t d\theta \int_{\mathbb{R}^n} \Gamma(t, x; \theta, \xi) f(\theta, \xi)\, d\xi + \int_{\mathbb{R}^n} \Gamma(t, x; \tau, \xi)\varphi(\xi)\, d\xi. \tag{4.2.1$'$}$$

4.2.2. Non-negative solutions. Let us consider a subclass of equations of the form (4.1.1) whose solutions have properties similar to those of solutions of second order parabolic differential equations.

Consider the equation (4.1.1) with $f(t,x) = 0$ and $1 \le \gamma \le 2$. Assume the conditions of Section 4.1.1. In particular, if $\gamma = 2$, then

$$-A = \sum_{j,l=1}^n a_{jl}(t,x)\partial_{x_j}\partial_{x_l}$$

is a second-order elliptic differential operator whose coefficients will be assumed real-valued. Representing all the ΨDOs as HSIs we write (4.1.1) in the following form, which encompasses both the case $\gamma = 2$ and the case $\gamma < 2$:

$$
\begin{aligned}
Lu = \partial_t u(t,x) &- \beta \sum_{j,l=1}^{n} a_{jl}(t,x)\partial_{x_j}\partial_{x_l} u(t,x) \\
&+ \frac{1-\beta}{d_{n,l}(\gamma)} \int\limits_{\mathbb{R}^n} \Omega\left(t,x,\frac{h}{|h|}\right) \frac{\left(\Delta_h^l u\right)(t,x)}{|h|^{n+\gamma}}\, dh \\
&+ \sum_{k=1}^{m} \frac{1}{d_{n,l}(\gamma_k)} \int\limits_{\mathbb{R}^n} \Omega_k\left(t,x,\frac{h}{|h|}\right) \frac{\left(\Delta_h^l u\right)(t,x)}{|h|^{n+\gamma_k}}\, dh \\
&+ \sum_{j=1}^{n} b_j(t,x)\partial_{x_j} u(t,x) = 0,
\end{aligned}
\tag{4.2.17}
$$

where the coefficients $b_j(t,x)$ are real-valued, $\beta = 1$ if $\gamma = 2$, and $\beta = 0$ if $\gamma < 2$. The last sum in (4.2.17) is absent if $\gamma = 1$.

We also assume the following additional condition.

(A_{44}) The characteristics Ω and Ω_k are even (except, possibly, those Ω_k, for which $\gamma_k < 1$) and non-negative.

Let $u(t,x)$ be a solution of (4.2.17), jointly continuous on $\Pi_{[\tau,T]}$ and satisfying the initial condition

$$u(\tau,x) = \varphi(x) \tag{4.2.18}$$

where φ is a continuous bounded function. As earlier, HSIs are understood in the sense of the conditional convergence (4.1.11). The function $u(t,x)$ is assumed to be continuously differentiable with respect to x (if $\gamma > 1$) and t for $t > \tau$, and if $\gamma = 2$, also twice continuously differentiable with respect to x.

Lemma 4.7. *If $\varphi(x) \geq 0$ for all $x \in \mathbb{R}^n$ and $u(t,x) \to 0$ as $|x| \to \infty$, $t > \tau$, then $u(t,x) \geq 0$ for all $x \in \mathbb{R}^n$ and $t \in [\tau,T]$.*

Proof. It suffices to consider the case when u is real-valued. Let

$$\alpha = \inf_{(t,x)\in\Pi_{[\tau,T]}} u(t,x).$$

Suppose that $\alpha < 0$. We introduce the function

$$v(t,x) = u(t,x) + (t-\tau)\rho \tag{4.2.19}$$

where ρ is a fixed number with $0 < \rho < -\alpha/(T-\tau)$. Obviously,

$$\inf_{(t,x)\in\Pi_{[\tau,T]}} v(t,x) < 0.$$

Since $v(t,x)$ is jointly continuous on $\Pi_{[\tau,T]}$, $v(\tau,x)=\varphi(x)\geq 0$, and $v(t,x)\to (t-\tau)\rho\geq 0$ as $|x|\to\infty$, $t>\tau$, it follows that $v(t,x)$ has a negative global minimum at some point (t_0,x_0). Here $x_0\in\mathbb{R}^n$ and $\tau<t_0\leq T$. It is clear that $\nabla_x v(t_0,x_0)=0$, $\partial_t v(t_0,x_0)=0$ if $t_0<T$, and $\partial_t v(t_0,x_0)\leq 0$ if $t_0=T$. It is known [83] that

$$\sum_{j,l=1}^{n} a_{jl}(t_0,x_0)\partial_{x_i}\partial_{x_j}v(t_0,x_0)\geq 0.$$

All the HSIs in (4.2.17) have even characteristics and orders less than 2. Therefore we may set $l=1$, so that

$$\left(\mathbf{\Delta}_h^l v\right)(t_0,x_0)=v(t_0,x_0)-v(t_0,x_0+h)\leq 0,$$

which implies the inequalities

$$\left(\mathbf{D}_{\Omega,\varepsilon}^{\gamma}v\right)(t_0,x_0)\leq 0,\quad \left(\mathbf{D}_{\Omega_k,\varepsilon}^{\gamma_k}v\right)(t_0,x_0)\leq 0.$$

Passing to the limit as $\varepsilon\to 0$, we get similar inequalities for $\mathbf{D}_{\Omega}^{\gamma}v$ and $\mathbf{D}_{\Omega_k}^{\gamma_k}v$.

Thus, $(Lv)(t_0,x_0)\leq 0$. However in view of (4.2.19) $(Lv)(t_0,x_0)=\rho>0$, and we have come to a contradiction. □

Let $\Gamma(t,x;\tau,\xi)$ be a FSCP for the equation (4.2.17), and let

$$u(t,x)=\int_{\mathbb{R}^n}\Gamma(t,x;\tau,\xi)\varphi(\xi)\,d\xi. \tag{4.2.20}$$

Assume that the function φ has a compact support. It follows from (4.1.25) and (4.2.3) that $u(t,x)\to 0$ as $|x|\to\infty$, $t>\tau$. The function $u(t,x)$ is a solution of the Cauchy problem (4.2.17)–(4.2.18). The next result follows now from Lemma 4.7.

Theorem 4.2. *If the condition (A_{44}) is satisfied, then $\Gamma(t,x;\tau,\xi)\geq 0$ for all $x,\xi\in\mathbb{R}^n$, $t>\tau$.*

As a corollary to Lemma 4.7 we also have the following fact.

Lemma 4.8. *If the condition (A_{44}) is satisfied, then the problem* (4.2.17)–(4.2.18) *can have at most one solution tending to zero as $|x|\to\infty$.*

It is deduced in the standard way from Lemma 4.8 and the estimates (4.1.25) and (4.2.3) that for $\tau<\sigma<t\leq T$,

$$\Gamma(t,x;\tau,\xi)=\int_{\mathbb{R}^n}\Gamma(t,x;\sigma,y)\Gamma(\sigma,y;\tau,\xi)\,dy. \tag{4.2.21}$$

Lemma 4.9. *If the condition (A_{44}) is satisfied, then for all $x\in\mathbb{R}^n$, $t>\tau$,*

$$\int_{\mathbb{R}^n}\Gamma(t,x;\tau,\xi)\,d\xi=1. \tag{4.2.22}$$

Proof. We write the operator L in the right-hand side of (4.2.17) in the form $L = \partial_t + M$. Suppose that $\psi \in C_0^\infty(\mathbb{R}^n)$, $\psi(x) = 1$ for $|x| \le 1/2$, and $\psi(x) = 0$ for $|x| \ge 1$. Let $\varphi_\nu(x) = \psi(x/\nu)$, $\nu = 1, 2, \ldots$.

Consider the Cauchy problem

$$\partial_t u(t,x) + (Mu)(t,x) = f_\nu(t,x), \quad t > \tau; \tag{4.2.23}$$

$$u(\tau, x) = \varphi_\nu(x), \tag{4.2.24}$$

where $f_\nu(t,x) = (M\varphi_\nu)(t,x)$.

The HSI operators $\mathbf{D}_\Omega^\gamma$ and $\mathbf{D}_{\Omega_k}^{\gamma_k}$ act on the function $\varphi_\nu \in C_0^\infty(\mathbb{R}^n)$ as ΨDOs of the form (4.1.3). This implies that f_ν is bounded and Hölder continuous with respect to x (uniformly with respect to t). Therefore the solution of the problem (4.2.23)–(4.2.24) can be found from (4.2.1′):

$$u(t,x) = \int_\tau^t d\theta \int_{\mathbb{R}^n} \Gamma(t,x;\theta,\xi) f_\nu(\theta,\xi)\, d\xi + \int_{\mathbb{R}^n} \Gamma(t,x;\tau,\xi)\varphi_\nu(\xi)\, d\xi.$$

According to (4.1.13),

$$|f_\nu(t,x)| \le C_\nu (1 + |x|)^{-n-\alpha}, \quad \alpha = \min(\gamma, \gamma_k) > 0.$$

Using this estimate and the inequalities (4.1.25) and (4.2.3), we see that $u(t,x) \to 0$ as $|x| \to \infty$, $t > \tau$.

On the other hand, the function $w(t,x) \equiv \varphi_\nu(x)$ also satisfies (4.2.23)–(4.2.24). By Lemma 4.8, $w(t,x) = u(t,x)$, so that

$$\varphi_\nu(x) = \int_\tau^t d\theta \int_{\mathbb{R}^n} \Gamma(t,x;\theta,\xi) f_\nu(\theta,\xi)\, d\xi + \int_{\mathbb{R}^n} \Gamma(t,x;\tau,\xi)\varphi_\nu(\xi)\, d\xi. \tag{4.2.25}$$

We have just used the estimate of f_ν for a fixed ν. Now we study its behaviour as $\nu \to \infty$. Obviously,

$$\begin{aligned}\left(\mathbf{\Delta}_h^2 \varphi_\nu\right)(x) &= \sum_{k=0}^{2} (-1)^k C_2^k \varphi_\nu(x - kh) \\ &= \sum_{k=0}^{2} (-1)^k C_2^k \psi((x - kh)/\nu) \\ &= \left(\mathbf{\Delta}_{h/\nu}^2 \psi\right)(x/\nu),\end{aligned}$$

so that

$$\left(\mathbf{D}_{\Omega_k}^{\gamma_k} \varphi_\nu\right)(x,t) = \frac{\nu^{-\gamma_k}}{d_{n,2}(\gamma_k)} \int_{\mathbb{R}^n} \Omega_k\left(t, x, \frac{s}{|s|}\right) \frac{\left(\mathbf{\Delta}_s^2 \psi\right)(x/\nu)}{|s|^{n+\gamma_k}}\, ds.$$

According to (4.1.10),

$$\left|\left(\mathbf{\Delta}_s^2\psi\right)(x/\nu)\right| \le C|s|^2 \sup_{|\kappa|=2} \sup_{y\in\mathbb{R}^n} |D^\kappa\psi(y)| \le C|s|^2.$$

From this we have

$$\left|\left(\mathbf{D}_{\Omega_k}^{\gamma_k}\varphi_\nu\right)(t,x)\right| \le C\nu^{-\gamma_k}\left[\int\limits_{|s|\le 1}\frac{ds}{|s|^{n+\gamma_k-2}} + \int\limits_{|s|>1}\frac{ds}{|s|^{n+\gamma_k}}\right] \le C\nu^{-\gamma_k},$$

where C is independent of ν, x, and t.

If $\gamma < 2$, then the proof that $\left|\left(\mathbf{D}_\Omega^\gamma\varphi_\nu\right)(t,x)\right| \le C\nu^{-\gamma}$ is similar. It is also obvious that $|D^\kappa\varphi_\nu(x)| \le C\nu^{-|\kappa|}$.

Thus, $|f_\nu(t,x)| \le C\nu^{-\alpha}$ for all $x \in \mathbb{R}^n$, $t > \tau$. It follows from (4.1.25), (4.2.3), and the dominated convergence theorem that

$$\int\limits_{\mathbb{R}^n}\Gamma(t,x;\tau,\xi)\varphi_\nu(\xi)\,d\xi \longrightarrow \int\limits_{\mathbb{R}^n}\Gamma(t,x;\tau,\xi)\,d\xi$$

as $\nu \to \infty$. Passing to the limit in (4.2.25) as $\nu \to \infty$, we get (4.2.22). □

Let B be a compact subset of $\mathbb{R}^n$.

Lemma 4.10. *The following relations hold:*

1)
$$\lim_{\nu\to\infty}\sup_{|x|\ge\nu,0\le\tau<t\le T}\int\limits_B\Gamma(t,x;\tau,\xi)\,d\xi = 0.$$

2) *For any* $\varepsilon > 0$,

$$\lim_{\delta\to 0}\sup_{x\in B,0\le\tau<t<t+\delta}\int\limits_{|\xi-x|>\varepsilon}\Gamma(t,x;\tau,\xi)\,d\xi = 0.$$

The *proof* follows easily from (4.1.25) and (4.2.3). □

Suppose now that the operator L is defined for all $t \ge 0$, and all the conditions of Sections 4.2.1 and 4.2.2 are satisfied for all $T > 0$. Using Theorems 4.1 and 4.2, the identity (4.2.21), Lemmas 4.8–4.10, and general theorems about Markov processes [40], we arrive at the following result.

Theorem 4.3 *The FSCP* $\Gamma(t,x;\tau,\xi)$ *is a transition density of a bounded non-terminating strict Markov process without second kind discontinuities.*

4.3 On qualitative properties of solutions of some equations with constant symbols

4.3.1. Introduction. In this section we use the estimates of the FSCP for an equation with a constant (that is, independent of t, x) homogeneous symbol obtained in Section 4.1, in order to find conditions for the stabilization of solutions of the Cauchy problem, and also to prove a version of the Liouville theorem. The proofs are similar to the ones for parabolic differential equations [45, 71, 72, 75, 197].

First we repeat, in a form convenient for subsequent applications, some of the above definitions and results.

We consider the equation

$$\partial_t u(t,x) + (Au)(t,x) = 0 \tag{4.3.1}$$

where A is a pseudo-differential operator with a symbol $a(\xi)$, homogeneous of a degree $\gamma \geq 1$, satisfying the conditions:

(A_{45}) $\qquad \operatorname{Re} a(\xi) \geq a_0 > 0, \quad \xi \in \mathbb{R}^n, \ |\xi| = 1;$

(A_{46}) $\quad a(\xi)$ has N continuous derivatives in ξ, $\xi \neq 0$, and

$$\left|\partial_\xi^\varkappa a(\xi)\right| \leq C_N |\xi|^{\gamma - |\varkappa|}$$

for all $\varkappa$ with $|\varkappa| \leq N$, $N \geq 2n + 2[\gamma] + 1$.

The simplest (but also the most important) example is the equation (4.3.1) with the symbol $a_0(\xi) = |\xi|^\gamma$, for which the condition (A_{46}) is fulfilled for any $\varkappa$.

We shall investigate properties of solutions of the equation (4.3.1) represented by the Poisson integral

$$u(t,x) = \int\limits_{\mathbb{R}^n} Z_0(t, x-y)\varphi(y)\,dy \tag{4.3.2}$$

and defined on the half-space $\mathbb{R}_+^{n+1} = \{(t,x) : \ 0 \leq t < \infty, x \in \mathbb{R}^n\}$.

The function $Z_0(t, x-y)$ is given by the formula (4.1.17), that is

$$Z_0(t, x-y) = (2\pi)^{-n} \int\limits_{\mathbb{R}^n} \exp\{i(x-y,\sigma) - a(\sigma)t\}\,d\sigma. \tag{4.3.3}$$

It will be convenient to rewrite this function as

$$Z_0(t, x-y) = t^{-n/\gamma} G((x-y)t^{-1/\gamma}) \tag{4.3.4}$$

where

$$G(z) = (2\pi)^{-n} \int\limits_{\mathbb{R}^n} \exp\{iz\sigma - a(\sigma)\}\,d\sigma.$$

It follows from the estimates (4.1.25) for Z_0 that

$$|D_z^{\varkappa} G(z)| \leq C_{\varkappa} t^{-|\varkappa|/\gamma}(1+|z|)^{-\gamma-|\varkappa|}, \quad |\varkappa| \leq N-2n-[\gamma]. \tag{4.3.5}$$

We shall use also the equality (see (4.1.29))

$$\int_{\mathbb{R}^n} Z_0(t, x-y)\, dy = 1, \quad t>0, x \in \mathbb{R}^n, \tag{4.3.6}$$

whence

$$\int_{\mathbb{R}^n} G(z)\, dz = 1. \tag{4.3.7}$$

4.3.2. Stabilization. Basic definitions and results. We consider solutions of the Cauchy problem (4.3.1) represented by the Poisson integral (4.3.2).

Definition. *A function $u(t,x)$ defined on the half-space $\mathbb{R}_+^{n+1} = \{(t,x) \mid 0 \leq t < \infty, x \in \mathbb{R}^n\}$ is said to stabilize to a function $v(x)$, as $t \to \infty$,*

1) uniformly on every compact subset of $\mathbb{R}^n$, *if $u(t,x) \to v(x)$ uniformly with respect to $x \in K$, for every compact subset $K \subset \mathbb{R}^n$;*
2) uniformly, *if $u(t,x) \to v(x)$ uniformly with respect to $x \in \mathbb{R}^n$;*
3) *at a point $x^0 \in \mathbb{R}^n$, if $u(t,x^0) \to v(x^0)$.*

Let us formulate a simple theorem giving sufficient conditions of the stabilization, uniform on every compact set. Denote by R_e one of the coordinate angles of $\mathbb{R}^n$, that is the set of points $x \in \mathbb{R}^n$ whose coordinates satisfy the inequalities $\varepsilon_1 x_1 \geq 0, \dots, \varepsilon_n x_n \geq 0$, where $e = (\varepsilon_1, \dots, \varepsilon_n)$ is a fixed vector with the components equal to 1 or -1. Let R_{ea} be the parallelepiped $\{\varepsilon_1 x_1 \leq a_1, \dots, \varepsilon_n x_n \leq a_n, a_i > 0, i = 1, \dots, n\}$. R_{eab} will denote the parallelepiped $\{b_s \leq x_s \leq b_s + a_s, \text{if } \varepsilon_s = 1;\ b_s - a_s \leq x_s \leq b_s, \text{if } \varepsilon_s = -1,\ s = 1, \dots, n\}$. Let $A = \prod_{s=1}^{n} a_s$.

Theorem 4.4. *Let the symbol a satisfy the conditions (A_{45}), (A_{46}), while the initial function φ is bounded and continuous. Suppose that one of the following conditions is fulfilled:*

a) *the function φ has an angular limit mean value, that is the sequences $\frac{1}{A} \int_{R_{ea}} \varphi(y)\, dy$, in which $a_1, \dots, a_n$ tend to infinity independently of each other, have a finite limit l, the same for all the vectors e;*
b) *the symbol a is an even function of the arguments $\xi_1, \dots, \xi_n$, and the function φ has the centrally-symmetric mean value, that is a finite limit l of the sequences*

$$\frac{1}{2A} \int_{R_{ea} \cup R_{-ea}} \varphi(y)\, dy,$$

as the coordinates of the vector a tend to infinity independently of each other, and the limit is the same for all the vectors e;

c) *the symbol a is an even function, separately in each of the arguments $\xi_1, \ldots, \xi_n$, and the function φ has the limit mean value, that is the expression*

$$\frac{1}{2^n A} \int_{-a_1}^{a_1} \cdots \int_{-a_n}^{a_n} \varphi(y)\, dy$$

has a finite limit l, as $a_1, \ldots, a_n$ tend to infinity independently of each other.

Then $u(t,x) \to l$, as $t \to \infty$, uniformly in every finite parallelepiped of the space $\mathbb{R}^n$.

Proof. Suppose that the condition a) is fulfilled. Note first of all that it is sufficient to consider the case $l = 0$, since

$$u(t,x) - l = \int_{\mathbb{R}^n} Z_0(t, x-y)[\varphi(y) - l]\, dy$$

and

$$\frac{1}{A} \int_{R_{ea}} \varphi(y)\, dy - l = \frac{1}{A} \int_{R_{ea}} [\varphi(y) - l]\, dy \longrightarrow 0,$$

as $a_1, \ldots, a_n \to \infty$, since A is the volume of R_{ea}.

Secondly, if a function φ has the angular limit mean value l, then

$$\frac{1}{A} \int_{R_{eab}} \varphi(y)\, dy \longrightarrow l,$$

as $a_1, \ldots, a_n \to \infty$, uniformly with respect to such vectors b that $|b_k| \le M$, $1 \le k \le n$. Indeed,

$$\frac{1}{A} \int_{R_{eab}} \varphi(y)\, dy - l = \frac{1}{A} \int_{R_{eab}} [\varphi(y) - l]\, dy = S_{a,b} + \frac{1}{A} \int_{R_{ea}} [\varphi(y) - l]\, dy,$$

where $S_{a,b}$ is a sum, in which every summand contains integrals of the form

$$\frac{1}{a_k} \int_{b_k}^{0} \quad \text{or} \quad \frac{1}{a_k} \int_{\varepsilon_k a_k}^{b_k + \varepsilon_k a_k}$$

tending to zero, as $a_1, \ldots, a_n \to \infty$, uniformly with respect to b_k, $|b_k| \le M$.

Writing our solution as

$$u(t,x) = t^{-n/\gamma} \int_{\mathbb{R}^n} G((x-y)t^{-1/\gamma}) \varphi(y)\, dy$$

we make the change of variables $y = x + t^{1/\gamma}\beta$. Then

$$u(t,x) = \int_{\mathbb{R}^n} G(\beta)\varphi(x + t^{1/\gamma}\beta)\, d\beta$$

$$= \int_{\mathbb{R}^n} G(\beta) \left(\partial_{\beta_1} \dots \partial_{\beta_n} \int_0^{\beta_1} \dots \int_0^{\beta_n} \varphi(x + t^{1/\gamma}\bar{\beta})\, d\bar{\beta} \right) d\beta. \tag{4.3.8}$$

Integrating by parts in (4.3.8) we obtain the equality

$$u(t,x) = (-1)^n \int_{\mathbb{R}^n} \partial_{\beta_1} \dots \partial_{\beta_n} \left(\int_0^{\beta_1} \dots \int_0^{\beta_n} \varphi(x + t^{1/\gamma}\bar{\beta})\, d\bar{\beta} \right) d\beta. \tag{4.3.9}$$

Here we used the estimates (4.3.5) and the boundedness of the function φ.

We have to prove that $u(t,x) \to 0$ uniformly with respect to x, $|x| \leq K$, where K is an arbitrary positive number. In fact we shall show that for any $\varepsilon > 0$ and $K > 0$ there exists such T that $|u(t,x)| \leq \varepsilon$ for arbitrary $t > T$ and $|x| \leq K$. Let us write the right-hand side of (4.3.9) as a sum of three integrals

$$u(t,x) = (-1)^n \left\{ \int_{\Omega_1} + \int_{\Omega_2} + \int_{\Omega_3} \right\} \equiv I_1 + I_2 + I_3. \tag{4.3.10}$$

Here

$$\Omega_1 = \left\{ (\beta_1, \dots, \beta_n) \middle| \, \exists s \in \{1, \dots, n\}, |\beta_s| > B_s > 0 \right\},$$
$$\Omega_2 = \left\{ (\beta_1, \dots, \beta_n) \middle| \, \forall s \in \{1, \dots, n\}, 0 < h_s < |\beta_s| \leq B_s > 0 \right\},$$
$$\Omega_3 = \left\{ (\beta_1, \dots, \beta_n) \middle| \, \exists s \in \{1, \dots, n\}, |\beta_s| \leq h_s \right\}.$$

The estimate (4.3.5) with $|\varkappa| = n$ and the assumption a) show that the integrals I_k, $k = 1, 2, 3$, exist and are bounded, uniformly with respect to t. Therefore, given $\varepsilon > 0$, one can choose B_s large enough, and h_s, $s = 1, \dots, n$ small enough, depending on ε, in such a way that for any t, x,

$$|I_1| < \frac{\varepsilon}{3}, \quad |I_3| < \frac{\varepsilon}{3}.$$

Let us give an estimate of I_2.

Consider the function

$$g_t(\beta) = \int_0^{\beta_1} \dots \int_0^{\beta_n} \varphi(x + t^{1/\gamma}\bar{\beta})\, d\bar{\beta}$$

$$= t^{-n/\gamma} \prod_{i=1}^n \beta_i \frac{1}{\prod\limits_{i=1}^n (t^{1/\gamma}\beta_i)} \int_{x_1}^{x_1 + t^{1/\gamma}\beta_1} \cdots \int_{x_n}^{x_n + t^{1/\gamma}\beta_n} \varphi(a)\, da.$$

Since φ has zero angular limit mean value, for any $\delta > 0$ there exist such N_1, ..., N_n that, as soon as $|t^{1/\gamma}\beta_s| > N_s$, $s = 1,\dots,n$, for any x with $|x_j| \le K$, $j = 1,\dots,n$, we have

$$\left| \frac{1}{\prod_{i=1}^{n}(t^{1/\gamma}\beta_i)} \int_{x_1}^{x_1+t^{1/\gamma}\beta_1} \cdots \int_{x_n}^{x_n+t^{1/\gamma}\beta_n} \varphi(a)\,da \right| < \delta. \tag{4.3.11}$$

Now we choose the number

$$\delta = \frac{\varepsilon}{3C_n}\left(\prod_{i=1}^{n} B_i \int_{\mathbb{R}^n} (1+|\beta|)^{-(n+\gamma)}\,d\beta\right)^{-1}$$

(C_n is taken from the inequality (4.3.5)), and the number $T > 0$ in such a way that

$$T^{1/\gamma} \ge \max_{1\le i\le n}\left\{N_i h_i^{-1}\right\}.$$

Using the estimates for derivatives of the function G, and the inequality (4.3.11), we find that for any $t > T$, $|x| \le K$,

$$\begin{aligned} |I_2| \le C_n & \int_{\{h_i\le|\beta_i|\le B_i, i=1,2,\dots,n\}} (1+|\beta)^{-(n+\gamma)} \left|\prod_{i=1}^{n} B_i\right| \\ & \times \left| \frac{1}{\prod_{i=1}^{n}(t^{1/\gamma}\beta_i)} \int_{x_1}^{x_1+t^{1/\gamma}\beta_1} \cdots \int_{x_n}^{x_n+t^{1/\gamma}\beta_n} \varphi(a)\,da \right| d\beta \le \frac{\varepsilon}{3}. \end{aligned}$$

Combining the estimates obtained for I_1, I_2, I_3, we see that $|u(t,x)| < \varepsilon$ for $t > T$ and $|x_j| \le K$. Thus, for the case a) the assertion has been proved.

In the case b) we use the representation

$$\begin{aligned} u(t,x) = \frac{1}{2} & \int_{\mathbb{R}^n} G(\beta)\partial_{\beta_1}\dots\partial_{\beta_n} \\ & \times \left[\int_0^{\beta_1}\cdots\int_0^{\beta_n} \varphi(x+t^{1/\gamma}\bar{\beta})\,d\bar{\beta} + \int_{-\beta_1}^{0}\cdots\int_{-\beta_n}^{0} \varphi(x+t^{1/\gamma}\bar{\beta})\,d\bar{\beta} \right] d\beta, \end{aligned} \tag{4.3.12}$$

while in the case c), in which $G(\beta_1,\dots,\beta_s,\dots,\beta_n) = G(\beta_1,\dots,-\beta_s,\dots,\beta_n)$, $s = 1,\dots,n$,

$$u(t,x) = \frac{1}{2^n}\int_{\mathbb{R}^n} G(\beta)\left(\partial_{\beta_1}\dots\partial_{\beta_n} \int_{-\beta_1}^{\beta_1}\cdots\int_{-\beta_n}^{\beta_n} \varphi(x+t^{1/\gamma}\bar{\beta})\,d\bar{\beta}\right) d\beta. \tag{4.3.13}$$

The representations (4.3.12) and (4.3.13) make it possible to repeat the reasoning, used above for the case a), in the situations where we assume the existence of the centrally-symmetric mean value and the limit mean value respectively. □

4.3.3. Necessary and sufficient conditions for uniform stabilization.

Theorem 4.5. *A solution* (4.3.2) *of the equation* (4.3.1) *with a bounded continuous initial function* φ *has the property of uniform stabilization, if and only if* φ *has a uniform limit mean value, that is*

$$S(N,x) = \frac{1}{(2N)^n} \int\limits_{V_N^x} \varphi(y)\,dy \longrightarrow l,$$

as $N \to \infty$, *uniformly with respect to* x. *Here* $V_N^x = \{\beta|\, |\beta_i - x_i| < N, i = 1,\dots,n\}$.

Proof. It is sufficient to consider the case $l = 0$.

1) *Sufficiency.* Let φ have the uniform limit mean value 0, that is for any $\varepsilon > 0$ there exists such $N_0(\varepsilon)$ that for all $N > N_0(\varepsilon)$ and any x,

$$\left| \frac{1}{(2N)^n} \int\limits_{V_N^x} \varphi(y)\,dy \right| < \varepsilon.$$

Using the arbitrariness of x, we can establish a similar inequality, with the integrals taken not over cubes but over parallelepipeds $V^x_{N_1,\dots,N_n} = \{\beta :\ |\beta_i - x_i| < N_i, i = 1,\dots,n\}$ where N_i tend to infinity independently of each other. This implies the existence of the zero angular limit mean values of φ, and (by Theorem 4.4) the stabilization of $u(t,x)$ to zero. The arguments from the proof of Theorem 4.4 show that in this case the stabilization is uniform with respect to x.

2) *Necessity.* Suppose that $u(t,x) \to 0$ uniformly with respect to x whereas the initial function does not possess the uniform limit mean value. This means that there exists such $\varepsilon_0 > 0$ that, for any positive N_0, we can find such $N > N_0$ and such a point $\eta = (\eta_1,\dots,\eta_n)$ that

$$|S(N,\eta)| > \varepsilon_0. \tag{4.3.14}$$

Consider a cube $V_B = \{|(z_1,\dots,z_n)| < B\}$. Using (4.3.7) we write

$$1 = \int\limits_{\mathbb{R}^n} G(z)\,dz = \int\limits_{V_B} G(z)\,dz + \int\limits_{\mathbb{R}^n \setminus V_B} G(z)\,dz$$

and choose B in such a way that

$$\int\limits_{\mathbb{R}^n \setminus V_B} |G(z)|\,dz < \frac{\varepsilon_0}{4}, \quad \int\limits_{V_B} |G(z)|\,dz \geq 1 - \frac{\varepsilon_0}{4}. \tag{4.3.15}$$

After the change of variables $y = x + zt^{1/\gamma}$ the representation (4.3.2) takes the form

$$u(t,x) = \int_{V_B} G(z)\varphi(x + zt^{1/\gamma})\,dz + \int_{\mathbb{R}^n\setminus V_B} G(z)\varphi(x + zt^{1/\gamma})\,dz = I_1 + I_2.$$

Suppose for simplicity that $|\varphi(x)| \le 1$. Then $|I_2| \le \dfrac{\varepsilon_0}{4}$.

The subsequent reasoning will be based on (4.3.15). Let us take a sequence $N_0^{(k)} \to \infty$. By (4.3.14) we can find a sequence $N_k \to \infty$, and points $\eta^{(k)}$, such that $|S(N_k, \eta^{(k)})| \ge \varepsilon_0$. In fact, we may assume that $S(N_k, \eta^{(k)}) \ge \varepsilon_0$. Let us define a sequence of time instants $t_k = \left(\dfrac{\varepsilon_0}{8nB} N_k\right)^\gamma \to \infty$. We have

$$\begin{aligned}
&\left| \frac{1}{(2N_k)^n} \int_{V_{N_k}^{\eta^{(k)}}} u(t_k, x)\,dx \right| \\
&= \left| \frac{1}{(2N_k)^n} \int_{V_{N_k}^{\eta^{(k)}}} \left\{ \int_{V_B} G(z)\varphi(x + zt^{1/\gamma})\,dz + \int_{\mathbb{R}^n\setminus V_B} G(z)\varphi(x + zt^{1/\gamma})\,dz \right\} dx \right| \\
&\ge \left| \frac{1}{(2N_k)^n} \int_{V_{N_k}^{\eta^{(k)}}} \left\{ \int_{V_B} G(z)\varphi(x + zt^{1/\gamma})\,dz \right\} dx \right| \\
&\quad - \left| \frac{1}{(2N_k)^n} \int_{V_{N_k}^{\eta^{(k)}}} \left\{ \int_{\mathbb{R}^n\setminus V_B} G(z)\varphi(x + zt^{1/\gamma})\,dz \right\} dx \right|.
\end{aligned}$$

We give an upper bound for the subtrahend in the last difference:

$$\begin{aligned}
&\left| \frac{1}{(2N_k)^n} \int_{V_{N_k}^{\eta^{(k)}}} \left\{ \int_{\mathbb{R}^n\setminus V_B} G(z)\varphi(x + zt^{1/\gamma})\,dz \right\} dx \right| = \left| \frac{1}{(2N_k)^n} \int_{V_{N_k}^{\eta^{(k)}}} I_2(x)\,dx \right| \\
&\le \frac{1}{(2N_k)^n} \int_{V_{N_k}^{\eta^{(k)}}} |I_2(x)|\,dx \le \frac{\varepsilon_0}{4} \frac{1}{(2N_k)^n} \int_{V_{N_k}^{\eta^{(k)}}} dx = \frac{\varepsilon_0}{4}.
\end{aligned}$$

Let us give a lower estimate of the minuend. We have

$$\left| \int\limits_{V_B} G(z)\,dz \frac{1}{(2N_k)^n} \int\limits_{V_{N_k}^{\eta(k)}} \varphi(x)\,dx \right.$$

$$\left. + \int\limits_{V_B} G(z)\,dz \frac{1}{(2N_k)^n} \left[\int\limits_{V_{N_k}^{\eta(k)}} \varphi(x + zt^{1/\gamma})\,dx - \int\limits_{V_{N_k}^{\eta(k)}} \varphi(x)\,dx \right] dz \right|$$

$$\geq \left| \int\limits_{V_B} G(z)\,dz \right| \cdot \left| \frac{1}{(2N_k)^n} \int\limits_{V_{N_k}^{\eta(k)}} \varphi(x)\,dx \right|$$

$$- \int\limits_{V_B} |G(z)| \left| \frac{1}{(2N_k)^n} \left[\int\limits_{V_{N_k}^{\eta(k)}} \varphi(x + zt^{1/\gamma})\,dx - \int\limits_{V_{N_k}^{\eta(k)}} \varphi(x)\,dx \right] \right| dz$$

$$\geq \varepsilon_0 \left(1 - \frac{\varepsilon_0}{4}\right) - \frac{\varepsilon_0}{4}.$$

Hence,

$$\left| \frac{1}{(2N_k)^n} \int\limits_{V_{N_k}^{\eta(k)}} u(t_k, x)\,dx \right| \geq \varepsilon_0 \left(1 - \frac{\varepsilon_0}{4}\right) - \frac{\varepsilon_0}{4} - \frac{\varepsilon_0}{4} = \frac{\varepsilon_0}{2} - \frac{\varepsilon_0^2}{4}. \tag{4.3.16}$$

We may assume that $\frac{\varepsilon_0}{2} - \frac{\varepsilon_0^2}{4} \geq \frac{\varepsilon_0}{4}$.

It follows from the inequality (4.3.16) that each $V_{N_k}^{\eta(k)}$ contains at least one point $x^{(k)}$, in which $u(t_k, x^{(k)}) \geq \frac{\varepsilon_0}{4}$. Since $t_k \to \infty$, this contradicts our initial assumption. □

4.3.4. Necessary and sufficient conditions for pointwise stabilization. The problem of finding necessary and sufficient conditions for pointwise stabilization is among the most subtle problems in the stabilization theory for parabolic equations and systems. In the problem of uniform stabilization (Theorem 4.5) the class of volumes, such that the existence of limit mean values of an initial function over them guarantees the stabilization of the solution, is wide enough and loosely connected with the equation under consideration. In contrast to that, necessary and sufficient conditions for pointwise stabilization are formulated in terms of the limit mean values over volumes restricted by the level surfaces of the fundamental solution.

We consider the Cauchy problem for the equation

$$\partial_t u(t,x) + (A_0 u)(t,x) = 0, \qquad (4.3.17),$$

in which the pseudo-differential operator A_0 has the symbol $|\xi|^\gamma$, $\gamma > 1$.

We shall prove two theorems giving necessary and sufficient conditions for pointwise stabilization of solutions of the equation (4.3.17).

Theorem 4.6. *A solution $u(t,x)$ of the equation* (4.3.17) *with $1 < \gamma < \infty$, given by the Poisson integral*

$$u(t,x) = \int_{\mathbb{R}^n} Z_0(t, x-y)\varphi(y)\,dy, \qquad (4.3.18)$$

stabilizes at a point x^0 to a constant l if and only if the bounded continuous initial function $\varphi(x)$ possesses a globular limit mean value, that is

$$\lim_{R\to\infty} \frac{1}{B_R^{x^0}} \int_{B_R^{x^0}} \varphi(x)\,dx = l \qquad (4.3.19)$$

where $B_R^{x^0}$ is a ball of radius R centered at x^0.

Theorem 4.7. *Let $u(t,x)$ be a solution of the equation* (4.3.17) *with $1 < \gamma < 2$, given by the Poisson-Stieltjes integral*

$$u(t,x) = \int_{\mathbb{R}^n} Z_0(t, x-y)\mu(dy), \qquad (4.3.20)$$

where μ is a real-valued measure defined on the Borel σ-algebra of $\mathbb{R}^n$. Suppose that

$$\int_{\mathbb{R}^n} (1+|x|^2)^{-\frac{n+\gamma}{2}} \mu(dx) < +\infty, \qquad (4.3.21)$$

and that μ is bounded from below, that is for every Borel set $M \subset \mathbb{R}^n$,

$$\mu(M) \geq -C|M| \qquad (4.3.22)$$

where $|M|$ is the n-dimensional Lebesgue measure of M, C is a constant independent of M.

The Poisson-Stieltjes integral (4.3.20) *stabilizes at a point x^0 to a constant l if and only if the measure μ has the limit mean value, that is*

$$\lim_{R\to\infty} \frac{\mu\left(B_R^{x^0}\right)}{\left|B_R^{x^0}\right|} = l.$$

Proof of Theorem 4.6. We shall use the following Wiener's Tauberian theorem (see [230]).

Let K_1 be a function from $L_1(\mathbb{R})$ whose Fourier transform is different from 0 everywhere on $\mathbb{R}$. Let K_2 belong to $L_1(\mathbb{R})$, and f be bounded on $\mathbb{R}$. If

$$\lim_{x\to\infty}\int_{-\infty}^{\infty} K_1(x-y)f(y)\,dy = A\int_{-\infty}^{\infty} K_1(x)\,dx,$$

then

$$\lim_{x\to\infty}\int_{-\infty}^{\infty} K_2(x-y)f(y)\,dy = A\int_{-\infty}^{\infty} K_2(x)\,dx.$$

Due to the convolution structure of the representation (4.3.2) it suffices to study the stabilization at the point $x = 0$. Let us consider the integral (4.3.3) with $a(\xi) = |\xi|^\gamma$. We have

$$\begin{aligned} Z_0(t,x) &= C_1 t^{-n/\gamma}\int_{\mathbb{R}^n} \exp\left\{i(xt^{-1/\gamma},\sigma) - |\sigma|^\gamma\right\} d\sigma \\ &= C_1 t^{-n/\gamma}\int_0^\infty \rho^{n-1}e^{-\rho^\gamma}\,d\rho\int_0^\pi \cos\left(irt^{-1/\gamma}\rho\cos\theta\right)\sin^{n-2}\theta\,d\theta, \end{aligned}$$

$r = |x|$. Then we use the representation (see [228] or [226])

$$J_\nu(z) = \frac{(z/2)^\nu}{\Gamma(\nu+1/2)\Gamma(1/2)}\int_0^\pi \cos(z\cos\theta)\sin^{2\nu}\theta\,d\theta,$$

with $\nu = \dfrac{n}{2} - 1$, $z = r(t)\rho$ where $r(t) = rt^{-1/\gamma}$. We obtain that

$$Z_0(t,x) = C_2 t^{-n/\gamma} G_1(r(t)) \tag{4.3.23}$$

where

$$G_1(s) = \int_0^\infty \rho^{n-1}e^{-\rho^\gamma}(s\rho)^{-\frac{n}{2}+1}J_{\frac{n}{2}-1}(s\rho)\,d\rho.$$

Now we can write

$$u(t,0) = C_3 t^{-1/\gamma}\int_0^\infty r(t)^{n-1}G_1(r(t))v_0(r)\,dr, \tag{4.3.24}$$

where $v_0(r)$ is the integral of φ over the sphere of the radius r centered at the origin.

Let us rewrite the initial problem in a form permitting the use of Wiener's Tauberian theorem. Consider the integral operators

$$I_1 v_0 = t^{-1/\gamma} \int_0^t v_0(r)\, dr,$$

$$I_2 v_0 = t^{-1/\gamma} \int_0^\infty r(t)^{n-1} G_1(r(t)) v_0(r)\, dr.$$

In both integrals we introduce new independent variables setting

$$r = e^y, \quad t^{1/\gamma} = e^\eta.$$

Then

$$I_m v_0 = \int_{-\infty}^\infty k_m(\eta - y) v_0(e^y)\, dy, \quad m = 1, 2,$$

where

$$k_1(\eta) = \begin{cases} e^{-\eta}, & \text{if } \eta \geq 0 \\ 0, & \text{if } \eta < 0 \end{cases}; \quad k_2(\eta) = e^{-n\eta} G_1(e^\eta).$$

Using the estimate for Z_0 we see that $k_m \in L_1(-\infty, \infty)$.

The final part of the proof of Theorem 4.6 consists of the verification of the fact that the Fourier transforms $\Phi_m(s)$ of the functions $k_m(\eta)$ are different from 0 for any real s. For the function Φ_1, that follows from a direct calculation. For Φ_2 we have

$$\Phi_2(s) = \int_{-\infty}^\infty e^{(is-n)\eta} \left(\int_0^\infty \rho^{n/2} e^{-\rho^\gamma} e^{(\frac{n}{2}-1)\eta} J_\nu(e^{-\eta}\rho)\, d\rho \right) d\eta$$

$$= \int_0^\infty \rho^{n/2} e^{-\rho^\gamma} \left(\int_{-\infty}^\infty e^{(is-n/2)\eta} J_\nu(e^{-\eta}\rho) e^{-\eta}\, d\eta \right) d\rho. \tag{4.3.25}$$

In the internal integral we introduce the new variable $\zeta = e^{-\eta}$. Then the formula (4.3.25) is rewritten as follows:

$$\Phi_2(s) = \int_0^\infty \rho^{n/2} e^{-\rho^\gamma} \left(\int_0^\infty \zeta^{\frac{n}{2}-is} J_\nu(\zeta\rho)\, d\zeta \right) d\rho$$

$$= \int_0^\infty \rho^{is} e^{-\rho^\gamma}\, d\rho \cdot \int_0^\infty \lambda^{\frac{n}{2}-is} J_{\frac{n}{2}-1}(\lambda)\, d\lambda.$$

Both integrals are calculated explicitly (in the first integral we can take ρ^γ for the new variable; the second integral is well known (see, for example, [194]). As a result we find that

$$\Phi_2(s) = \gamma^{-1} 2^{\frac{n}{2} - is} \frac{\Gamma(n - is)\Gamma(is\gamma^{-1})}{\Gamma(is)}. \tag{4.3.26}$$

Note that the functions $\Gamma(is\gamma^{-1})$ and $\Gamma(is)$ both have a simple pole at the point $s = 0$, with the residues $-i\gamma$ and $-i$ respectively, so that $\Phi_2(0) = 2^{n/2}\Gamma(n) = 2^{n/2}(n-1)!$ (as a Fourier transform of an integrable function, Φ_2 is continuous, and its value for $s = 0$ is a limit of $\Phi_2(s)$, as $s \to 0$).

It follows from (4.3.26) that $\Phi_2(s) \neq 0$ for all real values of s, as desired.

By Wiener's Tauberian theorem, $\lim\limits_{t \to 0} u(t, 0) = 0$ if and only if

$$\lim_{R \to \infty} R^{-1} \int_0^R v_0(r)\, dr = 0.$$

It remains to use the following elementary fact. If $v(r)$ is a bounded continuous function, then the limits $\lim\limits_{R \to \infty} nR^{-n} \int_0^R r^{n-1} v(r)\, dr$ and $\lim\limits_{R \to \infty} R^{-1} \int_0^R v(r)\, dr$ exist simultaneously and their values coincide. □

The *proof of Theorem* 4.7 is based on the generalized and improved version of Theorem 236 from Hardy's book [96] given in [75]. It is essential that (by Theorem 4.2) the function Z_0 is nonnegative.

A Tauberian type theorem. *Let $g(\rho)$ be a nonnegative function from $L_1(\mathbb{R}_+)$ whose Mellin transform*

$$\Phi(s) = \int_0^\infty g(\rho)\rho^{-is}\, d\rho$$

has no real zeroes, and λ be a nonnegative measure on $\mathbb{R}_+$. If

$$\tau^{-1} \int_0^\infty g\left(\frac{r}{\tau}\right) \lambda(dr) \longrightarrow 0 \text{ as } \tau \to \infty,$$

then

$$\lambda([0, R))/R \longrightarrow 0 \text{ as } R \to \infty.$$

Now we turn directly to the proof of Theorem 4.7. Our problem is essentially one-dimensional. Indeed, let us pass to the spherical coordinates in (4.3.20). If $E = [r_1, r_2)$, $C_E = \{\xi \in \mathbb{R}^n |\ |\xi| \in [r_1, r_2)\}$, then the set function

$$\lambda(E) = \int_{C_E} r^{n-1} \mu(d\xi), \quad r = |\xi|,$$

can be extended to a measure defined on the σ-algebra of Borel sets of the half-axis $\mathbb{R}_+$. Due to (4.3.20) and (4.3.21), we have

$$\int_0^\infty r^{n-1}(1+r^2)^{-\frac{n+\gamma}{2}}\lambda(dr) < \infty, \quad \lambda(E) \geq -C|E|. \tag{4.3.27}$$

Let us rewrite the Poisson-Stieltjes integral in a form convenient for applying the Tauberian theorem. Setting $t = \tau^\gamma$ we write (4.3.20) with $x = 0$ as

$$v(\tau) = u(t,0) = \tau^{-n}\int_0^\infty r^{n-1}G_2(r/\tau)\lambda(dr) \tag{4.3.28}$$

where

$$G_2(r/\tau) = G_1(\xi/\tau) = (2\pi)^{-n}\int_{\mathbb{R}^n} \exp\left\{i\left(-\frac{\xi}{\tau},\sigma\right) - |\sigma|^\gamma\right\}d\sigma, \quad \xi \in \mathbb{R}^n. \tag{4.3.29}$$

Turn the coordinate system in (4.3.29) in such a way that $\xi = (0,\dots,0,r)$, so that

$$G_2(r/\tau) = (2\pi)^{-n}\int_{\mathbb{R}^n} \exp\{i(-r/\tau)\,\sigma_n - |\sigma|^\gamma\}\,d\sigma,$$

$$v(\tau) = \frac{1}{\tau}\int_0^\infty \left(\frac{r}{\tau}\right)^{n-1} G_2(r/\tau)\lambda(dr). \tag{4.3.30}$$

By (4.3.30), the assertion of Theorem 4.7 is now as follows: in order that

$$\frac{1}{\tau}\int_0^\infty \left(\frac{r}{\tau}\right)^{n-1} G_2(r/\tau)\lambda(dr) \longrightarrow 0 \text{ as } \tau\to\infty,$$

it is necessary and sufficient that

$$R^{-n}\int_0^R r^{n-1}\lambda(dr) \longrightarrow 0 \text{ as } R\to\infty.$$

Necessity. Let us use the Tauberian type theorem setting $g(\rho) = \rho^{n-1}G_2(\rho)$. Since $|G_2(\rho)| \leq C(1+\rho)^{-(n+\gamma)}$, we have $|g(\rho)| \leq C(1+\rho)^{-1-\gamma}$, so that $g \in L_1(\mathbb{R}^1_+)$.

Consider the Mellin transform

$$\Phi(s) = \int_0^\infty g(\rho)\rho^{-is}\,d\rho = \int_0^\infty \rho^{n-1-is}G_2(\rho)\,d\rho.$$

As in (4.3.23), we get

$$G_2(\rho) = C\rho^{1-\frac{n}{2}} \int_0^\infty r^{n/2} e^{-r^\gamma} J_{\frac{n}{2}-1}(r\rho)\, dr$$

(here we assume that $n > 1$; the case $n = 1$ is much simpler), whence

$$\Phi(s) = C \int_0^\infty r^{n/2} e^{-r^\gamma}\, dr \int_0^\infty \rho^{\frac{n}{2}-is} J_{\frac{n}{2}-1}(r\rho)\, d\rho.$$

The function $\Phi(s)$ coincides, up to a multiplicative constant, with the function $\Phi_2(s)$ (see the formula (4.3.25)) from the proof of Theorem 4.6. Using (4.3.26) we see that the Tauberian type theorem is applicable, so that the relation $u(t, 0) \to 0$, $t \to \infty$, implies the relation

$$\lambda([0, R))/R \longrightarrow 0 \text{ as } R \to \infty.$$

It remains to use the fact (proved in [75]) that for any finite measure λ the limits of $\lambda([0, R))/R$ and $nR^{-n} \int_0^R r^{n-1}\lambda(dr)$, as $R \to \infty$, exist simultaneously, and are equal.

Sufficiency. We have to prove that the relation

$$\tau^{-n} \int_0^\tau x^{n-1}\lambda(dx) \longrightarrow 0 \text{ as } \tau \to \infty$$

implies $v(\tau) \to 0$ as $\tau \to \infty$. We carry out an estimate of v based on the inequality

$$|\partial_\rho G_2(\rho)| \le C(1 + \rho)^{-(n+\gamma+1)}.$$

We have by (4.3.28)

$$v(\tau) = \tau^{-n} \int_0^\infty r^{n-1} G_2(r/\tau)\lambda(dr) = \tau^{-n} \int_0^\infty r^{n-1} \left[- \int_{r/\tau}^\infty \partial_\rho G_2(\rho)\, d\rho \right] \lambda(dr)$$

$$= -\tau^{-n} \int_0^\infty \int_{r/\tau}^\infty r^{n-1} \partial_\rho G_2(\rho)\, d\rho\, \lambda(dr).$$

Changing the order of integration we obtain

$$v(\tau) = -\tau^{-n} \int_0^\infty \partial_\rho G_2(\rho)\, d\rho \int_0^{\rho\tau} r^{n-1}\lambda(dr)$$
$$= -\int_0^\infty \left[\rho^n \partial_\rho G_2(\rho)(\rho\tau)^{-n} \int_0^{\rho\tau} r^{n-1}\lambda(dr) \right] d\rho \equiv I_1 + I_2$$

where I_1 and I_2 correspond to the integration in ρ from 0 to δ and from δ to ∞ respectively.

Denote

$$C = \sup \left| (\rho\tau)^{-n} \int_0^{\rho\tau} r^{n-1}\lambda(dr) \right|, \quad K = \int_0^\infty \rho^n \left| \frac{\partial}{\partial\rho} G_2(\rho) \right| d\rho.$$

Given $\varepsilon > 0$, choose δ in such a way that

$$\int_0^\delta \rho^n \left| \partial_\rho G_2(\rho) \right| d\rho < \frac{\varepsilon}{2C},$$

and then choose t_0 so large that

$$\left| t^{-n} \int_0^t r^{n-1}\lambda(dr) \right| < \frac{\varepsilon}{2K},$$

if $t > t_0$. Now for $t > t_0/\delta$ we have $|I_1| < \varepsilon/2$, $|I_2| < \varepsilon/2$, as desired. □

Remark. The proof, based on the Tauberian theorems, of the fact that the stabilization of a solution of the Cauchy problem implies the existence of a limit of an appropriate average of the initial function, required an investigation of a certain reference function Φ. In the above situations, as well as in Section 4.3.6 below these functions are simple rational fractions of the Gamma function. Their poles and the corresponding residues are easily calculated.

4.3.5. A Liouville type theorem. Liouville type theorems determine a form of a function from a certain class provided its asymptotic behaviour is known. For example, the classical Liouville theorem states that an entire function of polynomial growth is a polynomial. A proof of this result is based on the representation by the Cauchy integral over a circle and the arbitrariness of the latter's radius. For a parabolic equation on a half-space $\{(t,x) | \, t \le T, x \in \mathbb{R}^n\}$ we have an integral representation with an arbitrary initial hyperplane which makes it possible to prove the following version of the Liouville theorem.

Theorem 4.8. *A solution of the equation* (4.3.1) *on a half-space* $Q_- = \{(t,x)|\ t \leq 0, x \in \mathbb{R}^n\}$ *represented by the Poisson integral*

$$u(t,x) - \int\limits_{\mathbb{R}^n} Z(t-t_0, x-\xi)\varphi(\xi)\,d\xi \tag{4.3.31}$$

and satisfying on Q_- *the inequality*

$$|u(t,x)| \leq C_0(1+|x|)^\beta \tag{4.3.32}$$

is a polynomial in x *whose degree does not exceed* $[\beta]$.

Proof. From (4.3.31) we obtain the equalities

$$\partial_x^\varkappa u(t,x) = \int\limits_{\mathbb{R}^n} \partial_x^\varkappa Z(t-t_0, x-\xi)u(t_0,\xi)\,d\xi.$$

Using the inequalities (4.3.5) and (4.3.32) we find that

$$|\partial_x^\varkappa u(t,x)|$$
$$\leq C_0 C_\varkappa \int\limits_{\mathbb{R}^n} (1+|\xi|)^\beta (t-t_0)^{-(n+|\varkappa|)\gamma^{-1}} \left[1+|x-\xi|(t-t_0)^{-1/\gamma}\right]^{-\gamma-|\varkappa|-n} d\xi.$$

After the standard change of variables $\xi = x + (t-t_0)^{1/\gamma}\alpha$, we obtain that

$$|\partial_x^\varkappa u(t,x)| \leq C_0 C_\varkappa \int\limits_{\mathbb{R}^n} (1+|x|+(t-t_0)^{1/\gamma}|\alpha|)^\beta (t-t_0)^{-|\varkappa|\gamma^{-1}} (1+|\alpha|)^{-\gamma-|\varkappa|-n} d\alpha$$
$$\leq C(\varkappa)(t-t_0)^{(\beta-|\varkappa|)/\gamma}.$$

If $|\varkappa| = [\beta]+1$, then, since t_0 is arbitrary, we have $D_x^\varkappa u(t,x) \equiv 0$, so that $u(t,x)$ is a polynomial in x whose degree does not exceed $[\beta]$. □

4.3.6. Examples. We shall consider three sufficiently simple examples. For the first example ($n = 1$, γ is arbitrary) we give as complete analysis of stabilization conditions as possible. As usual, this analysis is rather simple for the one-dimensional case. The second example (n is arbitrary, $\gamma = 1$) shows how the knowledge of an explicit formula for FSCP simplifies the investigation of stabilization. In the third example ($n = 1$, $\gamma = 1/2$) we look at the behaviour of solutions of an equation which does not belong to the class considered in the preceding sections (where it was assumed that $\gamma \geq 1$).

Below we assume that the initial functions are bounded and continuous, and that the solutions are represented by the Poisson integrals (4.3.2).

Example 1 ($n = 1$). Here the symbol $a(\xi)$ of the operator A from the equation (4.3.1) has the form

$$a(\xi) = |\xi|^\gamma a(\xi/|\xi|) = |\xi|^\gamma a(\theta) \tag{4.3.33}$$

where θ takes the values ± 1. Denote $a(-1) = a_-$, $a(1) = a_+$. By the condition (A_{45}) we have

$$\operatorname{Re} a_- > 0, \quad \operatorname{Re} a_+ > 0. \tag{4.3.34}$$

Now the function $Z_0(t,x)$ given by (4.3.3) can be written as follows:

$$Z_0(t,x) = (2\pi)^{-1} t^{-1/\gamma} \int_0^\infty \left[\exp\{iz\sigma - a_+\sigma^\gamma\} + \exp\{-iz\sigma - a_-\sigma^\gamma\}\right] d\sigma,$$

$$z = xt^{-1/\gamma}. \tag{4.3.35}$$

If $a(\xi) = |\xi|^\gamma$, then the formula (4.3.35) takes the form

$$Z_0(t,x) = \pi^{-1} t^{-1/\gamma} \int_0^\infty e^{-\sigma^\gamma} \cos\left(xt^{-1/\gamma}\sigma\right) d\sigma. \tag{4.3.36}$$

a) *Pointwise stabilization.* Here we give a specialization of Theorem 4.4 for the one-dimensional case.

Definition. *A function $\varphi(x)$ has the radial limit mean values at the origin, if there exist the limits*

$$\lim_{R\to\infty} \frac{1}{R} \int_0^R \varphi(x)\,dx, \quad \lim_{R\to\infty} \frac{1}{R} \int_{-R}^0 \varphi(x)\,dx,$$

called the right and left radial limit mean values respectively. A function $\varphi(x)$ has the limit mean value, if there exists the limit

$$\lim_{R\to\infty} \frac{1}{2R} \int_{-R}^R \varphi(x)\,dx.$$

Now Theorem 4.4 reads as follows.

Theorem 4.4′ $(n = 1)$.

1) *If the initial function $\varphi(x)$ has the left and right limit mean values at the origin, then the corresponding solution of the equation* (4.3.1), *with the symbol* (4.3.33) *satisfying* (4.3.34), *stabilizes to 0, that is*

$$\lim_{t\to\infty} u(t,x) = 0, \quad x \in \mathbb{R}.$$

2) *The solution of the Cauchy problem for the equation* (4.3.1) *with the symbol $a_0(\xi) = |\xi|^\gamma$, corresponding to an initial function having the limit mean value, stabilizes to 0.*

In both cases 1) *and* 2) *the stabilization is uniform on each finite interval* $[a,b] \subset \mathbb{R}$.

b) Let us show that for the case of the symbol $|\xi|^\gamma$ *the existence of the limit mean value is not only sufficient, but also necessary for the stabilization.* We do it using Wiener's Tauberian theorem.

By (4.3.36) we have

$$u(t,0) = \pi^{-1}t^{-1/\gamma} \int_{-\infty}^{\infty} \left[\int_0^\infty e^{-\rho^\gamma} \cos\left(\xi t^{-1/\gamma}\rho\right) d\rho \right] \varphi(\xi)\, d\xi.$$

Set $\xi = e^y$, $t^{1/\gamma} = e^\eta$. Then

$$u(t,0) = \pi^{-1} \int_{-\infty}^{\infty} \left[\int_0^\infty \rho^{-\gamma} \cos\left(e^{y-\eta}\rho\right) e^{y-\eta}\, d\rho \right] \varphi(e^y)\, dy.$$

Thus we have a representation

$$u(t,0) = \int_{-\infty}^{\infty} K_2(\eta - y)\varphi(e^y)\, dy$$

where

$$K_2(\eta) = \pi^{-1} e^{-\eta} \int_0^\infty \cos\left(e^{-\eta}\rho\right) e^{-\rho^\gamma}\, d\rho.$$

As before, $K_2 \in L_1(-\infty,\infty)$.

Now we compute the Fourier transform Φ of the kernel K_2:

$$\begin{aligned}\Phi(s) &= C_1 \int_{-\infty}^{\infty} e^{is\eta} \left[\int_0^\infty \cos\left(e^{-\eta}\rho\right) e^{-\rho^\gamma}\, d\rho \right] e^{-\eta}\, d\eta \\ &= C_2 \int_0^\infty \zeta^{-is} \left[\int_0^\infty \cos(\zeta\rho) e^{-\rho^\gamma}\, d\rho \right] d\zeta \\ &= I_1 I_2\end{aligned}$$

where

$$I_1 = \int_0^\infty e^{-\rho^\gamma} \rho^{is-1}\, d\rho, \quad I_2 = \int_0^\infty \zeta^{-is} \cos\zeta\, d\zeta.$$

We have $I_1 = C_3\Gamma(is\gamma^{-1})$, $I_2 = \Gamma(1-is)\cos\frac{\pi}{2}(-is+1) = -i\Gamma(1-is)\sinh\frac{\pi}{2}s$ (see [193], formula 2.5.3.10). Hence $\Phi(s) = C_4\Gamma(1-is)\Gamma(is\gamma^{-1})\sinh\frac{\pi}{2}s$. It is clear that $\Phi(s) \neq 0$ for $s \neq 0$. Noting that $\Gamma(is\gamma^{-1})$ has a first order pole at the origin, with the residue $-i\gamma$, while $\sinh\frac{\pi}{2}s$ has a first order zero there, we find that $\Phi(0) \neq 0$. □

c) *Uniform stabilization.* Let us reformulate Theorem 4.5 for the case $n = 1$.

Theorem 4.5′ ($n = 1$). *A solution* (4.3.2) *of the equation* (4.3.1) *with the symbol* (4.3.33) *satisfying* (4.3.34), *with a bounded continuous initial function* φ, *has the property of uniform stabilization to 0, if and only if* φ *has a uniform limit mean value 0, that is*

$$\frac{1}{2R}\int_{x-R}^{x+R} \varphi(r)\,dr \longrightarrow 0,$$

as $R \to \infty$, *uniformly with respect to* x.

Example 2. ($n > 1, \gamma = 1$). In this case the Poisson kernel can be written explicitly:

$$Z_0(t,x) = C_1 t^{-n}\left(1 + (t^{-1}|x|)^2\right)^{-\frac{n+1}{2}}$$

where $C_1 = \pi^{n/2}\left(\Gamma(\frac{n}{2}+1)\right)$. We shall show here that the explicit form makes it possible to prove Theorems 4.6 and 4.7 for the case $\gamma = 1$.

Proof of Theorem 4.6 ($n > 1, \gamma = 1$). As before, it suffices to prove that the control function $\Phi(s)$ has no real zeroes. In the earlier notation, we write

$$u(t,0) = C_1 t^{-n}\int_0^\infty r^{n-1}\left(1 + (rt^{-1})^2\right)^{-(n+1)/2} v_0(r)\,dr$$

where $v_0(r)$ is the integral of φ on the sphere in $\mathbb{R}^n$ of the radius r centered at the origin. After the change of variables $r = e^y$, $t = e^\eta$, we represent $u(t,0)$ by the convolution formula with the kernel

$$K_2(\eta) = C_1\left(1 + e^{-2\eta}\right)^{-(n+1)/2} e^{-n\eta}.$$

Calculating its Fourier transform, we get

$$\begin{aligned}\Phi(s) &= C_2\int_{-\infty}^{\infty} e^{i\eta s - n\eta}\left(1 + e^{-2\eta}\right)^{-(n+1)/2} d\eta\\ &= C_2\int_0^\infty \zeta^{-is+(n-1)}(1+\zeta^2)^{-(n+1)/2}\,d\zeta\\ &= C_3\int_0^\infty \beta^{\frac{-is+n}{2}-1}(1+\beta)^{-(n+1)/2}\,d\beta.\end{aligned}$$

Since

$$\int_0^\infty x^{a-1}(1+x)^{-(a+b)}\,dx = B(a,b)$$

(see [228]), we find that

$$\Phi(s) = B\left(\frac{n-is}{2}, \frac{1+is}{2}\right) = \Gamma\left(\frac{n-is}{2}\right)\Gamma\left(\frac{1+is}{2}\right)\left[\Gamma\left(\frac{n+1}{2}\right)\right]^{-1},$$

so that $\Phi(s) \neq 0$ for real values of s. □

Proof of Theorem 4.7 ($n > 1, \gamma = 1$). First we repeat the initial fragment of the above proof of Theorem 4.7. We pass to the spherical coordinates in (4.3.20). If $E = [r_1, r_2)$, $C_E = \{\xi \in \mathbb{R}^n \mid |\xi| \in [r_1, r_2)\}$, then the set function

$$\lambda(E) = \int_{C_E} r^{n-1}\mu(d\xi), \quad r = |\xi|,$$

can be extended to a measure defined on the σ-algebra of Borel sets of the half-axis $\mathbb{R}_+$, and

$$\int_0^\infty r^{n-1}(1+r^2)^{-\frac{n+1}{2}}\lambda(dr) < \infty, \quad \lambda(E) \geq -C|E|.$$

We have

$$u(t,0) = t^{-1}\int_0^\infty G(r/t)\lambda(dr)$$

where $G(\rho) = \rho^{n-1}(1+\rho^2)^{-(n+1)/2}$. Thus Theorem 4.7 (for $n > 1, \gamma = 1$) takes the following form: in order that

$$t^{-1}\int_0^\infty G(r/t)\lambda(dr) \longrightarrow 0 \text{ as } t \to \infty,$$

it is necessary and sufficient that

$$R^{-n}\int_0^R r^{n-1}\lambda(dr) \longrightarrow 0 \text{ as } t \to \infty.$$

The proof of the sufficiency is the same as the one given above for the case $\gamma > 1$. In order to prove the necessity we may use the above Tauberian type theorem with $g = G$. We see that $|g(\rho)| \leq \text{const} \cdot (1+\rho)^{-2}$, so that $g \in L_1(0,\infty)$. The appropriate control function

$$\Phi(s) = \int_0^\infty g(\rho)\rho^{-is}\,d\rho$$

coincides, up to a multiplicative constant, with the function Φ from the above proof of Theorem 4.6 for the case $n > 1, \gamma = 1$. Using the Tauberian type theorem we complete the proof, just as before.

Example 3. $(n = 1, \gamma = 1/2)$. Consider the equation

$$\partial_t u = \mathfrak{A} u \tag{4.3.37}$$

where $\mathfrak{A}$ is a pseudo-differential operator with the symbol $a(\xi) = |\xi|^{1/2}$. It was shown by Schneider [203] that the Poisson kernel for this case equals

$$Z_0(t,x) = \begin{cases} \frac{1}{2\sqrt{\pi}} t^{-2}(xt^{-2})^{-3/2} e^{-t^2/4x}, & \text{if } x > 0, \\ 0, & \text{if } x \le 0. \end{cases}$$

If φ is a bounded continuous function on $\mathbb{R}$, then the Cauchy problem for the equation (4.3.37) with the initial condition $u(0,x) = \varphi(x)$ has a solution

$$u(t,x) = \int_{-\infty}^{x} Z_0(t, x-\xi)\varphi(\xi)\, d\xi. \tag{4.3.38}$$

Theorem 4.9. *The solution* (4.3.38) *stabilizes to 0 at the origin, if and only if there exists the limit*

$$\lim_{t\to\infty} t^{-2} \int_0^{t^2} \varphi(-\xi)\, d\xi = 0.$$

Proof. It follows from (4.3.38) that

$$u(t,0) = \int_0^{\infty} G(\xi t^{-2})\varphi(-\xi)\, d\xi$$

where $G(z) = z^{-3/2}\exp(-1/4z)$. We shall use Wiener's Tauberian theorem. After the traditional change of variables $t^2 = e^{\eta}$, $\xi = e^{y}$, we come to the convolution operator with the kernel

$$K_2(\eta) = C_1 e^{-3\eta/2} e^{-\frac{1}{4}e^{\eta}} e^{-\eta}.$$

For the control function we have

$$\Phi(s) = C_2 \int_{-\infty}^{\infty} e^{i\eta s + \frac{3}{2}\eta - \frac{1}{4}e^{\eta}} e^{-\eta}\, d\eta = C_2 \int_0^{\infty} \zeta^{is - \frac{1}{2}} e^{-\zeta/4}\, d\zeta = C_3 \cdot 2^{2is-1}\Gamma\left(is + \frac{1}{2}\right),$$

so that $\Phi(s) \neq 0$ for real values of s. □

4.4 Comments

Theorems 4.1, 4.2, and 4.3 were proved by Kochubei [128]. Earlier the parametrix for the equation (4.1.1) was constructed by Ya. Drin' [31]. Ya. Drin' and Eidelman [33] obtained (by a different method) its precise estimates for the case $n = 1$. In [30] the Cauchy problem for such equations in the Hölder classes was studied.

The example from Section 4.1.5 is a special case of the results by Schneider [203] who obtained an H-function representation for densities of stable distributions, which immediately yields their asymptotic properties. By a different method, the asymptotics of the fundamental solution of the Cauchy problem (4.1.1)–(4.1.2) for equations with constant symbols (depending only on ξ) was found by Fedoryuk [80].

Lemma 4.6 is contained essentially in [200].

The results of Section 4.3 are adaptations of well-known stabilization theorems for parabolic partial differential equations [45, 71, 72, 75, 197]. Theorems 4.4–4.6 were proved by Eidelman and Ya. Drin' [32], and Theorem 4.7 is a result of R. Drin'. Theorem 4.9, proved by Eidelman, is published here for the first time.

For (partial) generalizations of main results of this chapter see Kolokoltsov [137] (equations with Lévy generators), Ya. Drin' and Eidelman [34], and R. Drin' [29] (systems of equations), Portenko and Podolynny [192, 183, 182] (singular perturbations).

An essential feature of pseudo-differential operators considered here is the finite smoothness of symbols at the point $\xi = 0$, a consequence of their homogeneity essential for probabilistic applications. Meanwhile there exists a theory of parabolic pseudo-differential equations with C^∞-symbols based on the standard calculus of pseudo-differential operators. See, for example, [164, 174, 211, 222, 223, 199].

Chapter 5

Fractional Diffusion Equations

5.1 Fractional derivatives

5.1.1. The Riemann-Liouville fractional integrals and derivatives. There exist many definitions of fractional derivatives and integrals, which are meaningful for different classes of functions, inspired by various applications etc. See [202] for a general survey and a detailed exposition of many aspects of the fractional calculus. Here we touch only a few of them necessary for the sequel. For the proofs see [202].

Suppose that a function φ belongs to $L_1(0,T)$. Let $\alpha \in (0,1)$ be a fixed number. *The Riemann-Liouville fractional integral* of the order α is given by the expression

$$\left(I_{0+}^{\alpha}\varphi\right)(t) = \frac{1}{\Gamma(\alpha)} \int_0^t (t-\tau)^{\alpha-1}\varphi(\tau)\,d\tau.$$

The Riemann-Liouville fractional derivative of the order α is defined as

$$\left(D_{0+}^{\alpha}\varphi\right)(t) = \frac{1}{\Gamma(1-\alpha)} \frac{d}{dt} \int_0^t (t-\tau)^{-\alpha}\varphi(\tau)\,d\tau,$$

that is $\left(D_{0+}^{\alpha}\varphi\right)(t) = \frac{d}{dt}\left(I_{0+}^{1-\alpha}\varphi\right)(t)$, provided the fractional integral $I_{0+}^{1-\alpha}\varphi$ is an absolutely continuous function. If φ is defined on $(0,\infty)$, then $I_{0+}^{\alpha}\varphi$ and $D_{0+}^{\alpha}\varphi$ are similarly defined on the whole half-axis $(0,\infty)$.

If φ is absolutely continuous on $[0,T]$, then $D_{0+}^{\alpha}\varphi$ exists almost everywhere and belongs to $L_p(0,T)$ for $1 \le p < \alpha^{-1}$. In this case

$$\left(D_{0+}^{\alpha}\varphi\right)(t) = \frac{1}{\Gamma(1-\alpha)} \left[t^{-\alpha}\varphi(0) + \int_0^t (t-\tau)^{-\alpha}\varphi'(\tau)\,d\tau \right]. \tag{5.1.1}$$

In fact the Riemann-Liouville fractional derivative exists for much wider classes of functions, in particular for some functions with singularities at the origin.

For example, if $\varphi(t) = t^{\beta-1}$, $\beta > 0$, then

$$\left(D_{0+}^{\alpha}\varphi\right)(t) = \frac{\Gamma(\beta)}{\Gamma(\beta-\alpha)} t^{\beta-\alpha-1}, \tag{5.1.2}$$

so that $D_{0+}^{\alpha}\varphi = 0$ if $\varphi(t) = t^{\alpha-1}$.

The Riemann-Liouville fractional integration and differentiation are inverse to each other in the following sense. If $\varphi \in L_1(0,T)$, then $D_{0+}^{\alpha} I_{0+}^{\alpha}\varphi = \varphi$. The equality $I_{0+}^{\alpha} D_{0+}^{\alpha}\varphi = \varphi$ holds under the stronger assumption that $\varphi = I_{0+}^{\alpha}\psi$ with some $\psi \in L_1(0,T)$. The latter is equivalent to the conditions of absolute continuity of $I_{0+}^{1-\alpha}\varphi$ on $[0,T)$ and the equality $\left(I_{0+}^{1-\alpha}\varphi\right)(0) = 0$.

The problem of boundedness of I_{0+}^{α} as an operator between certain function spaces has been investigated in great detail (see [202]). We shall use only one result of this kind – if $\varphi \in C[0,T]$, then $I_{0+}\varphi$ belongs to the space of Hölder continuous functions $H^{\alpha}[0,T]$.

Denote by $H_{\mu}^{\alpha+\lambda}[0,T]$ the class of such functions φ, for which the function $t^{\mu}\varphi(t)$ belongs to the Hölder class $H^{\alpha+\lambda}[0,T]$. If $\varphi \in H_{\mu}^{\alpha+\lambda}[0,T]$, $t^{\mu}\varphi(t) \to 0$ as $t \to 0$, where $0 < \lambda < 1-\alpha$, $0 \le \mu < \lambda + 1$, then the fractional derivative $\left(D_{0+}^{\alpha}\varphi\right)(t)$ exists for all $t \in [0,T]$ and can be represented in the form

$$\left(D_{0+}^{\alpha}\varphi\right)(t) = \frac{\varphi(t)}{t^{\alpha}\Gamma(1-\alpha)} + \frac{\alpha}{\Gamma(1-\alpha)} \lim_{\varepsilon\to+0} \psi_{\varepsilon}(t), \tag{5.1.3}$$

where $\psi_{\varepsilon}(t) = 0$ for $0 < t \le \varepsilon$,

$$\psi_{\varepsilon}(t) = \int_0^{t-\varepsilon} \frac{\varphi(t)-\varphi(\tau)}{(t-\tau)^{1+\alpha}}\,d\tau \text{ for } \varepsilon < t \le T.$$

The expression in the right-hand side of (5.1.3) is called *the Marchaud fractional derivative.*

5.1.2. The regularized fractional derivative. Suppose that a function φ is continuous on $[0,T]$ and possesses the Riemann-Liouville fractional derivative. The function

$$\begin{aligned}\left(\mathbb{D}^{(\alpha)}\varphi\right)(t) &= \frac{1}{\Gamma(1-\alpha)}\left[\frac{d}{dt}\int_0^t (t-\tau)^{-\alpha}\varphi(\tau)\,d\tau - t^{-\alpha}\varphi(0)\right] \\ &= \left(D_{0+}^{\alpha}\varphi\right)(t) - \frac{1}{t^{\alpha}\Gamma(1-\alpha)}\varphi(0) \end{aligned}\tag{5.1.4}$$

is called *the regularized fractional derivative* of the function φ, or its Caputo derivative (in honour of M. Caputo [22]).

If ϕ is absolutely continuous on $[0,T]$, then by (5.1.1)

$$\left(\mathbb{D}^{(\alpha)}\varphi\right)(t) = \frac{1}{\Gamma(1-\alpha)}\int_0^t (t-\tau)^{-\alpha}\varphi'(\tau)\,d\tau. \tag{5.1.5}$$

The form (5.1.5) is often used in physical literature. This fractional derivative is a special case of general fractional-differential operators investigated by Dzhrbashyan and Nersessyan [43, 44]. However the form (5.1.4) is more useful, since it makes sense for a wider class of functions.

The essential difference between the Riemann-Liouville and regularized fractional derivatives emerges in the study of corresponding "ordinary" differential equations. A solution of the equation

$$\mathbb{D}^{(\alpha)}\varphi = \lambda\varphi, \quad \lambda \in \mathbb{R}, \tag{5.1.6}$$

with the initial condition $\varphi(0) = 1$ is

$$\varphi(t) = E_\alpha(\lambda t^\alpha) \tag{5.1.7}$$

where E_α is the Mittag-Leffler function:

$$E_\alpha(z) = \sum_{k=0}^{\infty} \frac{z^k}{\Gamma(\alpha k + 1)}. \tag{5.1.8}$$

This formula [44] can be verified using (5.1.2). Note that the initial condition for the equation (5.1.6) is formulated like the one for a conventional first order differential equation.

If, instead of (5.1.6), we consider the equation with the Riemann-Liouville derivative

$$D^\alpha_{0+}\varphi = \lambda\varphi,$$

then its general solution is given by the formula

$$\varphi(t) = \text{const} \cdot t^{\alpha-1} E_{\alpha,\alpha}(\lambda t^\alpha) \tag{5.1.9}$$

containing the generalized Mittag-Leffler function:

$$E_{\alpha,\beta}(z) = \sum_{k=0}^{\infty} \frac{z^k}{\Gamma(\alpha k + \beta)}.$$

We see from (5.1.9) that in this case the solution has a singularity at the origin. Thus, in order to fix a specific solution, one should impose an initial condition not on the solution itself, but on its fractional integral. In modeling physical processes, where the initial state of the system is usually well-defined, it is natural to use equations with the regularized fractional derivative.

The Mittag-Leffler function (5.1.8) has much in common with the exponential function e^z, formally corresponding to the case $\alpha = 1$. For example, if z is real and $z \to +\infty$, then

$$E_\alpha(z) \sim \frac{1}{\alpha} \exp(t^{1/\alpha}).$$

The function $E_\alpha(-z)$ is completely monotone on the half-axis $z \geq 0$, that is

$$(-1)^n \frac{d^n E_\alpha(-z)}{dz^n} \geq 0, \quad z \geq 0.$$

An important property of E_α contrasting the case of the exponential, is the asymptotic behaviour at $-\infty$. Namely,

$$E_\alpha(z) \sim -\frac{z^{-1}}{\Gamma(1-\alpha)}, \quad z \to -\infty. \tag{5.1.10}$$

As we shall see, this property leads to serious differences between properties of the fractional diffusion equations and classical parabolic equations.

For further details regarding the Mittag-Leffler function see [10, 42]. In fact many problems of analysis connected with the exponential function and the operator of differentiation have their counterparts related to $\mathbb{D}^{(\alpha)}$ and based essentially on properties of the Mittag-Leffler function. Dzhrbashyan and Nersessyan [43] developed a theory of quasi-analytic classes corresponding to $\mathbb{D}^{(\alpha)}$ and more general fractional differential operators. The following theorem is a special case of their results. Let $\beta > \alpha$.

Theorem 5.1. *There exists a function $f \in C^\infty[0,T]$, $f(t) \not\equiv 0$, possessing the following property. Let $f_0(t) = f(t)$, $f_{n+1}(t) = \left(\mathbb{D}^{(\alpha)} f_n\right)(t)$, $n = 0,1,2,\dots$. Then $f_n(0) = 0$,*

$$\sup_{0 \leq t \leq T} |f_n(t)| \leq C B^{n\alpha} n^{n\beta}, \quad C, B > 0 \ (n = 0,1,2,\dots). \tag{5.1.11}$$

5.1.3. Asymptotics of some integrals. Dealing with fractional integrals and derivatives one has often to use asymptotic properties of some typical integrals. Here we give two results of this kind. The first of them is a special case of an asymptotic theorem from [202].

Lemma 5.1. *Suppose that $\varphi(t)$ is locally integrable on $(0,\infty)$ and*

$$\varphi(t) \sim a t^{-1-\beta}, \quad 0 < \beta \leq 1, \ \textit{as } t \to \infty,$$

where $a \neq 0$. Then

$$\left(I_{0+}^\alpha \varphi\right)(t) \sim \frac{b}{\Gamma(\alpha)} t^{\alpha-1} + \frac{\Gamma(-\beta)}{\Gamma(\alpha-\beta)} a t^{\alpha-\beta-1}, \quad t \to \infty,$$

where $b = \int\limits_0^\infty \varphi(t)\, dt$.

In the proof of the next lemma we follow [198]. Consider the integral

$$I(z) = \int\limits_0^1 (1-\eta)^{-\alpha} \exp\left(-z^{-s}\eta^{-s}\right) d\eta, \quad s > 0.$$

Lemma 5.2. *As $z \to 0$,*

$$I(z) \sim \Gamma(1-\alpha)s^{\alpha-1}z^{s(1-\alpha)}\exp\left(-z^{-s}\right). \tag{5.1.12}$$

Proof. Let $\sigma = z^{-s}$. After the substitution $u = \eta^{-s} - 1$ we get

$$\begin{aligned} I(z) &= e^{-\sigma}\int_0^1 \exp\left(-\sigma\left(\eta^{-s}-1\right)\right)(1-\eta)^{-\alpha}\,d\eta \\ &= \frac{1}{s}e^{-\sigma}\int_0^\infty e^{-\sigma u}\frac{\left[(1+u)^{1/s}-1\right]^{-\alpha}}{(1+u)^{\frac{1-\alpha}{s}+1}}\,du. \end{aligned}$$

As $u \to 0$,

$$\frac{\left[(1+u)^{1/s}-1\right]^{-\alpha}}{(1+u)^{\frac{1-\alpha}{s}+1}} \sim s^\alpha u^{-\alpha},$$

and Watson's lemma for the Laplace integrals (see, e.g., [176]) implies (5.1.12). □

5.2 Fundamental solution of the Cauchy problem

5.2.1. Main assumptions. In this chapter we consider the Cauchy problem

$$\left(\mathbb{D}_t^{(\alpha)}u\right)(t,x) - Bu(t,x) = f(t,x), \quad (t,x)\in\Pi_{(0,T]}, \tag{5.2.1}$$

$$u(0,x) = u_0(x), \quad x\in\mathbb{R}^n, \tag{5.2.2}$$

where $0<\alpha<1$,

$$B = \sum_{j,l=1}^n a_{jl}(x)\partial_{x_j}\partial_{x_l} + \sum_{j=1}^n b_j(x)\partial_{x_j} + c(x),$$

there exists such a constant $\delta>0$ that for any $x,\xi\in\mathbb{R}^n$

$$\sum_{j,l=1}^n a_{jl}(x)\xi_j\xi_l \ge \delta|\xi|^2; \tag{5.2.3}$$

the coefficients of the operator B are bounded and Hölder continuous.

We shall construct and study a *Green matrix* for the problem (5.2.1)–(5.2.2), that is such a pair

$$\{Z(t,x;\xi), Y(t,x;\xi)\}, \quad t\in(0,T],\ x,\xi\in\mathbb{R}^n,$$

that for any bounded continuous function u_0 (locally Hölder continuous, if $n>1$) and any bounded function f, jointly continuous in (t,x) and locally Hölder continuous in x, there exists a classical solution of the problem (5.2.1)–(5.2.2) of the form

$$u(t,x)=\int_{\mathbb{R}^n} Z(t,x;\xi)u_0(\xi)\,d\xi+\int_0^t d\lambda\int_{\mathbb{R}^n} Y(t-\lambda,x,;y)f(\lambda,y)\,dy. \tag{5.2.4}$$

We call a function $u(t,x)$ a classical solution if:

(i) $u(t,x)$ is twice continuously differentiable in x for each $t>0$;

(ii) for each $x\in\mathbb{R}^n$ $u(t,x)$ is continuous in t on $[0,T]$, and its fractional integral

$$\left(I_{0+}^{1-\alpha}u\right)(t,x)=\frac{1}{\Gamma(1-\alpha)}\int_0^t (t-\tau)^{-\alpha}u(\tau,x)\,d\tau$$

is continuously differentiable in t for $t>0$.

(iii) $u(t,x)$ satisfies (5.2.1) and (5.2.2).

5.2.2. A model equation. We begin with the case where $B=\Delta$ is the Laplacian. Suppose first that $f(t,x)\equiv 0$. Let us find (at first, formally) the FSCP Z, which in this case will depend on the difference $x-\xi$, so that $Z(t,x;\xi)=Z_0(t,x-\xi)$.

Applying the Fourier transform in x, for the function $\widehat{Z}_0(t,\xi)$ we obtain the Cauchy problem

$$\left(\mathbb{D}_t^{(\alpha)}\widehat{Z}_0\right)(t,\xi)=-|\xi|^2\widehat{Z}_0(t,\xi),\quad \widehat{Z}_0(0,\xi)=1$$

(as usual, the FSCP satisfies formally the initial condition with the δ-function).

By (5.1.7), $\widehat{Z}_0(t,\xi)=E_\alpha(-|\xi|^2t^\alpha)$. Let us represent the Mittag-Leffler function of a negative argument by Fox's H-function (see Appendix, formula (A.3)). Now

$$\widehat{Z}_0(t,\xi)=H_{12}^{11}\left[|\xi|^2t^\alpha\,\middle|\,\begin{matrix}(0,1)\\(0,1),\quad(0,\alpha)\end{matrix}\right],$$

so that

$$Z_0(t,x)=(2\pi)^{-\frac{n}{2}}|x|^{1-\frac{n}{2}}\int_0^\infty r^{\frac{n}{2}}H_{12}^{11}\left[r^2t^\alpha\,\middle|\,\begin{matrix}(0,1)\\(0,1),\quad(0,\alpha)\end{matrix}\right]J_{\frac{n}{2}-1}(r|x|)\,dr$$

(by the formula for the Fourier transform of a radial function; see [218]).

Next,

$$\int_0^\infty r^{\frac{n}{2}}H_{12}^{11}\left[r^2t^\alpha\,\middle|\,\begin{matrix}(0,1)\\(0,1),\quad(0,\alpha)\end{matrix}\right]J_{\frac{n}{2}-1}(r|x|)\,dr$$
$$=2^{\frac{n}{2}}|x|^{-1-\frac{n}{2}}H_{32}^{12}\left[4t^\alpha|x|^{-2}\,\middle|\,\begin{matrix}(1-\frac{n}{2},1),\quad(0,1),\quad(0,1)\\(0,1),\quad(0,\alpha)\end{matrix}\right]$$

(see [195]). Using the transformation formulas (A.4) and (A.5) we obtain the expression

$$Z_0(t,x) = \pi^{-\frac{n}{2}}|x|^{-n} H_{12}^{20}\left[\frac{1}{4}t^{-\alpha}|x|^2 \,\middle|\, \begin{matrix}(1,\alpha) \\ (\frac{n}{2},1), \quad (1,1)\end{matrix}\right]. \tag{5.2.5}$$

It will follow from the estimates given below that $Z_0(t,\cdot) \in L_1(\mathbb{R}^n)$. Then it will be possible to check the formula (5.2.5) rigorously, by calculating $\widehat{Z}_0$ using the same formulas.

Note that in [206] a different derivation of the expression (5.2.5) (given there in an equivalent form) is proposed. It is based on the Mellin transform in t, the explicit expression for the Green function of the Laplacian, the inverse Mellin transform, and integration formulas for H-functions.

Denote $R = t^{-\alpha}|x|^2$. Below the letters C, σ will denote various positive constants.

Proposition 5.1.

(i) *If* $R \geq 1$, *then*

$$|\partial_x^m Z_0(t,x)| \leq Ct^{-\frac{\alpha(n+m)}{2}} \exp\left\{-\sigma t^{-\frac{\alpha}{2-\alpha}}|x|^{\frac{2}{2-\alpha}}\right\}, \quad |m| \leq 3, \tag{5.2.6}$$

$$\left|\mathbb{D}^{(\alpha)} Z_0(t,x)\right| \leq Ct^{-\frac{\alpha(n+2)}{2}} \exp\left\{-\sigma t^{-\frac{\alpha}{2-\alpha}}|x|^{\frac{2}{2-\alpha}}\right\}; \tag{5.2.7}$$

(ii) *If* $R \leq 1$, $x \neq 0$, *then*

$$|\partial_x^m Z_0(t,x)| \leq Ct^{-\alpha}|x|^{-n+2-|m|}, \quad |m| \leq 3, \tag{5.2.8}$$

if $n \geq 3$, *or* $n = 2$, $m \neq 0$;

$$|Z_0(t,x)| \leq Ct^{-\alpha}[|\log(t^{-\alpha}|x|^2)| + 1], \tag{5.2.9}$$

if $n = 2$;

$$|\partial_x^m Z_0(t,x)| \leq Ct^{-\frac{(m+1)\alpha}{2}}, \tag{5.2.10}$$

if $n = 1$.

(iii) *If* $R \leq 1$, $x \neq 0$, *then*

$$\left|\mathbb{D}^{(\alpha)} Z_0(t,x)\right| \leq \begin{cases} Ct^{-2\alpha}|x|^{-n+2}, & \text{if } n \geq 3, \\ Ct^{-\alpha}\left[\left|\log(t^{-\alpha}|x|^2)\right| + 1\right], & \text{if } n = 2, \\ Ct^{-3\alpha/2}, & \text{if } n = 1. \end{cases} \tag{5.2.11}$$

The *proof* follows from the differentiation formulas (A.7), (A.8), (A.11), and the asymptotic representations (A.14), (A.18)–(A.21). Note that in some cases we have to use several terms of the asymptotic expansions, or estimates of the remainders, since main terms appearing in expressions for derivatives of Z_0 cancel each other. For example, let $n = 1$. Then

$$\partial_x Z_0(t,x) = -\pi^{-1/2}|x|^{-2}\operatorname{sign}(x)\left\{ H_{12}^{20}\left[\frac{1}{4}t^{-\alpha}|x|^2 \left| \begin{matrix} (1,\alpha) \\ (\frac{1}{2},1), \quad (1,1) \end{matrix} \right. \right] \right.$$
$$\left. +2H_{23}^{30}\left[\frac{1}{4}t^{-\alpha}|x|^2 \left| \begin{matrix} (1,\alpha), \quad (0,1) \\ (1,1), \quad (\frac{1}{2},1), \quad (1,1) \end{matrix} \right. \right] \right\}.$$

As $x \to 0$,

$$\partial_x Z_0(t,x) \sim -\pi^{-1/2}|x|^{-2}\operatorname{sign}(x)$$
$$\times\left\{ \frac{\sqrt{\pi}}{2\Gamma(1-\frac{\alpha}{2})}t^{-\alpha/2}|x| - \frac{\sqrt{\pi}}{2\Gamma(1-\alpha)}t^{-\alpha}x^2 - \frac{\sqrt{\pi}}{2\Gamma(1-\frac{\alpha}{2})}t^{-\alpha/2}|x| + \frac{\sqrt{\pi}}{\Gamma(1-\alpha)}t^{-\alpha}x^2 \right\}$$
$$= -\frac{\operatorname{sign}(x)}{2\Gamma(1-\alpha)}t^{-\alpha},$$

and we obtain the estimate (5.2.10) with $m = 1$. Here we used two terms of the asymptotics (A.18) and (A.19). A similar reasoning for higher derivatives requires three terms and the remainder estimates. □

In contrast to the parabolic differential equations, the FSCP $Z_0(t,x)$ has, if $n > 1$, a singularity not only at $t = 0$, but also at $x = 0$ (the orders of singularity given in Proposition 5.1 are precise, since the estimates are based on the asymptotics of H-functions). The fact of its dependence on the space dimension is natural – recall the expression of $\widehat{Z}_0$ via the Mittag-Leffler function and its asymptotics (5.1.10) on the negative half-axis.

The function $Y = Y_0(t-\lambda, x-y)$ appearing in (5.2.4) for this case has the form

$$Y_0(t,x) = \pi^{-\frac{n}{2}}|x|^{-n/2}t^{\alpha-1}H_{12}^{20}\left[\frac{1}{4}t^{-\alpha}|x|^2 \left| \begin{matrix} (\alpha,\alpha) \\ (\frac{n}{2},1), \quad (1,1) \end{matrix} \right. \right]. \tag{5.2.12}$$

$Y_0(t,x)$ is the Riemann-Liouville derivative of $Z_0(t,x)$ in t, of the order $1-\alpha$ (for $x \neq 0$, $Z_0(t,x) \to 0$ as $t \to 0$, so that the Riemann-Liouville derivative coincides in this case with the regularized fractional derivative).

The fact that (Z_0, Y_0) is a Green matrix for the problem (5.2.1)–(5.2.2) (with $B = \Delta$) will be proved later, as a special case of our investigation of general problems of this kind. Here we present only estimates of the function Y_0 and its derivatives, which follow from the asymptotic relations (A.14), (A.22)–(A.25). The (rather tedious) calculations are based on canceling of lower terms in asymptotics of various H-functions appearing in the expressions for higher-order derivatives of Y_0.

Proposition 5.2.

(i) *If* $R \geq 1$, *then*

$$|\partial_x^m Y_0(t,x)| \leq Ct^{-\frac{\alpha(n+m)}{2}-1+\alpha}\exp\left\{-\sigma t^{-\frac{\alpha}{2-\alpha}}|x|^{\frac{2}{2-\alpha}}\right\}, \quad |m| \leq 3. \tag{5.2.13}$$

(ii) *If $R \leq 1$, $x \neq 0$, $n > 4$, then*

$$|\partial_x^m Y_0(t,x)| \leq Ct^{-\alpha-1}|x|^{-n+4-|m|}, \quad |m| \leq 3. \tag{5.2.14}$$

(iii) *If $R \leq 1$, $x \neq 0$, $n = 4$, then*

$$|Y_0(t,x)| \leq Ct^{-\alpha-1}[|\log(t^{-\alpha}|x|^2)|+1], \tag{5.2.15}$$
$$|\partial_x Y_0(t,x)| \leq Ct^{-\frac{3\alpha}{2}-1}, \tag{5.2.16}$$
$$|\partial_x^m Y_0(t,x)| \leq Ct^{-2\alpha-1}[|\log(t^{-\alpha}|x|^2)|+1], \quad |m| = 2, \tag{5.2.17}$$
$$|\partial_x^m Y_0(t,x)| \leq Ct^{-2\alpha-1}|x|^{-1}[|\log(t^{-\alpha}|x|^2)|+1], \quad |m| = 3. \tag{5.2.18}$$

(iv) *If $R \leq 1$, $x \neq 0$, $n = 3$, then*

$$|Y_0(t,x)| \leq Ct^{-\frac{\alpha}{2}-1}, \tag{5.2.19}$$
$$|\partial_x Y_0(t,x)| \leq Ct^{-\alpha-1}, \tag{5.2.20}$$
$$|\partial_x^m Y_0(t,x)| \leq Ct^{-\alpha-1}|x|^{-1}, \quad |m| = 2, \tag{5.2.21}$$
$$|\partial_x^m Y_0(t,x)| \leq Ct^{-\alpha-1}|x|^{-2}, \quad |m| = 3. \tag{5.2.22}$$

(v) *If $R \leq 1$, $x \neq 0$, $n = 2$, then*

$$|Y_0(t,x)| \leq Ct^{-1}, \tag{5.2.23}$$
$$|\partial_x Y_0(t,x)| \leq Ct^{-\frac{\alpha}{2}-1}, \tag{5.2.24}$$
$$|\partial_x^m Y_0(t,x)| \leq Ct^{-\alpha-1}[|\log(t^{-\alpha}|x|^2)|+1], \quad |m| = 2, \tag{5.2.25}$$
$$|\partial_x^m Y_0(t,x)| \leq Ct^{-\alpha-1}|x|^{-1}[|\log(t^{-\alpha}|x|^2)|+1], \quad |m| = 3. \tag{5.2.26}$$

(vi) *If $R \leq 1$, $x \neq 0$, $n = 1$, then*

$$|\partial_x^m Y_0(t,x)| \leq Ct^{-\frac{(m-1)\alpha}{2}-1}, \quad m = 0,1,2,3. \tag{5.2.27}$$

5.2.3. The parametrix. If we consider the equation (5.2.1), in which the operator B has constant coefficients, and only the leading terms are present, that is $b_j = c = 0$, then the functions Z_0 and Y_0 for that case can be obtained from (5.2.5) and (5.2.12) by a change of variables:

$$Z_0(t,x-\xi) = \frac{\pi^{-n/2}}{(\det A)^{1/2}}\left[\sum_{j,l=1}^{n} A^{(jl)}(x_j-\xi_j)(x_l-\xi_l)\right]^{-n/2}$$
$$\times H_{12}^{20}\left[\frac{1}{4}t^{-\alpha}\sum_{j,l=1}^{n} A^{(jl)}(x_j-\xi_j)(x_l-\xi_l)\,\middle|\,\begin{matrix}(1,\alpha) \\ (\frac{n}{2},1), \quad (1,1)\end{matrix}\right], \tag{5.2.28}$$

$$Y_0(t,x-\xi) = \frac{\pi^{-n/2}}{(\det A)^{1/2}} \left[\sum_{j,l=1}^{n} A^{(ij)}(x_j-\xi_j)(x_l-\xi_l) \right]^{-n/2} t^{\alpha-1}$$

$$\times H_{12}^{20} \left[\frac{1}{4} t^{-\alpha} \sum_{j,l=1}^{n} A^{(jl)}(x_j-\xi_j)(x_l-\xi_l) \middle| \begin{matrix} (\alpha,\alpha) \\ (\frac{n}{2},1), \quad (1,1) \end{matrix} \right], \quad (5.2.29)$$

where $A = \left(A^{(ij)}\right)$ is the matrix inverse to (a_{ij}).

The parametrix kernels $Z_0(t,x-\xi;\zeta)$ and $Y_0(t,x-\xi;\zeta)$ associated with the general equation (5.2.1) are defined by (5.2.28) and (5.2.29). Now A corresponds, as above, to the matrix $(a_{ij}(\zeta))$ of the leading coefficients "frozen" at the parametric point $\zeta \in \mathbb{R}^n$. The estimates of Propositions 5.1 and 5.2 hold in this case too, with constants independent of ζ.

Since properties of Z_0 are different for $n \geq 2$ and $n = 1$, it is convenient to treat these cases separately. Therefore we assume in this and the next sections that $n \geq 2$. The results for the case $n = 1$ will be summarized in Section 5.2.5.

The behavior of Z_0 for $R \geq 1$ and $R < 1$ is described in Proposition 5.1 separately. However it is possible to write equivalent unified estimates:

$$|\partial_x^m Z_0(t,x-\xi;\zeta)| \leq Ct^{-\alpha}|x-\xi|^{-n+2-|m|} \exp\{-\sigma\rho(t,x;0,\xi)\},$$
$$|m| \leq 3, \quad (5.2.30)$$

$$\left|\mathbb{D}^{(\alpha)} Z_0(t,x-\xi;\zeta)\right| \leq Ct^{-\alpha}|x-\xi|^{-n} \exp\{-\sigma\rho(t,x;0,\xi)\},$$
$$|m| \leq 3, \quad (5.2.31)$$

where

$$\rho(t,x;\tau,\xi) = \left(\frac{|x-\xi|}{(t-\tau)^{\alpha/2}} \right)^{\frac{2}{2-\alpha}}$$

(this corresponds to the notation of Section 1.3.5 with $\beta = \alpha/2$). If $n = 2$, then (5.2.30) is valid only for $|m| \neq 0$. For the opposite case

$$|Z_0(t,x-\xi;\zeta)| \leq Ct^{-\alpha} \left[\left|\log\left(t^{-\alpha}|x-\xi|^2\right)\right| + 1 \right] \exp\{-\sigma\rho(t,x;0,\xi)\}. \quad (5.2.32)$$

The constants $C,\sigma > 0$ in (5.2.30)–(5.2.32) do not depend on ζ. Of course, the constants may be different from the ones in (5.2.6)–(5.2.11).

The next proposition gives similar estimates for differences of values of Z_0 and its derivatives corresponding to different values of the parameter ζ.

Proposition 5.3. *For any $y,\zeta',\zeta'' \in \mathbb{R}^n$, $0 < t \leq T$,*

$$|\partial_x^m Z_0(t,y;\zeta') - \partial_x^m Z_0(t,y;\zeta'')|$$
$$\leq Ct^{-\alpha}|\zeta'-\zeta''|^{\gamma}|y|^{-n+2-|m|} \exp\{-\sigma \left(t^{-\alpha/2}|y|\right)^{\frac{2}{2-\alpha}}\}, \quad |m| \leq 2. \quad (5.2.33)$$

Proof. Let $m = 0$. Denote

$$\mathfrak{A}(y,\zeta) = \sum_{i,j=1}^{n} A^{(ij)}(\zeta) y_i y_j.$$

By our assumptions

$$C_1|y|^2 \le \mathfrak{A}(y,\zeta) \le C_2|y|^2,$$
$$|\mathfrak{A}(y,\zeta') - \mathfrak{A}(y,\zeta'')| \le C|\zeta' - \zeta''|^\gamma |y|^2,$$
$$\left|[\det A(\zeta')]^{1/2} - [\det A(\zeta'')]^{1/2}\right| \le C|\zeta' - \zeta''|^\gamma.$$

According to (5.2.28), we have to use the estimate (A.15) for the function H_{12}^{20}, and also to find an estimate for a difference of such functions. Let

$$\varphi(s) = s^{-n/2} H_{12}^{20}\left[\omega s \middle| \begin{matrix} (1,\alpha) \\ (\frac{n}{2},1), \quad (1,1) \end{matrix}\right], \quad \omega = \frac{1}{4}t^{-\alpha}.$$

By (A.7),

$$\varphi'(s) = -\frac{n}{2}s^{-\frac{n}{2}-1} H_{12}^{20}\left[\omega s \middle| \begin{matrix} (1,\alpha) \\ (\frac{n}{2},1), \quad (1,1) \end{matrix}\right]$$
$$- s^{-\frac{n}{2}-1} H_{23}^{30}\left[\omega s \middle| \begin{matrix} (1,\alpha), \quad (0,1) \\ (\frac{n}{2},1), \quad (1,1), \quad (1,1) \end{matrix}\right].$$

For large s, (A.14) and (A.15) give the estimate

$$|\varphi'(s)| \le C s^{\frac{n+2}{2(2-\alpha)} - \frac{n}{2} - 1} \omega^{\frac{n+2}{2(2-\alpha)}} \exp\left\{-\sigma |\omega s|^{\frac{1}{2-\alpha}}\right\},$$

which implies the bound

$$|\varphi(\mathfrak{A}(y,\zeta')) - \varphi(\mathfrak{A}(y,\zeta''))| \le C t^{-\frac{n\alpha}{2}} |\zeta' - \zeta''|^\gamma \exp\left\{-\sigma t^{-\frac{\alpha}{2-\alpha}} |y|^{\frac{2}{2-\alpha}}\right\}.$$

For small s, we have

$$|\varphi'(s)| \le C\omega s^{-\frac{n}{2}}$$

by (A.18) and (A.19); note that the logarithmic terms emerging for $n = 2$ are cancelled.

As a result, we find that

$$|Z_0(t, x-\xi; \zeta') - Z_0(t, x-\xi, \zeta'')|$$
$$\le \begin{cases} C|\zeta' - \zeta''|^\gamma t^{-\frac{n\alpha}{2}} \exp\left\{-\sigma t^{-\frac{\alpha}{2-\alpha}} |y|^{\frac{2}{2-\alpha}}\right\} & \text{if } t^{-\alpha}|y| \ge 1, \\ C|\zeta' - \zeta''|^\gamma t^{-\alpha} |y|^{-n+2}, & \text{if } t^{-\alpha}|y| \le 1, \end{cases}$$

which is equivalent to the estimate (5.2.33) with $m = 0$.

For the first and second derivatives the proof is similar, though somewhat cumbersome – one has to use the asymptotics (A.14)–(A.16), (A.18)–(A.20) for the H-functions; again for $n = 2$ the logarithmic terms are cancelled. □

Just as for the function Z_0, it is desirable to obtain unified estimates of Y_0 and its derivatives which are valid for all values of the independent variables. Here the problem is a little more complicated because the behavior of Y_0 is different for different values of the dimension n.

In particular, for $n = 2$ we have

$$|\partial_x^m Y_0(t-\tau, x-\xi, \zeta)| \le C(t-\tau)^{-\frac{\alpha}{2}(2+|m|)+\alpha-1} \exp\{-\sigma\rho(t,x;\tau,\xi)\},$$
$$|m| = 0,1; \tag{5.2.34}$$

$$|\partial_x^m Y_0(t-\tau, x-\xi;\zeta)|$$
$$\le C(t-\tau)^{-\alpha-1}|x-\xi|^{2-|m|}\left[\left|\log\frac{|x-\xi|^2}{(t-\tau)^\alpha}\right|+1\right]\exp\{-\sigma\rho(t,x;\tau,\xi)\},$$
$$|m| = 2,3. \tag{5.2.35}$$

These estimates follow immediately from (5.2.13), (5.2.23)–(5.2.26).

For $n = 3$,

$$|\partial_x^m Y_0(t-\tau, x-\xi;\zeta)| \le C(t-\tau)^{-\frac{\alpha}{2}(3+|m|)+\alpha-1} \exp\{-\sigma\rho(t,x;\tau,\xi)\},$$
$$|m| = 0,1, \tag{5.2.36}$$

and

$$|\partial_x^m Y_0(t-\tau, x-\xi;\zeta)| \le C(t-\tau)^{-\alpha-1}|x-\xi|^{1-|m|}\exp\{-\sigma\rho(t,x;\tau,\xi)\},$$
$$m| = 2,3. \tag{5.2.37}$$

Similarly, for $n = 4$,

$$|\partial_x^m Y_0(t-\tau, x-\xi;\zeta)|$$
$$\le C(t-\tau)^{-\alpha-\frac{|m|\alpha}{2}-1}\left[\left|\log\frac{|x-\xi|^2}{(t-\tau)^\alpha}\right|+1\right]\exp\{-\sigma\rho(t,x;\tau,\xi)\},$$
$$|m| = 0,1,2, \tag{5.2.38}$$

$$|\partial_x^m Y_0(t-\tau, x-\xi;\zeta)|$$
$$\le C(t-\tau)^{-2\alpha-1}|x-\xi|^{-1}\left[\left|\log\frac{|x-\xi|^2}{(t-\tau)^\alpha}\right|+1\right]\exp\{-\sigma\rho(t,x;\tau,\xi)\},$$
$$|m| = 3. \tag{5.2.39}$$

Finally, if $n > 4$, then

$$|\partial_x^m Y_0(t-\tau, x-\xi;\zeta)| \le C(t-\tau)^{-\alpha-1}|x-\xi|^{-n-|m|+4}\exp\{-\sigma\rho(t,x;\tau,\xi)\},$$
$$|m| \le 3. \tag{5.2.40}$$

Just as in Proposition 5.3, the estimates for the differences $\partial_x^m Y_0(t-\tau, x-\xi;\zeta') - \partial_x^m Y_0(t-\tau, x-\xi;\zeta'')$ have the following form: the right-hand sides of

(5.2.34)–(5.2.40) are multiplied by $|\zeta' - \zeta''|^\gamma$. For example, if $n > 4$, then

$$\begin{aligned}&|\partial_x^m Y_0(t-\tau, x-\xi; \zeta') - \partial_x^m Y_0(t-\tau, x-\xi;\zeta'')|\\ &\le C(t-\tau)^{-\alpha-1}|\zeta'-\zeta''|^\gamma |x-\zeta|^{-n-|m|+4}\exp\{\ \sigma\rho(t,x;\tau,\xi)\},\\ &\qquad |m|\le 3.\end{aligned} \tag{5.2.41}$$

Let us compute the integrals of Z_0 and Y_0 in spatial variables. It follows from the construction of the function Z_0 that

$$\int_{\mathbb{R}^n} Z_0(t, x-\xi;\zeta)\, d\xi = 1. \tag{5.2.42}$$

Next, from (5.2.29) we see that

$$\begin{aligned}&\int_{\mathbb{R}^n} Y_0(t-\tau, x-\xi;\zeta)\, d\xi\\ &\quad = (t-\tau)^{\alpha-1}\pi^{-n/2}\int_{\mathbb{R}^n} |y|^{-n} H_{12}^{20}\left[\frac{1}{4}(t-\tau)^{-\alpha}|y|^2 \,\middle|\, \begin{matrix}(\alpha,\alpha) & \\ (\frac{n}{2},1), & (1,1)\end{matrix}\right] dy\\ &\quad = C(t-\tau)^{\alpha-1}\int_0^\infty r^{-1} H_{12}^{20}\left[\frac{1}{4}(t-\tau)^{-\alpha}r^2 \,\middle|\, \begin{matrix}(\alpha,\alpha) & \\ (\frac{n}{2},1), & (1,1)\end{matrix}\right] dr\\ &\quad = C_1(t-\tau)^{\alpha-1}\int_0^\infty s^{-1} H_{12}^{20}\left[\frac{1}{4}(t-\tau)^{-\alpha}s \,\middle|\, \begin{matrix}(\alpha,\alpha) & \\ (\frac{n}{2},1), & (1,1)\end{matrix}\right] ds\\ &\quad = -C_1(t-\tau)^{\alpha-1}\lim_{s\to\infty} H_{23}^{30}\left[\frac{1}{4}(t-\tau)^{-\alpha}s \,\middle|\, \begin{matrix}(\alpha,\alpha), & (1,1) & \\ (0,1), & (\frac{n}{2},1), & (1,1)\end{matrix}\right],\end{aligned}$$

so that

$$\int_{\mathbb{R}^n} Y_0(t-\tau, x-\xi,\zeta)\, d\xi = 0. \tag{5.2.43}$$

We have used the integration formula (A.13), the identity connecting H_{23}^{21} and H_{23}^{30} (A.6), and the asymptotics (A.14) of H_{23}^{30}.

Below we need also the first time derivative of the function $Z_0(t,x;\zeta)$. We find from (5.2.28) and (A.7) that

$$\begin{aligned}\partial_t Z_0(t,x;\zeta) &= \frac{\alpha\pi^{-n/2}t^{-1}}{(\det A)^{1/2}}\left[\sum_{j,l=1}^n A^{(jl)}x_j x_l\right]^{-n/2}\\ &\quad\times H_{23}^{30}\left[\frac{1}{4}t^{-\alpha}\sum_{j,l=1}^n A^{(jl)}x_j x_l \,\middle|\, \begin{matrix}(1,\alpha), & (0,1) & \\ (\frac{n}{2},1), & (1,1), & (1,1)\end{matrix}\right].\end{aligned}$$

Using the asymptotics (A.14) and (A.19) of the function H^{30}_{23} we can obtain an estimate of $\partial_t Z_0$. For example, if $n \geq 3$, then

$$|\partial_t Z_0(t,x;\zeta)| \leq \begin{cases} Ct^{-1-\frac{\alpha n}{2}} \exp\left\{-\sigma(t^{-\alpha/2}|x|)^{\frac{2}{2-\alpha}}\right\}, & \text{if } t^{-\alpha/2}|x| \geq 1, \\ Ct^{-\alpha-1}|x|^{-n+2}, & \text{if } t^{-\alpha/2}|x| < 1. \end{cases} \tag{5.2.44}$$

We can also use a roughened unified estimate

$$|\partial_t Z_0(t,x;\zeta)| \leq Ct^{-1}|x|^{-n} \exp\left\{-\sigma(t^{-\alpha/2}|x|)^{\frac{2}{2-\alpha}}\right\}. \tag{5.2.45}$$

As before, we get also an estimate for the difference $|\partial_t Z_0(t,x;\zeta') - \partial_t Z_0(t,x;\zeta'')|$ whose upper bound is the expression in the right-hand side of (5.2.44) or (5.2.45) multiplied by $|\zeta' - \zeta''|^\gamma$. Since

$$\int_{\mathbb{R}^n} \partial_t Z_0(t, x-\xi;\zeta')\, d\xi = 0$$

by virtue of (5.2.42), the above estimates imply also the estimate

$$\left| \int_{\mathbb{R}^n} \partial_t Z_0(t,x-\xi;\xi)\, d\xi \right| \leq Ct^{-1+\frac{\alpha\gamma}{2}} \tag{5.2.46}$$

obtained by subtracting the (zero) integral of $\partial_t Z_0(t,x;x)$ and using the estimate for the difference of the derivatives.

5.2.4. The Levi method ($n \geq 2$). We look for the functions Y, Z appearing in (5.2.4) assuming the following integral representations:

$$Z(t,x;\xi) = Z_0(t,x-\xi;\xi) + \int_0^t d\lambda \int_{\mathbb{R}^n} Y_0(t-\lambda, x-y;y) Q(\lambda,y;\xi)\, dy; \tag{5.2.47}$$

$$Y(t,x;\xi) = Y_0(t-\tau,x-\xi;\xi) + \int_\tau^t d\lambda \int_{\mathbb{R}^n} Y_0(t-\lambda,x-y;y)\Psi(\lambda,y;\xi)\, dy; \tag{5.2.48}$$

the functions Z_0, Y_0 were examined in detail in Section 5.2.3.

For the functions Q and Ψ we assume the integral equations

$$Q(t,x;\xi) = M(t,x;\xi) + \int_0^t d\lambda \int_{\mathbb{R}^n} K(t-\lambda,x;y) Q(\lambda,y;\xi)\, dy, \tag{5.2.49}$$

$$\Psi(t,x;\xi) = K(t,x;\xi) + \int_\tau^t d\lambda \int_{\mathbb{R}^n} K(t-\lambda,x;y)\Psi(\lambda,y;\xi)\, dy, \tag{5.2.50}$$

where

$$M(t,x;\xi)=\sum_{j,l=1}^{n}\left\{[a_{jl}(x)-a_{jl}(\xi)]\partial_{x_j}\partial_{x_l}Z_0(t,x-\xi;\xi)\right\}$$
$$+\sum_{j=1}^{n}b_j(x)\partial_{x_j}Z_0(t,x-\xi;\xi)+c(x)Z_0(t,x-\xi;\xi),$$
$$K(t,x;\xi)=\sum_{j,l=1}^{n}\left\{[a_{jl}(x)-a_{jl}(\xi)]\partial_{x_j}\partial_{x_l}Y_0(t,x-\xi;\xi)\right\}$$
$$+\sum_{j=1}^{n}b_j(x)\partial_{x_j}Y_0(t,x-\xi;\xi)+c(x)Y_0(t,x-\xi;\xi).$$

Using the estimates (5.2.30)–(5.2.32) we find that

$$|M(t,x;\xi)|\le Ct^{-\alpha}|x-\xi|^{-n+\gamma}\exp\{-\sigma\rho(t,x;0,\xi)\}. \tag{5.2.51}$$

In order to obtain estimates for K, we have to use estimates for Y_0, different for different values of n, and for different domains, and then to roughen the resulting estimates into less exact but unified bounds convenient for the Levi method.

Proposition 5.4. *For any* $n\ge 2$,

$$|K(t,x;\xi)|\le Ct^{(\gamma-\eta)\frac{\alpha}{2}-1}|x-\xi|^{-n+\eta}\exp\{-\sigma\rho(t,x;0,\xi)\}, \tag{5.2.52}$$

with $0<\eta<\gamma$.

Proof. Using Proposition 5.2 we write the estimate for $|x-\xi|\ge t^{\alpha/2}$:

$$|K(t,x;\xi)|\le Ct^{-\frac{\alpha n}{2}+\frac{\alpha\gamma}{2}-1}\exp\{-\sigma\rho(t,x;0,\xi)\}. \tag{5.2.53}$$

For $|x-\xi|\le t^{\alpha/2}$ we consider various cases separately. If $n=3$, or $n>4$, then

$$|K(t,x;\xi)|\le Ct^{-\alpha+\frac{\alpha\gamma}{2}-1}|x-\xi|^{-n+2}\exp\{-\sigma\rho(t,x;0,\xi)\}. \tag{5.2.54}$$

Since

$$|x-\xi|^2=|x-\xi|^{\eta}|x-\xi|^{2-\eta}\le|x-\xi|^{\eta}t^{\frac{\alpha}{2}(2-\eta)},$$

(5.2.53) and (5.2.54) imply (5.2.52). Note that transforming (5.2.53) we change the constant σ.

If $n=2$ or $n=4$, the initial bound for K is

$$|K(t,x;\xi)|\le Ct^{-\frac{\alpha n}{2}+\frac{\alpha\gamma}{2}-1}\left(\left|\log\frac{|x-\xi|^2}{t^{\alpha}}\right|+1\right)\exp\{-\sigma\rho(t,x;0,\xi)\}. \tag{5.2.55}$$

We roughen the estimate (5.2.55) replacing the factor $\left|\log\frac{|x-\xi|^2}{t^{\alpha}}\right|+1$ with the factor $\left(\frac{|x-\xi|}{t^{\alpha/2}}\right)^{-n+\eta}$, which results in (5.2.52). □

Next we have to study the increments

$$\Delta_x M(t,x;\xi) = M(t,x;\xi) - M(t,x';\xi),$$
$$\Delta_x K(t,x;\xi) = K(t,x;\xi) - K(t,x';\xi),$$

$x, x' \in \mathbb{R}^n$.

We can write $\Delta_x M(t,x;\xi) = M_1 + M_2 + M_3$ where

$$M_1 = \sum_{j,l=1}^{n} \Delta_x a_{jl}(x)\partial_{x_j}\partial_{x_l} Z_0(t, x-\xi;\xi),$$

$$M_2 = \sum_{j,l=1}^{n} [a_{jl}(x') - a_{jl}(\xi)]\Delta_x \partial_{x'_j}\partial_{x'_l} Z_0(t, x'-\xi;\xi),$$

and the term M_3 contains lower order derivatives. Since M_3 has a weaker singularity and does not influence the estimates, we omit its detailed description and consider only M_1 and M_2.

An estimate for M_1 is given directly:

$$|M_1| \le C_1|x-x'|^\gamma t^{-\alpha}|x-\xi|^{-n}\exp\{-\sigma\rho(t,x;0,\xi)\}. \tag{5.2.56}$$

Using (5.2.30) we find that

$$\begin{aligned}
&\left|\Delta_x\partial_{x'_j}\partial_{x'_l}Z_0(t,x'-\xi;\xi)\right| \le \left|\Delta_x\partial_{x'_j}\partial_{x'_l}Z_0\right|^\nu \left(\left|\partial_{x_j}\partial_{x_l}Z_0\right| + \left|\partial_{x'_j}\partial_{x'_l}Z_0\right|\right)^{1-\nu}\\
&\le C_3^\nu t^{-\alpha\nu}|\widetilde{x}-\xi|^{-n\nu-\nu}\exp\{-\nu\sigma\rho(t,\widetilde{x};0,\xi)\}|x-x'|^\nu C_2^{1-\nu}t^{-\alpha(1-\nu)}\\
&\times\left[|x-\xi|^{-n}\exp\{-\sigma\rho(t,x;0,\xi)\} + |x'-\xi|^{-n}\exp\{-\sigma\rho(t,x';0,\xi)\}\right]^{1-\nu}
\end{aligned} \tag{5.2.57}$$

where $\widetilde{x} = x + \theta(x'-x)$, $0 < \theta < 1$, $0 < \nu < \gamma$, and we assumed that $\min\{|x-\xi|, |x'-\xi|\} \ne 0$. From (5.2.56) and (5.2.57) we find that

$$\begin{aligned}
|\Delta_x M(t,x;\xi)| \le C_4 t^{-\alpha}\Big[&|x-x'|^\gamma|x-\xi|^{-n}\exp\{-\sigma\rho(t,x;0,\xi)\}\\
&+ |x-x'|^\nu|x'-\xi|^\gamma\widetilde{M}(t,x;\xi)\Big]
\end{aligned} \tag{5.2.58}$$

where

$$\begin{aligned}
\widetilde{M}(t,x;\xi) = |\widetilde{x}-\xi|^{-n\nu-\nu}\exp\{-\nu\sigma\rho(t,\widetilde{x};0,\xi)\}\,&\big[|x-\xi|^{-n}\exp\{-\sigma\rho(t,x;0,\xi)\}\\
&+|x'-\xi|^{-n}\exp\{-\sigma\rho(t,x';0,\xi)\}\big]^{1-\nu}.
\end{aligned}$$

We shall use also the following estimate which is a direct consequence of the definition of $\Delta_x M$:

$$\begin{aligned}
|\Delta_x M(t,x;\xi)| \le Ct^{-\alpha}\big(&|x-\xi|^{-n+\gamma}\exp\{-\sigma\rho(t,x;0,\xi)\}\\
&+|x'-\xi|^{-n+\gamma}\exp\{-\sigma\rho(t,x';0,\xi)\}\big).
\end{aligned} \tag{5.2.59}$$

Let us consider two possible cases.

a) Suppose that $|x - x'| > At^{\alpha/2}$. It follows from (5.2.59) that

$$|\Delta_x M(t,x;\xi)| \le C_5(A) t^{(\gamma-\varepsilon)\frac{\alpha}{2}-\alpha} \left[|x-\xi|^{-n+\varepsilon} \exp\{-\sigma_1 \rho(t,x;0,\xi)\} \right. \\ \left. + |x'-\xi|^{-n+\varepsilon} \exp\{-\sigma_1 \rho(t,x';0,\xi)\} \right], \quad \varepsilon > 0,$$

whence

$$|\Delta_x M(t,x;\xi)| \le C_6 t^{-\alpha} |x-x'|^{\gamma-\varepsilon} \left[|x-\xi|^{-n+\varepsilon} \exp\{-\sigma_1 \rho(t,x;0,\xi)\} \right. \\ \left. + |x'-\xi|^{-n+\varepsilon} \exp\{-\sigma_1 \rho(t,x';0,\xi)\} \right], \quad \varepsilon > 0. \quad (5.2.60)$$

b) Let $|x - x'| \le At^{\alpha/2}$ with some $A > 0$. If $|x - x'| > \eta \max(|\xi - x|, |\xi - x'|)$ with some $\eta > 0$, then we obtain again the estimate (5.2.60). Thus, we have now to consider the most complicated subcase, for which

$$|x - x'| \le \eta \max(|\xi - x|, |\xi - x'|), \quad \eta > 0.$$

Here we use the estimate (5.2.57).

Without restricting generality, we may assume that $|\xi - x'| \le |\xi - x|$. Then

$$\begin{aligned} &|x-x'| \le \eta|\xi - x|, \\ &|\widetilde{x} - \xi| = |x - \xi + \theta(x'-x)| \\ &\qquad \ge |x-\xi| - \theta\eta|x-\xi| \ge (1-\eta)|x-\xi|. \end{aligned}$$

Note that the function

$$r \mapsto r^{-n} \exp\left\{ -\sigma \left(\frac{r}{t^{\alpha/2}} \right)^{\frac{2}{2-\alpha}} \right\}, \quad 0 < \alpha < 1,$$

is monotone decreasing. This implies the estimate

$$\left| \widetilde{M}(t,x;\xi) \right| \le C_7 \Big[|x-\xi|^{-n-\nu} \exp\{-\sigma_1 \rho(t,x;0,\xi)\} \\ + |x'-\xi|^{-n-\nu} \exp\{-\sigma \rho(t,x';0,\xi)\} \Big].$$

Setting $\varepsilon = \gamma - \nu$ and using (5.2.59), (5.2.60), we come to the estimate contained in the following proposition.

Proposition 5.5. *If x'' is one of the points x, x' for which* $|x'' - \xi| = \min(|x - \xi|, |x' - \xi|)$, *then*

$$|\Delta_x M(t,x;\xi)| \le C t^{-\alpha} |x-x'|^{\gamma-\varepsilon} |x''-\xi|^{-n+\varepsilon} \exp\{-\sigma\rho(t,x'';0,\xi)\}, \quad (5.2.61)$$

$$|\Delta_x K(t,x;\xi)| \le C t^{-1} |x-x'|^{\gamma-\varepsilon} |x''-\xi|^{-n+\varepsilon} \exp\{-\sigma\rho(t,x'';0,\xi)\}. \quad (5.2.62)$$

The *proof* of (5.2.62) is similar to that of (5.2.61), since the estimates for Y_0 have a structure similar to those for Z_0. □

Now we are ready to consider the integral equations (5.2.49) and (5.2.50).

Proposition 5.6. *The integral equations* (5.2.49) *and* (5.2.50) *have the solutions* $Q(t,x;\xi)$, $\Psi(t,x;\xi)$, *which are continuous for* $x \neq \xi$ *and satisfy the estimates*

$$|Q(t,x;\xi)| \le Ct^{-\alpha}|x-\xi|^{-n+\gamma}\exp\{-\sigma\rho(t,x;0,\xi)\}, \tag{5.2.63}$$

$$|\Delta_x Q(t,x;\xi)| \le Ct^{-\alpha}|x-x'|^{\gamma-\varepsilon}|x''-\xi|^{-n+\varepsilon}\exp\{-\sigma\rho(t,x'';0,\xi)\}, \tag{5.2.64}$$

$$|\Psi(t,x;\xi)| \le Ct^{-1}|x-\xi|^{-n+\gamma}\exp\{-\sigma\rho(t,x;0,\xi)\}, \tag{5.2.65}$$

$$|\Delta_x \Psi(t,x;\xi)| \le Ct^{-1}|x-x'|^{\gamma-\varepsilon}|x''-\xi|^{-n+\varepsilon}\exp\{-\sigma\rho(t,x'';0,\xi)\}, \tag{5.2.66}$$

with an arbitrary $\varepsilon \in (0,\gamma)$ *and the constants depending on* ε. *If* $n=2$ *or* $n=4$, *then for any* $\mu > 0$,

$$|Q(t,x;\xi)| \le Ct^{-\alpha-\frac{\mu\alpha}{2}}|x-\xi|^{-n+\varepsilon-\mu}\exp\{-\sigma\rho(t,x;0,\xi)\}, \tag{5.2.67}$$

$$|\Delta_x Q(t,x;\xi)| \le Ct^{-\alpha-\frac{\mu\alpha}{2}}|x-x'|^{\gamma-\varepsilon}|x''-\xi|^{-n+\varepsilon-\mu}\exp\{-\sigma\rho(t,x'';0,\xi)\}, \tag{5.2.68}$$

$$|\Psi(t,x;\xi)| \le Ct^{-1-\frac{\mu\alpha}{2}}|x-\xi|^{-n+\varepsilon-\mu}\exp\{-\sigma\rho(t,x;0,\xi)\}, \tag{5.2.69}$$

$$|\Delta_x \Psi(t,x;\xi)| \le Ct^{-1-\frac{\mu\alpha}{2}}|x-x'|^{\gamma-\varepsilon}|x''-\xi|^{-n+\varepsilon-\mu}\exp\{-\sigma\rho(t,x'';0,\xi)\}. \tag{5.2.70}$$

Here the constants depend on ε *and* μ.

Proof. The equations (5.2.49) and (5.2.50) are solved by iterations. The estimates (5.2.63) and (5.2.67), (5.2.65) and (5.2.69) follow from Lemma 1.14.

Let us prove (5.2.64). We have $\Delta_x Q = \Delta_x M + J$ where

$$J = \int_0^t d\lambda \int_{\mathbb{R}^n} \Delta_x K(t-\lambda,x;y)Q(\lambda,y;\xi)\,dy, \tag{5.2.71}$$

$$|J| \le C|x-x'|^{\gamma-\varepsilon}\int_0^t (t-\lambda)^{-1}\lambda^{-\alpha}d\lambda$$
$$\times\Bigg[\int_{\mathbb{R}^n}|x-y|^{-n+\varepsilon}|y-\xi|^{-n+\gamma}\exp\{-\sigma\rho(t,x;\lambda,y)-\sigma\rho(\lambda,y;0,\xi)\}\,dy$$
$$+\int_{\mathbb{R}^n}|x'-y|^{-n+\varepsilon}|y-\xi|^{-n+\gamma}\exp\{-\sigma\rho(t,x';\lambda,y)-\sigma\rho(\lambda,y;0,\xi)\}\,dy\Bigg]$$

$$
\leq C|x-x'|^{\gamma-\varepsilon} \int\limits_0^t (t-\lambda)^{-1+\frac{\alpha\varepsilon}{4}} \lambda^{(1-\alpha)-1}\, d\lambda
$$

$$
\times \left[\int\limits_{\mathbb{R}^n} |x-y|^{-n+\frac{\varepsilon}{2}} |y-\xi|^{-n+\gamma} \exp\{-\sigma_1\rho(t,x;\lambda,y) - \sigma_1\rho(\lambda,y;0,\xi)\}\, dy \right.
$$

$$
\left. + \int\limits_{\mathbb{R}^n} |x'-y|^{-n+\frac{\varepsilon}{2}} |y-\xi|^{-n+\gamma} \exp\{-\sigma_1\rho(t,x';\lambda,y) - \sigma_1\rho(\lambda,y;0,\xi)\}\, dy \right].
$$

By Lemma 1.12, the integral (5.2.71) is estimated as follows:

$$
|J| \leq C|x-x'|^{\gamma-\varepsilon} t^{-\alpha+\frac{\alpha\varepsilon}{4}} \left[|x-\xi|^{-n+\gamma+\frac{\varepsilon}{2}} \exp\{-\sigma_2\rho(t,x;0,\xi)\} \right.
$$
$$
\left. + |x'-\xi|^{-n+\gamma+\frac{\varepsilon}{2}} \exp\{-\sigma_2\rho(t,x;0,\xi)\} \right].
$$

Together with (5.2.61), this implies (5.2.64) with an appropriate $\sigma > 0$.

The proof of the inequalities (5.2.66) and (5.2.70) is similar. □

5.2.5. The one-dimensional case. In the case $n = 1$ the parametrix and the fundamental solution have no singularity in the spatial variable. This simplifies the situation greatly and makes it similar to the theory of conventional parabolic equations of an arbitrary order. Note however that even this case is more complicated than the classical study of a second order parabolic differential equation. In particular, the estimates of iterated kernels in the Levi method should still be performed in two stages (see Lemma 1.15).

Here we give the main estimates for this case; their proofs, simplified versions of those given the preceding section, are omitted. The scheme of the Levi method and the main notation is as above; see (5.2.47)–(5.2.50).

Proposition 5.7. *If $n = 1$, then*

$$
|M(t,x;\xi)| \leq Ct^{-(3-\gamma)\alpha/2} \exp\{-\sigma\rho(t,x;0,\xi)\},
$$
$$
|K(t,x;\xi)| \leq Ct^{-1-(1-\gamma)\alpha/2} \exp\{-\sigma\rho(t,x;0,\xi)\},
$$
$$
|Q(t,x;\xi)| \leq Ct^{-(3-\gamma)\alpha/2} \exp\{-\sigma\rho(t,x;0,\xi)\},
$$
$$
|\Psi(t,x;\xi)| \leq Ct^{-1-(1-\gamma)\alpha/2} \exp\{-\sigma\rho(t,x;0,\xi)\}.
$$

The estimates for differences of the above kernels are collected in the next proposition.

Proposition 5.8. *If $n = 1$, then*

$$
|\Delta_x M(t,x;\xi)| \leq C|x-x'|^{\gamma-\varepsilon} t^{-(3-\varepsilon)\alpha/2} \exp\{-\sigma\rho(t,x'';0,\xi)\},
$$
$$
|\Delta_x K(t,x;\xi)| \leq C|x-x'|^{\gamma-\varepsilon} t^{-1-(1-\varepsilon)\alpha/2} \exp\{-\sigma\rho(t,x'';0,\xi)\},
$$
$$
|\Delta_x Q(t,x;\xi)| \leq C|x-x'|^{\gamma-\varepsilon} t^{-(3-\varepsilon)\alpha/2} \exp\{-\sigma\rho(t,x'';0,\xi)\},
$$
$$
|\Delta_x \Psi(t,x;\xi)| \leq C|x-x'|^{\gamma-\varepsilon} t^{-1-(1-\varepsilon)\alpha/2} \exp\{-\sigma\rho(t,x'';0,\xi)\},
$$
$$
0 < \varepsilon < \gamma.
$$

5.3 The Cauchy problem: Existence and representation of solutions

5.3.1. The Green matrix. Let us formulate the main result regarding the existence and properties of the Cauchy problem (5.2.1)–(5.2.2) for the general operator B. See Section 5.2.1 for the main assumptions and definitions.

In order to describe regularity properties of solutions of the equation (5.2.1) with respect to the variable t, it is convenient to introduce the class $H_\mu^{\alpha+\lambda}[0,T]$ of such functions $\varphi(t)$ that $t^\mu\varphi(t)$ is Hölder continuous on $[0,T]$ with the exponent $\alpha+\lambda$, $\lambda>0$. It will be shown below that the problem (5.2.1)–(5.2.2) cannot have more than one bounded classical solution belonging (for each fixed $x\in\mathbb{R}^n$) to $H_\mu^{\alpha+\lambda}[0,T]$, if $0<\lambda<1-\alpha$, $0\le\mu<\lambda+1$.

Below we denote by γ various Hölder exponents with respect to spatial variables (without restricting generality they will be assumed equal).

Theorem 5.2.
a) *There exists a Green matrix* $\{Z(t,x;\xi),Y(t,x;\xi)\}$ *for the problem* (5.2.1)–(5.2.2), *of the form*

$$Z(t,x;\xi)=Z_0(t,x-\xi;\xi)+V_Z(t,x;\xi),$$
$$Y(t,x;\xi)=Y_0(t,x-\xi;\xi)+V_Y(t,x;\xi),$$

where (Z_0,Y_0), *the Green matrix of the Cauchy problem for the equation obtained by "freezing" the coefficients* a_{ij} *at the parameter point* ξ *and setting other coefficients equal to zero, satisfies the estimates listed in Propositions* 5.1 *and* 5.2, *with the constants independent of* $\xi\in\mathbb{R}^n$. *The functions* V_Z,V_Y *satisfy the estimates*

(i) *If* $n=1$, *then*

$$|\partial_x^m V_Z(t,x;\xi)|\le Ct^{(\gamma-m-1)\alpha/2}\exp\left\{-\sigma t^{-\frac{\alpha}{2-\alpha}}|x-\xi|^{\frac{2}{2-\alpha}}\right\},$$
$$|\partial_x^m V_Y(t,x;\xi)|\le Ct^{\alpha-1+(\gamma-m-1)\alpha/2}\exp\left\{-\sigma t^{-\frac{\alpha}{2-\alpha}}|x-\xi|^{\frac{2}{2-\alpha}}\right\},$$

$m=0,1,2.$

(ii) *If* $n=2$, *then*

$$|V_Z(t,x;\xi)|\le Ct^{\frac{\gamma\alpha}{2}-\alpha}\exp\left\{-\sigma t^{-\frac{\alpha}{2-\alpha}}|x-\xi|^{\frac{2}{2-\alpha}}\right\},$$
$$|\partial_x V_Z(t,x;\xi)|\le Ct^{\frac{\gamma_0\alpha}{2}-\alpha}|x-\xi|^{-1+\gamma-\gamma_0}\exp\left\{-\sigma t^{-\frac{\alpha}{2-\alpha}}|x-\xi|^{\frac{2}{2-\alpha}}\right\},$$
$$|V_Y(t,x;\xi)|\le Ct^{\frac{\gamma\alpha}{2}-1}\exp\left\{-\sigma t^{-\frac{\alpha}{2-\alpha}}|x-\xi|^{\frac{2}{2-\alpha}}\right\},$$
$$|\partial_x V_Y(t,x;\xi)|\le Ct^{\frac{\gamma_0\alpha}{2}-1}|x-\xi|^{-1+\gamma-\gamma_0}\exp\left\{-\sigma t^{-\frac{\alpha}{2-\alpha}}|x-\xi|^{\frac{2}{2-\alpha}}\right\},$$

where $\gamma_0<\gamma$ *is an arbitrary fixed positive constant (here and below in the formulas containing* γ_0 *the constant* C *depends on* γ_0*).*

(iii) *If $n = 3$, then*

$$|\partial_x^m V_Z(t,x;\xi)| \le Ct^{\frac{\gamma_0\alpha}{2}-\alpha}|x-\xi|^{-1-|m|+\gamma-\gamma_0}\exp\left\{-\sigma t^{-\frac{\alpha}{2-\alpha}}|x-\xi|^{\frac{2}{2-\alpha}}\right\},$$
$$|\partial_x^m V_Y(t,x;\xi)| \le Ct^{(\gamma_0+\gamma)\frac{\alpha}{4}-1}|x-\xi|^{-1-|m|+(\gamma-\gamma_0)/2}\exp\left\{-\sigma t^{-\frac{\alpha}{2-\alpha}}|x-\xi|^{\frac{2}{2-\alpha}}\right\},$$

$|m| = 0, 1$.

(iv) *If $n = 4$, then*

$$|\partial_x^m V_Z(t,x;\xi)| \le Ct^{\frac{(\gamma-\gamma_0)\alpha}{2}-\alpha}|x-\xi|^{-2-|m|+\gamma-2\gamma_0}\exp\left\{-\sigma t^{-\frac{\alpha}{2-\alpha}}|x-\xi|^{\frac{2}{2-\alpha}}\right\},$$
$$|\partial_x^m V_Y(t,x;\xi)| \le Ct^{\frac{(\gamma-\gamma_0)\alpha}{4}-1}|x-\xi|^{-2-|m|+\gamma-2\gamma_0}\exp\left\{-\sigma t^{-\frac{\alpha}{2-\alpha}}|x-\xi|^{\frac{2}{2-\alpha}}\right\},$$

$|m| = 0, 1;\ \gamma_0 < \gamma/2$.

(v) *If $n \ge 5$, then*

$$|\partial_x^m V_Z(t,x;\xi)| \le Ct^{\frac{\gamma_0\alpha}{2}-\alpha}|x-\xi|^{2-n-|m|+\gamma-\gamma_0}\exp\left\{-\sigma t^{-\frac{\alpha}{2-\alpha}}|x-\xi|^{\frac{2}{2-\alpha}}\right\},$$
$$|\partial_x^m V_Y(t,x;\xi)| \le Ct^{(\gamma_0+\gamma)\frac{\alpha}{4}-1}|x-\xi|^{2-n-|m|+(\gamma-\gamma_0)/2}\exp\left\{-\sigma t^{-\frac{\alpha}{2-\alpha}}|x-\xi|^{\frac{2}{2-\alpha}}\right\},$$

$|m| = 0, 1$.

(vi) *If $n \ge 2, |m| = 2$, then*

$$|\partial_x^m V_Z(t,x;\xi)| \le Ct^{\gamma_1-\alpha}|x-\xi|^{-n+\gamma_2}\exp\left\{-\sigma t^{-\frac{\alpha}{2-\alpha}}|x-\xi|^{\frac{2}{2-\alpha}}\right\},$$
$$|\partial_x^m V_Y(t,x;\xi)| \le Ct^{\gamma_1-1}|x-\xi|^{-n+\gamma_2}\exp\left\{-\sigma t^{-\frac{\alpha}{2-\alpha}}|x-\xi|^{\frac{2}{2-\alpha}}\right\},$$

where γ_1, γ_2 are some positive constants.

(vii) *The estimates for the fractional derivatives $\mathbb{D}^{(\alpha)}V_Z$, $\mathbb{D}^{(\alpha)}V_Y$ of the functions Z, Y are the same as the above estimates for their second order derivatives.*

b) *Let $u(t,x)$ be the solution* (5.2.4) *of the problem* (5.2.1)–(5.2.2) *with a bounded continuous function u_0 (locally Hölder continuous if $n > 1$) and a bounded jointly continuous and locally Hölder continuous, in x, function f. If $0 < \lambda < 1-\alpha$, $\alpha+\lambda < \nu$, then for any $x \in \mathbb{R}^n$ $u(\cdot,x) \in H_\nu^{\alpha+\lambda}[0,T]$.*

c) *The functions Z and Y are nonnegative.*

The rest of this section will be devoted to the proof of Theorem 5.2. Note that it can be generalized in various directions. In particular, the initial function may have a certain exponential growth (see [130] for the case $B = \Delta$). All the coefficients except the leading ones may depend also on t. Moreover, if we consider only the case of the zero initial functions, that is we look only for the function Y, then we can include, without any changes, the general case of time-dependent coefficients. The same is true for the case $n = 1$ which resembles the classical theory of parabolic equations. In fact, it seems probable that the stationarity assumption can be dropped completely. However, then the machinery of Fox's H-functions would not be available, and we would have to deal directly with contour integral representations of solutions of fractional diffusion equations with coefficients depending only on t. Their study is a complicated task in itself, which has not been fulfilled so far..

5.3.2. Heat potentials. In order to verify that the functions (5.2.47) and (5.2.48) indeed solve the Cauchy problem via the representation (5.2.4), we have to study the heat potential

$$W(t,x) = \int_0^t d\lambda \int_{\mathbb{R}^n} Y_0(t-\lambda, x-y; y) f(\lambda, y)\, dy \tag{5.3.1}$$

for two situations. Firstly, we must consider the case of a bounded, locally Hölder continuous (in x) function f, jointly continuous in $(t,x) \in \Pi_{[0,T]}$. This is, of course, dictated by the formula (5.2.4) containing a term which solves the inhomogeneous equation with the zero initial condition. Note that Y_0 is the main singular part of the function Y appearing in (5.2.4). Secondly, heat potentials appear also in (5.2.47) and (5.2.48), where the role of f is played by the functions Q and Ψ which are much more singular (see Proposition 5.6).

Let us consider the potential (5.3.1) with a bounded, locally Hölder continuous f. The existence of the integral (5.3.1) and the possibility to find its first order derivatives in x by differentiating under the integral's symbol follow directly from the estimates (5.2.34), (5.2.36), (5.2.38), and (5.2.40). For studying the second order derivatives we consider the function

$$W(t,x) = \int_0^{t-h} d\lambda \int_{\mathbb{R}^n} Y_0(t-\lambda, x-y; y) f(\lambda, y)\, dy, \quad h > 0.$$

As $t - \lambda \geq h$, the function Y_0 and its first and second derivatives have integrable singularities at $x = y$. Therefore we may differentiate under the integral's symbol, so that

$$\partial_{x_j}\partial_{x_l} W_h(t,x) = J_1 + J_2$$

where

$$J_1 = \int_0^{t-h} d\lambda \int_{\mathbb{R}^n} \partial_{x_j}\partial_{x_l} Y_0(t-\lambda, x-y; y)[f(\lambda, y) - f(\lambda, x)]\, dy, \tag{5.3.2}$$

$$J_2 = \int_0^{t-h} f(\lambda, x)\, d\lambda \int_{\mathbb{R}^n} \partial_{x_j}\partial_{x_l} Y_0(t-\lambda, x-y; y)\, dy. \tag{5.3.3}$$

We decompose further, $J_1 = J_1^{(1)} + J_1^{(2)}$, where $J_1^{(1)}$ corresponds to integration over the set $\Pi^{(1)} = \{y \in \mathbb{R}^n \big| \, |x-y| \le (t-\lambda)^{\alpha/2}\}$ while for $J_1^{(2)}$ the domain of integration is $\Pi^{(2)} = \{y \in \mathbb{R}^n \big| \, |x-y| > (t-\lambda)^{\alpha/2}\}$. Suppose that, for example, $n > 4$; other cases are treated similarly. If $y \in \Pi^{(1)}$, then

$$\left|\partial_{x_j}\partial_{x_l} Y_0(t-\lambda, x-y; y)\right| \le C(t-\lambda)^{-\alpha-1}|x-y|^{-n+2},$$

so that the $\Pi^{(1)}$ part of the integral in (5.3.2) is majorized by

$$I_1^{(1)} = C \int\limits_0^{t-h} (t-\lambda)^{-\alpha-1} d\lambda \int\limits_{\Pi^{(1)}} |x-y|^{-n+2+\gamma}\, dy,$$

and the change of variables $\widehat{y} = (t-\lambda)^{-\alpha/2}(y-x)$ gives

$$I_1^{(1)} \le C \int\limits_0^{t-h} (t-\lambda)^{-1+\frac{\alpha\gamma}{2}}\, d\lambda \int\limits_{|\widehat{y}|\le 1} |\widehat{y}|^{-n+2+\gamma}\, d\widehat{y}. \tag{5.3.4}$$

Here $\gamma > 0$ is the Hölder exponent of the function $f(\lambda, x)$ in x.

For $y \in \Pi^{(2)}$ we use the inequality

$$\left|\partial_{x_j}\partial_{x_l} Y_0(t-\lambda, x-y; y)\right| \le C(t-\lambda)^{-\frac{\alpha n}{2}-1} \exp\left\{-\sigma\left(\frac{|x-y|}{(t-\lambda)^{\alpha/2}}\right)^{\frac{2}{2-\alpha}}\right\},$$

a consequence of the general estimate (5.2.40). A similar argument shows that the $\Pi^{(2)}$ part of the integral in J_1 is majorized by

$$C \int\limits_0^{t-h} (t-\lambda)^{-1+\frac{\alpha\gamma}{2}}\, d\lambda \int\limits_{|\widehat{y}|>1} |\widehat{y}|^{\gamma} \exp\left\{-\sigma|\widehat{y}|^{\frac{2}{2-\alpha}}\right\} d\widehat{y}.$$

We see the existence of the limit of J_1 as $h \to 0$.

Let us consider J_2. Using (4.14) we have

$$\begin{aligned} J_2 = &\int\limits_0^{t-h} f(\lambda, x)\, d\lambda \\ &\times \int\limits_{\mathbb{R}^n} \left[\partial_{x_j}\partial_{x_l} Y_0(t-\lambda, x-y; y) - \partial_{x_j}\partial_{x_l} Y_0(t-\lambda, x-y; \delta)\big|_{\delta=x}\right] dy. \end{aligned} \tag{5.3.5}$$

By (5.2.41), the expression in brackets is majorized by

$$C|x-y|^{-n+2+\gamma}(t-\lambda)^{-\alpha-1}\exp\{-\sigma\rho(t, x; \lambda, y)\}.$$

The same change of variables (used above) shows that the integrand in (5.3.5) is majorized by

$$(t-\lambda)^{-1+\frac{\alpha\gamma}{2}}|\widehat{y}|^{-n+2+\gamma}\exp\left\{-\sigma|\widehat{y}|^{\frac{2}{2-\alpha}}\right\}.$$

This means the existence of the limit of J_2 as $h \to 0$. Simultaneously we have proved the formula

$$\begin{aligned}\partial_{x_j}\partial_{x_l}W(t,x) &= \int_0^t d\lambda \int_{\mathbb{R}^n} \partial_{x_j}\partial_{x_l}Y_0(t-\lambda,x-y;y)[f(\lambda,y)-f(\lambda,x)]\,dy \\ &\quad + \int_0^t f(\lambda,x)\,d\lambda \int_{\mathbb{R}^n} \partial_{x_j}\partial_{x_l}Y_0(t-\lambda,x-y;y)\,dy. \qquad (5.3.6)\end{aligned}$$

Now we have to carry out a similar procedure for the potential (5.3.1) with $f(\lambda,y) = Q(\lambda,y;\xi)$ with a fixed ξ (the case of $\Psi(\lambda,y;\tau,\xi)$ is quite similar). We use the estimates (5.2.63) and (5.2.64) for the function Q and its increment, and also the following roughened estimate:

$$\left|\partial_{x_j}\partial_{x_l}Y_0(t-\lambda,x-y;\xi)\right| \le C(t-\lambda)^{-1}|x-y|^{-n}\exp\{-\sigma\rho(t,x;\lambda,y)\}. \quad (5.3.7)$$

In fact we have to prove convergence of the integrals

$$I_1' = \int_0^t d\lambda \int_{\mathbb{R}^n} \left|\partial_{x_j}\partial_{x_l}Y_0(t-\lambda,x-y;y)\right| |Q(\lambda,x;\xi)-Q(\lambda,y;\xi)|\,dy,$$

$$\begin{aligned}I_2' &= \int_0^t |Q(\lambda,x;\xi)|\,d\lambda \\ &\quad \times \int_{\mathbb{R}^n} \left|\partial_{x_j}\partial_{x_l}Y_0(t-\lambda,x-y;y) - \partial_{x_j}\partial_{x_l}Y_0(t-\lambda,x-y;\delta)\big|_{\delta=x}\right|\,dy.\end{aligned}$$

Let $a>0$ be a small positive constant. Changing σ we can rewrite (5.3.7) as

$$\left|\partial_{x_j}\partial_{x_l}Y_0(t-\lambda,x-y;\xi)\right| \le C(t-\lambda)^{-1+\frac{a\alpha}{2}}|x-y|^{-n-a}\exp\{-\sigma\rho(t,x;\lambda,y)\}.$$

Now we get by Lemma 1.12 and Lemma 1.13 that for $0<\varepsilon<\gamma$,

$$\begin{aligned}I_1' &\le C\int_0^t (t-\lambda)^{-1+\frac{a\alpha}{2}}\lambda^{-\alpha}\,d\lambda \\ &\quad\times\Bigg[\int_{\mathbb{R}^n} |x-y|^{-n+\gamma-\varepsilon-a}|x-\xi|^{-n+\varepsilon}\exp\{-\sigma\rho(t,x;\lambda,y)-\sigma\rho(\lambda,x;0,\xi)\}\,dy \\ &\quad + \int_{\mathbb{R}^n} |x-y|^{-n+\gamma-\varepsilon-a}|y-\xi|^{-n+\varepsilon}\exp\{-\sigma\rho(t,x;\lambda,y)-\sigma\rho(\lambda,y;0,\xi)\}\,dy\Bigg]\end{aligned}$$

$$= C|x-\xi|^{-n+\varepsilon}\exp\{-\sigma\rho(t,x;0,\xi)\}$$
$$\times \int_0^t (t-\lambda)^{-1+\frac{a\alpha}{2}}\lambda^{-\alpha}d\lambda \int_{\mathbb{R}^n} |x-y|^{-n+\gamma-\varepsilon-a}\exp\{-\sigma\rho(t,x;\lambda,y)\}\,dy$$
$$+ C\int_0^t (t-\lambda)^{-1+\frac{a\alpha}{2}}\lambda^{-\alpha}d\lambda \int_{\mathbb{R}^n} |x-y|^{-n+\gamma-\varepsilon-a}|y-\xi|^{-n+\varepsilon}$$
$$\times \exp\{-\sigma\rho(t,x;\lambda,y)-\sigma\rho(\lambda,y;0,\xi)\}\,dy$$
$$\le C|x-\xi|^{-n+\varepsilon}\exp\{-\sigma\rho(\lambda,x;0,\xi)\}\int_0^t (t-\lambda)^{\frac{(\gamma-\varepsilon)\alpha}{2}-1}\lambda^{-\alpha}d\lambda$$
$$+ Ct^{-\alpha+\frac{a\alpha}{2}}|x-\xi|^{-n+\gamma-a}\exp\{-\sigma\rho(t,x;0,\xi)\}.$$

As before, the letters σ, C mean various positive constants; this abuse of notation will be convenient in the sequel too.

In order to obtain a bound for I_2', we use the estimate

$$\left|\partial_{x_j}\partial_{x_l}Y_0(t-\lambda,x-y;y)-\partial_{x_j}\partial_{x_l}Y_0(t-\lambda,x-y;\delta)\big|_{\delta=x}\right|$$
$$\le C|x-y|^{\gamma-n}(t-\lambda)^{-1}\exp\{-\sigma\rho(t,x;\lambda,y)\}$$
$$\le C|x-y|^{-n+\frac{\gamma}{2}}(t-\lambda)^{-1+\frac{\alpha\gamma}{4}}\exp\{-\sigma\rho(t,x;\lambda,y)\}$$

(with a different σ). Then by Lemma 1.13

$$I_2' \le C|x-\xi|^{-n+\gamma}\exp\{-\sigma\rho(t,x;0,\xi)\}$$
$$\times \int_0^t (t-\lambda)^{-1+\frac{\alpha\gamma}{4}}\lambda^{-\alpha}\,d\lambda \int_{\mathbb{R}^n} |x-y|^{-n+\frac{\gamma}{2}}\exp\{-\sigma\rho(t,x;\lambda,y)\}$$
$$\le C|x-\xi|^{-n+\gamma}\exp\{-\sigma\rho(t,x;0,\xi)\}\int_0^t (t-\lambda)^{-1+\frac{\alpha\gamma}{2}}\lambda^{-\alpha}\,d\lambda$$
$$\le Ct^{\frac{\alpha\gamma}{2}-\alpha}|x-\xi|^{-n+\gamma}\exp\{-\sigma\rho(t,x;0,\xi)\}.$$

Thus, we have proved (5.3.6) for $f(\lambda,y)=Q(\lambda,y;\xi)$.

5.3.3. The fractional derivative. Our next task is to study the fractional derivative of the heat potential (5.3.1), again for the above two cases.

Suppose that $f(t,x)$ is bounded, locally Hölder continuous in x, and jointly continuous in (t,x). We have $\mathbb{D}^{(\alpha)}W=\partial_t v$ where $v=I_{0+}^{1-\alpha}W$ is the Riemann-Liouville fractional integral, that is

$$v(t,x)=\frac{1}{\Gamma(1-\alpha)}\int_0^t (t-\theta)^{-\alpha}\,d\theta\int_0^\theta d\lambda\int_{\mathbb{R}^n} Y_0(\theta-\lambda,x-y;y)f(\lambda,y)\,dy$$

$$= \frac{1}{\Gamma(1-\alpha)} \int_0^t d\lambda \int_\lambda^t (t-\theta)^{-\alpha}\, d\theta \int_{\mathbb{R}^n} Y_0(\theta-\lambda, x-y; y) f(\lambda, y)\, dy$$

$$= \int_0^t d\lambda \int_{\mathbb{R}^n} Z_0(t-\lambda, x-y; y) f(\lambda, y)\, dy,$$

by the definition of Y_0.

Let h be a small positive number,

$$v_h(t,x) = \int_0^{t-h} d\lambda \int_{\mathbb{R}^n} Z_0(t-\lambda, x-y; y) f(\lambda, y)\, dy.$$

Then $v_h \to v$ pointwise as $h \to 0$,

$$\partial_t v_h = v_h^{(1)}(t,x) + v_h^{(2)}(t,x)$$

where

$$v_h^{(1)}(t,x) = \int_{\mathbb{R}^n} Z_0(h, x-y; y) f(t-h, y)\, dy,$$

$$v_h^{(2)}(t,x) = \int_0^{t-h} d\lambda \int_{\mathbb{R}^n} \partial_t Z_0(t-\lambda, x-y; y) f(\lambda, y)\, dy.$$

We have

$$v_h^{(1)}(t,x) = \int_{\mathbb{R}^n} [Z_0(h, x-y; y) - Z_0(h, x-y; x)] f(t-h, y)\, dy$$
$$+ \int_{\mathbb{R}^n} Z_0(h, x-y; x)[f(t-h, y) - f(t-h, x)]\, dy + f(t-h, x), \quad (5.3.8)$$

due to (5.2.42). By Proposition 5.3, the absolute value of the first integral in the right-hand side of (5.3.8) does not exceed

$$Ch^{-\alpha} \int_{\mathbb{R}^n} |x-y|^{-n+2+\gamma} \exp\left\{ -\sigma \left(h^{-\alpha/2} |x-y| \right)^{\frac{2}{2-\alpha}} \right\} dy = \text{const} \cdot h^{\frac{\gamma\alpha}{2}} \to 0,$$

as $h \to 0$. Similarly, using (5.2.30) we show that the second integral in (5.3.8) tends to zero as $h \to 0$. Thus

$$v_h^{(1)}(t,x) \longrightarrow f(t,x), \quad (5.3.9)$$

as $h \to 0$.

Turning to $v_h^{(2)}$ we write

$$v_h^{(2)}(t,x) = \int\limits_0^{t-h} d\lambda \int\limits_{\mathbb{R}^n} \partial_t Z_0(t-\lambda, x-y; y)[f(\lambda,y) - f(\lambda,x)]\, dy$$
$$+ \int\limits_0^{t-h} f(\lambda,x)\, d\lambda \int\limits_{\mathbb{R}^n} \partial_t Z_0(t-\lambda, x-y; y)\, dy.$$

It follows from (5.2.44) that

$$\left| \int\limits_0^{t-h} d\lambda \int\limits_{\mathbb{R}^n} \partial_t Z_0(t-\lambda, x-y; y)[f(\lambda,y) - f(\lambda,x)]\, dy \right| \le C(t-\lambda)^{-1+\frac{\alpha\gamma}{2}}.$$

Together with (5.2.46) this shows that

$$v_h^{(2)}(t,x) \longrightarrow \int\limits_0^{t} d\lambda \int\limits_{\mathbb{R}^n} \partial_t Z_0(t-\lambda, x-y; y)[f(\lambda,y) - f(\lambda,x)]\, dy$$
$$+ \int\limits_0^{t} f(\lambda,x)\, d\lambda \int\limits_{\mathbb{R}^n} \partial_t Z_0(t-\lambda, x-y; y)\, dy. \qquad (5.3.10)$$

By the definition of Y_0, $\partial_t Z_0 = \mathbb{D}^{(\alpha)} Y_0$, and it follows from (5.3.9) and (5.3.10) that the fractional derivative $\mathbb{D}^{(\alpha)} W$ exists and can be represented as follows:

$$\left(\mathbb{D}^{(\alpha)} W\right)(t,x) = f(t,x) + \int\limits_0^{t} d\lambda \int\limits_{\mathbb{R}^n} \partial_t Z_0(t-\lambda, x-y; y)[f(\lambda,y) - f(\lambda,x)]\, dy$$
$$+ \int\limits_0^{t} f(\lambda,x)\, d\lambda \int\limits_{\mathbb{R}^n} \partial_t Z_0(t-\lambda, x-y; y)\, dy. \qquad (5.3.11)$$

The case, in which $f(\lambda,y) = Q(\lambda,y;\xi)$, is considered similarly, on the basis of the estimates (5.2.45), (5.2.46), (5.2.63), (5.2.64).

5.3.4. The homogeneous equation. Now, in order to verify that the equation (5.2.1) is satisfied, it remains (due to the whole construction of the Levi method) to study the integral

$$\int\limits_{\mathbb{R}^n} Z_0(t, x-y; \xi) u_0(y)\, dy,$$

and to show that it represents a solution of the equation (5.2.1) with the coefficients of the operator B "frozen" at the point ξ.

Equivalently (up to a change of variables), we have to prove that the function

$$v(t,x)=\int_{\mathbb{R}^n} Z_0(t,x-y)u_0(y)\,dy, \tag{5.3.12}$$

satisfies the model equation $\mathbb{D}_t^{(\alpha)}v=\Delta v$. Here the kernel Z_0 is given by (5.2.5).

Let us fix $t>0$, $x^0\in\mathbb{R}^n$, $k\in\{1,\dots,n\}$, and compute $\partial^2_{x_k}v(t,x^0)$. Suppose that $n\geq 2$. Using the estimates from Proposition 5.1 we can find the first derivative by differentiating under the symbol of the integral. Taking into account (5.2.42) we have

$$\partial_{x_k}v(t,x^0)=\int_{\mathbb{R}^n}\frac{\partial_{x_k}Z_0(t,x^0-\xi)}{\partial x_k}\left[u_0(\xi)-u_0(x^0)\right]\,d\xi. \tag{5.3.13}$$

Let us decompose the domain of integration in (5.3.13) into the two sets:

$$\Omega_1=\left\{\xi\in\mathbb{R}^n\,\middle|\,|\xi-x^0|\geq t^\alpha\right\},\quad \Omega_2=\mathbb{R}^n\setminus\Omega_1.$$

Correspondingly, the integral becomes the sum of two integrals, $v_1(t,x^0)+v_2(t,x^0)$.

If x lies in a small neighbourhood of the point x^0, and $\xi\in\Omega_1$, then $x-\xi|$ is separated from zero. Therefore

$$\partial_{x_k}v_1(t,x^0)=\int_{\Omega_1}\partial^2_{x_k}Z_0(t,x^0-\xi)\left[u_0(\xi)-u_0(x^0)\right]\,d\xi. \tag{5.3.14}$$

Let $d>0$, $\tilde{d}=(0,\dots,d,\dots 0)$ (d stands in the k-th place). Then

$$\begin{aligned}
&\frac{1}{d}\left[v_2(t,x^0+\tilde{d})-v_2(t,x^0)\right]-\int_{\Omega_2}\partial^2_{x_k}Z_0(t,x^0-\xi)\left[u_0(\xi)-u_0(x^0)\right]\,d\xi\\
&=\frac{1}{d}\int_{|x^0-\xi|\leq 2d}\partial_{x_k}Z_0(t,x^0+\tilde{d}-\xi)\left[u_0(\xi)-u_0(x^0)\right]\,d\xi\\
&\quad-\frac{1}{d}\int_{|x^0-\xi|\leq 2d}\partial_{x_k}Z_0(t,x^0-\xi)\left[u_0(\xi)-u_0(x^0)\right]\,d\xi\\
&\quad-\int_{|x^0-\xi|\leq 2d}\partial^2_{x_k}Z_0(t,x^0-\xi)\left[u_0(\xi)-u_0(x^0)\right]\,d\xi\\
&\quad+\int_{2d<|x^0-\xi|\leq t^{\alpha/2}}\left\{\frac{1}{d}\left[\partial_{x_k}Z_0(t,x^0+\tilde{d}-\xi)-\partial_{x_k}Z_0(t,x^0-\xi)\right]\right.\\
&\quad\left.-\partial^2_{x_k}Z_0(t,x^0-\xi)\right\}\left[u_0(\xi)-u_0(x^0)\right]\,d\xi
\end{aligned}$$

(the integrals converge since u_0 is locally Hölder continuous). Each of the integrals on the right tends to zero as $d \to 0$. This follows from Proposition 5.1 and the local Hölder continuity of u_0; in the last integral we use the Taylor formula

$$\frac{1}{d}\left[\partial_{x_k} Z_0(t, x^0 + \tilde{d} - \xi) - \partial_{x_k} Z_0(t, x^0 - \xi)\right] - \partial^2_{x_k} Z_0(t, x^0 - \xi) = \frac{d}{2}\partial^3_{x_k} Z_0(t, x' - \xi),$$

$x' = x^0 + \theta d$, $0 \le \theta \le 1$, and take into account (5.2.8) and the fact that the inequality $|x^0 - \xi| \ge 2d$ implies

$$|x' - \xi| \ge |\xi - x^0| - |x' - x^0| \ge |\xi - x^0| - d \ge \frac{1}{2}|\xi - x^0|,$$

so that the last integral is majorized by

$$Cd^\gamma \int\limits_{2 \le |\zeta| < t^{\alpha/2} d^{-1}} |\zeta|^{-n-1+\gamma}\, d\zeta.$$

Thus,

$$\partial_{x_k} v_2(t, x^0) = \int\limits_{\Omega_2} \partial^2_{x_k} Z_0(t, x^0 - \xi)\left[u_0(\xi) - u_0(x^0)\right] d\xi.$$

Together with (5.3.14) this gives

$$\partial^2_{x_k} v(t, x^0) = \int\limits_{\mathbb{R}^n} \partial^2_{x_k} Z_0(t, x^0 - \xi)\left[u_0(\xi) - u_0(x^0)\right] d\xi. \tag{5.3.15}$$

In the case $n = 1$ this formula is obtained directly, by differentiating under the symbol of the integral.

On the other hand, using the inequality (5.2.11) we find that

$$\left(\mathbb{D}_t^{(\alpha)} v\right)(t, x^0) = \int\limits_{\mathbb{R}^n} \left(\mathbb{D}_t^{(\alpha)} Z_0\right)(t, x^0 - \xi)\left[u_0(\xi) - u_0(x^0)\right] d\xi. \tag{5.3.16}$$

Let $Z_\alpha \in \mathcal{S}'(\mathbb{R}^n)$ be the result of the action of the operation $\mathbb{D}_t^{(\alpha)}$ upon the function Z understood as a distribution in x depending on the parameter t. Using (5.2.7) and (5.2.11) we find, for any $\psi \in \mathcal{S}(\mathbb{R}^n)$, that

$$\langle Z_\alpha, \psi\rangle = \mathbb{D}_t^{(\alpha)}\langle Z, \psi\rangle = D^\alpha_{0+}\langle Z, \psi\rangle - \frac{\psi(0)}{\Gamma(1-\alpha)t^\alpha} = \left\langle D^\alpha_{0+} Z, \psi\right\rangle - \frac{\psi(0)}{\Gamma(1-\alpha)t^\alpha}$$

whence

$$\left\langle \widehat{Z_\alpha}, \psi\right\rangle = \left\langle D^\alpha_{0+} Z, \widehat{\psi}\right\rangle - \frac{\widehat{\psi}(0)}{\Gamma(1-\alpha)t^\alpha}.$$

Next,

$$\left\langle \mathbb{D}_t^{(\alpha)} \widehat{Z}, \psi \right\rangle = D_{0+}^{\alpha} \left\langle \widehat{Z}, \psi \right\rangle - \frac{1}{\Gamma(1-\alpha)t^{\alpha}} \int_{\mathbb{R}^n} \psi(x)\, dx = \left\langle D_{0+}^{\alpha} Z, \widehat{\psi} \right\rangle - \frac{\widehat{\psi}(0)}{\Gamma(1-\alpha)t^{\alpha}},$$

that is $\widehat{Z_\alpha} = \mathbb{D}_t^{(\alpha)} \widehat{Z}$, and it follows from the construction of Z that $\widehat{Z_\alpha}$ coincides with the Fourier transform of the distribution ΔZ. Hence, $Z_\alpha = \Delta Z$, and

$$\left(\mathbb{D}_t^{(\alpha)} Z\right)(t,x) = \Delta Z(t,x), \quad x \neq 0.$$

Now it follows from (5.3.15) and (5.3.16) that $\left(\mathbb{D}_t^{(\alpha)} v\right)(t,x^0) = \Delta v(t,x^0)$, as desired.

5.3.5. The initial condition. Let us verify the initial condition (5.2.2). As above, we begin with the model equation $\mathbb{D}_t^{(\alpha)} v = \Delta v$ and consider its solution (5.3.12). Using the equality (5.2.42) we write

$$v(t,x) = \int_{\mathbb{R}^n} Z_0(t, x-\xi)[u_0(\xi) - u_0(x)]\, d\xi + u_0(x).$$

It follows from (5.2.5) that

$$Z_0(t, xt^{\alpha/2}) = t^{-\alpha n/2} Z_0(1,x). \tag{5.3.17}$$

Therefore

$$v(t,x) = \int_{\mathbb{R}^n} Z_0(1,\eta) \left[u_0(x - \eta t^{\alpha/2}) - u_0(x)\right] d\eta + u_0(x),$$

and the inequalities (5.2.6), (5.2.8)–(5.2.10), together with the dominated convergence theorem, show that $v(t,x) \to u_0(x)$, as $t \to 0$, for any $x \in \mathbb{R}^n$.

Let us consider the general case. By our construction,

$$Z(t,x;\xi) = Z_0(t, x-\xi, \xi) + V_Z(t,x;\xi),$$

where

$$V_Z(t,x;\xi) = \int_0^t d\lambda \int_{\mathbb{R}^n} Y_0(t-\lambda, x-y; y) Q(\lambda, y; \xi)\, dy$$

(see (5.2.47)). We have

$$\int_{\mathbb{R}^n} Z_0(t, x-\xi, \xi) u_0(\xi)\, d\xi = \int_{\mathbb{R}^n} Z_0(t, x-\xi, x) u_0(\xi)\, d\xi$$
$$+ \int_{\mathbb{R}^n} [Z_0(t, x-\xi, \xi) - Z_0(t, x-\xi, x)] u_0(\xi)\, d\xi.$$

As we have just shown, the first summand tends to $u_0(x)$, as $t \to 0$. By Proposition 5.3, the second summand is majorized by

$$Ct^{-\alpha} \int_{\mathbb{R}^n} |x-\xi|^{-n+2+\gamma} \exp\left\{-\sigma \left(t^{-\alpha/2}|x-\xi|\right)^{\frac{2}{2-\alpha}}\right\} d\xi$$
$$= Ct^{\gamma\alpha} \int_{\mathbb{R}^n} |\eta|^{-n+2+\gamma} \exp\left\{-\sigma|\eta|^{\frac{2}{2-\alpha}}\right\} d\eta \longrightarrow 0,$$

as $t \to 0$.

Let us find an estimate for V_Z. Suppose, for example, that $n = 3$ (other cases are treated similarly). Then by (5.2.36) and (5.2.63)

$$|V_Z(t,x;\xi)|$$
$$\leq C\int_0^t (t-\lambda)^{-\frac{\alpha}{2}-1}\lambda^{-\alpha}\, d\lambda \int_{\mathbb{R}^3} |y-\xi|^{-3+\gamma} \exp\{-\sigma[\rho(t,x;\lambda,y)+\rho(\lambda,y;0,\xi)]\}\, dy.$$

Taking $0 < \gamma_0 < \gamma$ we can rewrite this as

$$|V_Z(t,x;\xi)| \leq C\int_0^t (t-\lambda)^{-\frac{\gamma_0\alpha}{2}-1}\lambda^{-\alpha}\, d\lambda$$
$$\times \int_{\mathbb{R}^3} \left(\frac{|x-y|}{(t-\lambda)^{\alpha/2}}\right)^{1+\gamma_0} |x-y|^{-1-\gamma_0}|y-\xi|^{-3+\gamma}$$
$$\times \exp\{-\sigma[\rho(t,x;\lambda,y)+\rho(\lambda,y;0,\xi)]\}\, dy$$
$$\leq C_1\int_0^t (t-\lambda)^{-\frac{\gamma_0\alpha}{2}-1}\lambda^{-\alpha}\, d\lambda \int_{\mathbb{R}^3} |x-y|^{-1-\gamma_0}|y-\xi|^{-3+\gamma}$$
$$\times \exp\{-\sigma_1[\rho(t,x;\lambda,y)+\rho(\lambda,y;0,\xi)]\}\, dy,$$

and now Lemma 1.12 implies the estimate (iii) from Theorem 5.2.

Other estimates given in the formulation of Theorem 5.2 are proved in a quite similar way (note that we roughen some estimates containing logarithmic terms replacing the logarithm by the power function with an arbitrarily small exponent). Returning now to the verification of the initial condition we obtain that

$$\left|\int_{\mathbb{R}^3} V_Z(t,x;\xi)u_0(\xi)\, d\xi\right| \leq Ct^{\frac{\gamma_0\alpha}{2}-\alpha} \int_{\mathbb{R}^3} |\xi|^{\gamma-\gamma_0-1} \exp\left\{-\sigma\left(t^{-\alpha/2}|\xi|\right)^{\frac{2}{2-\alpha}}\right\} d\xi$$
$$= Ct^{\frac{\gamma\alpha}{2}} \int_{\mathbb{R}^3} |\eta|^{\gamma-\gamma_0-1} \exp\left\{-\sigma|\eta|^{\frac{2}{2-\alpha}}\right\} d\eta \longrightarrow 0,$$

as $t \to 0$. Thus, the initial condition (5.2.2) has been verified.

5.3.6. The Hölder continuity in t. Let us prove the assertion b) of Theorem 5.2. Suppose that $0 < \nu < 1-\alpha$, $\alpha+\lambda < \nu$. We have to show that for any fixed $x \in \mathbb{R}^n$ $t^\nu u(t,x)$ is Hölder continuous in t with the exponent $\alpha+\lambda$. It is sufficient to prove the Hölder continuity near the origin $t=0$. Let us consider, for example, the case $n=3$; all other cases are treated similarly.

Let us write $u(t,x) = u_1(t,x) + u_2(t,x) + u_3(t,x) + u_4(t,x)$ where

$$u_1(t,x) = \int_{\mathbb{R}^3} Z_0(t, x-\xi; \xi) u_0(\xi)\, d\xi,$$

$$u_2(t,x) = \int_{\mathbb{R}^3} V_Z(t, x; \xi) u_0(\xi)\, d\xi,$$

$$u_3(t,x) = \int_0^t d\lambda \int_{\mathbb{R}^3} Y_0(t-\lambda, x-\xi; \xi) f(\lambda, \xi)\, d\xi,$$

$$u_4(t,x) = \int_0^t d\lambda \int_{\mathbb{R}^3} V_Y(t-\lambda, x; \xi) f(\lambda, \xi)\, d\xi.$$

Let $w(t,x) = t^\nu u_1(t,x)$. Below we prove that $D_{0+}^{\alpha+\lambda} w \in C[0,T]$ for any fixed x, if $\nu > \alpha+\lambda$. It is also clear that $\left(I_{0+}^{\alpha+\lambda} w\right)(0,x) = 0$. This will imply (see Section 5.1.1) that $w = I_{0+}^{\alpha+\lambda} D_{0+}^{\alpha+\lambda} w \in H_0^{\alpha+\lambda}[0,T]$, that is $u_1 \in H_\nu^{\alpha+\lambda}[0,T]$.

Denote

$$\mathfrak{A}(x,\xi;\zeta) = \sum_{i,j=1}^{3} A^{(ij)}(\zeta)(x_i - \xi_i)(x_j - \xi_j),$$

where, as before, $A = \left(A^{(ij)}(\zeta)\right)$ is the matrix inverse to $(a_{ij}(\zeta))$. By (5.2.28),

$$Z_0(t, x-\xi; \xi) = \frac{\pi^{-3/2}}{(\det A)^{1/2}} \left[\mathfrak{A}(x,\xi;\xi)\right]^{-3/2} H_{12}^{20}\left[\frac{1}{4} t^{-\alpha} \mathfrak{A}(x,\xi;\xi) \,\middle|\, \begin{matrix} (1,\alpha) \\ (\frac{3}{2},1), \quad (1,1) \end{matrix}\right].$$

Using the representation (A.10) of the H-function as a contour integral and applying the Riemann-Liouville derivative (as in (A.11) or in (A.12)) we find that

$$\left(D_{0+}^{\alpha+\lambda} w\right)(t,x) = \int_{\mathbb{R}^n} \Xi(t,x;\xi) u_0(\xi)\, d\xi = \int_{\mathbb{R}^n} \Xi(t,x;\xi)[u_0(\xi) - u_0(x)]\, d\xi,$$

where

$$\Xi(t, x-\xi; \xi) = \frac{\pi^{-3/2}}{(\det A)^{1/2}} \left[\mathfrak{A}(x,\xi;\xi)\right]^{-3/2} t^{\nu-\alpha-\lambda}$$
$$\times H_{23}^{30}\left[\frac{1}{4} t^{-\alpha} \mathfrak{A}(x,\xi;\xi) \,\middle|\, \begin{matrix} (1,\alpha), & (1+\nu-\alpha-\lambda, \alpha) & \\ (\frac{3}{2},1), & (1,1), & (1+\nu,\alpha) \end{matrix}\right].$$

By (A.14) and (A.26), we have

$$|\Xi(t,x-\xi;\xi)| \le Ct^{\nu-\alpha-\lambda-\frac{3\alpha}{2}} \exp\left\{-\mu_\alpha |x-\xi|^{\frac{2}{2-\alpha}}\right\}, \qquad \text{if } t^{-\alpha}|x-\xi|^2 \ge 1,$$

and

$$|\Xi(t,x-\xi;\xi)| \le Ct^{\nu-2\alpha-\lambda}|x-\xi|^{-1}, \qquad \text{if } t^{-\alpha}|x-\xi|^2 < 1.$$

These inequalities show, after a simple evaluation of integrals, that $D_{0+}^{\alpha+\lambda} w \in C[0,T]$ for each x. As we already know, this implies the Hölder continuity of u_1.

As we saw above, $|u_2(t,x)| \le Ct^{\alpha\gamma/2}$, so that $|t^\nu u_2(t,x)| \le Ct^{\frac{\alpha\gamma}{2}+\nu} < Ct^{\alpha+\lambda}$ for small values of t, since $\alpha+\lambda<\nu$.

Next, by (5.2.36)

$$\begin{aligned}|u_3(t,x)| &\le C\int_0^t (t-\lambda)^{-\frac{\alpha}{2}-1}\,d\lambda \int_{\mathbb{R}^3} \exp\left\{-\sigma\left(\frac{|x-\xi|}{(t-\lambda)^{\alpha/2}}\right)^{\frac{2}{2-\alpha}}\right\} d\xi \\ &\le C_1\int_0^t (t-\lambda)^{\alpha-1}\,d\lambda = C_2 t^\alpha,\end{aligned}$$

so that $|t^\nu u_3(t,x)| \le C_2 t^{\alpha+\nu} < C_2 t^{\alpha+\lambda}$ for small values of t.

Finally, from the estimate (iii) of Theorem 5.2 we find that

$$|t^\nu u_4(t,x)| \le Ct^{\frac{\alpha\gamma}{2}+\alpha+\nu} < Ct^{\alpha+\lambda}$$

for small values of t, as desired.

5.3.7. Nonnegativity. Let us prove that the functions Z and Y are nonnegative. It is sufficient to show the nonnegativity of the solution (5.2.4) of the problem (5.2.1), (5.2.2) for arbitrary nonnegative functions $u_0 \in C_0^\infty(\mathbb{R}^n)$, $f \in C_0^\infty([0,T]\times\mathbb{R}^n)$.

As we already know, the solution $u(t,x)$ is bounded (due to our estimates of Z and Y), say $|u(t,x)| \le M$, and belongs to $H_\nu^{\alpha+\lambda}[0,T]$, $0<\lambda<1-\alpha$, $\alpha+\lambda<\nu$, for each x. This regularity in t makes it possible to represent the regularized fractional derivative $\mathbb{D}_t^{(\alpha)} u$ via the Marchaud fractional derivative:

$$\left(\mathbb{D}_t^{(\alpha)} u\right)(t,x) = \frac{1}{\Gamma(1-\alpha)}\left[t^{-\alpha}u(t,x) - t^{-\alpha}u_0(x)\right] + \frac{\alpha}{\Gamma(1-\alpha)}\lim_{\varepsilon\to+0}\psi_\varepsilon(t,x)$$

where $\psi_\varepsilon(t,x)=0$ for $0<t\le\varepsilon$,

$$\psi_\varepsilon(t,x) = \int_0^{t-\varepsilon} \frac{u(t,x)-u(\tau,x)}{(t-\tau)^{1+\alpha}}\,d\tau, \quad \varepsilon<t\le T,$$

and the limit exists for all $(t,x)\in\Pi_{[0,T]}$ (see Section 5.1.1).

Denote $v(t,x) = \dfrac{u(t,x)}{E_\alpha(\beta t^\alpha)}$ where E_α is the Mittag-Leffler function, $\beta > 0$ will be chosen later. Using the fact that $\mathbb{D}^{(\alpha)} E_\alpha(\beta t^\alpha) = \beta E_\alpha(\beta t^\alpha)$ we obtain that

$$\mathbb{D}^{(\alpha)} u(t,x) = \beta E_\alpha(\beta t^\alpha) v(t,x) - \frac{t^{-\alpha}}{\Gamma(1-\alpha)} u_0(x) + (L_\beta v)(t,x)$$

where

$$(L_\beta v)(t,x) = \frac{1}{\Gamma(1-\alpha)} \left[v(t,x) t^{-\alpha} + \alpha \lim_{\varepsilon \to +0} \int_0^{t-\varepsilon} E_\alpha(\beta \tau^\alpha) \frac{v(t,x) - v(\tau,x)}{(t-\tau)^{1+\alpha}} \, d\tau \right].$$

Now the equation (5.2.1) takes the form

$$\frac{1}{E_\alpha(\beta t^\alpha)} (L_\beta v)(t,x) - (B - \beta) v(t,x) = g(t,x) \tag{5.3.18}$$

where

$$g(t,x) = \frac{1}{E_\alpha(\beta t^\alpha)} \left[\frac{t^{-\alpha}}{\Gamma(1-\alpha)} u_0(x) + f(t,x) \right].$$

Let $c_0 = \sup\limits_{x \in \mathbb{R}^n} |c(x)|$, $d > 0$, $\beta = c_0 + d$. Consider the function

$$F_R(t,x) = \frac{M}{R^2} \left(|x|^2 + \mu t^\alpha + 1 \right), \quad R, \mu > 0.$$

Since $F_R(t,x)$ and $E_\alpha(\beta t^\alpha)$ are monotone increasing in t, we have

$$(L_\beta F_R)(t,x) \geq \left(D_{0+}^\alpha F_R \right)(t,x) \geq \frac{M\mu}{R^2} D_{0+}^\alpha (t^\alpha) = \frac{M\mu \Gamma(1+\alpha)}{R^2}$$

due to (5.1.2). In estimating $L_\beta F_R$ we used the Marchaud form of D_{0+}^α.

On the other hand,

$$\begin{aligned}(B-\beta)F_R(t,x) &= \sum_{i=1}^n a_{ii}(x) \frac{2M}{R^2} + \sum_{j=1}^n b_j(x) \frac{2M}{R^2} x_j + [c(x) - c_0] F_R(t,x) - dF_R(t,x) \\ &\leq \frac{M}{R^2} \left(C_1 + C_2|x| - d|x|^2 - d\mu t^\alpha - d \right),\end{aligned}$$

so that

$$\begin{aligned}&\frac{1}{E_\alpha(\beta t^\alpha)} (L_\beta F_R)(t,x) - (B - \beta) F_R(t,x) \\ &\quad \geq \frac{M}{R^2} \left[\frac{\mu \Gamma(1+\alpha)}{E_\alpha(\beta t^\alpha)} - C_1 - C_2|x| + d|x|^2 + d\mu t^\alpha + d \right] \geq 0 \end{aligned} \tag{5.3.19}$$

for all $(t,x) \in \Pi_{[0,T]}$, if the number μ is large enough.

Let $G(t,x) = v(t,x) + F_R(t,x)$. Since $g(t,x) \geq 0$, it follows from (5.3.18) and (5.3.19) that

$$\frac{1}{E_\alpha(\beta t^\alpha)} \left(L_\beta G\right)(t,x) - (B-\beta)G(t,x) \geq 0. \tag{5.3.20}$$

If $|x| = R$, then

$$G(t,x) = v(t,x) + M + \frac{\mu M t^\alpha}{R^2} + \frac{M}{R^2} > 0.$$

Also $G(0,x) = u_0(x) + F_R(0,x) > 0$. Then $G(t,x) \geq 0$ for all (t,x) with $t \in [0,T]$, $|x| < R$. Indeed, otherwise the function G possesses a point (t^0, x^0) of its global minimum on the set $\{0 < t \leq T, |x| < R\}$, such that $G(t^0,x^0) < 0$. At this point $(B-\beta)G(t^0,x^0) \geq 0$ [83], and by (5.3.20) $\left(L_\beta G\right)(t^0,x^0) \geq 0$, which contradicts the definition of L_β.

Thus, if $|x| \leq R$, then

$$u(t,x) \geq -\frac{M}{R^2}\left(|x|^2 + \mu t^\alpha + 1\right) E_\alpha(\beta t^\alpha).$$

Since R is arbitrary, this means that $u(t,x) \geq 0$.

This completes the proof of Theorem 5.2. □

5.4 Uniqueness theorems

5.4.1. Bounded solutions. The proof of non-negativity of the components of the Green matrix (see the assertion c) of Theorem 5.2 and its proof in Section 5.3.7) actually did not use the representation (5.2.4), but only some properties of the resulting solution. If those properties are taken as a priori assumptions, we come to the uniqueness theorem for bounded solutions, which is actually valid in a more general situation. In particular, the uniform parabolicity condition (5.2.3) may be weakened to parabolicity:

$$\sum_{j,l=1}^{n} a_{jl}\xi_j\xi_l \geq 0;$$

it is sufficient to assume continuity and boundedness of coefficients of the operator B (the Hölder continuity is not required). Moreover, the coefficients may depend also on t.

Theorem 5.3. *Let $u(t,x)$ be a classical solution of the problem* (5.2.1)–(5.2.2) *with $u_0(x) \equiv 0$. Suppose that $|u(t,x)| \leq M$ for all $t \in [0,T]$, $x \in \mathbb{R}^n$, and that for each $x \in \mathbb{R}^n$ $u(\cdot,x) \in H_\mu^{\alpha+\lambda}[0,T]$ where $0 < \lambda < 1-\alpha$, $0 \leq \mu < \lambda + 1$. Then u equals zero identically.*

Proof. As in Section 5.3.7, we prove that $u(t,x) \geq 0$ for all t,x. Then, considering the solution $-u(t,x)$, we find that $u(t,x) \leq 0$, so that $u(t,x) \equiv 0$. □

5.4.2. Solutions of exponential growth. Let $n = 1$.

Theorem 5.4. *Let the conditions of Theorem 5.3 be satisfied, except that the solution $u(t,x)$ is not necessarily bounded, but*

$$|u(t,x)| \le M_1 \exp\left\{M_2|x|^{2/(2-\alpha)}\right\}, \quad M_1, M_2 > 0. \tag{5.4.1}$$

Then $u(t,x) \equiv 0$.

Proof. We follow again the scheme of Section 5.3.7, but instead of the function $F_R(t,x)$ use the function

$$F_R^{(1)}(t,x) = M_1 \exp\left\{M_2 R^{2/(2-\alpha)}\right\}[h(t,x-R) + h(t,x+R)], \quad -R < x < R,$$

where $h(t,x) = \exp\left\{-\rho t^{-\alpha/(2-\alpha)}|x|^{2/(2-\alpha)}\right\}$, while the constant $\rho \in (0,1)$ will be determined later.

As in Section 5.3.7, we reduce the equation (5.2.1) to the equation (5.3.18) with $g = 0$. This time we set $\beta = c_0$. Let us prove that for $|x| < R$, $0 < t \le T$,

$$\frac{1}{E_\alpha(\beta t^\alpha)}\left(L_\beta F_R^{(1)}\right)(t,x) - (B-\beta)F_R^{(1)}(t,x) \ge 0. \tag{5.4.2}$$

It is sufficient to show that

$$\frac{1}{E_\alpha(\beta t^\alpha)}\left(L_\beta h\right)(t,x) - (B-\beta)h(t,x) \ge 0, \tag{5.4.3}$$

if $x \ne 0$, $0 < t \le T$.

Evidently, if $x \ne 0$, then $h(0,x) = 0$, so that $\left(\mathbb{D}_t^{(\alpha)} h\right)(t,x) = \left(D_{0+}^\alpha h\right)(t,x)$. Next,

$$\partial_t h(t,x) = \frac{\alpha\rho}{2-\alpha} t^{-2/(2-\alpha)}|x|^{2/(2-\alpha)} \exp\left\{-\rho t^{-\alpha/(2-\alpha)}|x|^{2/(2-\alpha)}\right\} > 0,$$

so that, by (5.1.1) and (5.1.3),

$$\frac{1}{E_\alpha(\beta t^\alpha)}(L_\beta h)(t,x) \ge \frac{1}{E_\alpha(\beta T^\alpha)}\left(\mathbb{D}_t^{(\alpha)} h\right)(t,x)$$

$$= \frac{\alpha\rho|x|^{2/(2-\alpha)}}{E_\alpha(\beta T^\alpha)(2-\alpha)\Gamma(1-\alpha)} \int_0^t (t-\tau)^{-\alpha}\tau^{-2/(2-\alpha)} \exp\left\{-\rho\tau^{-\alpha/(2-\alpha)}|x|^{2/(2-\alpha)}\right\} d\tau.$$

After the change of variables $\tau = \rho^{(2-\alpha)/\alpha}|x|^{2/\alpha}\sigma$ we find that

$$\frac{1}{E_\alpha(\beta t^\alpha)}(L_\beta h)(t,x) \ge \frac{\alpha|x|^{-2}\rho^{\alpha-2}}{E_\alpha(\beta T^\alpha)(2-\alpha)} S\left(t\rho^{-(2-\alpha)/\alpha}|x|^{-2/\alpha}\right) \tag{5.4.4}$$

where

$$S(z)=\frac{1}{\Gamma(1-\alpha)}\int\limits_0^z (z-\sigma)^{-\alpha}\sigma^{-2/(2-\alpha)}\exp\left\{-\sigma^{-\alpha/(2-\alpha)}\right\}d\sigma.$$

According to Lemma 5.1, $S(z)\sim \nu z^{-\alpha}$, as $z\to\infty$, where

$$\nu=\frac{1}{\Gamma(1-\alpha)}\int\limits_0^\infty \sigma^{-2/(2-\alpha)}\exp\left\{-\sigma^{-\alpha/(2-\alpha)}\right\}d\sigma.$$

This means, in particular, that for $z\geq 1$,

$$S(z)\geq \nu_1 z^{-\alpha},\quad \nu_1>0. \tag{5.4.5}$$

Let $0<z\leq 1$. Then

$$\begin{aligned}S(z)&=\frac{1}{\Gamma(1-\alpha)}z^{1-\alpha-2/(2-\alpha)}\int\limits_0^1 (1-\eta)^{-\alpha}\eta^{-2/(2-\alpha)}\exp\left\{-(z\eta)^{-\alpha/(2-\alpha)}\right\}d\eta\\ &\geq \frac{1}{\Gamma(1-\alpha)}z^{1-\alpha-2/(2-\alpha)}\int\limits_0^1 (1-\eta)^{-\alpha}\exp\left\{-(z\eta)^{-\alpha/(2-\alpha)}\right\}d\eta.\end{aligned}$$

Now it follows from Lemma 5.2 that

$$S(z)\geq \nu_2 z^{-2\alpha/(2-\alpha)}\exp\left\{-z^{-\alpha/(2-\alpha)}\right\},\quad \nu_2>0, \tag{5.4.6}$$

for $z\leq 1$.

The inequalities (5.4.4)–(5.4.6) imply the lower bound

$$\frac{1}{E_\alpha(\beta t^\alpha)}\left(L_\beta h\right)(t,x)\geq \begin{cases}\nu_3 t^{-\alpha},\\ \text{if } t\rho^{-(2-\alpha)/\alpha}|x|^{-2/\alpha}\geq 1,\\[2ex] \nu_4\rho^\alpha |x|^{2\alpha/(2-\alpha)}t^{-2\alpha/(2-\alpha)}\exp\left\{-\rho|x|^{2/(2-\alpha)}t^{-\alpha/(2-\alpha)}\right\},\\ \text{if } t\rho^{-(2-\alpha)/\alpha}|x|^{-2/\alpha}\leq 1,\end{cases} \tag{5.4.7}$$

where $\nu_3,\nu_4>0$ do not depend on t,x,ρ.

Calculating $(B-\beta)h(t,x)$ we come (after tedious but quite elementary computations) to the upper bound

$$(B-\beta)h(t,x) \le \begin{cases} \nu_5\rho^{1-\frac{\alpha}{2}}t^{-\alpha}, \\ \text{if } t\rho^{-(2-\alpha)/\alpha}|x|^{-2/\alpha} \ge 1, \\ \\ \nu_6\rho^{3\alpha/2}|x|^{2\alpha/(2-\alpha)}t^{-2\alpha/(2-\alpha)}\exp\left\{-\rho|x|^{2/(2-\alpha)}t^{-\alpha/(2-\alpha)}\right\}, \\ \text{if } t\rho^{-(2-\alpha)/\alpha}|x|^{-2/\alpha} \le 1; \end{cases} \tag{5.4.8}$$

ν_5, ν_6 do not depend on t, x, ρ.

If ρ is taken small enough, the bounds (5.4.7) and (5.4.8) imply (5.4.3). Hence, we have proved the inequality (5.4.2).

If $x = \pm R$, then

$$\begin{aligned} F_R^{(1)}(t,x) &= M_1\exp\left\{M_2R^{2/(2-\alpha)}\right\}[1+h(t,\pm 2R)] \\ &\ge M_1\exp\left\{M_2R^{2/(2-\alpha)}\right\} \ge u(t,x) \ge v(t,x) \end{aligned}$$

(we follow the notation of Section 5.3.7) because $E_\alpha(\beta T^\alpha) \ge 1$.

If $-R < x < R$, then $F_R^{(1)}(0,x) = v(0,x) = 0$. Now the reasoning of Section 5.3.7 yields the inequality $v(t,x) \le F_R^{(1)}(t,x)$ for all $t \in [0,T]$, $x \in (-R,R)$.

Let us fix $t \in (0,\varepsilon]$, $x \in \mathbb{R}$, where

$$0 < \varepsilon < \min\left(\left(\frac{M_2}{\rho}\right)^{-(2-\alpha)/\alpha}\cdot 2^{-2/\alpha}, T\right).$$

If $R > 2|x|$, then

$$v(t,x) \le F_R^{(1)}(t,x) \le M_1\exp\left\{\left(M_2 - \rho\cdot 2^{2/(2-\alpha)}t^{-\alpha/(2-\alpha)}\right)R^{2/(2-\alpha)}\right\}.$$

Since R is arbitrary, this means that $v(t,x) \le 0$, so that $u(t,x) \le 0$. Similarly we get, again for any $t \in (0,\varepsilon]$, $x \in \mathbb{R}$, that $u(t,x) \ge 0$.

In order to consider larger values of t, we introduce the function

$$w(t,x) = u(t+\varepsilon,x), \quad t > 0, x \in \mathbb{R}.$$

We have $w(0,x) = 0$,

$$\begin{aligned} \left(\mathbb{D}_t^{(\alpha)}w\right)(t,x) = \left(D_{0+}^\alpha w\right)(t,x) &= \frac{1}{\Gamma(1-\alpha)}\int_\varepsilon^{t+\varepsilon}(t+\varepsilon-\sigma)^{-\alpha}\partial_\sigma u(\sigma,x)\,d\sigma \\ &= \frac{1}{\Gamma(1-\alpha)}\int_0^{t+\varepsilon}(t+\varepsilon-\sigma)^{-\alpha}\partial_\sigma u(\sigma,x)\,d\sigma \\ &= \left(\mathbb{D}_t^{(\alpha)}u\right)(t+\varepsilon,x) = Bu(t+\varepsilon,x) = Bw(t,x), \end{aligned}$$

so that the function w also satisfies the equation (5.2.1). By the above result, $w(t,x) = 0$ as $t \le \varepsilon$. Thus, $u(t,x) = 0$ for $0 \le t \le \min(2\varepsilon, T)$. Repeating this argument, after a finite number of steps we complete the proof. □

5.4.3. A counter-example. Let us show that Theorem 5.4 cannot be improved in the sense that the exponent $2/(2-\alpha)$ cannot be increased: if $p > 2/(2-\alpha)$, then there exists a non-trivial solution $u(t,x)$ of the problem (5.2.1)–(5.2.2) with $B = \partial_x^2$, $u_0 = 0$, such that

$$|u(t,x)| \le C_1 \exp\{C_2|x|^p\} \tag{5.4.9}$$

for all $t \in [0,T]$, $x \in \mathbb{R}$.

Indeed, let $\beta = 2 - \dfrac{2}{p}$. Then $\beta > \alpha$. By Theorem 5.1 there exists such a smooth function $f(t)$, different from the identical zero, that the sequence $f_0(t) = f(t)$, $f_{n+1}(t) = (\mathbb{D}^{(\alpha)} f_n)(t)$, $n = 0,1,2,\ldots$, satisfies the inequalities (5.1.11), and $f_n(0) = 0$ for all n. Let

$$u(t,x) = \sum_{n=0}^{\infty} \frac{f_n(t)}{(2n)!} x^{2n}. \tag{5.4.10}$$

A direct calculation shows that $u(t,x)$ is a solution of the above Cauchy problem. With respect to the variable x, $u(t,x)$ is an entire function, and its exponential growth order q satisfies the inequality

$$q \le 2 \limsup_{n\to\infty} \frac{n \log n}{\log \frac{(2n)!}{|f_n(t)|}}$$

(see Chapter 8 in [221]). Using the Stirling formula and the inequalities (5.1.11) we find that $q \le \dfrac{2}{2-\beta} = p$, which implies the required estimate (5.4.9).

Note that the conditions of Theorem 5.3 are similar to those of the classical uniqueness theorems for second order parabolic differential equations [83]. Classically [83], an example like (5.4.10) is constructed on the basis of the theory of conventional quasi-analytic classes.

5.5 Comments

The expression (5.2.5) for the FSCP Z_0 was found independently by Schneider and Wyss [206] and Kochubei [130]. The non-negativity of Z_0 was first noticed in [206], though the proofs of the uniqueness Theorems 5.3 and 5.4 given in [130] actually contained a proof of the non-negativity. Theorem 5.2 was proved by Eidelman and Kochubei [63].

The results of this chapter demonstrate a number of interesting features of the equation (5.2.1), which represent a peculiar union of properties typical for second order parabolic differential equations (like a kind of the maximum principle; see the proofs of Theorems 5.3, 5.4) and general parabolic equations and systems. It is well known [45, 83, 146] that the latter are characterized by the parabolic weight $2b$ ($b \geq 1$), so that the differentiation in t has the same "force" as the differentiation of the order $2b$ with respect to the spatial variables. In other words, the differentiation in spatial variables has the weight $1/2b$ with respect to ∂_t. For the equation (5.2.1) we have $\alpha/2$ instead of $1/2b$, and this formal analogy goes through all the constructions and estimates including the description of uniqueness and correctness classes.

In contrast to classical parabolic equations, the fundamental solution Z of the Cauchy problem (5.2.1)–(5.2.2) (even for $B = \Delta$) has, if $n > 1$, a singularity not only at $t = 0$, but also at the "diagonal" $x = \xi$ with respect to the spatial variables. This causes serious complications for the implementation of the Levi method. The estimates of the iterated kernels require a series of additional regularization procedures absent in the classical case. Moreover, while the classical heat kernel has a simple expression in elementary functions, in the fractional case we have to deal with Fox's H-functions and their asymptotics, which makes even simple (in principle) transformations and estimates difficult technical tasks. Note also that, in order to consider an inhomogeneous equation (5.2.1), we have to develop Levi's method twice – separately for Z and Y.

In this book we do not touch boundary value problems for fractional diffusion equations. So far mainly model equations have been considered (mostly in the physical literature). In fact there exists a general approach giving some results both for the Cauchy problem and boundary value problems. That is the study of operator-differential equations

$$\mathbb{D}^{(\alpha)}u = Bu, \quad u(0) = u_0, \tag{5.5.1}$$

where u is a function on $[0, T]$ with values from a Banach space X, B is a closed linear operator on X. The study of such equations was initiated by Kochubei [129].

In [129] a solution of the problem (5.5.1) is understood as a strongly continuous function on $[0, T]$, such that $u(t)$ belongs to the domain $D(B)$ of the operator B, if $t > 0$, the fractional integral $I_{0+}^{1-\alpha}u$ is strongly continuously differentiable for $t > 0$, and u satisfies the equation and the initial condition. This setting resembles the weakened abstract Cauchy problem for first order evolution equations [141].

Let $R(\lambda) = (B - \lambda I)^{-1}$ be the resolvent of the operator B. It is proved in [129] that the problem (5.5.1) with $u_0 \in D(B)$ has a unique solution if the operator-function $R(\lambda^\alpha)$ exists for $\operatorname{Re}\lambda > \omega$ ($\omega > 0$), and

$$\|R(\lambda^\alpha)\| \leq C(1 + |\operatorname{Im}\lambda|)^{-\beta}, \quad 0 < \beta < 1. \tag{5.5.2}$$

The uniqueness of a solution is proved under much weaker conditions.

If B is an operator on some function space corresponding to a boundary value problem for a uniformly elliptic equation, or a uniformly elliptic operator over $\mathbb{R}^n$, then the resolvent $R(\lambda)$ has been thoroughly investigated, the condition (5.5.2) can be easily verified, and we obtain unique solvability results for fractional diffusion equations. Of course, the analytic methods of Chapter 5, where they are applicable, give much more detailed information about solutions.

The investigation of the abstract Cauchy problem (5.5.1) was continued by Bazhlekova [12] (who developed a version of the theory of the uniformly correct Cauchy problem), El-Sayed [79], Bauemer and Meerschaert [11], and others. The abstract Cauchy problem with the Riemann-Liouville fractional derivative (and an appropriately modified initial condition) was studied by Kostin [140].

Another interesting direction is the investigation of evolution equations with the fractional derivative of an order $\alpha \in (1, 2)$. The first results [206, 84] show that in this case solutions possess an interesting combination of properties typical for parabolic and hyperbolic equations. Note also some recent papers on space-time fractional diffusion equations containing a fractional derivative in t and pseudo-differential operators with respect to spatial variables; see [153, 169].

Appendix
Fox's H-Functions

The H-functions introduced by Fox [81] provide a very useful framework for investigating asymptotic properties of solutions of various equations. In many cases a solution can be expressed in terms of H-functions, and then its asymptotic behaviour can be found from general results about H-functions.

Let μ, ν, p, q be integers satisfying the conditions $0 \le \nu \le p$, $1 \le \mu \le q$. Suppose we have also the complex parameters $c_1, \ldots, c_p$ and $d_1, \ldots, d_q$, and positive real parameters $\gamma_1, \ldots, \gamma_p$ and $\delta_1, \ldots, \delta_q$, such that $P_1 \cap P_2 = \varnothing$ where

$$P_1 = \{s = -(d_j + k)/\delta_j,\ j = 1, \ldots, \mu;\ k = 0, 1, 2, \ldots\},$$
$$P_2 = \{s = (1 - c_j + k)/\gamma_j,\ j = 1, \ldots, \nu;\ k = 0, 1, 2, \ldots\}.$$

We will need only the case where

$$\rho := \sum_{k=1}^{q} \delta_k - \sum_{k=1}^{p} \gamma_k > 0. \tag{A.1}$$

The H-function

$$H_{pq}^{\mu\nu}(z)\left[z\left|\begin{matrix}[c_p,\gamma_p]\\ [d_q,\delta_q]\end{matrix}\right.\right] = H_{pq}^{\mu\nu}\left[z\left|\begin{matrix}(c_1,\gamma_1), & \ldots, & (c_\nu,\gamma_\nu); & (c_{\nu+1},\gamma_{\nu+1}), & (c_p,\gamma_p)\\ (d_1,\delta_1), & \ldots, & (d_\mu,\delta_\mu); & (d_{\mu+1},\delta_{\mu+1}), & (d_q,\delta_q)\end{matrix}\right.\right],$$
$$[c_p,\gamma_p] = (c_1,\gamma_1),\ldots,(c_p,\gamma_p); \quad [d_q,\delta_q] = (d_1,\delta_1),\ldots,(d_q,\delta_q),$$

is defined by the contour integral

$$H_{pq}^{\mu\nu}(z)\left[z\left|\begin{matrix}[c_p,\gamma_p]\\ [d_q,\delta_q]\end{matrix}\right.\right] = \frac{1}{2\pi i}\int_L \frac{C(s)D(s)}{E(s)F(s)} z^{-s}\, ds \tag{A.2}$$

with

$$C(s) = \prod_{k=1}^{\mu} \Gamma(d_k + \delta_k s), \qquad D(s) = \prod_{k=1}^{\nu} \Gamma(1 - c_k - \gamma_k s),$$
$$E(s) = \prod_{k=\mu+1}^{q} \Gamma(1 - d_k - \delta_k s), \qquad F(s) = \prod_{k=\nu+1}^{p} \Gamma(c_k + \gamma_k s)$$

(P_1 and P_2 are the sets of poles for $C(s)$ and $D(s)$ respectively). The integration contour L is an infinite loop running between $s = -\infty - i\sigma$ and $s = -\infty + i\sigma$ where $\sigma > \max\limits_{1\le j\le\mu} \{|\operatorname{Im} d_j|/\delta_j\}$ in such a way that P_1 lies to the left of L, and P_2 to the right of L.

Under the above assumptions the function $H^{\mu\nu}_{pq}(z)\left[z\left|\begin{matrix}[c_p,\gamma_p]\\ [d_q,\delta_q]\end{matrix}\right.\right]$ is holomorphic in a certain sector containing the positive real half-axis.

Nearly all classical special functions can be represented as H-functions with appropriate parameters. In particular, Meijer's G-function corresponds to the case where all the numbers δ_k and γ_k are equal to 1; this case includes all the elementary functions, the Bessel functions, various functions of hypergeometric type, classical orthogonal polynomials, and others. The Mittag-Leffler function

$$E_\alpha(z) = \sum_{k=0}^{\infty} \frac{z^k}{\Gamma(\alpha k + 1)}, \quad \alpha > 0,$$

on the negative real half-axis can be written as

$$E_\alpha(-x) = H^{11}_{12}\left[x\left|\begin{matrix}(0,1) & \\ (0,1), & (0,\alpha)\end{matrix}\right.\right]. \tag{A.3}$$

The theory of the H-function including its analytic properties, asymptotics, various relations, is expounded in [19, 195, 217]. Below we formulate all the results we need in this book.

a) *Transformation formulas.*

$$H^{\mu\nu}_{pq}(z)\left[z\left|\begin{matrix}[c_p,\gamma_p] & \\ [d_{q-1},\delta_{q-1}], & (c_1,\gamma_1)\end{matrix}\right.\right] = H^{\mu,\nu-1}_{p-1,q-1}(z)\left[z\left|\begin{matrix}(c_2,\gamma_2), & \dots, & (c_p,\gamma_p)\\ [d_{q-1},\delta_{q-1}] & & \end{matrix}\right.\right]. \tag{A.4}$$

$$H^{\mu\nu}_{pq}(z)\left[z\left|\begin{matrix}[c_p,\gamma_p]\\ [d_q,\delta_q]\end{matrix}\right.\right] = H^{\nu\mu}_{qp}(z)\left[z^{-1}\left|\begin{matrix}[1-d_q,\delta_q]\\ [1-c_p,\gamma_p]\end{matrix}\right.\right]. \tag{A.5}$$

$$H^{\mu\nu}_{pq}(z)\left[z\left|\begin{matrix}(c,\gamma), & [c_{p-1},\gamma_{p-1}]\\ [d_{q-1},\delta_{q-1}], & (c\pm l,\gamma)\end{matrix}\right.\right] = (-1)^l H^{\mu+1,\nu-1}_{pq}(z)\left[z\left|\begin{matrix}[c_{p-1},\gamma_{p-1}], & (c,\gamma)\\ (c\pm l,\gamma), & [d_{q-1},\delta_{q-1}]\end{matrix}\right.\right] \tag{A.6}$$

b) *Differentiation and integration formulas.* We write them explicitly for some cases we need.

$$\frac{d}{dz} H^{20}_{12}\left[z\left|\begin{matrix}(c_1,\gamma_1) & \\ (d_1,\delta_1), & (d_2,\delta_2)\end{matrix}\right.\right] = -z^{-1} H^{30}_{23}\left[z\left|\begin{matrix}(c_1,\gamma_1), & (0,1) & \\ (d_1,\delta_1), & (d_2,\delta_2), & (1,1)\end{matrix}\right.\right]. \tag{A.7}$$

$$\frac{d}{dz} H_{23}^{30}\left[z \left| \begin{matrix} (c_1,\gamma_1), & (c_2,\gamma_2) & \\ (d_1,\delta_1), & (d_2,\delta_2), & (d_3,\delta_3) \end{matrix} \right.\right]$$

$$= -z^{-1} H_{34}^{40}\left[z \left| \begin{matrix} (c_1,\gamma_1), & (c_2,\gamma_2), & (0,1) & \\ (d_1,\delta_1), & (d_2,\delta_2), & (d_3,\gamma_3) & (1,1) \end{matrix} \right.\right]. \quad \text{(A.8)}$$

In a similar way $\frac{d}{dz} H_{34}^{40}$ is expressed via H_{45}^{50} etc.

By (A.2),

$$H_{12}^{20}\left[z \left| \begin{matrix} (1,\alpha) & \\ (\frac{n}{2},1), & (1,1) \end{matrix} \right.\right] = \frac{1}{2\pi i} \int_L \frac{\Gamma(\frac{n}{2}+s)\Gamma(1+s)}{\Gamma(1+\alpha s)} z^{-s}\, ds. \quad \text{(A.9)}$$

We need this function with $z = \omega(t-\tau)^{-\alpha}$, $\omega > 0$, $t > \tau$. We have

$$u(t) = H_{12}^{20}\left[\omega(t-\tau)^{-\alpha} \left| \begin{matrix} (1,\alpha) & \\ (\frac{n}{2},1), & (1,1) \end{matrix} \right.\right]$$

$$= \frac{1}{2\pi i} \int_L \frac{\Gamma(\frac{n}{2}+s)\Gamma(1+s)}{\Gamma(1+\alpha s)} \omega^{-s} (t-\tau)^{\alpha s}\, ds. \quad \text{(A.10)}$$

In order to calculate the Riemann-Liouville fractional derivative

$$\left(D_{\tau+}^{\alpha} u\right)(t) = \frac{1}{\Gamma(1-\alpha)} \partial_t \int_\tau^t (t-s)^{-\alpha} u(s)\, ds,$$

we note that $D_{\tau+}^{\alpha}$ transforms $(t-\tau)^{\alpha s}$ into $\frac{\Gamma(1+\alpha s)}{\Gamma(1-\alpha+\alpha s)}(t-\tau)^{\alpha s-\alpha}$ (see [202]). Now we find from (A.10) that

$$\left(D_{\tau+}^{\alpha} u\right)(t) = (t-\tau)^{-\alpha} H_{12}^{20}\left[\omega(t-\tau)^{-\alpha} \left| \begin{matrix} (1-\alpha,\alpha) & \\ (\frac{n}{2},1), & (1,1) \end{matrix} \right.\right]. \quad \text{(A.11)}$$

Similarly

$$\left(D_{\tau+}^{1-\alpha} u\right)(t) = (t-\tau)^{\alpha-1} H_{12}^{20}\left[\omega(t-\tau)^{-\alpha} \left| \begin{matrix} (\alpha,\alpha) & \\ (\frac{n}{2},1), & (1,1) \end{matrix} \right.\right]. \quad \text{(A.12)}$$

Next,

$$\int_0^t s^{-1} H_{12}^{20}\left[as \left| \begin{matrix} (\alpha,\alpha) & \\ (\frac{n}{2},1), & (1,1) \end{matrix} \right.\right] ds = H_{23}^{21}\left[at \left| \begin{matrix} (1,1), & (\alpha,\alpha) & \\ (\frac{n}{2},1), & (1,1), & (0,1) \end{matrix} \right.\right]. \quad \text{(A.13)}$$

c) *Asymptotics at infinity.* The asymptotic behavior of H-functions for $z \to \infty$ has been thoroughly investigated. We will need only two results of this kind, for specific classes of H-functions, for a real positive argument z, with only the leading term of the asymptotic expansion.

If $\nu = 0$ and $q = \mu$, then the H-function has an exponential decrease for $z \to +\infty$:

$$H_{p\mu}^{\mu 0}\left[z\left|\begin{matrix}(c_1,\gamma_1), & \dots, & (c_p,\gamma_p)\\ (d_1,\delta_1), & \dots, & (d_\mu,\delta_\mu)\end{matrix}\right.\right] \sim \text{const} \cdot z^{(1-a)/\rho} \exp\left\{-b^{1/\rho}\rho z^{1/\rho}\right\} \quad (A.14)$$

where

$$a = \sum_{k=1}^{p} c_k - \sum_{k=1}^{\mu} d_k + \frac{1}{2}(\mu - p + 1), \quad b = \prod_{k=1}^{p} \gamma_k^{\gamma_k} \prod_{l=1}^{\mu} \delta_l^{-\delta_l},$$

and ρ is given by (A.1).

In particular, from (A.14) we get estimates of some specific H-functions for $|z| \geq 1$:

$$\left|H_{12}^{20}\left[z\left|\begin{matrix}(1,\alpha) & \\ (\frac{n}{2},1), & (1,1)\end{matrix}\right.\right]\right| \leq C|z|^{\frac{n}{2(2-\alpha)}} \exp\left\{-\sigma|z|^{\frac{1}{2-\alpha}}\right\}; \quad (A.15)$$

$$\left|H_{12}^{20}\left[z\left|\begin{matrix}(\alpha,\alpha) & \\ (\frac{n}{2},1), & (1,1)\end{matrix}\right.\right]\right| \leq C|z|^{\frac{n+2-2\alpha}{2(2-\alpha)}} \exp\left\{-\sigma|z|^{\frac{1}{2-\alpha}}\right\}. \quad (A.16)$$

Similar estimates hold for functions appearing in expressions for derivatives of the above H-functions. In all cases the estimates contain the same exponentially decreasing factor while the degrees of positive powers of $|z|$ are different and specific for each case.

For another situation (used in Chapter 4), the H-function has a power-like decrease:

$$H_{22}^{11}\left[z\left|\begin{matrix}\left(1-\frac{1}{\gamma},\frac{1}{\gamma}\right), & \left(\frac{1}{2},\frac{1}{2}\right)\\ (0,1), & \left(\frac{1}{2},\frac{1}{2}\right)\end{matrix}\right.\right] \sim \frac{\gamma}{\pi}\Gamma(1+\gamma)\sin\frac{\pi\gamma}{2} \cdot z^{-1-\gamma}, \quad z \to +\infty, \quad (A.17)$$

where $1 < \gamma < 2$.

d) *Asymptotics near the origin.* In order to obtain, for the above class of H-functions, an asymptotic expansion near the origin, we write an H-function as

$$H_{p\mu}^{\mu 0}\left[z\left|\begin{matrix}(c_1,\gamma_1), & \dots, & (c_p,\gamma_p)\\ (d_1,\delta_1), & \dots, & (d_\mu,\delta_\mu)\end{matrix}\right.\right] = \sum_{s\in P_1} \operatorname{Res}\left(\frac{C(s)}{F(s)}z^{-s}\right).$$

In specific cases, writing P_1 explicitly and using well-known properties of the function Gamma, we find a required number of terms in asymptotic expansions.

In particular, for $z \to +0$,

$$H_{12}^{20}\left[z\,\middle|\,\begin{matrix}(1,\alpha)\\ (\frac{n}{2},1), \quad (1,1)\end{matrix}\right] = \begin{cases} \frac{\Gamma(\frac{n}{2}-1)}{\Gamma(1-\alpha)} z + O\left(z^{\min(\frac{n}{2},2)}\right), & \text{if } n \geq 3, \\ -\frac{z\log z}{\Gamma(1-\alpha)} + O(z), & \text{if } n = 2, \\ \frac{\sqrt{\pi}}{\Gamma(1-\frac{\alpha}{2})} z^{1/2} - \frac{2\sqrt{\pi}}{\Gamma(1-\alpha)} z + \frac{2\sqrt{\pi}}{\Gamma(1-\frac{3\alpha}{2})} z^{3/2} + O(z^2), & \text{if } n = 1 \end{cases} \tag{A.18}$$

(here and below we present those terms of the asymptotic expansions which are actually used in this book),

$$H_{23}^{30}\left[z\,\middle|\,\begin{matrix}(1,\alpha), \quad (0,1), \\ (\frac{n}{2},1), \quad (1,1), \quad (1,1)\end{matrix}\right] = \begin{cases} -\frac{\Gamma(\frac{n}{2}-1)}{\Gamma(1-\alpha)} z + O\left(z^{\min(\frac{n}{2},2)}\right), & \text{if } n \geq 3, \\ \frac{z\log z}{\Gamma(1-\alpha)} + O(z), & \text{if } n = 2, \\ -\frac{\sqrt{\pi}}{2\Gamma(1-\frac{\alpha}{2})} z^{1/2} + \frac{2\sqrt{\pi}}{\Gamma(1-\alpha)} z - \frac{3\sqrt{\pi}}{\Gamma(1-\frac{3\alpha}{2})} z^{3/2} + O(z^2), & \text{if } n = 1, \end{cases} \tag{A.19}$$

$$H_{34}^{40}\left[z\,\middle|\,\begin{matrix}(1,\alpha), \quad (0,1), \quad (0,1) \\ (\frac{n}{2},1), \quad (1,1), \quad (1,1), \quad (1,1)\end{matrix}\right] = \begin{cases} \frac{\Gamma(\frac{n}{2}-1)}{\Gamma(1-\alpha)} z + O\left(z^{\min(\frac{n}{2},2)}\right), & \text{if } n \geq 3, \\ -\frac{z\log z}{\Gamma(1-\alpha)} + O(z), & \text{if } n = 2, \\ \frac{\sqrt{\pi}}{4\Gamma(1-\frac{\alpha}{2})} z^{1/2} - \frac{2\sqrt{\pi}}{\Gamma(1-\alpha)} z & \\ +\frac{9\sqrt{\pi}}{2\Gamma(1-\frac{3\alpha}{2})} z^{3/2} + O(z^2), & \text{if } n = 1, \end{cases} \tag{A.20}$$

$$H_{45}^{50}\left[z\,\middle|\,\begin{matrix}(1,\alpha), \quad (0,1), \quad (0,1), \quad (0,1) \\ (\frac{n}{2},1), \quad (1,1), \quad (1,1), \quad (1,1), \quad (1,1)\end{matrix}\right] = \begin{cases} -\frac{\Gamma(\frac{n}{2}-1)}{\Gamma(1-\alpha)} z + O\left(z^{\min(\frac{n}{2},2)}\right), & \text{if } n \geq 3, \\ \frac{z\log z}{\Gamma(1-\alpha)} + O(z), & \text{if } n = 2, \\ -\frac{\sqrt{\pi}}{8\Gamma(1-\frac{\alpha}{2})} z^{1/2} + \frac{2\sqrt{\pi}}{\Gamma(1-\alpha)} z & \\ -\frac{27\sqrt{\pi}}{4\Gamma(1-\frac{3\alpha}{2})} z^{3/2} + O(z^2), & \text{if } n = 1. \end{cases} \tag{A.21}$$

Next we give the expansions for the function appearing in (5.2.12) and those emerging in the course of differentiating (5.2.12). A different character of their asymptotics is caused by the fact that in this case the pole $s = 1$ appears both in the numerator and denominator of the fraction in (A.2) and cancel each other.

We have

$$H_{12}^{20}\left[z\,\middle|\,\begin{matrix}(\alpha,\alpha)\\ (\frac{n}{2},1), & (1,1)\end{matrix}\right]
=\begin{cases}
-\frac{\Gamma(\frac{n}{2}-2)}{\Gamma(-\alpha)}z^2+O\left(z^{\min(\frac{n}{2},3)}\right), & \text{if } n\geq 5,\\
\frac{z^2\log z}{\Gamma(-\alpha)}+\gamma_1 z^2+O(z^3\log z), & \text{if } n=4,\\
-\frac{2\sqrt{\pi}}{\Gamma(-\frac{\alpha}{2})}z^{3/2}+\frac{2\sqrt{\pi}}{\Gamma(-\alpha)}z^2+O(z^{5/2}), & \text{if } n=3,\\
\alpha z-\frac{z^2\log z}{\Gamma(-\alpha)}+\gamma_2 z^2+O(z^3\log z), & \text{if } n=2,\\
\frac{\sqrt{\pi}}{\Gamma(\frac{\alpha}{2})}z^{1/2}+\frac{2\sqrt{\pi}}{\Gamma(-\frac{\alpha}{2})}z^{3/2}+O(z^2), & \text{if } n=1,
\end{cases} \tag{A.22}$$

where γ_1, γ_2 are certain constants. Next,

$$H_{23}^{30}\left[z\,\middle|\,\begin{matrix}(\alpha,\alpha), & (0,1)\\ (\frac{n}{2},1), & (1,1), & (1,1)\end{matrix}\right]
=\begin{cases}
\frac{2\Gamma(\frac{n}{2}-2)}{\Gamma(-\alpha)}z^2+O\left(z^{\min(\frac{n}{2},3)}\right), & \text{if } n\geq 5,\\
-2\frac{z^2\log z}{\Gamma(-\alpha)}-2\gamma_1 z^2+O(z^3\log z), & \text{if } n=4,\\
\frac{3\sqrt{\pi}}{\Gamma(-\frac{\alpha}{2})}z^{3/2}-\frac{4\sqrt{\pi}}{\Gamma(-\alpha)}z^2+O(z^{5/2}), & \text{if } n=3,\\
-\alpha z-2\frac{z^2\log z}{\Gamma(-\alpha)}-2\gamma_2 z^2+O(z^3\log z), & \text{if } n=2,\\
-\frac{\sqrt{\pi}}{2\Gamma(\frac{\alpha}{2})}z^{1/2}-\frac{3\sqrt{\pi}}{\Gamma(-\frac{\alpha}{2})}z^{3/2}+O(z^2), & \text{if } n=1,
\end{cases} \tag{A.23}$$

$$H_{34}^{40}\left[z\,\middle|\,\begin{matrix}(\alpha,\alpha), & (0,1), & (0,1)\\ (\frac{n}{2},1), & (1,1), & (1,1), & (1,1)\end{matrix}\right]
=\begin{cases}
-\frac{4\Gamma(\frac{n}{2}-2)}{\Gamma(-\alpha)}z^2+O\left(z^{\min(\frac{n}{2},3)}\right), & \text{if } n\geq 5,\\
4\frac{z^2\log z}{\Gamma(-\alpha)}+4\gamma_1 z^2+O(z^3\log z), & \text{if } n=4,\\
-\frac{9\sqrt{\pi}}{2\Gamma(-\frac{\alpha}{2})}z^{3/2}+\frac{8\sqrt{\pi}}{\Gamma(-\alpha)}z^2+O(z^{5/2}), & \text{if } n=3,\\
\alpha z-4\frac{z^2\log z}{\Gamma(-\alpha)}+4\gamma_2 z^2+O(z^3\log z), & \text{if } n=2,\\
\frac{\sqrt{\pi}}{4\Gamma(\frac{\alpha}{2})}z^{1/2}+\frac{9\sqrt{\pi}}{2\Gamma(-\frac{\alpha}{2})}z^{3/2}+O(z^2), & \text{if } n=1.
\end{cases} \tag{A.24}$$

Finally,

$$H_{45}^{50}\left[z\,\middle|\,\begin{matrix}(\alpha,\alpha), & (0,1), & (0,1), & (0,1)\\ (\frac{n}{2},1), & (1,1), & (1,1), & (1,1), & (1,1)\end{matrix}\right]
=\begin{cases}
\frac{8\Gamma(\frac{n}{2}-2)}{\Gamma(-\alpha)}z^2+O\left(z^{\min(\frac{n}{2},3)}\right), & \text{if } n\geq 5,\\
-8\frac{z^2\log z}{\Gamma(-\alpha)}-8\gamma_1 z^2+O(z^3\log z), & \text{if } n=4,\\
\frac{27\sqrt{\pi}}{4\Gamma(-\frac{\alpha}{2})}z^{3/2}-\frac{16\sqrt{\pi}}{\Gamma(-\alpha)}z^2+O(z^{5/2}), & \text{if } n=3,\\
-\alpha z+8\frac{z^2\log z}{\Gamma(-\alpha)}-8\gamma_2 z^2+O(z^3\log z), & \text{if } n=2,\\
-\frac{\sqrt{\pi}}{8\Gamma(\frac{\alpha}{2})}z^{1/2}-\frac{27\sqrt{\pi}}{4\Gamma(-\frac{\alpha}{2})}z^{3/2}+O(z^2), & \text{if } n=1.
\end{cases} \tag{A.25}$$

In Chapter 5 we meet also the function

$$H(z) = H_{23}^{30}\left[z \,\middle|\, \begin{matrix} (1,\alpha), & (1+\nu-\alpha-\lambda,\alpha) & \\ (\frac{n}{2},1), & (1,1), & (1+\nu,\alpha) \end{matrix}\right],$$

$$0 < \lambda < 1-\alpha, \nu > \alpha+\lambda.$$

For this case we need only an estimate for small z's, without explicit values of coefficients of the asymptotic expansion:

$$|H(z)| \leq \begin{cases} C|z|, & \text{if } n \geq 3, \\ C|z|(|\log z|+1), & \text{if } n = 2, \\ C|z|^{1/2}, & \text{if } n = 1. \end{cases} \tag{A.26}$$

Bibliography

[1] M.S. Agranovich and M.I. Vishik, Elliptic problems with parameter and general parabolic problems, *Usp. Mat. Nauk*, **19**, No. 3 (1964), 53–161 (Russian).

[2] L.N. Androsova and S.D. Ivasyshen, On integral representation and initial values of solutions of certain degenerate parabolic equations, *Dokl. AN Ukr. SSR, Ser. A*, No. 1 (1989), 16–19 (Russian).

[3] L.N. Androsova and S.D. Ivasyshen, Integral representation of solutions of a class of degenerate parabolic Kolmogorov equations, *Diff. Equations*, **27** (1991), 341–347.

[4] V.V. Anh and N.N. Leonenko, Spectral analysis of fractional kinetic equations with random data, *J. Statist. Phys.*, **104** (2001), 1349–1387.

[5] D.G. Aronson, The fundamental solution of a linear parabolic equation containing a small parameter, *Ill. J. Math.*, **3** (1959), 580–619.

[6] D.G. Aronson, On the initial value problem for parabolic systems of differential equations, *Bull. Amer. Math. Soc.*, **65** (1959), 310–318.

[7] D.G. Aronson, Uniqueness of solutions of the initial value problem for parabolic systems of differential equations, *J. Math. Mech.*, **11** (1962), 403–420.

[8] D.G. Aronson and P. Besala, Parabolic equations with unbounded coefficients, *J. Diff. Equat.*, **3** (1967), 1–14.

[9] H. Bateman, A. Erdélyi, *Higher Transcendental Functions, Vol. 1*, McGraw-Hill, New York, 1953.

[10] H. Bateman, A. Erdélyi, *Higher Transcendental Functions, Vol. 3*, McGraw-Hill, New York, 1955.

[11] B. Baeumer and M. Meerschaert, Stochastic solutions for fractional Cauchy problems, *Fract. Calc. Appl. Anal.* **4** (2001), 481–500.

[12] E. Bazhlekova, The abstract Cauchy problem for fractional evolution equation, *Fract. Calc. Appl. Anal.* **1** (1998), 255–270.

[13] L.P. Berezan, Integral representation of solutions of the generalized Cauchy problem for a $\overrightarrow{2b}$-parabolic system, strongly degenerate on the initial hyperplane, *Nauk. Visnyk Chernivtsi University*, No. 46 (1999), 13–18 (Ukrainian).

[14] L.P. Berezan, Some properties of the fundamental matrix of solutions of the Cauchy problem for $\overrightarrow{2b}$-parabolic systems with degeneration on the initial hyperplane*Nauk. Visnyk Chernivtsi University*, No. 76 (2000), 5–10 (Ukrainian).

[15] L.P. Berezan and S.D. Ivasyshen, Fundamental matrix of solutions of the Cauchy problem for $\overrightarrow{2b}$-parabolic systems with degeneration on the initial hyperplane, *Dopovidi NAN Ukrainy*, No. 12 (1998), 7–12 (Ukrainian).

[16] L.P. Berezan and S.D. Ivasyshen, On $\overrightarrow{2b}$-parabolic systems, strongly degenerate on the initial hyperplane, *Visn. Nat. Univ. Lvivska Polytechnika*, No. 337 (1998), 73–76 (Ukrainian).

[17] P. Besala, On the existence of a fundamental solution for a parabolic differential equation with unbounded coefficients, *Ann. Polon. Math.*, **29** (1975), 403–409.

[18] A.T. Bharucha-Reid, *Elements of the Theory of Markov Processes and Their Applications*, McGraw-Hill, New York, 1960.

[19] B.L.J. Braaksma, Asymptotic expansions and analytic continuation for a class of Barnes integrals, *Compositio Math.* **15** (1964), 239–341.

[20] M. Bramanti, M.C. Cerutti, and M. Manfredini, L^p estimates for some ultraparabolic operators with discontinuous coefficients, *J. Math. Anal. Appl.*, **200** (1996), 332–354.

[21] S.Z. Bruk, Fundamental solutions of systems of differential equations of the parabolic type, *Dokl. AN SSSR*, **60** (1948), 9–12 (Russian).

[22] M. Caputo. Linear model of dissipation whose Q is almost frequency independent. II, *Geophys. J. R. Astr. Soc.*, **13** (1967), 529–539.

[23] A. Cavalluci, Sulle proprieta differenziali delle soluzioni delle equazioni quasi-ellittiche, *Ann. Mat. Pura ed Appl.*, **67** (1965), 143–167.

[24] J. Chabrowski, Representation theorems for parabolic systems, *J. Austral. Math. Soc.*, **32** (1982), 246–288.

[25] S. Chandrasekar, Stochastic problems in physics and astronomy, *Rev. Mod. Phys.*, **15** (1943), 1–110.

[26] J.L. Doob, The Brownian motion and stochastic equations, *Ann. Math.*, **43** (1942), 351–369.

[27] J.L. Doob, *Stochastic Processes*, Wiley, New York, 1953.

[28] F. Dressel, The fundamental solution of a parabolic equation. I, II, *Duke Math. J.*, **7**, No. 4 (1940), 186–203; **13**, No. 1 (1946), 61–70.

[29] R.Ya. Drin', Stabilization of solutions of the Cauchy problem for systems of parabolic pseudo-differential equations with non-smooth symbols. In: *Integral Transforms and their Applications to Boundary Value Problems*, Institute of Mathematics, Kiev, 1997, pp. 88–102.

[30] Ya.M. Drin', Investigation of a class of parabolic pseudo-differential operators on classes of Hölder continuous functions, *Dopovidi AN Ukr. SSR, Ser. A*, No. 1 (1974), 19-22 (Ukrainian).

[31] Ya.M. Drin', The fundamental solution of the Cauchy problem for a class of parabolic pseudo-differential equations, *Dokl. AN Ukr. SSR, Ser. A*, No. 3 (1977), 198–203 (Russian).

[32] Ya.M. Drin' and S.D. Eidelman, Necessary and sufficient conditions for stabilization of solutions of the Cauchy problem for parabolic pseudo-differential equations. In: *Approximate Methods of Mathematical Analysis*, Kiev Gos. Ped. Inst., Kiev, 1974, 60–69 (Russian).

[33] Ya.M. Drin' and S.D. Eidelman, Construction and investigation of classical fundamental solution of the Cauchy problem for uniformly parabolic pseudo-differential equations, *Mat. Issled.*, **63** (1981), 18–33 (Russian).

[34] Ya.M. Drin' and S.D. Eidelman, On the theory of systems of parabolic pseudo-differential equations, *Dokl. AN Ukr. SSR, Ser. A*, No.4 (1989), 35–37.

[35] V.S. Dron', On the maximum principle for degenerate parabolic equations of the Kolmogorov type. In: *Integral Transforms and their Applications to Boundary Value Problems. Collection of Papers, Vol. 12*, Institute of Mathematics, Kiev, 1996, 272–277 (Ukrainian).

[36] V.S. Dron', On correct solvability of the Cauchy problem for an ultra-parabolic equation of the Kolmogorov type, *Mat. Metody Fiz.-Mech. Polya*, **42**, No. 3 (1999), 52–55 (Ukrainian).

[37] V.S. Dron', On correct solvability in weighted Hölder spaces for the Cauchy problem for a class of degenerate parabolic equations of the Kolmogorov type, *Nauk. Visnyk Chernivtsi University*, No. 76 (2000), 32–41 (Ukrainian).

[38] V.S. Dron' and S.D. Ivasyshen, Properties of fundamental solutions and uniqueness theorems for solutions of the Cauchy problem for a class of ultraparabolic equations, *Ukrainian Math. J.*, **50** (1998), 1692–1709.

[39] V.S. Dron' and S.D. Ivasyshen, On properties of the volume potential and correct solvability of the Cauchy problem for a model ultraparabolic equation, *Nauk. Visnyk Chernivtsi University*, No. 46 (1999), 36–43 (Ukrainian).

[40] E.B. Dynkin, *Foundations of the Theory of Markov Processes*, Prentice-Hall, Englewood Cliffs, NJ, 1961.

[41] E.B. Dynkin, *Markov Processes I, II*, Springer, Berlin, 1965.

[42] M.M. Djrbashian (Dzhrbashyan), *Harmonic Analysis and Boundary Value Problems in Complex Domain*, Birkhäuser, Basel, 1993.

[43] M.M. Dzhrbashyan and A.B. Nersessyan, Some integro-differential operators and quasi-analytic classes of functions associated with them, *Izv. AN Arm. SSR, Phys. and Math. Sci.*, **11**, No. 1 (1958), 107–120 (Russian).

[44] M.M. Dzhrbashyan and A.B. Nersessyan, Fractional derivatives and the Cauchy problem for differential equations of fractional order, *Izv. AN Arm. SSR. Mathematics*, **3**, No. 1 (1968), 3–29 (Russian).

[45] S.D. Eidelman, *Parabolic Systems*, North-Holland, Amsterdam, 1969 (Russian edition, Nauka, Moscow, 1964).

[46] S.D. Eidelman, Parabolic equations. In: *Encyclopaedia of Mathematical Sciences. Vol. 63. Partial Differential Equations VI*, Springer, Berlin, 1994, 203–316.

[47] S.D. Eidelman, On fundamental solutions of parabolic systems. I, *Mat. Sbornik*, **38**, No. 1 (1956), 51–92 (Russian).

[48] S.D. Eidelman, On some applications of fundamental matrices of solutions of parabolic systems. In: *Theoretical and Applied Mathematics. Vol. 1*, Lviv, 1958, pp. 99–149.

[49] S.D. Eidelman, Fundamental matrices of solutions of general parabolic systems, *Dokl. AN SSSR*, **120** (1958), 980–983 (Russian).

[50] S.D. Eidelman, On the Cauchy problem for systems with increasing coefficients, *Dokl. AN SSSR*, **127** (1959), 760–763 (Russian).

[51] S.D. Eidelman, On a class of parabolic systems, *Soviet Math. Dokl.*, **1** (1960), 815–818.

[52] S.D. Eidelman, On fundamental solutions of parabolic systems. II. *Mat. Sbornik* **53** (1961), 73–136 (Russian).

[53] S.D. Eidelman and S.D. Ivasyshen, $\overrightarrow{2b}$-parabolic systems, *Trudy Sem. Funkt. Anal. Kiev, Institute of Mathematics*, **1** (1968), 3–175, 271–273 (Russian).

[54] S.D. Eidelman and S.D. Ivasyshen, On a new class of degenerate parabolic equations, *Usp. Mat. Nauk*, **51**, No. 5 (1996), 227 (Russian).

[55] S.D. Eidelman and S.D. Ivasyshen, On fundamental solution of the Cauchy problem for a new class of degenerate parabolic pseudo-differential equations, *Dopovidi NAN Ukrainy*, No. 6 (1997), 18–23 (Ukrainian).

[56] S.D. Eidelman and S.D. Ivasyshen, $\overrightarrow{2b}$-parabolic equations with degeneration in some of the variables, *Dokl. Math.*, **57** (1998), 384–386.

[57] S.D. Eidelman and S.D. Ivasyshen, On fundamental solutions of the Cauchy problem for degenerate equations of the Kolmogorov type with $\overrightarrow{2b}$-parabolic part with respect to the main group of variables, *Diff. Equations*, **34** (1998), 1537–1546.

[58] S.D. Eidelman and S.D. Ivasyshen, On the Cauchy problem for degenerate equations of Kolmogorov type with a $\overrightarrow{2b}$-parabolic part with respect to the main group of variables, *Diff. Equations*, **36** (2000), 587–599.

[59] S.D. Eidelman and S.D. Ivasyshen, On the integral representation of solutions of Kolmogorov-type degenerate equations with a $\overrightarrow{2b}$-parabolic part with respect to the main group of variables, *Diff. Equations*, **36** (2000), 717–727.

[60] S.D. Eidelman and S.D. Ivasyshen, On solutions of parabolic equations from families of Banach spaces depending on time. In: *Operator Theory: Advances and Applications, Vol. 117*, Birkhäuser, Basel, 2000, 111–125.

[61] S.D. Eidelman, S.D. Ivasyshen, and A.P. Malitskaya, A modified Levi method: development and application, *Dopovidi NAN Ukrainy*, No. 5 (1998), 14–19.

[62] S.D. Eidelman, S.D. Ivasyshen, and L.M. Tychinskaya, Fundamental solution of the Cauchy problem for a class of second order ultra-parabolic equations, *Dopovidi AN Ukr. SSR, Ser. A*, No. 5 (1990), 6–8 (Ukrainian).

[63] S.D. Eidelman and A.N. Kochubei, Cauchy problem for fractional diffusion equations, *J. Diff. Equat.*, **199** (2004), 211–255.

[64] S.D. Eidelman and A.P. Malitskaya, On equations of diffusion with inertia and increasing coefficients, *Dopovidi AN Ukr. SSR, Ser. A*, No. 2 (1974), 106–110 (Russian).

[65] S.D. Eidelman and A.P. Malitskaya, On fundamental solutions and stabilization of a solution of the Cauchy problem for a class of degenerate parabolic equations, *Diff. Equations*, **11** (1976), 988–998.

[66] S.D. Eidelman and M.I. Matiychuk, Parabolic systems with coefficients satisfying Dini's condition, *Soviet Math.Dokl.*, **6** (1965), 1461–1464.

[67] S.D. Eidelman and M.I. Matiychuk, Cauchy problem for parabolic systems whose coefficients have a little smoothness, *Ukrainian Math. J.*, **22** (1970), 18–30.

[68] S.D. Eidelman and F.O. Porper, The behavior of solutions of second order parabolic equations with dissipation, *Diff. Uravneniya*, **7** (1971), 1684–1695 (Russian).

[69] S.D. Eidelman and F.O. Porper, Investigation of the behaviour of L_2-norms of solutions for strongly parabolic systems with dissipation, *Izv. AN SSSR. Ser. Mat.*, **37** (1973), 676–690 (Russian).

[70] S.D. Eidelman and F.O. Porper, Two-sided estimates of fundamental solutions of second-order parabolic equations, and some applications, *Russ. Math. Surv.*, **39** (1984), No.3, 119–178.

[71] S.D. Eidelman and V.D. Repnikov, Necessary and sufficient conditions for the establishment of a solution of the Cauchy problem, *Soviet Math. Dokl.*, **7** (1966), 388–391.

[72] S.D. Eidelman and V.D. Repnikov, A new proof of the theorem on the stabilization of the solution of the Cauchy problem for the heat equation, *Math. USSR Sb.*, **2** (1967), 135–139.

[73] S.D. Eidelman and L.M. Tychinskaya, Construction of fundamental solutions for some degenerate parabolic equations of an arbitrary order, *Dopovidi AN Ukr. SSR, Ser. A*, No. 11 (1979), 896–899 (Ukrainian).

[74] S.D. Eidelman and L.M. Tychinskaya, Cauchy problem for some equations of diffusion with inertia and increasing coefficients, *Dopovidi AN Ukrainy*, No. 6 (1993), 18–21 (Russian).

[75] S.D. Eidelman and Yu.N. Valitsky, A necessary and sufficient condition for the stabilization of positive solutions of the heat equation, *Sib. Mat. Zh.*, **17** (1976), 744–756 (Russian).

[76] S.D. Eidelman and N.V. Zhitarashu, *Parabolic Boundary Value Problems* (Operator Theory: Advances and Applications, Vol. 101), Birkhäuser, Basel, 1998.

[77] A. Einstein, *Investigations on the Theory of the Brownian Movement*, Dover, New York, 1956.

[78] A. Einstein and M. von Smoluchowski, *Brownsche Bewegung*, Verlag Harri Deutsch, Frankfurt am Main, 1997.

[79] A.M. El-Sayed, Fractional order evolution equations, *J. Fract. Calc.* **7** (1995), 89–100.

[80] M.V. Fedoryuk, Asymptotic properties of Green's function of a parabolic pseudodifferential equation, *Diff. Equations*, **14** (1978), 923–927.

[81] C. Fox, The G and H functions as symmetrical Fourier kernels, *Trans. Amer. Math. Soc.*, **98** (1961), 395–429.

[82] J. Friberg, *Estimates for Partially Hypoelliptic Differential Operators*, Lund, 1963.

[83] A. Friedman, *Partial Differential Equations of Parabolic Type*, Prentice-Hall, Englewood Cliffs, NJ, 1964.

[84] Y. Fujita, Integrodifferential equation which interpolates the heat equation and the wave equation. I, II, *Osaka J. Math.*, **27** (1990), 309–321, 797–804.

[85] V.I. Fushchich, A.S. Galitsyn, and A.S. Polubinskii, A new mathematical model of heat conduction processes, *Ukrainian Math. J.*, **42** (1990), 210–216.

[86] I.M. Gelfand and G.E. Shilov, Fourier transforms of rapidly increasing functions, *Usp. Mat. Nauk*, **8**, No. 6 (1953), 3–54 (Russian).

[87] I.M. Gelfand and G.E. Shilov, *Generalized Functions. Vol. 1: Properties and Operations*, Academic Press, New York, 1964.

[88] I.M. Gelfand and G.E. Shilov, *Generalized functions. Vol. 3. Theory of differential equations*, Academic Press, New York, 1967.

[89] T.G. Genchev, Ultraparabolic equations, *Soviet Math. Dokl.*, **4** (1963), 979–982.

[90] A.V. Glushak and S.D. Shmulevich, On some correct problems for parabolic equations of high order degenerating in the time variable, *Diff. Uravneniya*, **22** (1986), 1065–1068 (Russian).

[91] V.P. Glushko, Degenerate linear differential equations. II, *Diff. Uravneniya*, **4** (1968), 1956–1966 (Russian).

[92] R. Gorenflo, F. Mainardi, D. Moretti, and P. Paradisi, Time fractional diffusion: A discrete random walk approach, *Nonlinear Dynamics* **29** (2002), 129–143.

[93] V.V. Gorodetsky and I.V. Zhitaryuk, On solutions of the Cauchy problem for equations of the parabolic type with degeneracy, *Dopovidi AN Ukr. SSR, Ser. A*, No. 12 (1989), 5–8 (Ukrainian).

[94] V.V. Gorodetsky and I.V. Zhitaryuk, Solvability of the Cauchy problem for parabolic-type evolution equations with degeneracy in certain spaces, *Diff. Equations*, **28** (1992), 1118–1125.

[95] E. Goursat, *Cours d'Analyse Mathématique*, Dover, New York, 1959.

[96] G.H. Hardy, *Divergent Series*, Chelsea, New York, 1991.

[97] L. Hörmander, On the regularity of solutions of boundary problems, *Acta Math.*, **99** (1958), 225–264.

[98] L. Hörmander, *Linear Partial Differential Operators*, Springer, Berlin, 1963.

[99] L. Hörmander, Hypoelliptic second order differential equations, *Acta Math.*, **119** (1967), 147–171.

[100] L. Hörmander, *The Analysis of Linear Partial Differential Operators, III*, Springer, Berlin, 1985.

[101] A.M. Il'in, On a class of ultra-parabolic equations, *Soviet Math. Dokl.*, **5** (1965), 1673–1676.

[102] A.M. Il'in, A.S. Kalashnikov, and O.A. Olejnik, Linear second order equations of parabolic type, *Russ. Math. Surv.*, **17**, No.3, (1962), 1–146.

[103] A. M. Il'in and R.Z. Khasminsky, On equations of Brownian motion, *Theor. Probab. Appl.*, **9** (1964), 421–444.

[104] K. Ito, *On Stochastic Differential Equation*, Mem. Amer. Math. Soc., **4** (1951).

[105] S.D. Ivasyshen, On integral representations and the Fatou property for solutions of parabolic systems, *Usp. Mat. Nauk*, **41**, No. 4 (1986), 173–174. (Russian).

[106] S.D. Ivasyshen, On initial values of solutions of ultra-parabolic equations, *Usp. Mat. Nauk*, **43**, No. 4 (1988), 188–189 (Russian).

[107] S.D. Ivasyshen, *Green Matrices of Parabolic Boundary Value Problems*, Vyshcha Shkola, Kiev, 1990 (Russian).

[108] S.D. Ivasyshen, Integral representation and initial values of solutions of $\overrightarrow{2b}$-parabolic systems, *Ukrainian Math. J.*, **42** (1990), 443–448.

[109] S.D. Ivasyshen and O.S. Kondur, On the Green matrix of the Cauchy problem and the characterization of certain classes of solutions for $\overrightarrow{2b}$-parabolic systems of an arbitrary order, *Mat. Studii*, **14**, No. 1 (2000), 73–84 (Ukrainian).

[110] S.D. Ivasyshen and I.P. Medynsky, On the correct solvability of parabolic systems with degeneration on the initial hyperplane, *Nauk. Visnyk Chernivtsi University*, No. 76 (2000), 71–76 (Ukrainian).

[111] S.D. Ivasyshen and I.P. Medynsky, Properties of integrals which have the type of derivatives of volume potentials for parabolic systems with degeneration on the initial hyperplane, *Mat. Studii*, **13**, No. 1 (2000), 33–46.

[112] S.D. Ivasyshen and I.P. Medynsky, Properties of integrals which have the type of derivatives of volume potentials for $\overrightarrow{2b}$-parabolic systems with degeneration on the initial hyperplane, *Mat. Metody Fiz.-Mech. Polya*, **45**, No. 4 (2002), 76–86 (Ukrainian).

[113] S.D. Ivasyshen and I.P. Medynsky, Cauchy problem for $\overrightarrow{2b}$-parabolic systems with degeneration on the initial hyperplane, *Mat. Metody Fiz.-Mech. Polya*, **46**, No. 3 (2003), 15–24 (Ukrainian).

[114] S.D. Ivasyshen and G.S. Pasichnyk, On the fundamental matrix of solutions of the Cauchy problem for dissipative $\overrightarrow{2b}$-parabolic systems with degeneration on the initial hyperplane, *Dopovidi NAN Ukrainy*, No. 6 (1999), 18–22 (Ukrainian).

[115] S.D. Ivasyshen and G.S. Pasichnyk, On the Cauchy problem for $\overrightarrow{2b}$-parabolic systems with increasing coefficients, *Ukrainian Math. J.*, 52 (2000), 1691–1705.

[116] S.D. Ivasyshen and G.S. Pasichnyk, On the fundamental matrix of solutions of the Cauchy problem for a class of parabolic systems with unbounded coefficients and degenerations on the initial hyperplane, *Nauk. Visnyk Chernivtsi University*, No. 76 (2000), 82–91 (Ukrainian).

[117] S.D. Ivasyshen and G.S. Pasichnyk, On solvability of the Cauchy problem for parabolic systems with increasing coefficients, *Nonlinear Boundary Value Problems* (Donetsk), No. 12 (2002), 84–91 (Ukrainian).

[118] S.D. Ivasyshen and O.G. Voznyak, The Cauchy problem for parabolic systems with degeneration on the initial hyperplane, *Dopovidi AN Ukrainy*, No. 6 (1994), 7–11 (Ukrainian).

[119] S.D. Ivasyshen and O.G. Voznyak, Fundamental solutions of the Cauchy problem for a class of parabolic equations, and their applications, *Dopovidi NAN Ukrainy*, No. 10 (1996), 11–16 (Ukrainian).

[120] S.D. Ivasyshen and O.G. Voznyak, On fundamental solutions of the Cauchy problem for a class of degenerate parabolic equations, *Mat. Metody Fiz.-Mech. Polya*, **41**, No. 2 (1998), 13–19 (Ukrainian).

[121] S.D. Ivasyshen and O.G. Voznyak, Unique solvability and property of localization of solutions of the Cauchy problem for a class of degenerate equations with generalized initial data, *Mat. Metody Fiz.-Mech. Polya*, **44**, No. 4 (2001), 27–39 (Ukrainian).

[122] N. Jacob, *Pseudo-Differential Operators and Markov Processes*, Akademie-Verlag, Berlin, 1996.

[123] N. Jacob and R.L. Schilling, Lévy-type processes and pseudo-differential operators. In: *Lévy Processes – Theory and Applications*, Birkhäuser, Boston, 2001, pp. 139–168.

[124] A.S. Kalashnikov, Increasing solutions of linear second-order equations with non-negative characteristic form, *Math. Notes*, **3** (1968), 110–114.

[125] A.S. Kalashnikov, The problem without initial conditions in classes of increasing functions for certain second-order linear degenerate parabolic systems. I, II, *Vestn. Moscow Univ. Math. Mech.*, No. 2 (1971), 42–48; No. 3 (1971), 3–9 (Russian).

[126] Y. Kato, The hypoellipticity of degenerate parabolic differential operators, *J. Funct. Anal.*, **7** (1971), 116–131.

[127] A.N. Kochubei, Singular parabolic equations and Markov processes, *Math. USSR Izvestiya*, **24** (1985), 73–97.

[128] A.N. Kochubei, Parabolic pseudodifferential equations, hypersingular integrals, and Markov processes, *Math. USSR Izvestiya*, **33** (1989), 233–259.

[129] A.N. Kochubei, A Cauchy problem for evolution equations of fractional order, *Diff. Equations*, **25** (1989), 967–974.

[130] A.N. Kochubei, Fractional-order diffusion, *Diff. Equations*, **26** (1990), 485–492.

[131] A.N. Kochubei, Parabolic equations over the field of p-adic numbers, *Math. USSR Izvestiya*, **39** (1992), 1263–1280.

[132] A.N. Kochubei, *Pseudo-Differential Equations and Stochastics over Non-Archimedean Fields*, Marcel Dekker, New York, 2001.

[133] A.N. Kolmogorov, Ueber die analytischen Methoden in der Wahrscheinlichkeitsrechnung, *Math. Ann.*, **104** (1931), 415–458.

[134] A.N. Kolmogorov, Zur Theorie der stetigen zufälligen Prozesse, *Math. Ann.*, **108** (1933), 149–160.

[135] A.N. Kolmogorov, Zufällige Bewegungen (zur Theorie der Brownschen Bewegung), *Ann. Math.*, **35** (1934), 116–117.

[136] A.N. Kolmogorov, *Probability Theory and Mathematical Statistics*, Kluwer, Dordrecht, 1992.

[137] V. Kolokoltsov, Symmetric stable laws and stable-like jump-diffusions, *Proc. London Math. Soc.*, **80** (2000), 725–768.

[138] T. Kolsrud, On a class of probabilistic integrodifferential equations. In: *Ideas and Methods in Mathematics and Physics. Memorial Volume Dedicated to Raphael Høegh-Krohn, Vol. 1*, Cambridge University Press, 1992, pp. 168–172.

[139] T. Komatsu, Pseudo-differential operators and Markov processes, *J. Math. Soc. Japan*, **36** (1984), 387–418.

[140] V.A. Kostin, Cauchy problem for an abstract differential equation with fractional derivatives, *Russian Acad. Sci. Dokl. Math.*, **46** (1993), 316–319.

[141] S.G. Krein, *Linear Differential Equations in Banach Space*, American Mathematical Society, Providence, 1972.

[142] L.P. Kuptsov, On fundamental solutions of a class of second order elliptic-parabolic equations, *Diff. Uravneniya*, **8** (1972), 1649–1660 (Russian).

[143] L.P. Kuptsov, Mean value theorem and a maximum principle for Kolmogorov's equation, *Math. Notes*, **15** (1974), 280–286.

[144] L.P. Kuptsov, Fundamental solutions of certain degenerate second-order parabolic equations, *Math. Notes*, **31** (1982), 283–289.

[145] L. P. Kuptsov, A mean property for a generalized Kolmogorov equation, *Diff. Equations*, **19** (1983), 222-230.

[146] O.A. Ladyzhenskaya, V.A. Solonnikov, and N.N. Uraltseva, *Linear and Quasilinear Equations of Parabolic Type*, American Mathematical Society, Providence, 1968.

[147] E. Lanconelli and F. Lascialfari, A boundary value problem for a class of quasilinear operators of Fokker-Planck type, *Ann. Univ. Ferrara. Sez. VII (N.S.)*, **41**, Suppl. (1996), 65–84.

[148] E. Lanconelli and S. Polidoro, On a class of hypoelliptic evolution operators, *Rend. Sem. Mat. Univ. Politec. Torino*, **52**, No. 1 (1994), 29–63.

[149] F. Lascialfari and D. Morbidelli, A boundary value problem for a class of ultraparabolic equations, *Comm. Part. Diff. Equat.*, **23** (1998), 847–868.

[150] C. Lemoine, Fourier transforms of homogeneous distributions, *Ann. Scuola Norm. Sup. Pisa*, **26** (1972), 117–149.

[151] J.-P. Lepeltier and B. Marchal, Problème des martingales et équations différentielles stochastiques associées à un opérateur intégro-differentiel, *Ann. Inst. H. Poincaré, Sect. B*, **12** (1976), 43–103.

[152] E.E. Levi, Sulle equazioni lineari totalmente ellittiche alle derivate parziali, *Rend. Circ. Mat. Palermo*, **24** (1907), 275–317.

[153] F. Mainardi, Yu. Luchko and G. Pagnini, The fundamental solution of the space-time fractional diffusion equation, *Fract. Calc. Appl. Anal.*, **4** (2001), 153–192.

[154] A.P. Malitskaya, Fundamental solutions for a class of degenerate parabolic equations. In: *Approximate Methods of Integration of Differential and Integral Equations*, Kiev Gos. Ped. Inst., Kiev, 1973, 109–130 (Russian).

[155] A.P. Malitskaya, Construction of a fundamental solution for a class of high-order degenerate parabolic equations, *Ukrainian Math. J.*, **32** (1980), 508–514.

[156] A.P. Malitskaya, First mixed problem in the half space for the Sonin equation, *Ukrainian Math. J.*, **35** (1983), 274–279.

[157] A.P. Malitskaya, The first boundary value problem in a half space for a degenerate parabolic equation, *Diff. Uravneniya*, **19** (1983), 1821–1823 (Russian).

[158] A.P. Malitskaya, Structure of the fundamental solutions of some ultraparabolic equations of high order, *Ukrainian Math. J.*, **37** (1985), 582–587.

[159] A.P. Malitskaya, Degenerate parabolic equations whose coefficients are increasing, *Ukrainian Math. J.*, **41** (1989), 158–163.

[160] A.P. Malitskaya, The third mixed problem for the Sonin equation in a half space, *Ukrainian Math. J.*, **45** (1993), 1236–1243.

[161] A.P. Malitskaya, On the maximum principle for ultraparabolic equations, *Ukrainian Math. J.*, **48** (1996), 220–227.

[162] M. Manfredini, The Dirichlet problem for a class of ultraparabolic equations, *Adv. Differ. Equat.*, **2** (1997), 831–866.

[163] M. Manfredini and S. Polidoro, Interior regularity for weak solutions of ultraparabolic equations in divergence form with discontinuous coefficients, *Boll. Unione Mat. Ital. B*, **1** (1998), 651–675.

[164] V.P. Maslov and I.A. Shishmarev, On symbols of one-parameter semigroups e^{tA} for $t \to 0$, *Usp. Mat. Nauk*, **29**, No. 4 (1974), 177 (Russian).

[165] M.I. Matiychuk, Fundamental matrices of solutions of general $\overrightarrow{2b}$-parabolic and $\overrightarrow{2b}$-elliptic systems whose coefficients satisfy an integral Hölder condition, *Dopovidi AN Ukr. SSR*, No. 8 (1964), 1010–1013 (Ukrainian).

[166] I.P. Medynsky, A priori estimates for solutions of parabolic systems with degeneration, *Visn. Nat. Univ. Lvivska Polytechnika*, No. 337 (1998), 133–136 (Ukrainian).

[167] I.P. Medynsky, On properties of the fundamental matrix of solutions of the Cauchy problem for a parabolic system with degeneration on the initial hyperplane, *Visn. Nat. Univ. Lvivska Polytechnika*, No. 364 (1999), 298–307 (Ukrainian).

[168] I.P. Medynsky, On a priori estimates for solutions of parabolic systems with degeneration on the initial hyperplane, *Visn. Nat. Univ. Lvivska Polytechnika*, No. 407 (2000), 185–194 (Ukrainian).

[169] M.M. Meerschaert, D.A. Benson, H.P. Scheffler and B. Baeumer, Stochastic solutions of space-time fractional diffusion equations, *Phys. Rev. E* **65** (2002), 1103–1106.

[170] R. Metzler and J. Klafter, The random walk's guide to anomalous diffusion: a fractional dynamics approach, *Physics Reports*, **339** (2000), 1–77.

[171] V.P. Mikhailov, On the first boundary value problem for a class of hypoelliptic equations, *Mat. Sbornik*, 63 (1964), 238–264 (Russian).

[172] D. Morbidelli, Formule di media e disuguaglianza di Harnack per una classe di operatori parabolici degeneri. In: *Seminario di Analisi Matematica*, Dip. Mat. Univ. Bologna, Technoprint, Bologna, 1994-1995, pp. 191–211.

[173] P.P. Mosolov, On the generalized first boundary value problem for a class of differential operators, *Mat. Sbornik*, **57** (1962), 334–374; **59** (1962, suppl.), 165–188 (Russian).

[174] M. Nagase, On the Cauchy problem for parabolic pseudo-differential equations, *Osaka J. Math.*, **11** (1974), 239–264.

[175] U. Neri, Spherical harmonics, *Lecture Notes Math.*, **200** (1971), 1–272.

[176] F. W. J. Olver, *Asymptotics and Special Functions*, Academic Press, New York, 1974.

[177] R. Paley and N. Wiener, *Fourier Transforms in the Complex Domain*, American Mathematical Society, New York, 1934.

[178] G.S. Pasichnyk, On solvability of the Cauchy problem for $\overrightarrow{2b}$-parabolic systems with increasing coefficients, *Mat. Metody Fiz.-Mech. Polya*, **42**, No. 3 (1999), 61–65 (Ukrainian).

[179] G.S. Pasichnyk, On the fundamental matrix of solutions of the Cauchy problem for dissipative $\overrightarrow{2b}$-parabolic systems, *Visnyk Lviv Univ. Ser. Mech. Mat.*, No. 54 (1999), 140–151 (Ukrainian).

[180] I. Petrowsky, Über das Cauchysche Problem fur ein System linearer partieller Differentialgleichungen im Gebiete der nichtanalytischen Funktionen, *Bull. Univ. Etat. Moscou, Ser. Int.*, Sect. A, Math. et Mecan. **1**, Fasc. 7 (1938), 1–74.

[181] N.S. Piskunov, Boundary value problems for an equation of the elliptic-parabolic type, *Mat. Sbornik*, **7** (1940), 385–424 (Russian).

[182] S.I. Podolynny, Perturbation of homogeneous parabolic pseudo-differential equations by locally unbounded vector fields, *Ukr. Math. J.*, **49** (1997), 1756–1762.

[183] S.I. Podolynny and N.I. Portenko, On multidimensional stable processes with locally unbounded drift, *Random Oper. Stoch. Equ.* **3**, No.2 (1995), 113–124.

[184] W. Pogorzelski, Étude de la solution fondamentale de l'equation parabolique, *Ricerche di Mat.*, **5** (1956), 25–57.

[185] W. Pogorzelski, Étude de la matrice des solution fondamentales du système parabolique d'équations aux dérivees partielles, *Ricerche di Mat.*, **7** (1958), 153–185.

[186] W. Pogorzelski, Propriétés de solution du système parabolique d'équations aux dérivees partielles, *Math. Scand.*, **6** (1958), 237–262.

[187] S. Polidoro, On a class of ultraparabolic operators of Kolmogorov-Fokker-Planck type, *Le Matematiche*, **49** (1994), 53–105.

[188] S. Polidoro, Uniqueness and representation theorems for solutions of Kolmogorov-Fokker-Planck equations, *Rend. di Mat. VII*, **15** (1995), 535–560.

[189] S. Polidoro, A global lower bound for the fundamental solution of Kolmogorov-Fokker-Planck equations, *Arch. Rat. Mech. Anal.*, **137** (1997), 321–340.

[190] S. Polidoro and M.A. Ragusa, Sobolev-Morrey spaces related to an ultraparabolic equation, *Manuscripta Math.*, **96** (1998), 371–392.

[191] S. Polidoro and M.A. Ragusa, Hölder regularity for solutions of an ultraparabolic equation in divergence form, *Potential Anal.*, **14** (2001), 341–350.

[192] N.I. Portenko, On some perturbations of symmetric stable processes. In: *Probability Theory and Mathematical Statistics. Proceedings of the Seventh Japan-Russia Symposium*, World Scientific, Singapore, 1996, pp. 414–422.

[193] A.P. Prudnikov, Yu.A. Brychkov, and O.I. Marichev, *Integrals and Series. Vol. 1: Elementary Functions*, Gordon and Breach, New York, 1986.

[194] A.P. Prudnikov, Yu.A. Brychkov, and O.I. Marichev, *Integrals and Series. Vol. 2: Special Functions*, Gordon and Breach, New York, 1986.

[195] A.P. Prudnikov, Yu.A. Brychkov, and O.I. Marichev, *Integrals and Series. Vol. 3: More Special Functions*, Gordon and Breach, New York, 1990.

[196] S.G. Pyatkov, Solvability of boundary value problems for a certain ultraparabolic equation. In: *Non-Classical Equations and Equations of Mixed Type*, Novosibirsk, 1990, pp. 182–197 (Russian).

[197] V.D. Repnikov, Some theorems concerning stabilization of the solution of the Cauchy problem for parabolic equations, *Soviet Math. Dokl.* **4** (1963), 137–140.

[198] E. Riekstiņš, Asymptotic expansions of some types of convolution integrals, *Latviysky Mat. Ezhegodnik*, **8** (1970), 223–239 (Russian).

[199] V.L. Roitburd, On the representation of the solution of a Cauchy problem by a functional integral, *Soviet Math. Dokl.*, **12** (1971), 1691–1694.

[200] S.G. Samko, Generalized Riesz potentials and hypersingular integrals with homogeneous characteristics, their symbols and inversion, *Proc. Steklov Inst. Math.*, **156** (1983), 173–243.

[201] S.G. Samko, *Hypersingular Integrals and Their Applications*, Taylor and Francis, London, 2001.

[202] S.G. Samko, A.A. Kilbas, and O.I. Marichev, *Fractional Integrals and Derivatives: Theory and Applications*, Gordon and Breach, New York, 1993.

[203] W.R. Schneider, Stable distributions: Fox function representation and generalization, *Lecture Notes Phys.*, **262** (1986), 497–511.

[204] W.R. Schneider, Fractional diffusion, *Lecture Notes Phys.*, **355** (1990), 276–286.

[205] W.R. Schneider, Grey noise. In: *Ideas and Methods in Mathematics and Physics. Memorial Volume Dedicated to Raphael Høegh-Krohn, Vol. 1*, Cambridge University Press, 1992, pp. 261–282.

[206] W.R. Schneider, W. Wyss, Fractional diffusion and wave equations, *J. Math. Phys.*, **30** (1989), 134–144.

[207] V. Scornazzani, Sulle soluzioni non negative dell'equazione di Kolmogorov, *Rend. mat. e appl.*, **2** (1982), 689–709.

[208] Ya.I. Shatyro, Inner estimates for solutions of a class of ultra-parabolic equations, *Mat. Zapiski Uralsk. Univ.*, **7**, No. 4 (1970), 131–154 (Russian).

[209] Ya.I. Shatyro, On smoothness of solutions of some degenerate second order equations, *Mat. Zametki*, **10**, No. 1 (1971), 101–111 (Russian).

[210] Ya.I. Shatyro, The first boundary value problem for an ultraparabolic equation, *Differ. Uravneniya*, **7** (1971), 1089–1096 (Russian).

[211] R. Shinkai, On symbols of fundamental solutions of parabolic systems, *Proc. Japan Acad.*, **50** (1974), 337–341.

[212] A.V. Skorokhod, *Studies in the Theory of Random Processes*, Addison-Wesley, Reading, 1965.

[213] L.N. Slobodeckij, The fundamental solution and the Cauchy problem for parabolic systems, *Amer. Math. Soc. Transl.*, **41** (1964), 121–154.

[214] S.L. Sobolev, *Some Applications of Functional Analysis in Mathematical Physics*, AMS, Providence, 1991.

[215] V.A. Solonnikov, On boundary value problems for linear parabolic systems of differential equations of general form, *Proc. Steklov Inst. Math.*, **83** (1965), 1–184.

[216] I.M. Sonin, On a class of degenerate diffusion processes, *Theor. Probab. Appl.*, **12** (1967), 490–496.

[217] H.M. Srivastava, K.C. Gupta, and S.P. Goyal, *The H-Functions of One and Two variables with Applications*, South Asian Publishers, New Dehli, 1982.

[218] E.M. Stein and G. Weiss, *Introduction to Fourier Analysis on Euclidean Spaces*, Princeton University Press, Princeton, 1971.

[219] D.W. Stroock, Diffusion processes associated with Lévy generators, *Z. Wahrsch. Verw. Geb.*, **32** (1975), 209–244.

[220] V.I. Tikhonov and N.K. Kulman, *Nonlinear Filtration and Quasi-Coherent Reception of Signals*, Sovetskoye Radio, Moscow, 1975 (Russian).

[221] E.C. Titchmarsh, *The Theory of Functions*, Oxford University Press, Oxford, 1939.

[222] C. Tsutsumi, The fundamental solution for a degenerate parabolic pseudo-differential operator, *Proc. Japan. Acad.*, **50** (1974), 11–15.

[223] C. Tsutsumi, The fundamental solution for a parabolic pseudo-differential operator and parametrices for degenerate operators, *Proc. Japan. Acad.*, **51** (1975), 103–108.

[224] G.E. Uhlenbeck and L.S. Ornstein, On the theory of Brownian motion, *Phys. Rev.*, **36** (1930), 823–841.

[225] L.R. Volevich, Local properties of solutions of quasi-elliptic systems, *Mat. Sbornik*, **59** (1962, suppl.), 3–52 (Russian).

[226] G.N. Watson, *The Theory of Bessel Functions*, Cambridge University Press, Cambridge, 1922.

[227] M. Weber, The fundamental solution of a degenerate partial differential equation of parabolic type, *Trans. Amer. Math. Soc.*, **71** (1951), 24–37.

[228] E.T. Whittaker and G.N. Watson, *A Course of Modern Analysis*, Cambridge University Press, Cambridge, 1927.

[229] D.V. Widder, Positive temperatures on an infinite rod, *Trans. Amer. Math. Soc.*, **55** (1944), 85–95.

[230] N. Wiener, *The Fourier Integral and Certain of its Applications*, Cambridge University Press, Cambridge, 1933.

[231] W. Wyss, The fractional diffusion equation, *J. Math. Phys.*, **27** (1986), 2782–2785.

[232] M. Yor, W. Schneider's grey noise and fractional Brownian motion. In: *Easter Meeting on Probability (Edinburgh, April 10–14, 1989).*

[233] Ya.I. Zhitomirsky, On the Cauchy problem for a second order parabolic equation with variable coefficients, *Dokl. AN SSSR*, 116 (1957), 913–916 (Russian).

[234] Ya.I. Zhitomirsky, Cauchy problem for parabolic systems of linear partial differential equations with increasing coefficients, *Izv. VUZ. Matematika*, No. 1 (1959), 55–74 (Russian).

Notation

In this book we use without comments the standard notations, like $\mathbb{N}, \mathbb{R}, \mathbb{R}^n, \mathbb{C}, \mathbb{C}^n$, i (the imaginary unit), δ_{jl} (the Kronecker symbol). We denote by $x_1, \dots, x_n$ the coordinates of a point $x \in \mathbb{R}^n$ or $x \in \mathbb{C}^n$. The coordinates of a point $(t, x) \in \mathbb{R}^{n+1}$ are denoted by $t, x_1, \dots, x_n$. Some other notation is listed below. Of course, this is not a complete list (for example, it does not cover the difference operators used in Chapter 4).

$\lvert x \rvert$	$:= \left(\sum\limits_{j=1}^{n} x_j^2 \right)^{1/2}$, if $x \in \mathbb{R}^n$
$\mathbb{C}_{mr}$	the set of all $m \times r$ matrices with complex entries; $\mathbb{C}_{1r} = \mathbb{C}^r$
$\lvert A \rvert$	$:= \left(\sum\limits_{j=1}^{m} \sum\limits_{l=1}^{r} \lvert a_{jl} \rvert^2 \right)^{1/2}$, if $A \in \mathbb{C}_{mr}$
A'	the transpose of a square matrix A
I	the unit matrix
$\mathbb{Z}_+^n$	the set of all n-dimensional multi-indices $k := (k_1, \dots, k_n)$
$\lvert k \rvert$	$:= \sum\limits_{j=1}^{n} k_j$, if $k \in \mathbb{Z}_+^n$
$\lVert k \rVert$	$:= \sum\limits_{j=1}^{n} m_j k_j$, if $k \in \mathbb{Z}_+^n$ and $m_1, \dots, m_n$ are natural numbers specified for each particular case
∂_t, ∂_y	the first order differentiation operations in t and y respectively
∂_y^l	the differentiation in y of an order $l > 1$
∂_x^k	$:= \partial_{x_1}^{k_1} \dots \partial_{x_n}^{k_n}$, if $x \in \mathbb{R}^n$ and $k \in \mathbb{Z}_+^n$
x^k	$:= x_1^{k_1} \dots x_n^{k_n}$, if $x \in \mathbb{R}^n$ and $k \in \mathbb{Z}_+^n$

$\Delta_t^\tau f(t,\cdot)$	$:= f(t,\cdot) - f(\tau,\cdot)$
$\Delta_x^y f(\cdot,x,\cdot)$	$:= f(\cdot,x,\cdot) - f(\cdot,y,\cdot)$
$\Delta_{t,x}^{\tau,y} f(t,x,\cdot)$	$:= f(t,x,\cdot) - f(\tau,y,\cdot)$
T	a fixed positive number
Π_H	$:= \{(t,x) \in \mathbb{R}^{n+1} \vert t \in H, x \in \mathbb{R}^n\}$, if $H \subset \mathbb{R}$
$B(x,R)$	$:= \{y \in \mathbb{R}^n \vert d(x,y) \leq R\}$, the ball of a radius R centered at $x \in \mathbb{R}^n$, with respect to a distance d (specified for each particular case)
B_R	$:= B(0,R)$

We often denote by the same letter various constants whose specific values are of no importance.

Index

MIX
Papier aus verantwortungsvollen Quellen
Paper from responsible sources
FSC® C105338

If you have any concerns about our products,
you can contact us on
ProductSafety@springernature.com

In case Publisher is established outside the EU,
the EU authorized representative is:
Springer Nature Customer Service Center GmbH
Europaplatz 3, 69115 Heidelberg, Germany

Printed by Libri Plureos GmbH
in Hamburg, Germany